Critical Values for Student's t

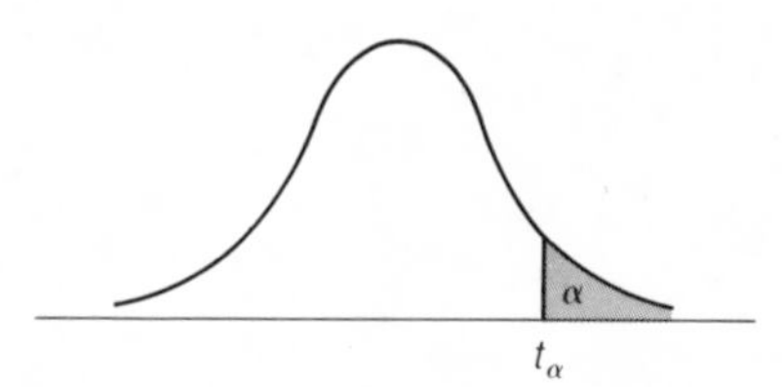

DEGREES OF FREEDOM	$t_{.100}$	$t_{.050}$	$t_{.025}$	$t_{.010}$	$t_{.005}$
1	3.078	6.314	12.706	31.821	63.657
2	1.886	2.920	4.303	6.965	9.925
3	1.638	2.353	3.182	4.541	5.841
4	1.533	2.132	2.776	3.747	4.604
5	1.476	2.015	2.571	3.365	4.032
6	1.440	1.943	2.447	3.143	3.707
7	1.415	1.895	2.365	2.998	3.499
8	1.397	1.860	2.306	2.896	3.355
9	1.383	1.833	2.262	2.821	3.250
10	1.372	1.812	2.228	2.764	3.169
11	1.363	1.796	2.201	2.718	3.106
12	1.356	1.782	2.179	2.681	3.055
13	1.350	1.771	2.160	2.650	3.012
14	1.345	1.761	2.145	2.624	2.977
15	1.341	1.753	2.131	2.602	2.947
16	1.337	1.746	2.120	2.583	2.921
17	1.333	1.740	2.110	2.567	2.898
18	1.330	1.734	2.101	2.552	2.878
19	1.328	1.729	2.093	2.539	2.861
20	1.325	1.725	2.086	2.528	2.845
21	1.323	1.721	2.080	2.518	2.831
22	1.321	1.717	2.074	2.508	2.819
23	1.319	1.714	2.069	2.500	2.807
24	1.318	1.711	2.064	2.492	2.797
25	1.316	1.708	2.060	2.485	2.787
26	1.315	1.706	2.056	2.479	2.779
27	1.314	1.703	2.052	2.473	2.771
28	1.313	1.701	2.048	2.467	2.763
29	1.311	1.699	2.045	2.462	2.756
∞	1.282	1.645	1.960	2.326	2.576

Source: From M. Merrington, "Table of Percentage Points of the t-Distribution," *Biometrika*, 1941, *32*, 300. Reproduced by permission of E. S. Pearson.

Statistics by Example

Statistics by Example

Terry Sincich
University of Florida

Dellen Publishing Company
San Francisco and Santa Clara, California

On the cover: The work on the cover is a collage and acrylic painting on canvas by Robert Natkin. His work is in many private and public collections, among them the Museum of Modern Art, the Guggenheim Museum, the Metropolitan Museum of Art, and the Los Angeles County Museum. Natkin's work may be seen at the Tortue Gallery in Santa Monica, California, and at the Gimpel and Weitzenhoffer Gallery in New York City.

Printed in the United States of America

10 9 8 7 6 5 4 3 2 1

ISBN 0-89517-037-X

Contents

Preface

This introductory college statistics text is designed for a one semester or one quarter course for students who have only a high school background in mathematics as a prerequisite. It differs from most other texts in two ways:

1. Explanations of basic statistical concepts and methodology are based on and motivated by the use of real data sets.
2. Concepts and statistical methods are explained in examples. These examples arise as questions posed about the data sets.

We think that this practical orientation helps the student to relate statistics to real-life problems and, hopefully, will develop a pattern of thought that will persist after the student leaves the academic environment.

The text contains four data sets; the first two are heavily used as instructional vehicles. These data sets are:

Appendix A. The set of actual starting salaries, majors, and colleges of 948 bachelor's degree graduates of the University of Florida during the period June 1980–March 1981.

Appendix B. The starting salaries (extracted from Appendix A) of graduates of the Colleges of Business Administration, Education, Engineering, Liberal Arts, and Sciences.

Appendix C. Supermarket customer checkout times for mechanical and automated checkers.

Appendix D. Length, weight, and DDT measurements for various species of fish collected from the Tennessee River, Alabama, and its creek tributaries.

Although all of the data sets are used to develop the notion of a population and a sample, the starting salaries of college graduates (Appendices A and B) are used to demonstrate the need for data description, to develop the notion of a sampling distribution, and to motivate the inferential methods commonly studied in an introductory statistics course.

In addition to teaching via data sets and by example, this text contains the following features:

1. Definitions are boxed.
2. Steps for constructing histograms, performing statistical calculations, and conducting statistical tests are listed and boxed for each procedure.
3. Key words, which must be added to a student's vocabulary, are listed (and boxed) at the end of each chapter.
4. Warnings, indicating situations where a student might misuse a statistical technique, are presented in boxed form. The student is directed to specific alternative methods.
5. The basic concepts of probability and their relation to statistical inference are presented in an easy-to-understand manner and are developed around the game of blackjack. Problem solving for the sake of problem solving is avoided.
6. The use of computer program packages is introduced in the presentation of the analysis of variance (Chapter 11) and multiple regression analysis (Chapter 13). The computer printouts for three different program packages, Minitab, SAS, and SPSS, are presented for the analyses of identical sets of data.
7. Case studies that detail specific current events are used at the end of each chapter to pose questions for the student. These case studies, extracted from news articles and journals, demonstrate to the student the relevance of statistics to the solution of current practical problems.
8. The data sets can be used by the instructor to illustrate the concept of a sampling distribution and the concepts of estimation and tests of hypotheses.
9. The data sets can be entered into computer storage and can be accessed by students for sampling and statistical inference. The student can then access the data sets for the demonstration of statistical concepts and for realistic statistical exercises.

In addition to the features described above, the text is accompanied by the following supplementary material:

1. A student's exercise solutions manual that presents the solutions for half of the exercises contained in the text.
2. A study guide for the student which provides additional worked examples.

I wish to acknowledge the many individuals who provided their invaluable assistance to this project. Their efforts are much appreciated. In particular, I thank the following reviewers for providing helpful suggestions and advice on the writing of the manuscript: John S. Bowdidge, Southwest Missouri State University; John Cameron, Rockhurst College; Geoffrey B. Holmewood, Hudson Valley Community College; Thomas B. Laase, University of Southern Colorado; James T. McClave, University of Florida; William Mendenhall, University of Florida; Susan Reiland, North Carolina State University; and John B. Rushton, Metropolitan State College. Susan Reiland deserves special recognition for her excellent line-by-line reviews, during both the writing and the production of the text. I am very grateful to the following for providing the data sets and accompanying background information: Maurice Mayberry (Director, Career Resources Center, University of Florida), Jim Sullivan (Water and Air Research), Info Tech, and Venus Wong (who patiently spent long hours at the supermarket recording customer checkout times). Also, I thank Robert Fordham and Jim Yucha for preparing answers to the exercises, and my typist, Brenda Dobson, who did a remarkable job of converting my hand-scribbled notes into immaculate type.

Finally, I owe very special thanks to William Mendenhall and James T. McClave who, together, suggested the concept of this text and then provided me with the opportunity to write the manuscript. Without their guidance and encouragement this text would never have been completed.

Statistics by Example

One Introduction

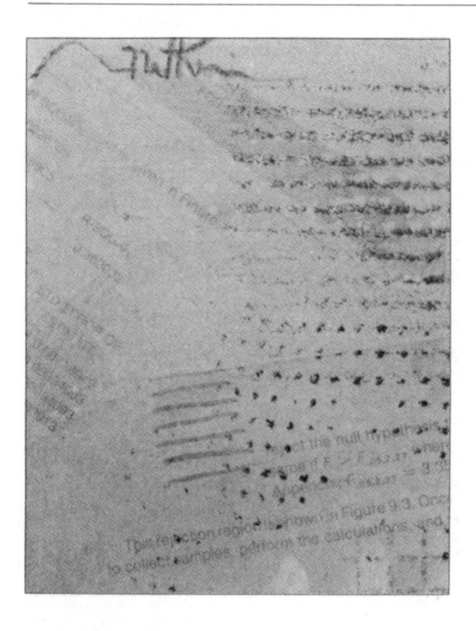

Have you ever thought of the difficulty that you would encounter if you tried to audit the inventory of a large hospital? The inventory shown in computerized records will be larger than the actual inventory because of poor record keeping, loss or theft of supplies, etc. Think of the massive task of actually counting and recording the values of all items in supply, and the attendant errors associated with that procedure. Then think of sampling! How you can use sampling and the science of statistics to solve this and other problems is the subject of this chapter and this text. The hospital inventory problem is discussed in greater detail in Case Study 1.3.

Contents

1.1 What Is Statistics?

You know from contact with the news media and with literature in your chosen field of study that statistics involves numbers (e.g., George Brett's batting average, the number of traffic deaths during a holiday period, monthly indicators summarizing the state of our national economy, the Nielsen television ratings, and the results of a political poll), data description, and sampling. Consequently, a good way to develop an understanding of the word *statistics* is to examine a real data set and the phenomenon which it purports to describe. We have chosen data that should be of interest to students of various academic backgrounds—namely data concerned with the annual starting salaries of college graduates.

For our purposes, we have chosen to examine the starting salaries of graduates of the University of Florida. Appendix A provides information on annual starting salaries for 948 graduates who earned a bachelor's degree during the period June 1980 to March 1981.* Each row of the data set, Appendix A, pertains to a single graduate and gives the following:

1. Date of Graduation
2. Sex
3. College or School
4. Major
5. Starting Salary

The sex and starting salaries of graduates in five colleges, Business Administration, Education, Engineering, Liberal Arts, and Sciences, were extracted from the data set, Appendix A, and are given in Appendix B.† These disciplines were chosen because they represent the most popular major fields of study among today's college students. Within a given discipline, the starting salaries are, for the most part, commensurate. Engineering and Sciences graduates tend to receive higher starting salaries and are mostly male. Education and Liberal Arts graduates are generally female with lower starting salaries. Business Administration graduates are well represented with both males and females and their salary structure appears to be the most diversified. The data presented in Appendix B will enable you to compare starting salaries for graduates of these five colleges.

EXAMPLE 1.1 What practical phenomena are characterized by the data sets, Appendices A and B?

Solution Before we attempt to answer this question, it is important that we understand how the data of Appendix A were collected. A few weeks after completion of an academic quarter, the Career Resource Center (CRC) at the University of Florida mails out questionnaires pertaining to the employment status and, if employed, the starting salary of all students graduating that particular quarter. The Career Resource Center estimates that about 48% of the questionnaires are returned, and of these, only half

*The data were obtained from the Career Resource Center, University of Florida, with permission granted by Maurice Mayberry, Director.

†Liberal Arts includes English Literature, History, Sociology, and Psychology majors. Sciences include Biology, Chemistry, Physics, Mathematics, and Statistics majors.

of the graduates indicate that they had secured a job as of the date of graduation. The salaries of Appendix A, then, are the annual starting salaries of those graduates who (1) have returned the Center's questionnaire and (2) have indicated that they had secured a job at the time of graduation. Thus, the data set, Appendix A, is not a complete listing of the annual starting salaries of all June 1980–March 1981 University of Florida bachelor's degree graduates.

Keeping this in mind, two of the many possible phenomena characterized by the data sets are as follows: The annual starting salaries, Column 6 of Appendix A, characterize the annual financial compensations of the June 1980–March 1981 University of Florida graduates during their first year in the job market—at least the graduates who returned the CRC questionnaires. Each of the five data sets, Appendix B, characterizes the annual starting salaries of graduates (who returned the questionnaire) in one of the five colleges.

When you examine a data set in the course of your studies or as part of your future employment, you will be doing so because the data characterize some phenomenon of interest to you. The data set which is the target of your interest is called a ***population.*** Notice that a population does not refer to people; it refers to a collection of data.

DEFINITION 1.1

A ***population*** is a collection (or set) of data that describes some phenomenon of interest to you.

The objects upon which measurements are made are called ***experimental units.*** For example, each of the college graduates who returned the CRC questionnaire indicating they had secured a job is an experimental unit. Measurements taken on this collection of units produced the populations of Appendices A and B. It is not uncommon for people to use the word *population* in two ways: to characterize the collection of experimental units upon which the measurements are made, as well as to characterize the data set itself. The particular meaning attached to the word will be clear from the context in which it is used.

Now that you have learned what we mean by the word *population,* let us return to the notion that a population characterizes some phenomenon of interest to you. Who defines a population? You do! Consequently, if you are interested in some phenomenon, you must be sure that your population really does characterize it. For example, if you take the complete set of starting salaries, Appendix A, as a measure of the financial compensations of June 1980–March 1981 University of Florida graduates during their first year in the job market, you should be certain that this data set characterizes the graduates' financial compensations. To illustrate:

EXAMPLE 1.2 Do the annual starting salary data, Appendix A, characterize the financial compensations of June 1980–March 1981 University of Florida graduates during their first year in the job market?

Solution To a certain extent they do characterize financial compensation but they do not do so completely. To properly characterize this phenomenon, we would need to know the starting salaries of all June 1980–March 1981 graduates, not only those who returned the CRC questionnaires but also those who never bothered to return the questionnaire even though they had secured a job by the date of graduation. This population, i.e., the set of starting salaries for all graduates who had secured a job by the date of graduation, would (in our opinion) characterize the first year financial compensations of June 1980–March 1981 University of Florida graduates.

EXAMPLE 1.3 Refer to Example 1.2. Why is the set of starting salaries, Appendix A, inadequate for characterizing the first year financial compensations of June 1980–March 1981 University of Florida graduates?

Solution The data set, Appendix A, represents only a subset of the starting salaries for all June 1980–March 1981 graduates. Thus, it represents a ***sample*** selected from the totality of all starting salaries. If this sample possesses characteristics that are very close to those of the complete set of all starting salaries, you might be safe in using it to characterize financial compensations. But this situation is unlikely. The graduates who return the CRC questionnaire are most likely comfortable with their starting salary and thus are willing to reveal it. Many others will treat their starting salary as confidential information no matter how large their compensation. Not all major fields of study may appear in your sample. For example, many Education majors who graduated in December or in March will not have secured a job until late August or early September, when grade schools, high schools, colleges, and universities begin the new academic year. Consequently, they will indicate their employment status on the CRC questionnaire as "negotiating for a job" or "still interviewing" and their eventual starting salary will not be known. Thus, the distribution of starting salaries in the sample may vary substantially from the distribution of starting salaries of all graduates.

DEFINITION 1.2

A ***sample*** is a subset of data selected from a population.

EXAMPLE 1.4 Can the data set, Appendix A, be either a sample or a population?

Solution Remember, you are the one who defines the population. Presumably, it is a data set that characterizes some phenomenon of interest to you. If you are interested only in the starting salaries of June 1980–March 1981 graduates who have returned the CRC questionnaire, then the data set, Appendix A, characterizes this phenomenon and it would be the population. In contrast, if you are interested in characterizing the first-year financial compensations of all June 1980–March 1981 graduates, then, as explained in Example 1.3, the data set, Appendix A, is a sample selected from the complete set of starting salaries of June 1980–March 1981 graduates who have secured a job. Whether a data set is a sample or a population depends upon the phenomenon that you wish to characterize.

EXAMPLE 1.5 Research* by faculty at the Virginia Polytechnic Institute suggests that the 2-drink lunch decreases the output of laborers. Particularly, a team of researchers found that a normal person needs about 19% more time to accomplish a relatively simple task if the person's blood alcohol level reaches 0.07%. This level is produced (approximately) by two mixed drinks or two 16-ounce glasses of beer. The sample data produced by this research included the times required for a group of persons to assemble a common water tap under two conditions, at alcohol blood levels of 0% and 0.07%, and the percentage increase in the length of assembly time.

a. Describe the population of data from which this sample was selected.
b. What phenomenon does the population characterize?

Solution

a. The sample consists of a set of percentages corresponding to the increase in the lengths of times for a group of volunteer persons to assemble a common water tap. Consequently, the population is conceptual (existing in our minds) and consists of the percentage increases for the collection of all workers similar to those who volunteered for the study and who could be required to assemble a common water tap.
b. The population, part a, characterizes the percentage increase in time to assemble a common water tap for assemblers with physical abilities (coordination, strength, etc.) and experience comparable to the volunteers employed in the researchers' experiment and who have had a 2-drink lunch. Most of us would like to extrapolate and imagine that the population characterizes the percentage increases in assembly time for any and all assembly workers and for any and all assembly operations. This, of course, is not the case, but it is easy to believe that similar experiments with almost any type of worker in any assembly operation would produce data that are similar to that obtained by the researchers at the Virginia Polytechnic Institute. The average percentage increase in assembly times induced by an alcohol level of 0.07% in the blood of assembly workers may vary, depending on the type of worker and on the assembly operation, but we would expect the percentage change in assembly time to be positive and sufficiently large to affect worker productivity.

1.2 How Can Statistics Be of Value in Your Field of Study?

The data sets, Appendices A and B, and the preceding discussion identify the two ways that the body of knowledge called *statistics* can be of value to economists, sociologists, biologists, psychologists, government workers, etc. We will assume that you are interested in some phenomenon associated with some aspect of your field of study and that you have identified a population of data that characterizes this phenomenon. Then statistics can be of value in two ways:

*Reported in the Orlando *Sentinel Star*, May 18, 1980.

1. If you have the population in hand, i.e., if you have every measurement in the population, then statistical methodology can help you describe this typically large set of data. That is, we will find graphical and numerical ways to make sense out of a large mass of data. The branch of statistics devoted to this application is called ***descriptive statistics.***

DEFINITION 1.3

The branch of statistics devoted to the summarization and description of data sets is called ***descriptive statistics.***

2. It may be too expensive to obtain every measurement in the population or, as in the case of the starting salaries of college graduates, it may be impossible to acquire them. Then we will wish to select a sample of data from the population and use the sample to infer the nature of the population.

DEFINITION 1.4

The branch of statistics concerned with using sample data to make an inference about a population of data is called ***inferential statistics.***

A careful analysis of most data sets will reveal that they are samples from larger data sets that are really the object of our interest. Consequently, most applications of statistics involve sampling and using the information in the sample to make inferences about the sampled population. To illustrate, if the annual starting salaries, Appendix A, really comprise a representative* sample of the starting salaries of all June 1980–March 1981 University of Florida graduates, then we could use the characteristics of the sample to infer the characteristics of the population. The "average" starting salary in the sample would be an estimate of the "average" starting salary of all June 1980–March 1981 graduates. The spread of the sample starting salaries would be a rough indicator of the dispersion of the population starting salaries.

EXAMPLE 1.6 Would the "average" of the sample starting salaries equal the "average" starting salary for all June 1980–March 1981 graduates?

Solution The answer is "not likely," and this pinpoints the major contribution of statistics to modern society. ***Anyone can examine a sample and calculate an estimate of some population characteristic. But statistical methodology enables you to go one step further. When the sample is selected in a specified way from the population, statistical methodology will enable you to say how accurate your sample estimate***

*We use this term loosely. We will have more to say about sampling in later discussions.

will be. Statistical methodology not only tells you how to make an inference about a population based upon sample data, but it also tells you how reliable your inference will be.

To illustrate the value of possessing a measure of reliability for a particular inference, consider the following situation:

EXAMPLE 1.7 Before investing in real estate you would like to know the "fair" value of a particular residential property, i.e., you would like to know the price at which the property will be sold. Although two real estate appraisers are available for assessing the property's value, you can hire only one. Both appraisers are supposed to be "good," but the properties of appraiser #2's appraisal errors are known—records show that this appraiser is within $2,000 of the sale price 99% of the time. You have no information on the appraisal errors of appraiser #1. Which appraiser would you choose?

Solution We think that you would choose appraiser #2 because you would attach a very real value to knowing how much reliability you could place in the appraisal. The appraisals of appraiser #1 may be better or worse than those of appraiser #2, i.e., they might fall within $500 of the sale prices of appraised properties—or they might fall within $5,000—but, either way, it is a great disadvantage not to know how large this error is apt to be.

This example points to the major contribution of statistics, one that a nonstatistical method does not possess: statistical methods provide measures of reliability for each inference obtained from a sample.

The key facts to remember in this section are:

THE OBJECTIVE OF STATISTICS

1. To describe data sets
2. To use sample data to make inferences about a population

THE MAJOR CONTRIBUTION OF STATISTICS

Provides a measure of reliability for every statistical inference

1.3 Summary

This chapter identified the types of problems for which statistical procedures are useful, describing data sets and using sample data to make inferences about a sampled population. Basic to the application of these techniques is the identification of a population of data that truly characterizes the phenomenon of interest to you.

Most statistical problems involve sampling and using a sample to make inferences about the sampled population. For example, the starting salaries of University of Florida graduates for the five colleges, Appendix B, could be viewed as samples of the starting salaries of University of Florida graduates for their respective colleges. Do these sample values suggest a difference in the distributions of the starting salaries among the five colleges? Statistical methods (to be covered later) will help us answer this question and will provide us with a measure of reliability for our decision.

The remainder of this course will examine some basic statistical procedures for describing data sets and giving them meaning. More important, we will learn how to use sample data to infer the nature of the sampled population and to do so with a known degree of reliability.

KEY WORDS

Population
Sample
Inference
Reliability

EXERCISES

1.1 Appendix C contains the checkout times for 500 grocery shoppers at each of two supermarkets, supermarket A and supermarket B. *Customer checkout time* is defined here as the total length of time required for service personnel to check the prices of the customer's food items, total the prices, accept payment, and return change.

The checking of food items at supermarket A is conducted in the usual manner; the cashier searches for the price marked on each item and manually punches in the price on the cash register. Supermarket B, however, employs automated checkers. With automated checkers, the cashier need only brush the item across a scanning window located on the counter. A laser beam is then activated which reads the price code several times, verifies its accuracy, and transmits the price to the checkstand to be printed on the receipt tape. Thus, the data of Appendix C provide us with an opportunity to compare customer checkout times at supermarkets using manual and automated checkers.

a. Suppose that you were to view the two sets of customer checkout times as two different populations. What phenomenon do the supermarket A checkout times characterize? The supermarket B checkout times?

b. Do the data sets actually represent the populations you described in part a, or do they represent samples selected, respectively, from the totality of all customer checkout times at supermarket A and supermarket B? Explain.

c. Suppose that you were to use the average of the 500 supermarket A customer checkout times to estimate the average checkout time of all customers who shop at supermarket A. Would the sample average equal the average for the population? Explain.

1.2 The slope of a river delta region can sometimes be accurately predicted from knowledge of the typical size of stones found there. With this in mind, a geographer studying South America would like to estimate the average size of stones found in the delta region of the Amazon River. In order to obtain this estimate, the geographer collects 50 stones and measures the diameter of each.

a. What is the population of interest to the geographer?
b. What phenomenon does the population characterize?
c. Describe the sample in this problem.
d. Suppose that the average diameter of the 50 stones is 7.2 inches. Do you believe that this sample average will equal the average for the population? Explain.

1.3 Postpartum depression is the term used to describe the usually short-lived period of emotional sensitivity which many women suffer following childbirth. Studies have indicated that nearly 90% of all mothers experience some symptoms of postpartum blues (Gainesville *Sun,* April 26, 1981). However, new evidence shows that men, too, can suffer from postpartum depression. Suppose a developmental psychologist wishes to estimate the proportion of fathers who suffer from postpartum blues. Fifty men who have recently fathered a child are interviewed and observed in the home and the number experiencing some form of postpartum depression is recorded.

a. What is the population of interest to the developmental psychologist?
b. What phenomenon does the population characterize?
c. Describe the sample in this problem.
d. Suppose that 31 of the 50 men are diagnosed as having postpartum blues. Thus, the psychologist estimates that 62% of all fathers experience postpartum depression. Do you believe that this estimate is equal to the proportion for the entire population? Explain.

1.4 A successful discount clothing store, in business for 30 years, must order Levi's blue jeans from the manufacturer one month in advance. In order to avoid large losses, the store needs to be able to predict the monthly demand for blue jeans for each month of the year. Suppose that the store has ready access to the monthly sales records (i.e., the number of blue jeans sold by the store during each month) for the past 10 years. This information will be used to project monthly demand.

a. If the discount store views the monthly blue jean sales data over the past 10 years as a population, what phenomenon does it characterize?
b. If the discount store views the monthly blue jean sales data over the past 10 years as a sample selected from a population, describe the population.
c. Suggest a way in which the discount clothing store could use the monthly sales records over the past 10 years to project monthly blue jean demand.

1.5 The high cost of fuel oil and the diversion of low-cost natural gas from power plants to home heating are often cited as the principal reasons for the record high January 1981 utility bills registered across the country.* To partially ease the burden

---

*Gainesville *Sun,* February 4, 1981.

of payment on the customer, many utility companies allowed their customers to pay only half of January's bill initially, and then to pay the balance over the next 90 days. For example, customers of Gainesville (Florida) Regional Utilities whose January 1981 electric bills were twice as much or more than in January 1980 were declared eligible for a deferred-payment plan. Suppose that the director of utilities in a city which had a January 1981 deferred-payment plan wished to estimate the true proportion of the utility's residential customers who qualified for the plan (i.e., whose January 1981 bill was at least two times as much as their January 1980 bill). To obtain this estimate, the director randomly sampled 200 customers, compared their January 1980 and January 1981 utility bills, and recorded the number who qualified for the deferred-payment plan.

a. What is the population of interest to the director?
b. What is the sample?
c. What type of inference does the director desire to make?

1.6 Refer to Exercise 1.5. Suppose that the utility director also wished to estimate the difference between the average January 1981 utility bill and the average January 1980 utility bill of the utility's customers. Then the problem is to compare the averages of two populations, where the first population, say population A, is the collection of all January 1981 utility bill amounts issued by the utility, and the second population, say population B, is the collection of all January 1980 utility bill amounts issued by the utility.

a. What phenomenon does population A characterize?
b. What phenomenon does population B characterize?
c. Suggest how the director could use the information collected on the January utility bills for the sample of 200 customers to estimate the difference between the averages of the populations.
d. Would you expect the difference in the sample averages to equal the difference in the population averages? Explain.

1.7 Manufacturers of low-tar, low-nicotine cigarettes claim that smokers who switch to their brands will be better off physically than those who continue to smoke regular cigarettes. However, in his annual survey of smoking, the U.S. Surgeon General reported* that low-yield brands reduce the risk of developing lung cancer only slightly, and reduce the risk of heart disease, emphysema, and bronchitis not at all. In addition, some of the additives (e.g., cocoa) which manufacturers have been using to enhance the weaker flavor of the low-tar, low-nicotine cigarettes turn into cancer-causing substances when burned. In view of these facts, let us suppose that a manufacturer of low-yield cigarettes is interested in determining the fraction of smokers of low-tar, low-nicotine cigarettes who are aware of the U.S. Surgeon General's report.

a. What is the population of interest to the manufacturer?
b. What phenomenon does it characterize?

**Time,* January 26, 1981.

c. Do you believe that the manufacturer could actually determine the true fraction of smokers of low-yield cigarettes who are aware of the report? Explain.
d. Suggest a way in which the manufacturer could obtain an estimate of the true fraction.

1.8 A research cardiologist is interested in the average amount of time spent per week on vigorous exercise by Russian athletes. The cardiologist contacted the 20 members of the visiting Soviet hockey team and asked each (through an interpreter) to state the number of hours they spend in vigorous exercise per week.
a. What is the population of interest to the cardiologist?
b. What phenomenon does the population characterize?
c. What is the sample in this problem?
d. Do you think the average for the sample will be an adequate estimate of the average for the population? Explain.

1.9 To evaluate the current status of the dental health of school children, the American Dental Association conducted a survey to estimate the average number of cavities per child in grade school in the United States. One thousand school children from across the country were selected, examined by a dentist, and the number of cavities for each was recorded.
a. Identify the population of interest to the American Dental Association.
b. Identify the sample.
c. How could the American Dental Association use the sample information to estimate the average number of cavities per child in grade school? Will this estimate equal the average for the population? Explain.

1.10 Euthanasia has long been a dilemma of medical ethics. Euthanasia is the act of putting to death painlessly a person suffering from an incurable and painful disease or condition. Developments in medical technology have contributed to the dilemma: individuals who a few years ago would have died from their affliction can now be sustained beyond the point which even they themselves would desire. Suppose that you work for a major opinion pollster and you have been assigned the task of conducting a survey for the Concern for Dying (formerly the Euthanasia Society). The purpose of the survey is to estimate the proportion of American adults who support euthanasia.
a. Clearly define the population of interest to the Concern for Dying.
b. What phenomenon does the population characterize?
c. Do you think it is possible to obtain the entire population? Explain.
d. Why is it necessary that the sample you select for the survey be representative of the population?

CASE STUDY 1.1 Contamination of Fish in the Tennessee River

Chemical and manufacturing plants often discharge toxic waste materials into nearby rivers and streams. These toxicants have a detrimental effect on the plant and animal life inhabiting the river and the river's bank. One type of pollutant, commonly known as DDT, is especially harmful to fish and, indirectly, to people. The Food and Drug Administration (FDA) sets the limit for DDT content in individual fish at

5 parts per million (ppm). Fish with DDT content exceeding this limit are considered potentially hazardous to people if consumed. A study was undertaken recently to examine the DDT content of fish inhabiting the Tennessee River (Alabama) and its tributaries.

The Tennessee River flows in a west-east direction across the northern part of the state of Alabama, through Wheeler Reservoir, a national wildlife refuge. Ecologists fear that contaminated fish migrating from the mouth of the river to the reservoir could endanger other wildlife which prey on the fish. This concern is more than academic. A manufacturing plant was once located along Indian Creek, which enters the Tennessee River 321 miles upstream from the mouth. Although the plant has been inactive for over ten years, there is evidence that the plant discharged toxic materials into the creek, contaminating all the fish in the immediate area. Have the fish in the Tennessee River and its tributary creeks also been contaminated? And if so, how far upstream have the contaminated fish migrated? In order to answer these and other questions, a team of U.S. Army Corps of Engineers in the summer of 1980 collected fish samples at different locations along the Tennessee River and three tributary creeks: Flint Creek (which enters the river 309 miles upstream from the river's mouth), Limestone Creek (310 miles upstream), and Spring Creek (282 miles upstream). Each fish was first weighed (in grams) and measured (length in centimeters), then the filet of the fish was extracted and the DDT concentration (in parts per million) in the filet measured.

Appendix D contains these length, weight, and DDT measurements for a total of 144 sampled fish.* Notice that the data set also contains information on location (i.e., where the fish were captured) and species of the fish. Three species of fish were examined: channel catfish, large mouth bass, and small mouth buffalo. The different symbols for location are interpreted as follows: the first two characters represent the river or creek and the remaining characters represent the distance (in miles) from the mouth of the river or creek. For example, FCM5 indicates that the fish was captured in Firestone Creek (FC), 5 miles upstream from the mouth of the creek (M5). Similarly, TRM380 denotes a fish sample collected from the Tennessee River (TR), 380 miles upstream from the river's mouth (M380). These data provide us with an opportunity to compare the DDT content of fish at different locations and among the different species and to determine the relationship (if any) of length and weight to DDT content.

a. Suppose you were to view the totality of fish lengths (or weights) in Appendix D as a population. What phenomenon do they characterize?

b. Suppose you were to view the totality of fish DDT concentrations in Appendix D as a population. What phenomenon do they characterize?

c. Suppose you were to view the DDT concentrations in each species of fish as a distinct population. What phenomena do the three populations characterize?

d. Select ten DDT concentrations from the listing, Appendix D. Would this constitute a sample from the population you described in part b? Explain.

---

*Source: U.S. Army Corps of Engineers, Mobile District, Alabama.

e. Suppose you were to use the average value of the ten DDT concentrations (part d) to estimate the average DDT content of the 144 fish of Appendix D. Would the sample average equal the average for the population? Explain.

CASE STUDY 1.2 Cruising: How Foresters Estimate Timber Weights

Paper and lumber companies pay for timber by the weight per truckload of 16-foot logs. Consequently, an investor in forest land needs to estimate the total weight of logs that can be produced by a property. This is done by "cruising the property" and counting the total number* of trees capable of producing 16-foot logs. A random sample of trees (usually 10% of the total number) is selected from this group and the diameter at chest height and the number of logs per tree (a visual guess) are recorded for each. A forester can then use the diameter and logs-per-tree measurements to calculate the *approximate* weight for each tree.

For use in this and later case studies, we were able to obtain data of this type for an actual 40-acre tract of short-leaf pine timber located in western Arkansas.** Twenty ⅕-acre plots were sampled (i.e., a total of 4 acres were "cruised"), and the chest height diameters and logs per tree were measured for each of the 117 trees on these plots. The trees were then grouped according to diameter (10, 11, . . . , 15, or 16 inches), and an estimate of the weight per tree in each diameter group (based on the average of the logs-per-tree measurements) was calculated. The "grouped" data are reported in Table 1.1.

TABLE 1.1 Diameters and Estimated Weights for a Sample of 117 Short-Leaf Pine Trees

DIAMETER AT CHEST HEIGHT (INCHES)	ESTIMATED (AVERAGE) WEIGHT IN POUNDS PER TREE	NUMBER OF TREES
10	580	38
11	750	34
12	1,100	21
13	1,800	15
14	2,000	5
15	2,660	3
16	3,000	1
		117

a. Describe the population of measurements from which the sample of tree-weight measurements was selected.

b. What phenomenon does the population, part a, characterize?

c. What will the average of the sample of tree weights estimate? Do you think that this estimate will equal or be close to the quantity that you have estimated? Explain.

*This number is usually close to but not actually equal to the exact number of trees. For the purpose of this discussion, we will assume that it is an exact count.

**Timber data and information courtesy of Delton F. Price, Fort Smith, Arkansas.

CASE STUDY 1.3 Auditing Hospital Inventory

Consider the problem of auditing the inventory of a large hospital and assessing the value on hand of the very large number of items that are in daily use. Some items, such as drugs, adhesive tape, bandages, etc., are expended daily in large amounts while others, such as stethoscopes, scissors, typewriters, etc., are lost due to wear, damage, or theft.

Theoretically, auditing this inventory might appear to be easy. Thus, the hospital records (usually stored in a computer) should contain the number and dollar value on hand for each of this large number of items stocked by the hospital. In practice, however, the auditing problem is not so easy because the actual number and dollar value on hand for any given item will usually be less than the numbers shown in the hospital records. This discrepancy may be due to the failure to record the use or destruction of an item or, for many small disposable items (tape, gauze, etc.), it may be due to theft.

An auditor envisions two populations of data. The first population, population A (stored in the computer), is the collection of the recorded dollar values of all items held in inventory. The second population, population B (whose values are unknown to us), is the collection of the actual dollar values of all items held in inventory.

a. What phenomenon does population A characterize?

b. What phenomenon does population B characterize?

c. How is the goal of the auditor related to the two populations?

d. Suppose an auditor decided to base an estimate of the total value of inventory on a complete count of the number on hand and dollar value for each and all of the items listed in the hospital records. Aside from the fact that this procedure would be very costly, explain why the total dollar value of all items in inventory obtained by this method would be subject to error, i.e., differ from the true dollar value of the inventory.

e. Suppose that you were to conduct a complete count and calculate the dollar value on hand for a sample of 50 items selected from the complete list of items held in inventory. Can you suggest a way to use these sample values to obtain an estimate of the total value of the inventory on hand? Explain why this estimate would be subject to error.

REFERENCES

Careers in statistics. American Statistical Association and the Institute of Mathematical Statistics, 1974.

Tanur, J. M., Mosteller, F., Kruskal, W. H., Link, R. F., Pieters, R. S., & Rising, G. R. *Statistics: A guide to the unknown.* San Francisco: Holden-Day, 1978.

Two

Graphical Methods for Describing Data Sets

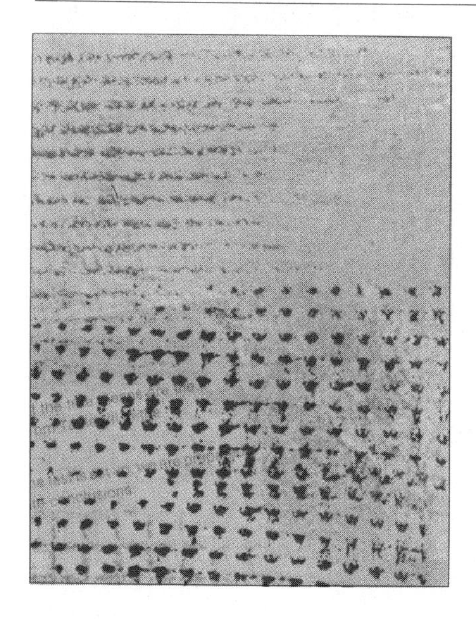

If you smoke, you are probably aware of the dangers of inhaling the excess tar and nicotine from a burning cigarette. Recently, however, the U.S. Surgeon General announced that breathing carbon monoxide may be just as hazardous to smokers. As a result of these findings, the Federal Trade Commission ranked 187 American cigarette brands according to their carbon monoxide, tar, and nicotine content. A graphical method for summarizing and making sense of this mass of data is the topic of Case Study 2.3. Graphical methods that rapidly convey information contained in a data set are discussed in this chapter.

Contents

2.1 The Objective of Data Description

The objective of data description is to summarize the characteristics of a data set. Ultimately, we wish to make the data set more comprehensible and meaningful. In this chapter we will show you how to construct charts and graphs that convey the nature of a data set. The procedure that we will use to accomplish this objective will depend upon the type of data that you wish to describe.

2.2 Types of Data

You will recall that Appendix B contains sex and starting salary information on June 1980–March 1981 University of Florida graduates who earned their bachelor's degree in one of five colleges. If you examine Appendix B, you will see that it contains two different types of data. To illustrate, we show in Table 2.1 data pertaining to each of 10 graduates randomly selected from Appendix B.

TABLE 2.1
A Sample of Data from Appendix B

OBSERVATION NUMBER	SEX	COLLEGE	STARTING SALARY
90	F	Business Administration	$12,700
320	M	Engineering	19,200
264	M	Engineering	18,800
454	F	Liberal Arts	8,000
95	M	Business Administration	14,300
257	M	Engineering	17,900
162	M	Education	12,700
216	M	Engineering	14,800
153	F	Education	7,100
80	M	Business Administration	36,500

The last piece of information (Starting salary) represents ***quantitative data,*** data which take values on a numerical scale. The first two (Sex and College) are examples of ***qualitative data.*** Qualitative data are those which take values that are nonnumerical; they can only be classified. For example, the observation that graduate #90 is female is a qualitative observation. Similarly, we can classify each graduate according to college but we cannot represent the College of Education as a numerical quantity.

DEFINITION 2.1

Quantitative data are observations measured on a numerical scale.

DEFINITION 2.2

Nonnumerical data which can only be classified into one of a group of categories are said to be ***qualitative.***

The three columns of data, Table 2.1, represent measurements (or observations) on variables. For example, starting salary varies from graduate to graduate. It is called a ***quantitative variable*** because the data obtained by observing starting salary are quantitative. For similar reasons, sex and college may vary from graduate to graduate and are called ***qualitative variables.***

EXAMPLE 2.1 Suppose the following types of data were available for the graduates, Appendix B. Classify the data according to whether they are quantitative or qualitative.

a. Age of a graduate
b. Race of a graduate
c. Grade point average (GPA) of a graduate

Solution Age and grade point average are measured on a numerical scale. Typical values for age would be 20 years, 21 years, 22 years, etc. Similarly, grade point average (on a 4-point scale) could take any value between 0.00 and 4.00. Consequently, both of these are quantitative variables. In contrast, race cannot be measured on a quantitative scale; it can only be classified (e.g., Caucasian, Black, Hispanic, etc.). Consequently, data on race are qualitative.

EXAMPLE 2.2 State whether the following variables are quantitative or qualitative.

a. Prime interest rate
b. Bacteria count in your drinking water
c. Security guard at a general hospital
d. Divorce rate
e. Brand of gasoline used by a race car driver

Solution The variables a, b, and d are quantitative because the values that the variables can assume are numerical, i.e., they are measured on a numerical scale. In contrast, the two variables c and e are qualitative. For example, suppose security at the general hospital operates on three shifts, using a different guard for each shift. Thus, the security guard varies from shift to shift, but the possible values of the variable, Security guard, cannot be quantified. We can identify a security guard only by giving his or her name. Similarly, Brand of gasoline may vary from race car driver to race car driver, but the values that this variable can assume cannot be quantified. We can identify them only as Shell, Texaco, Exxon, etc., but cannot locate these "values" on a numerical scale.

EXAMPLE 2.3 Many Americans are skeptical about the information they obtain from the news media. A recent Newsweek-Gallup poll of 760 randomly selected American adults revealed that 61% believe little or only some of the news. Surprisingly, only 13% believe that news reporters should always reveal their sources to readers or listeners.* Clearly, the purpose of this survey was to examine the opinions of American adults on the reliability of the news media. The experimental units, the objects upon which observations were taken, are the individual American adults. Suppose that for each American adult we record whether he or she feels news reporters should always reveal their sources. Describe the variable observed for each experimental unit and explain whether it is a quantitative variable.

Solution One way to view this situation is to note that the variable being measured, Opinion on public disclosure of news sources, can assume one of two conditions. An American adult will either believe that news sources should always be revealed or he or she will not. From this point of view, opinion on public disclosure of news sources is a qualitative variable.

A second way to view the response would be to convert the qualitative variable into a meaningful quantitative response by assigning a number, 1, to all American adults who believe that news sources should always be revealed and a 0 to those who do not. Then the sum of all the 0's and 1's in the sample will equal the total number of American adults in the sample who feel news reporters should always reveal their sources. (For example, if there were 5 adults in the sample with 2 who believe and 3 who do not believe news reporters should always reveal their sources, then the sample data would be 1, 1, 0, 0, 0. The sum of these observations would equal 2, the number of adults in the sample who believe news reporters should always reveal their sources.) Qualitative variables cannot always be converted into meaningful quantitative variables, but it can be done (as shown above) when the number of categories into which the observations fall is equal to 2. Then for a (0, 1) assignment to the two categories, the sum of the observations will equal the number of observations falling in the "1" category.

EXERCISES

2.1 Examine the following variables and state whether they are quantitative or qualitative.

a. Number of acres in a plot of land
b. Mode of transportation (to and from work) for a city employee
c. Type of residential water heating system
d. Time required for postoperative pain to be relieved in surgery patients

2.2 Many conservationists fear that Antarctic whales, particularly blue whales, will soon become extinct. State whether each of the variables relating to Antarctic blue whales listed below is quantitative or qualitative.

a. Length of an Antarctic blue whale
b. Cause of death of each blue whale that died last year
c. Number of blue whales sighted in the Antarctic on a particular day

---

*Source: Gainesville *Sun*, April 27–28, 1981.

2.3 List the variables that you consider before you purchase a new automobile. State whether each is qualitative or quantitative.

2.4 Classify the following variables as either quantitative or qualitative.

a. Political affiliation of a chief executive whose firm is listed in the *Fortune* 500
b. Geographical region with the highest unemployment rate in the United States
c. Gas mileage attained by an automobile powered by alcohol
d. Fee charged by an attorney to handle an uncontested divorce
e. Highest educational degree attained by members of the faculty at a community college

2.5 List the variables that your family physician considers while giving you a complete physical. State whether each is qualitative or quantitative.

2.3 Graphical Descriptions of Qualitative Data

Bar graphs and ***pie charts*** are two of the most widely used graphical methods for describing qualitative data sets. Essentially, they show how many observations fall in each qualitative category.

The observations for the qualitative variable Security guard on duty in part c of Example 2.2 could fall into one of a number of categories or ***classes.*** If three security guards were on rotating duty at the hospital, then the number of classes would equal 3. If the variable Security guard was observed on a number of shifts, we would find that security guard #1 was on duty a certain number of times, say n_1 times, security guard #2, n_2 times, and security guard #3, n_3 times.

The summary information that we seek about qualitative variables is either the number of observations falling in each class or the proportion of the total number of observations falling in each class. Bar graphs can be constructed to show either of these types of information. Pie charts usually show the proportions or percentages of the total number of measurements falling in the classes.

EXAMPLE 2.4 Figure 2.1 (see next page) is a bar graph that allows you to visually compare the frequencies of graduates who indicated they had secured a job on the questionnaire for the five colleges of Appendix B. Notice that the figure contains a rectangle or ***bar*** for each college, and that the height of a particular bar is proportional to the number of graduates for its college. You can rapidly compare the employed graduates for the five colleges by visually comparing the heights of the bars.

DEFINITION 2.3

The ***frequency*** for a particular class is the number of observations falling in that class.

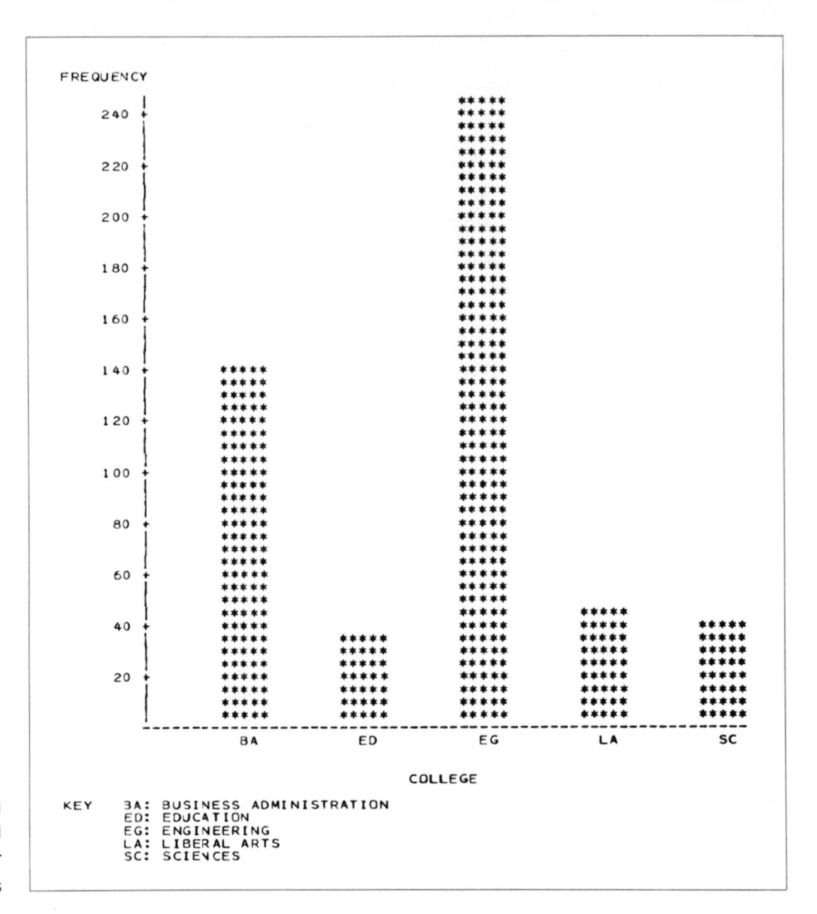

FIGURE 2.1
A Bar Graph Showing Frequency of Employed Graduates for Five Colleges

DEFINITION 2.4

The ***relative frequency*** for a particular class is equal to the class frequency divided by the total number of observations.

The class frequencies, the number of graduates who indicated they had secured a job for each of the five colleges, are shown in Table 2.2. The class relative frequencies, obtained by dividing each class frequency by the total number of graduates, 505, are also shown in Table 2.2.

The bar graph that permits a comparison of employed graduates can be constructed in several different ways. The heights of the bars can be measured in units of frequency (see Figure 2.1) or relative frequency (see Figure 2.2). It is also

TABLE 2.2
Frequencies and Relative Frequencies of Employed Graduates for the Five Colleges of Appendix B

COLLEGE	CLASS FREQUENCY	RELATIVE FREQUENCY
Business Administration	141	.279
Education	34	.068
Engineering	245	.485
Liberal Arts	45	.089
Sciences	40	.079
TOTALS	505	1.000

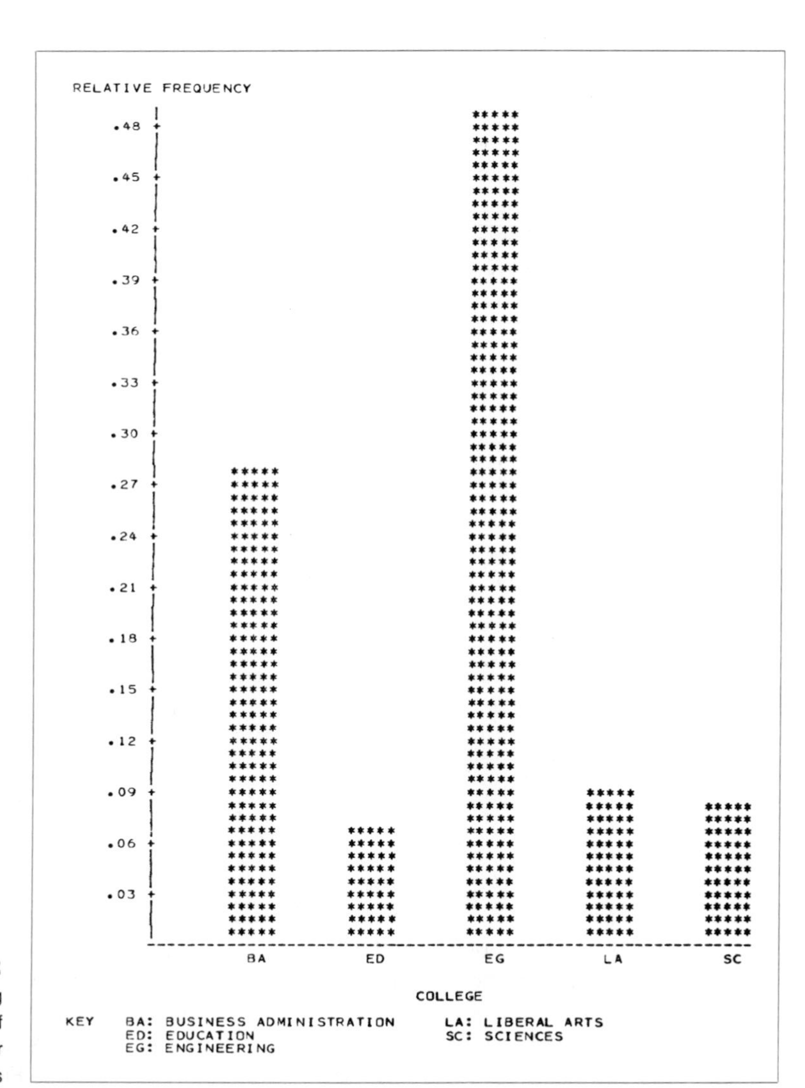

FIGURE 2.2
A Bar Graph Showing Relative Frequency of Employed Graduates for Five Colleges

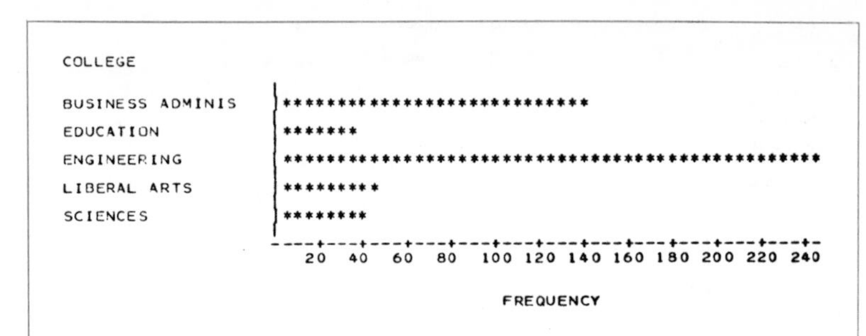

FIGURE 2.3
A Horizontal Bar Graph Showing Frequency of Employed Graduates for Five Colleges

common to reverse the axes and display the bars in a horizontal fashion as shown in Figure 2.3.

EXAMPLE 2.5 Figure 2.4 is a pie chart that conveys the same information as the bar chart in Figure 2.2. The total number of employed graduates for the five colleges, the pie, is split into five pieces. The size (angle) of the slice assigned to a college is proportional to the relative frequency for that college. It is common to show the percentage of measurements in each class on the pie chart as indicated.

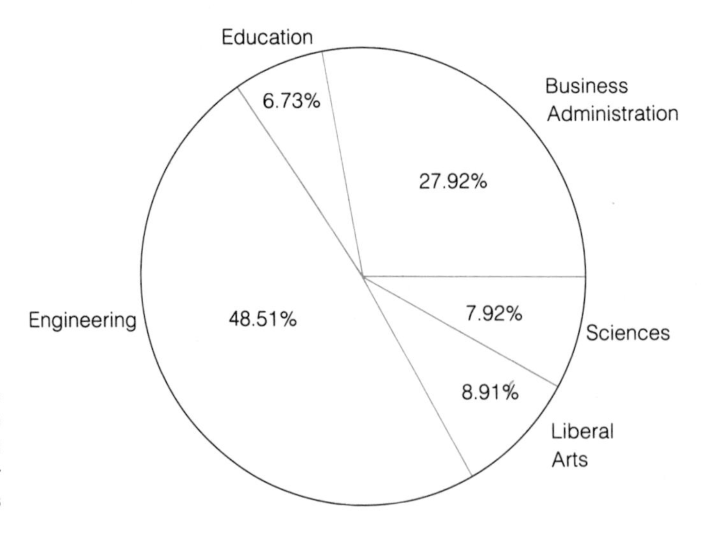

FIGURE 2.4
Pie Chart Showing the Percentages of Employed Graduates for Five Colleges

EXAMPLE 2.6 Beginning in 1790, the nation's population has been counted in each year ending in zero. The main purpose of this decennial census is to apportion seats in the U.S. House of Representatives among the states, according to their population. The U.S. Bureau of the Census tabulated the 1980 population data by various geographical regions (state, county, district, etc.) and by race. In 1980, Alachua County, located in

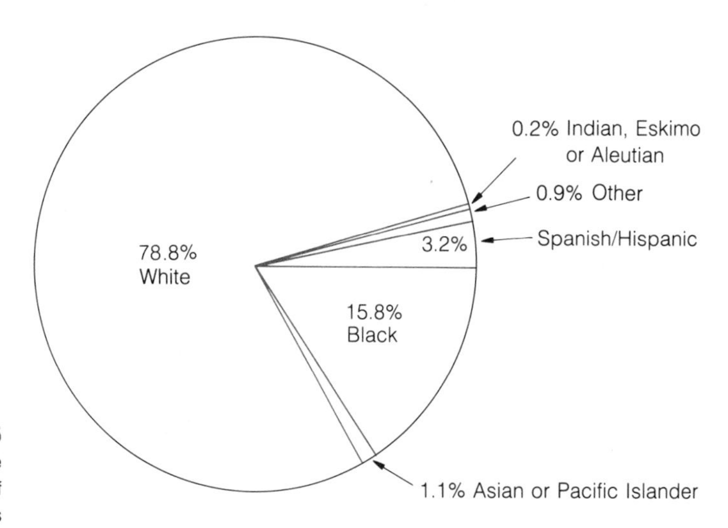

FIGURE 2.5 Pie Chart Showing the 1980 Proportion of Alachua County Residents by Race

Source: Bureau of the Census, U.S. Department of Commerce.

central Florida, had a total population count of 151,348. Figure 2.5 gives the breakdown of the 151,348 persons living in the county according to race. Interpret the figure.

Solution Each of the 151,348 persons living in Alachua County has been classified according to Race, a qualitative variable. The six types of race, or classes, are shown in the figure. The percentage given in each section of the pie (i.e., the percentage of all the measurements corresponding to each class) represents the proportion of the 151,348 persons who are of the respective race. For example, approximately 79% of the 151,348 Alachua County residents (i.e., 119,205 people) are White, while approximately 16% (23,875 people) are Black.

EXAMPLE 2.7 A recent advertisement claims that by virtue of their sensible design, unmatched safety record, and network of factory-owned service centers, the Cessna *Citation* has become the world's best-selling business jet. Cessna presents as evidence of the *Citation*'s leader status the bar graph shown in Figure 2.6 on page 24. Discuss the information provided by the graph. Does it lend support to the Cessna claim?

Solution Each of the 400 business jets delivered throughout the world in 1979 has been classified according to model type. The height of the vertical bar above each jet model is proportional to the number of worldwide deliveries for that particular model. Since the height of the bar above *Citation* is 140, we conclude that 140 of the 400 business jets delivered worldwide in 1979 were Cessna *Citations*. Cessna's nearest competitor is *Learjet,* whose 1979 worldwide deliveries were (approximately) 110. Figure 2.6 thus supports Cessna *Citation*'s claim to be the world's best-selling business jet, at least in 1979.

FIGURE 2.6
Frequency of 1979 Worldwide Deliveries of Business Jets (Total Number of Deliveries = 400)

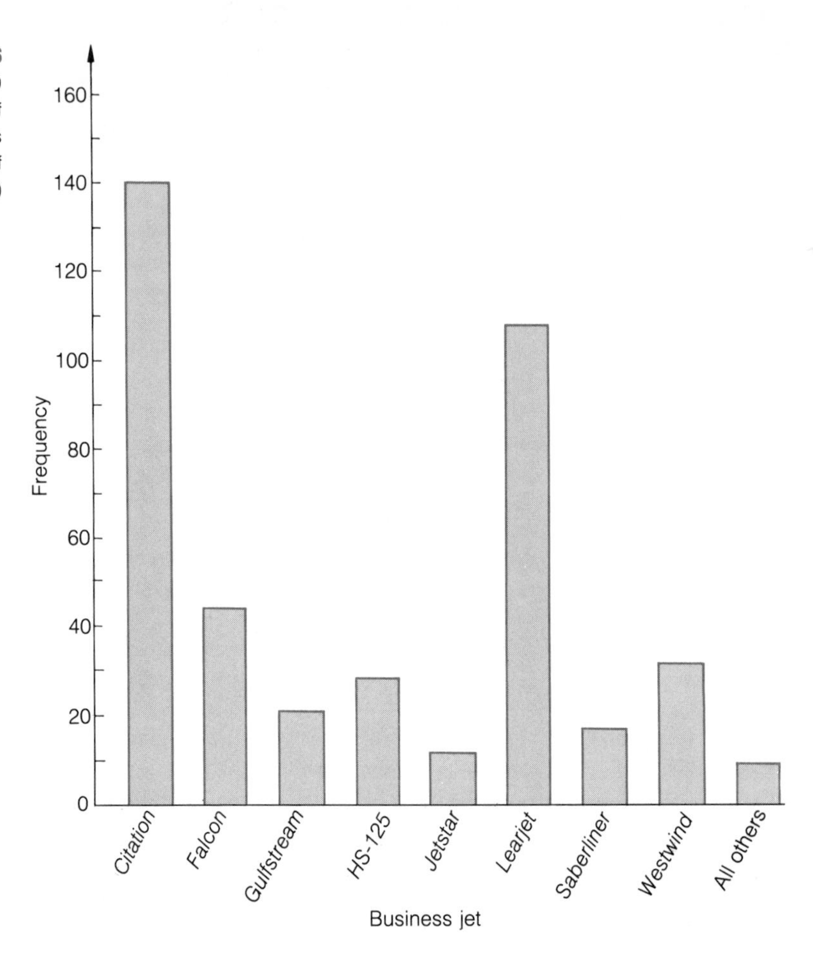

Source: *Weekly of Business Aviation,* January 14, 1980.

EXERCISES

2.6 During August, 1980, hundreds of thousands of Poland's workers walked off the job, protesting the country's poor labor conditions. One of their demands, eventually negotiated with the government through the independent Polish labor union Solidarity, was the reduction of a mandatory 6-day, 48-hour work week to a 5-day, 40-hour week. In the United States, many corporations are considering instituting a 4-day, 40-hour work week or a 3-day, 40-hour work week. Suppose that a company surveyed its employees concerning the type of work week they would prefer: a 6-day, 48-hour work week; a 5-day, 40-hour work week; a 4-day, 40-hour work week; or a 3-day, 40-hour work week. Twenty-five employees responded as shown in the table.

a. Construct a bar graph for the data on work-week preferences.

b. What is the frequency of workers who prefer a 5-day, 40-hour work week?

c. What is the relative frequency of workers who prefer a 3-day, 40-hour work week?

EMPLOYEE	WORK WEEK	EMPLOYEE	WORK WEEK
1	5-day, 40-hour	14	4-day, 40-hour
2	5-day, 40-hour	15	6-day, 48-hour
3	3-day, 40-hour	16	4-day, 40-hour
4	6-day, 48-hour	17	5-day, 40-hour
5	4-day, 40-hour	18	5-day, 40-hour
6	4-day, 40-hour	19	5-day, 40-hour
7	5-day, 40-hour	20	4-day, 40-hour
8	3-day, 40-hour	21	3-day, 40-hour
9	6-day, 48-hour	22	5-day, 40-hour
10	5-day, 40-hour	23	3-day, 40-hour
11	4-day, 40-hour	24	4-day, 40-hour
12	5-day, 40-hour	25	5-day, 40-hour
13	4-day, 40-hour		

2.7 Sea turtles (an endangered species) are often the subject of ecological research. However, little is known about the natural hatching success of sea turtles on nesting beaches since most studies of hatching success are conducted under artificial hatchery conditions. As part of an investigation of the natural survival rate of the green sea turtle, L. E. Fowler (1979) marked and monitored a total of 350 sea turtle nests on a 4-kilometer area of the Tortuguero beach located on the Caribbean coast of Costa Rica. The fate of each of the 350 marked nests was of prime interest. Fowler reported the results given in the table. Construct a bar graph for the data. Interpret your results.

NEST FATE	NUMBER
Undisturbed, young emerged (successful hatching)	148
Disturbed by predators, some young emerged	18
Destroyed by animal predators	122
Washed out by surf	20
Lost to human predators	23
Dead although undisturbed	19
TOTAL	350

Source: *Ecology,* October 1979, *60,* 946–955. Copyright 1979, the Ecological Society of America.

2.8 In order to gain insight on how well the law works to compensate the victims of automobile accidents, an elaborate study of auto accidents was undertaken. The resulting distribution of reparations is given in the table on the next page.

a. Illustrate the distribution of reparations with a pie chart.

b. Many auto drivers assume that liability law is overwhelmingly the most important source of compensation for accident victims. Does the pie chart (part a) support this belief?

SOURCE OF REPARATION TO INJURED PARTY	PERCENT OF TOTAL DOLLARS
Liability of third parties who had negligently caused the accident	55
Injured's own insurance:	
Accident	22
Hospital and medical	11
Life and burial	5
Social Security	2
Employer and Workmen's Compensation	1
Other	4
TOTAL	100%

2.9 *Time* magazine (January 26, 1981) reported that Jimmy Carter's "farewell" 1982 budget "does little to reverse the federal spending machine. Outlays in fiscal 1982 are slated to rise by 11.5%, to $739 billion, leaving a projected deficit of $27.5 billion" for President Ronald Reagan's administration. *Time* illustrated the areas of increased federal spending with pie charts which are reproduced here.

Source: *Time*, January 26, 1981. Reprinted by permission from *Time*, The Weekly Newsmagazine; Copyright Time Inc. 1981.

a. Compute the percentage of the federal budget allocated to each of the spending areas for both the 1981 and 1982 budgets.

b. Interpret the percentages you computed in part a.

c. Which area had been allocated the largest increase in federal spending, relative to the total 1982 budget?

2.10 Is alcoholism more prevalent among medical professionals? Does easy access to drugs tend to encourage their use? Is there a difference between the rates of addiction for nurses and physicians? In an attempt to answer these and other questions, Bissell and Jones (1981) conducted a survey of nurses and physicians who (1) considered themselves alcoholics, (2) were members of Alcoholics Anonymous (AA), and (3) had been completely abstinent for at least one calendar year immediately prior to being interviewed. One aspect of the study concerned the subjects' addiction to other drugs. Each of the subjects was asked if they had used a drug outside a hospital setting. If they had, they were asked if they had been addicted to it. Counting only those who responded in the affirmative, the results shown in the table were observed.

ADDICTION	NUMBER OF NURSES	NUMBER OF PHYSICIANS
Alcohol only	65	55
Both alcohol and narcotics	1	4
Both alcohol and nonnarcotic drugs	21	24
Alcohol, narcotics and nonnarcotic drugs	13	14
TOTALS	100	97

Source: *Nursing Outlook*, February 1981.

a. Construct either a bar graph or pie chart to describe addiction among the nurses interviewed.
b. Repeat part a for the group of physicians interviewed.
c. Compare the two figures, parts a and b. Does there appear to be a difference between the rates of addiction for the two groups of subjects? Explain.

2.11 Many worldwide industries currently use, or are planning to use, newly designed robots to perform certain assembly tasks which often require as many as ten people to complete. Information on the estimated number of industrial robots currently being utilized in each of 11 countries is given in the table.

COUNTRY	NUMBER OF INDUSTRIAL ROBOTS
Finland	130
France	200
Great Britain	185
Italy	500
Japan	10,000
Norway	200
Poland	360
Sweden	600
United States	3,000
U.S.S.R.	25
West Germany	850
TOTAL	16,050

Source: *Time*, December 8, 1980. Figures are 1979 estimates. Reprinted by permission from *Time*, The Weekly Newsmagazine; Copyright Time Inc. 1980.

a. Construct a bar graph to describe the distribution of industrial robots in the 11 countries.
b. Interpret the bar graph in part a.

2.4 Graphical Descriptions of Quantitative Data

You will recall that Appendix A contains the 948 starting salaries of graduates who earned a bachelor's degree at the University of Florida during June 1980–March 1981. Turn to Appendix A and examine the set of 948 starting salaries, and, as you do so, keep in mind that these salaries provide an indicator of the starting salaries of all University of Florida bachelor's degree graduates in this period. How could we provide a graphical description of this data set?

To answer this question, we shall extend the procedure employed in describing qualitative data sets. That is, we shall classify the starting salaries by forming intervals, say $10,000 to $12,499, $12,500 to $14,999, $15,000 to $17,499, etc. Then we shall count the starting salaries falling within each interval (or ***class***) and form a figure similar to the bar graph of Section 2.3. The bars can be constructed so that their heights are proportional to either the class ***frequencies*** or to the class ***relative frequencies.*** The corresponding graphs are called ***frequency distributions*** and ***relative frequency distributions,*** respectively.*

The ***relative frequency distribution*** that describes the 948 starting salaries is shown in Figure 2.7. The classes are marked off in intervals of $2,500 along the horizontal axis of the graph. Notice that the ***class intervals*** are of equal width and that we have essentially divided the starting salary axis into 16 equal class intervals. The heights of the bars constructed over the intervals are proportional to the relative frequencies of the respective classes.

Figure 2.7 provides a good graphical description of the 948 starting salaries. You can see how the starting salaries are ***distributed*** along the starting salary axis. The starting salaries tend to pile up near $15,000 (notice that the class, $13,750 to $16,250, has the greatest relative frequency). None of the starting salaries was less than $3,750, the ***lower class boundary*** of the lowest salary class, or larger than $43,750, the ***upper class boundary*** of the highest salary class.

Because the classes are of equal width, the area of the bar associated with a particular class is proportional to its class relative frequency. Consequently, we can visually guess the proportion of starting salaries falling within any particular starting salary interval by comparing the area of bars over that interval with the total area of all the bars.

EXAMPLE 2.8 Examine Figure 2.7 and visually estimate the proportion of the total number of starting salaries that lie between $8,750 and $18,750.

Solution Shade the bars lying above the interval, $8,750 to $18,750, as indicated in Figure 2.8. You can see that this shaded portion represents approximately .60 (actually .608) of the total area of the bars for the complete distribution. This tells us that approximately 60% of the starting salaries were in the interval, $8,750 to $18,750. A more precise (but less rapid) answer could be obtained by recording and summing the relative frequencies for the classes in the interval, $8,750 to $18,750.

---

*The proper term is ***histogram*** but the word ***distribution*** conveys a better notion of the functional use of the graph.

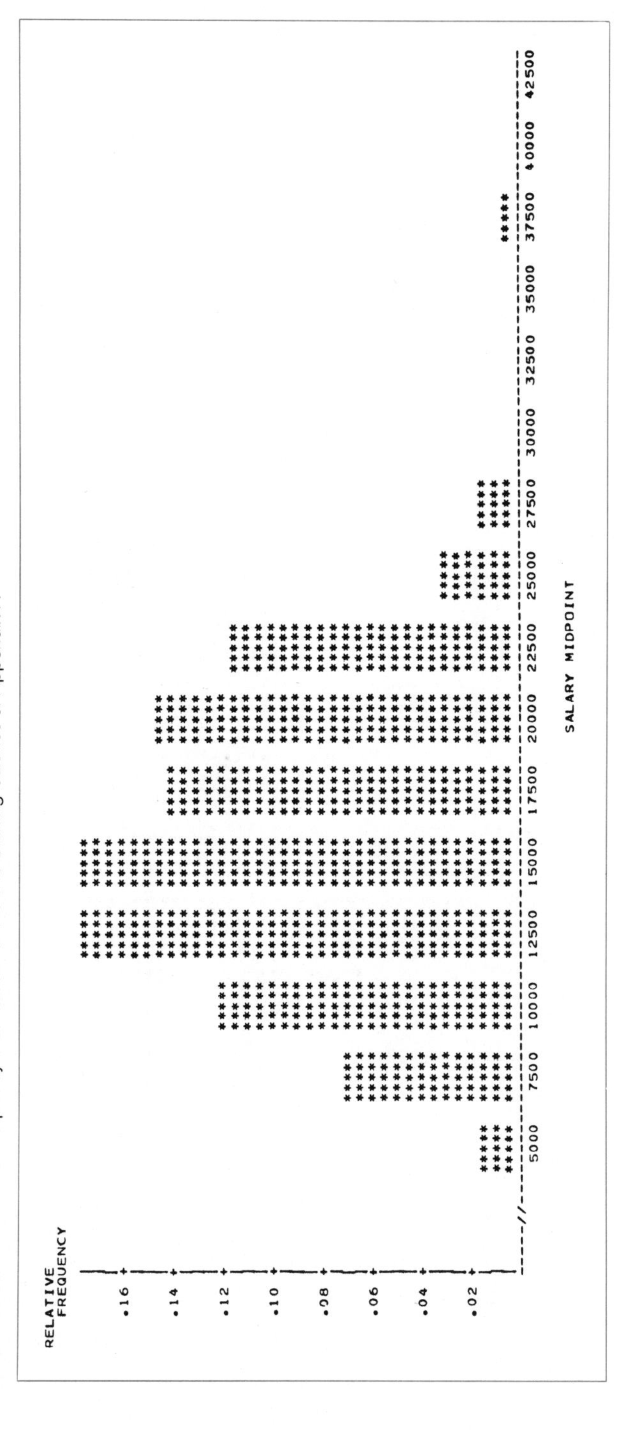

FIGURE 2.7 A Relative Frequency Distribution for the 948 Starting Salaries of Appendix A

FIGURE 2.8 Proportion of the 948 Starting Salaries That Lie between $8,750 and $18,750

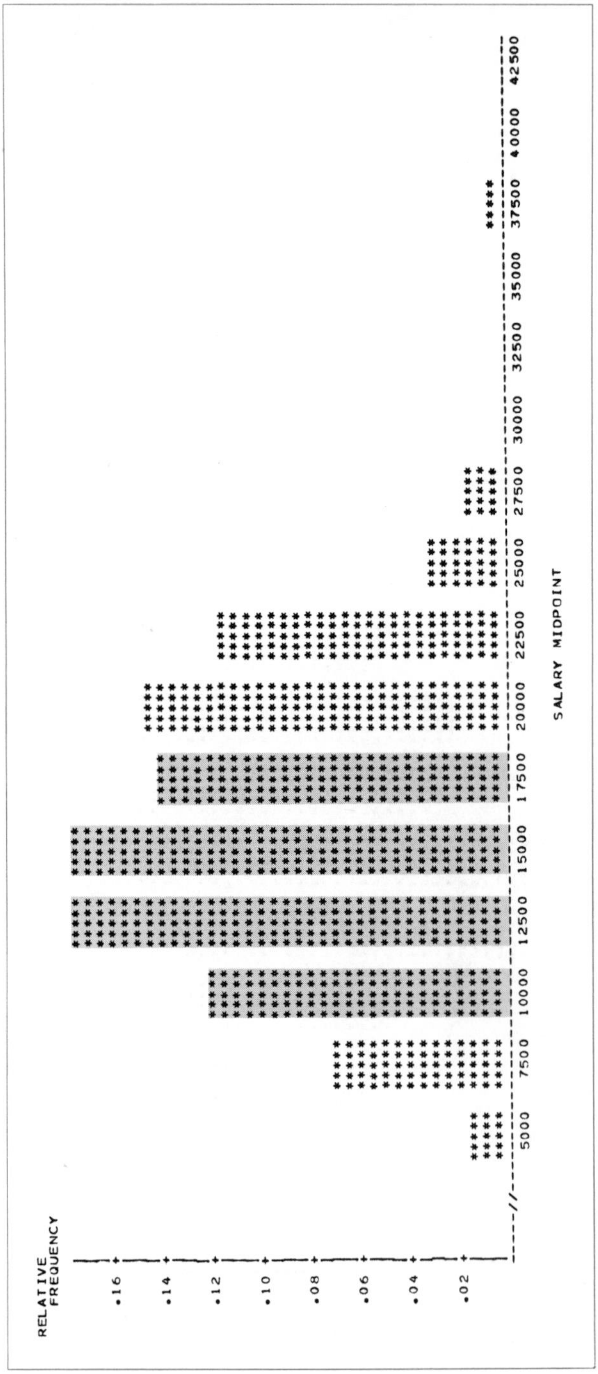

RELATIVE FREQUENCY

.16 .14 .12 .10 .08 .06 .04 .02

5000 7500 10000 12500 15000 17500 20000 22500 25000 27500 30000 32500 35000 37500 40000 42500

SALARY MIDPOINT

Now that we know what a relative frequency distribution is and the type of information it conveys, let us examine the details of its construction. We will use a sample of data from Appendix A to illustrate the procedure.

EXAMPLE 2.9 Table 2.3 gives the starting salaries for 50 graduates selected at random from the 948 starting salaries of Appendix A.

TABLE 2.3
The Starting Salaries for a Sample of 50 Graduates Selected from Appendix A

11,300	15,800	13,900	20,800	10,800
14,100	21,300	21,300	16,300	14,000
16,900	11,100	16,400	17,600	11,600
12,900	15,500	13,100	19,300	15,500
11,200	38,400	8,200	12,900	16,900
15,400	8,700	8,400	15,300	19,800
10,400	6,000	16,900	13,600	12,300
20,400	20,900	23,700	13,300	16,500
13,800	17,500	14,600	21,200	15,700
21,400	10,300	9,400	19,800	7,600

The first step in constructing the relative frequency distribution for this sample is to define the ***class intervals.*** To do this, we need to know the smallest and largest starting salaries in the data set. These salaries are \$6,000 and \$38,400, respectively. Since we want the smallest salary to fall in the lowest class interval and the largest salary to fall in the highest class interval, we shall want the class intervals to span starting salaries in the interval, \$6,000 to \$38,400.

The next step is to choose the ***class interval width,*** and this will depend upon how many intervals we wish to use to span the starting salary range. The starting salary range is equal to

$$\begin{aligned}\text{Range} &= \text{Largest measurement} - \text{Smallest measurement}\\ &= \$38{,}400 - \$6{,}000\\ &= \$32{,}400\end{aligned}$$

Suppose that we choose to use 11 class intervals.* Then the class interval width should approximately equal

$$\begin{aligned}\text{Class interval} &\approx \frac{\text{Range}}{\text{Number of class intervals}}\\ &\approx \frac{32{,}400}{11}\\ &\approx \$3{,}000\end{aligned}$$

We shall start the first class slightly below the smallest observation, \$6,000, and choose the point so that no observation can fall on a class boundary. Since starting salaries are recorded to the nearest hundred dollars, we can do this by

*We shall discuss the selection of an appropriate number of class intervals later in this section.

choosing the lower class boundary of the first class interval to be $5,550. [*Note:* We could just as easily have chosen $5,525, $5,650, $5,850 or any one of many other points below and near $6,000.] Then the class intervals would be $5,550 to $8,550, $8,550 to $11,550, etc. The 11 class intervals are shown in the second column of Table 2.4.

The next step in constructing a relative frequency distribution is to obtain each class frequency, i.e., the number of observations falling within each class. This is done by examining each starting salary in Table 2.3 and recording by tally (as shown in the third column of Table 2.4) the class in which it falls. The tally for each class gives the class frequencies shown in Column 4 of Table 2.4. Finally, we calculate the class relative frequency as

$$\text{Class relative frequency} = \frac{\text{Class frequency}}{\text{Total number of observations}}$$

$$= \frac{\text{Class frequency}}{50}$$

These values are shown in the fifth column of Table 2.4.

TABLE 2.4 Tabulation of Data for the 50 Starting Salaries, Table 2.3

CLASS	CLASS INTERVAL	TALLY	CLASS FREQUENCY	CLASS RELATIVE FREQUENCY
1	5,550— 8,550	////	4	.08
2	8,550—11,550	~~////~~ ///	8	.16
3	11,550—14,550	~~////~~ ~~////~~ /	11	.22
4	14,550—17,550	~~////~~ ~~////~~ ////	14	.28
5	17,550—20,550	~~////~~	5	.10
6	20,550—23,550	~~////~~ /	6	.12
7	23,550—26,550	/	1	.02
8	26,550—29,550		0	.00
9	29,550—32,550		0	.00
10	32,550—35,550		0	.00
11	35,550—38,550	/	1	.02
			50	1.00

The final step in constructing a relative frequency distribution is to draw the graph. Mark off the class intervals along a horizontal line, as shown in Figure 2.9.

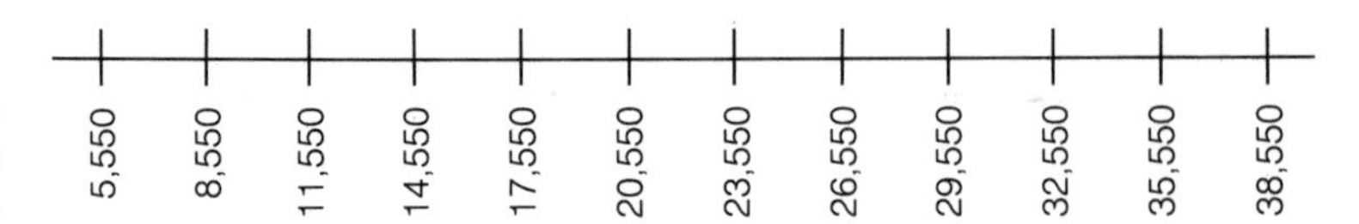

FIGURE 2.9 Class Intervals for the Data, Table 2.4

Then construct over each class interval a bar with the height proportional to the class relative frequency. The resulting relative frequency distribution is shown in Figure 2.10.

FIGURE 2.10 Relative Frequency Distribution for the Data, Table 2.3

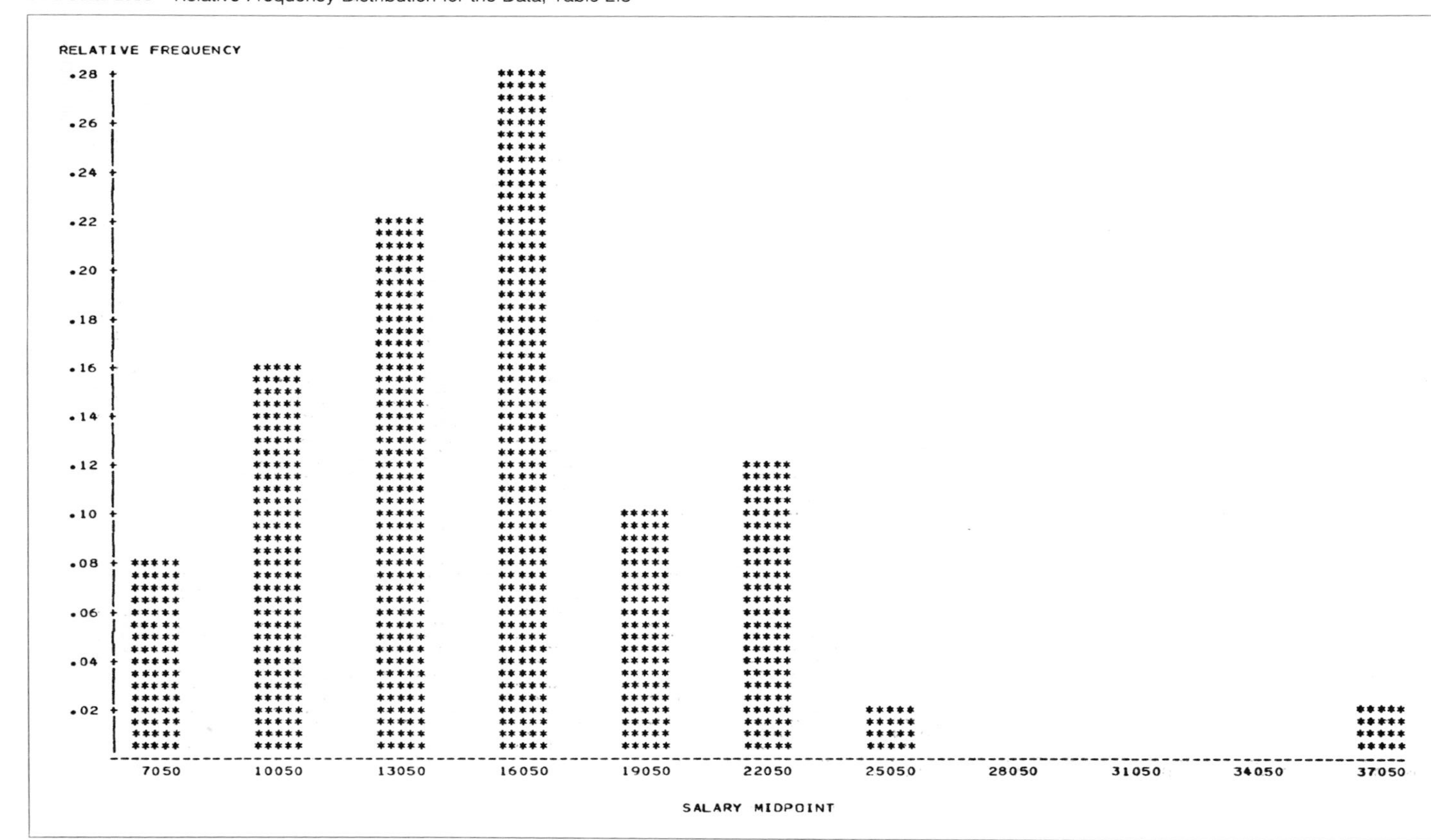

If you had any question about how to construct a relative frequency distribution, you would probably ask why we chose 11 class intervals for Figure 2.10. Why not 5, 10, 15, or 20? The answer is that there is no "best" number, but 11 seemed to be a good choice. Remember, the reason for constructing the graph is to obtain a figure that visually tells us how the observations are distributed along the starting salary axis. You can see at a glance that most of the graduates in the sample of 50 had starting salaries between \$8,550 and \$17,550; and you can obtain an approximate value for the proportion of starting salaries in this interval by comparing the area of the three bars over the interval, \$8,550 to \$17,550, to the total area of all the bars. This ratio is approximately ⅔. None of the starting salaries was less than \$5,550, but several high salaries caused the distribution to be ***skewed*** to the right. Suppose that you had spanned the range with 3 classes instead of 11. Almost all of the observations would have fallen in a single class and the resulting relative frequency distribution would not have been nearly as informative as Figure 2.10. Or suppose that you had chosen to span the interval, \$6,000 to \$38,400, with 50 classes. Then many classes would have contained only a few observations and others would have been empty. Again, this figure would not have been as informative as Figure 2.10.

The rule of thumb in deciding on the number of class intervals is to use a small number when you wish to describe a small amount of data, say 5 or 6 classes for up to 25 or 30 observations. You can increase the number of classes (as we did) for 50 observations, and you may wish to use 15 or 20 classes for large amounts of data. (See the relative frequency distribution, Figure 2.7, for the 948 observations of Appendix A.) Remember, the objective is to obtain a graph that rapidly conveys a visual picture of the data. If your first choice of class interval width is not satisfactory, you may wish to choose a different interval width and try again.

The steps employed in constructing a relative frequency distribution are summarized in the box on the next page.

The important concept to use in visually interpreting a relative frequency distribution is as follows:

INTERPRETING A RELATIVE FREQUENCY DISTRIBUTION

The percentage of the total number of measurements falling within a particular interval is proportional to the area of the bar that is constructed above the interval. If 30% of the area under the distribution lies over a particular interval, then 30% of the observations fell in that interval.

EXERCISES

2.12 Hospitals are required to file a yearly Cost Report in order to obtain reimbursement from the state for patient bills paid through the Medicare, Medicaid, and Blue Cross programs. Many factors contribute to the amount of reimbursement that

STEPS EMPLOYED IN CONSTRUCTING A RELATIVE FREQUENCY DISTRIBUTION

1. Examine the data to obtain the smallest and the largest measurement.
2. Divide the interval between the smallest and the largest measurement into between five and twenty equal subintervals called ***classes.*** These classes should satisfy the following requirement:

   Each measurement falls into one and only one subinterval.

   Note that this requirement implies that no measurement falls on a boundary of a subinterval. Although the choice of the number of classes is arbitrary, you will obtain a better description of the data if you use a small number of subintervals when you have a small amount of data and use a large number of subintervals for a large amount of data.
3. Compute the proportion (relative frequency) of measurements falling within each subinterval.*
4. Using a vertical axis of about three-fourths the length of the horizontal axis, plot each relative frequency as a rectangle over the corresponding subinterval.

the hospital receives. One important factor is bed size (i.e., the total number of beds available for patient use). The data below represent the bed sizes for 54 hospitals which were satisfied with their Cost Report reimbursements last year. Construct a relative frequency distribution for the bed size data using ten classes to span the range.

303	550	243	282	195	310	288	188	190	335
473	169	292	492	200	478	182	172	231	375
171	262	198	313	600	264	311	371	145	242
278	183	215	719	519	382	249	350	99	218
300	450	337	330	252	400	514	427	533	930
319	210	550	488						

2.13 Repeat Exercise 2.12, but use only three classes to span the range. Compare with the relative frequency distribution you constructed in that exercise. Which is more informative? Why does an inadequate number of classes limit the information conveyed by the relative frequency distribution?

2.14 Repeat Exercise 2.12, but use 25 classes. Comment on the information provided by this graph as compared to that of Exercise 2.12.

*Note that *frequencies* rather than relative frequencies may be used in constructing a frequency distribution. The frequency is the actual number of measurements in each interval.

2.15 The Community Attitude Assessment Scale (CAAS) measures citizens' attitudes towards fifteen life areas (e.g., education, employment, and health) on four dimensions—importance, influence, equality of opportunity, and satisfaction. In order to develop the CAAS, households in each of 35 communities were randomly selected and sent questionnaires. Because relatively low response rates suggest that there could be a substantial but unknown bias in the reported data, the percentage of the sample responding to the survey was determined in each community. The results are given below (in percent). Construct a relative frequency distribution for the data.

21	14	18	20	14	16	6	22	28
16	26	14	13	15	25	21	15	7
12	8	15	14	21	22	10	3	22
15	31	9	19	7	10	20	11	

2.16 In a study of a generality of response to pain, subjects were exposed to two different pain-producing stimuli. The objective of the study was to determine whether the subjects showed consistency in pain response. The following data represent a sample of pain-response scores from the study. Construct a relative frequency distribution for the data.

10	13	20	15	13	10	16	13	21
19	11	17	12	16	11	15	16	15

2.17 Examine the relative frequency distribution (Figure 2.7) for the starting salaries of the 948 University of Florida graduates of Appendix A. By visually comparing the area under the distribution to the left of \$13,750 to the total area under the distribution, estimate the proportion of starting salaries less than or equal to \$13,750.

2.5 Summary

The method for describing data sets of both qualitative and quantitative variables is based upon the same concept. Qualitative data, by their very nature, fall into specific categories or classes. Similarly, the values associated with quantitative data can be subdivided into classes. Then, to describe the data, we calculate the number of observations falling into each class and construct a bar graph where the height of the bar constructed for each class is proportional to the number of observations falling within that class. Alternatively, the height of the bar may be proportional to the proportion of the total number of observations falling within that class.

The difference between a bar graph and a relative frequency distribution is that the classes of a qualitative variable are unrelated. In contrast, the classes for a quantitative variable are intervals on a real line which are connected; the upper class boundary of one interval is the lower class boundary of the next.

If the bars are constructed so that they are of equal width, then the area of a bar, in comparison to the total area of all the bars, is proportional to the proportion of the total number of observations falling within that class interval. This enables you to examine a relative frequency distribution and visually estimate the proportion of the total number of measurements falling within specific intervals. In brief, bar graphs, pie charts, and relative frequency distributions provide for a rapid description of qualitative and quantitative data sets.

KEY WORDS

Quantitative data	Bar graph
Quantitative variable	Pie chart
Qualitative data	Class interval
Qualitative variable	Class boundaries (lower and upper)
Class	Frequency distribution
Class frequency	Relative frequency distribution
Class relative frequency	Skewed distribution

SUPPLEMENTARY EXERCISES

2.18 Classify each of the following variables as either quantitative or qualitative.

a. Number of 1982 federal income tax returns which are filed incorrectly with the Internal Revenue Service
b. Brand of cigarette with the highest nicotine content
c. Time (in seconds) before a subject reacts to a pain-producing stimulus
d. Neighborhood (of a city) with the lowest crime rate
e. Type of heating fuel used in a rural home
f. Number of animals treated in a veterinary clinic on a particular day
g. Price (per gallon) of unleaded gasoline at a self-service station
h. Professional sport with the highest average attendance per game

2.19 Many experts are predicting that energy, particularly in the form of liquid fuel, will be much more costly and in limited supply in the near future. Consumers are therefore being urged to conserve energy while new domestic supplies of liquid fuel are being developed. One seemingly popular form of energy conservation is carpooling. However, statistics compiled by the Bureau of the Census indicate that American motorists drive without anyone else in the car on 52% of their trips (*Scientific American,* May 1981), as shown in the table.

NUMBER OF CAR OCCUPANTS	RELATIVE FREQUENCY OF CAR TRIPS
1	.52
2	.28
3	.10
4	.06
5 or more	.04
	1.00

a. Use one of the graphical methods described in this chapter to summarize the data in the table.
b. What percentage of all automobile trips are made with no more than two people in the car?

2.20 "Possibly one of the single greatest sources of abuse and neglect of the elderly is the family, especially the . . . family in which the child has assumed a caretaking role and the parent is now in the dependent role," writes Suzanne K. Steinmetz (*Aging,* January–February 1981). In order to investigate the problem of elder abuse, Steinmetz interviewed a sample of 60 adult children caring for a dependent elderly parent (65 years or older). During the course of the interview, Steinmetz was able to ascertain various methods used by the adult children to control their elderly parents. The results are summarized below.

METHOD OF CONTROL	PROPORTION OF ADULT CHILDREN
Screamed and yelled	.40
Used physical restraint	.06
Forced feeding or medication	.06
Threatened to send to nursing home	.06
Threatened with physical force	.04
Hit or slapped	.03
Non-abusive method	.35
	1.00

a. State whether the variable of interest, Method of controlling elderly parents, is qualitative or quantitative.

b. Construct a pie chart for the data given in the table. Interpret your results.

2.21 The U.S. Environmental Protection Agency (EPA) performs fuel economy tests on all makes and models of automobiles each year. An important variable that is measured on each car is Miles per gallon (mpg). Listed below are the EPA highway estimates of mpg for thirty 1980 model automobiles. Construct a relative frequency distribution for the data using six class intervals.

AMC Spirit	18	Lincoln Continental	15
Buick LeSabre	17	Jaguar XJ	14
Buick Skylark	20	Olds Cutlass	20
Cadillac Eldorado	14	Ford Thunderbird	18
Chevy Malibu	19	Honda Civic	36
Chrysler Cordoba	17	Olds Omega	22
Datsun 310	31	Plymouth Arrow	29
Datsun 210SX	26	Plymouth Champ	37
Chevy Chevette	25	Pontiac Firebird	20
Audi 4000	22	Porsche 924	19
Dodge Aspen	17	Toyota Corolla	26
Dodge Colt	30	Triumph Spitfire	21
Fiat Brava	20	Rolls Royce/Bentley	10
Ford LTD	17	Plymouth Volare	17
Mazda GLC	27	VW Diesel Rabbit	42

Source: *The World Almanac & Book of Facts,* 1980 ed. Copyright © Newspaper Enterprise Association, 1981, New York, N.Y. 10166

2.22 Each year the National Soft Drink Association (NSDA) presents a review of the industry's sales performance based upon a survey of the association's members. The survey provides an estimate of annual industry growth that is reconciled with the U.S. Department of Commerce's Census of Manufacturers.

The NSDA reported the information given in the table on the 1977 market share of regular (nondiet) soft drink flavors.

SOFT DRINK FLAVOR	PROPORTION OF ALL PACKAGED SALES IN 1977
Cola	.622
Lemon Lime	.121
Orange	.035
Ginger Ale	.015
Root Beer	.030
Grape	.016
Other	.161
	1.000

Source: NSDA 1977 Sales Survey of the Soft Drink Industry.

a. Construct a bar graph for the 1977 market share data.
b. What was the predominant flavor of soft drink in terms of packaged sales in 1977?
c. Estimate the 1977 percentage of packaged sales of noncola soft drinks.

2.23 In the past decade the sunflower, grown mainly for the oil in its seeds, has become a major factor in U.S. agriculture (*Scientific American,* May 1981). The sunflower currently ranks second worldwide to the soybean as a source of vegetable oil. The rank of "sunoil" among the other sources of vegetable oil for the 1979–1980 crop year is shown in the figure.

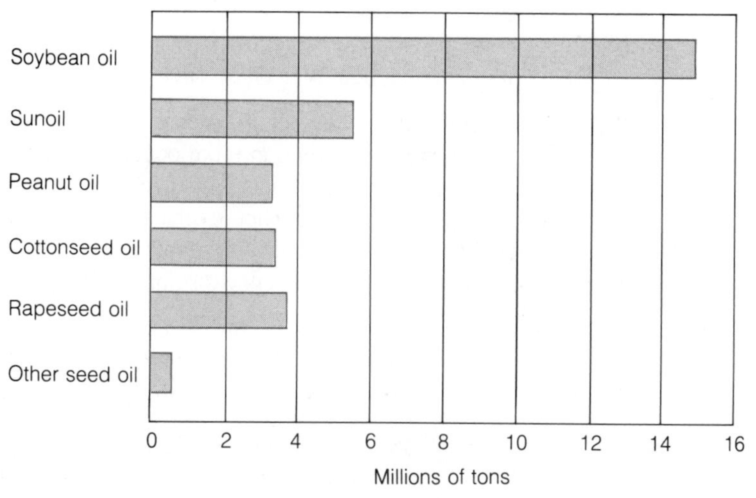

Source: B. H. Beard. "The sunflower crop." *Scientific American,* May 1981, *244,* 151–161.

a. What type of graphical method is displayed by the figure?
b. Approximately what proportion of the vegetable oil produced during the 1979–1980 crop year is sunoil?

2.24 Since 1972, the total number of business failures per year has slowly declined. The Business Economics Division of the U.S. Department of Labor classifies each business failure into one of the following eight categories: failure due to (1) neglect; (2) fraud; (3) lack of experience in the line; (4) lack of managerial experience; (5) unbalanced experience; (6) incompetence; (7) disaster; (8) unknown reasons. These classifications are based on the opinions of informed creditors and information in Business Economics Division reports.

A summary of 1,463 failures of construction enterprises in 1977 is given in the table.

UNDERLYING CAUSE	RELATIVE FREQUENCY
1. Neglect	.008
2. Fraud	.002
3. Lack of line experience	.076
4. Lack of managerial experience	.161
5. Unbalanced experience	.215
6. Incompetence	.477
7. Disaster	.004
8. Reason unknown	.057
	1.000

Source: *The Business Failure Record,* compiled by the Business Economics Division of the U.S. Department of Labor.

a. Use an appropriate graphical method to describe the 1977 construction business failures.
b. Visually estimate the percentage of construction businesses which failed due to inadequate experience or incompetence.

2.25 A traditional pulse rate of the economic health of the accommodations (hotel-motel) industry is the trend in room occupancy. (The ***room occupancy*** at a hotel or motel is determined by dividing the number of rooms occupied at the hotel or motel by the total number of inservice rooms available.) Given below are the room occupancies for 30 Miami, Florida, hotels (motels) during a randomly selected day in August. Construct a relative frequency distribution for the data.

.68	.93	.55	.70	.58	.60	.81	.48	.39	.43
.71	.80	.67	.52	.60	.92	.41	.59	.91	.85
.53	.77	.68	.66	.33	.62	.60	.82	.69	.67

2.26 The data on page 41 represent the numbers of 1979 graduates of Florida public high schools in each of the 67 counties in the state who entered a postsecondary institution.

1,452	741	86	507	1,019	1,243	258
214	447	42	443	1,228	5,705	206
1,340	12,474	165	131	452	3,238	93
211	154	105	48	518	526	2,236
3,918	102	158	1,002	444	325	114
8,505	5,717	180	1,482	1,874	542	255
100	2,607	380	1,223	204	931	176
359	84	489	203	4,962	1,775	
471	97	5,433	43	413	1,954	
1,017	473	198	177	4,285	248	

a. Classify the variable of interest, Number of high school graduates entering a postsecondary institution, as qualitative or quantitative.

b. Based on your answer to part a, use a graphical method to summarize the data.

2.27 The palatability of a new food product can often be determined by preliminary market taste tests. Experience has shown that having as few as 50 people taste and evaluate a new product under controlled conditions is adequate to reveal a major problem in consumer acceptability, if one exists. Suppose that 50 randomly selected individuals agreed to participate in a taste test for a new product, chocolate peanut butter. After tasting the product, each person was asked to mark a ballot rating overall acceptability on a scale from -3 to $+3$ (-3 = "terrible," -2 = "very poor," -1 = "poor," 0 = "average," $+1$ = "good," $+2$ = "very good," and $+3$ = "excellent"). The results are displayed in the graph.

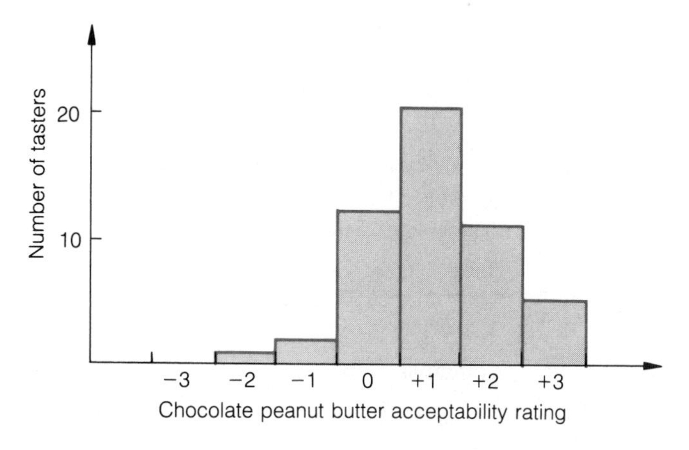

a. What type of graphical tool is used to describe the results of the taste test?

b. What information is conveyed by the graph?

c. What proportion of the 50 tasters rated the new chocolate peanut butter as "terrible"?

2.28 The pie chart (next page) gives the breakdown of each dollar spent by the U.S. federal government on elderly Americans (65 years or over) during the 1980 fiscal year.

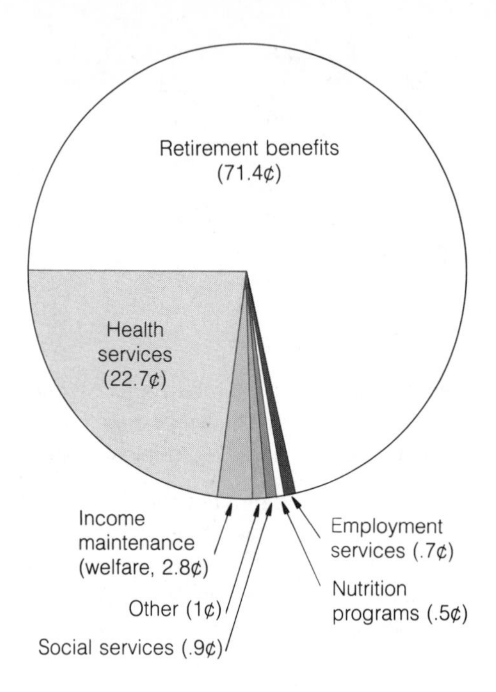

Source: U.S. House of Representatives, Select Committee on Aging, Subcommittee on Human Services, *Future Directions for Aging Policy: A Human Service Model,* Committee Publication No. 96-226 (Washington D.C., U.S. Government Printing Office, 1980) Figure 9.

a. What percentage of the federal government's 1980 expenditures on the elderly were allotted to health services?

b. What percentage of the federal government's 1980 expenditures on the elderly were allotted to retirement benefits?

c. What percentage of the federal government's 1980 expenditures on the elderly were allotted to nutrition programs and income maintenance (welfare)?

2.29 As part of a profile study of acute and chronic schizophrenics, a psychoanalyst classified each of 100 diagnosed schizophrenic patients according to age. The data are summarized in the table.

AGE (IN YEARS)	NUMBER OF SCHIZOPHRENICS
9.5–19.5	7
19.5–29.5	38
29.5–39.5	30
39.5–49.5	17
49.5–59.5	5
59.5–69.5	3
	100

a. Calculate the relative frequency of schizophrenics in each age class interval.

b. Use the relative frequencies of part a to construct a relative frequency distribution for the data. Interpret the figure.

2.30 During 1975, 15,759 felony offenders in Pennsylvania were sentenced to state prisons. The type of crime that each committed is summarized in the table. [*Note:* Although it is possible for an offender to be convicted for more than one crime, each felony offender was classified in only one of the crime type classes listed below.]

CRIME TYPE	NUMBER OF FELONY OFFENDERS CONVICTED
Murder	569
Rape	223
Robbery	1,903
Burglary	3,729
Drug offenses	5,153
Larceny/auto theft/stolen property	4,182
	15,759

Source: *Policy Sciences,* February 1981.

a. Classify the variable of interest, Crime type, as qualitative or quantitative.
b. What type of graphical method is appropriate for summarizing the data in the table?
c. Construct the graph of part b.
d. What proportion of the felony offenders in the state were convicted for drug offenses?
e. What proportion of the felony offenders in the state were convicted for robbery?

2.31 In his essay "Making Things Right," W. Edwards Deming considered the role of statistics in the quality control of industrial products.* In one example, Deming examined the quality control process for a manufacturer of steel rods. Rods produced with diameters smaller than 1 centimeter fit too loosely in their bearings and ultimately must be rejected (thrown out). To determine if the diameter setting of the machine which produces the rods is correct, 500 rods are randomly selected from the day's production and their diameters are recorded. The distribution of the 500 diameters for one day's production is shown in the figure on page 44. Note that the symbol LSL in the figure represents the 1-centimeter lower specification limit of the steel rod diameters.

a. What type of data, quantitative or qualitative, does the figure portray?
b. What type of graphical method is being used to describe the data?
c. Use the figure to estimate the proportion of rods with diameters between 1.0025 and 1.0045 centimeters.
d. There has been speculation that some of the inspectors are unaware of the trouble that an undersized rod diameter would cause later on in the manufacturing process. Consequently, these inspectors may be passing rods with

*From J. Tanur et al., eds. *Statistics: A guide to the unknown.* San Francisco: Holden-Day, 1978, pp. 279–281.

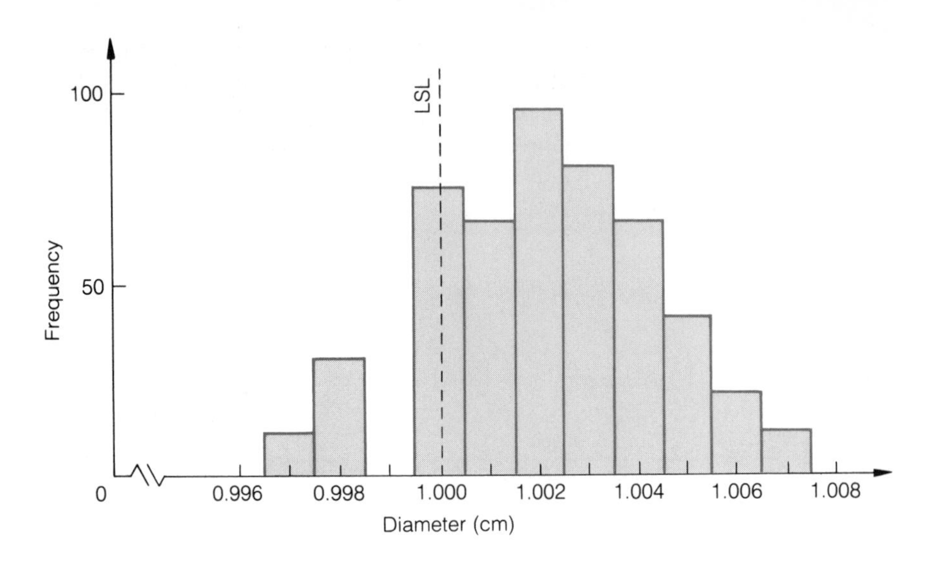

diameters that were barely below the lower specification limit, and recording them in the interval centered at 1.000 centimeter. According to the figure, is there any evidence to support this claim? Explain.

2.32 Refer to the supermarket customer checkout time data of Appendix C.

a. Classify the variable Supermarket customer checkout time as qualitative or quantitative.

b. Use one of the graphical methods described in this chapter to depict the distribution of checkout times for the 500 customers at supermarket A.

c. Repeat part b for the 500 customer checkout times at supermarket B. Place the figure on the same axes as the figure for part b so that the distribution of customer checkout times for the two supermarkets can be compared.

2.33 Refer to Exercise 2.32, part c. Explain how the manager of supermarket A could use these results as an aid in determining whether to authorize a switch from mechanical checking to automated checking. What is your recommendation if the manager is only interested in speeding up the customer checkout process at the supermarket?

CASE STUDY 2.1 The Most Serious Health Problem Facing Women

Has the public's general knowledge of breast cancer progressed since 1973? According to a National Cancer Institute (NCI) survey (*Nursing Outlook,* February 1981), 96% of women today have heard of the early detection technique of breast self-examination (BSE), an increase of 19% since 1973; and 40% of today's women—10% more than in 1973—use the technique at least once a month. The survey was conducted for the NCI by Opinion Research Corporation of Princeton, New Jersey, in the fall of 1979 and included a national sample of 1,580 adult women interviewed on a variety of health topics.

"Despite measurable progress," reports *Nursing Outlook,* "the findings from the NCI survey suggest a need for further public education about breast cancer.

Among all women in the . . . sample who said they examined their breasts, only 60 percent described at least three of six recommended steps in BSE . . . and . . . 10 percent were only able to describe one correct step." Surprisingly, half of the 1,580 respondents wrongly believed that a blow or other injury to the breast can cause cancer. Nevertheless, the survey results are encouraging to the NCI when one considers the responses to one question in particular: "In your opinion, what is the most serious health problem facing women?" The responses are summarized in Table 2.5.

TABLE 2.5

THE MOST SERIOUS HEALTH PROBLEM FOR WOMEN	RELATIVE FREQUENCY
1. Breast cancer	.44
2. Other cancers	.31
3. Emotional stress	.07
4. High blood pressure	.06
5. Heart trouble	.03
6. Other problems	.09
	1.00

Source: *Nursing Outlook,* February 1981.

a. Construct a bar graph for the data of Table 2.5.
b. What proportion of the respondents believe that high blood pressure or heart trouble is the most serious health problem for women?
c. Estimate the percentage of all women who feel that some type of cancer is the most serious health problem facing women.
d. Refer to your answer to part c. Comment on the reliability of your inference. Do you think that the sample of 1,580 responses are representative of the set of responses for the population of all (American) women? (You will be better able to answer this question after reading Chapters 7 and 8.)

CASE STUDY 2.2 Trouble beneath the Sidewalks of New York—Subways Deep in Debt

An article in *Railway Age* (May 12, 1980) describes the problems of New York's subways as follows:

> The nation's busiest passenger railroad, the 230-mile system that operates mainly beneath the sidewalks of New York, was back in business April 11 [1980] after a crippling 11-day strike by the Transit Workers Union. For New Yorkers, the return of their trains marked the end of a crisis—one that brought them not to their knees but to their feet, as millions walked, bicycled, and even roller-skated to work.
>
> But for the New York City Transit Authority (NYCTA), business as usual meant problems as usual.

Steven Kauffman, NYCTA's senior executive officer and general manager, hopes to solve their equipment shortage by acquiring hundreds of millions of dollars worth of new and rebuilt cars (in 1980), and is trying to correct the shortage of knowledgeable workers with a new personnel development program. However, the NYCTA is currently faced with "a deepening financial crisis that tends to shove [these and] all other problems to the back burner." Kauffman estimates that the NYCTA will need to spend $18 billion over the next 10 years to rebuild itself.

The severity of the NYCTA's debt for the fiscal year ending June 30, 1980, is illustrated by the pie charts shown in Figures 2.11(a) and 2.11(b).

FIGURE 2.11 NYCTA Funds for Fiscal Year Ending June 30, 1980 (Excludes Transit Police, Capital Engineering, Debt Service)

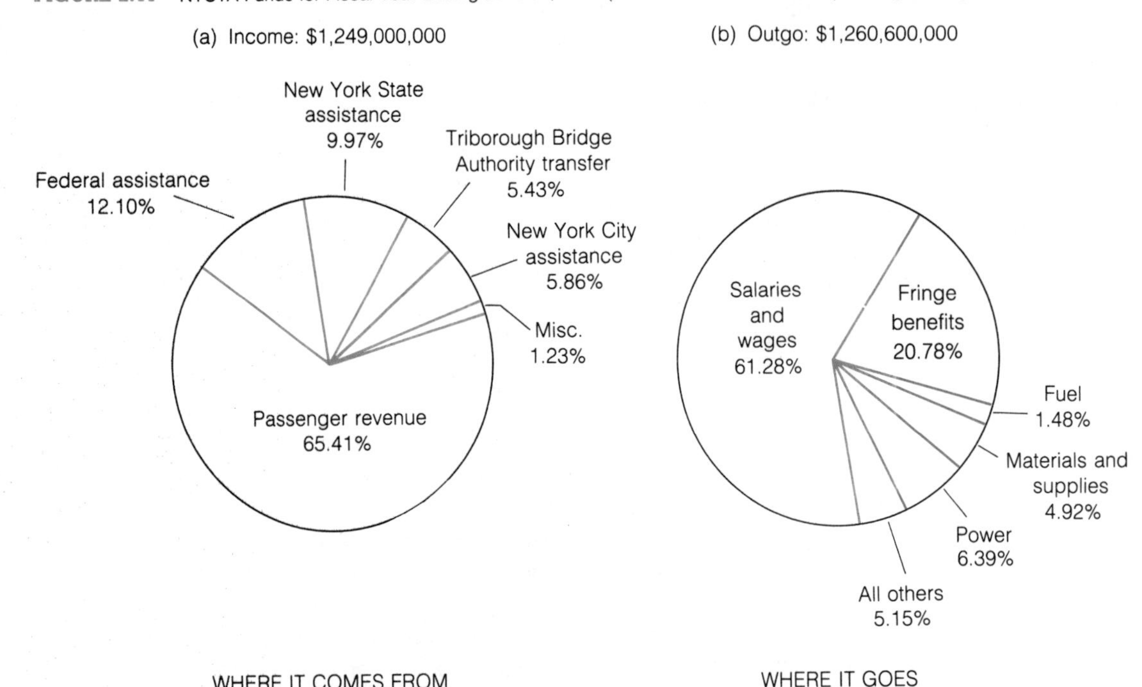

a. What proportion of incoming NYCTA funds were generated by passenger revenue during this fiscal year?
b. What proportion of incoming NYCTA funds were generated through the federal, New York State, or New York City assistance programs?
c. What proportion of outgoing NYCTA funds were allocated to employees' salaries, wages, or fringe benefits?

CASE STUDY 2.3 Carbon Monoxide: A Potential Hazard for Cigarette Smokers

An ongoing campaign of heavy advertising and warnings in the media by the U.S. Surgeon General has made both cigarette smokers and nonsmokers well aware of the dangers of inhaling the excess tar and nicotine from a burning cigarette. Now, however, the Surgeon General has added a third element to its list of hazardous substances that affect cigarette smokers—carbon monoxide. According to the Surgeon General and the Gainesville *Sun* (May 6, 1981), "Breathing carbon monoxide, a product of incomplete combustion, reduces the ability of blood to carry oxygen. For smokers, this occurs at the same time that inhaled nicotine is increasing the heart's oxygen needs. Research [conducted by the Surgeon General] has indicated carbon monoxide can be particularly hazardous to pregnant women smokers and heart disease patients."

As a result of these findings, the Federal Trade Commission (FTC), for the first time ever, has ranked 187 American cigarette brands in terms of the amount of carbon monoxide in their smoke. Table 2.6 shows the FTC's list, which is arranged

TABLE 2.6 Federal Trade Commission Listing of Carbon Monoxide (CO) Rankings for 187 Cigarette Brands

OBS	BRAND	LIGHT	FILTER	MENTHOL	PACK	LENGTH	TAR	NICOTINE	CO
1	ALPINE	R	F	M	SP	85	14.7	0.96	15.8
2	AMERICAN LIGHTS	L	F	NM	SP	120	7.7	0.67	8.6
3	AMERICAN LIFHTS	L	F	M	SP	120	8.9	0.82	9.8
4	ARTIC LIGHTS	L	F	M	SP	85	8.2	0.69	10.3
5	ARTIC LIGHTS	L	F	M	SP	100	8.4	0.74	10.9
6	ASPEN	R	F	M	SP	85	8.8	0.81	11.1
7	ASPEN	R	F	M	SP	100	8.6	0.83	10.3
8	BELAIR	R	F	M	SP	85	9.1	0.82	11.3
9	BELAIR	R	F	M	SP	100	8.2	0.65	10.2
10	BENSON & HEDGES	R	F	NM	SP	70	0.6	0.12	1.3
11	BENSON & HEDGES	R	F	NM	HP	85	15.6	1.31	14.5
12	BENSON & HEDGES 100	R	F	NM	HP	100	15.5	1.11	17.4
13	BENSON & HEDGES 100	R	F	M	HP	100	15.8	1.12	17.7
14	BENSON & HEDGES 100	R	F	NM	SP	100	15.9	1.13	17.6
15	BENSON & HEDGES 100	R	F	M	SP	100	15.7	1.15	18.0
16	BENSON & HEDGES	L	F	NM	SP	100	9.7	0.77	13.7
17	BENSON & HEDGES	L	F	M	SP	100	9.8	0.76	12.9
18	BROOKWOOD	R	F	M	SP	85	7.3	0.71	8.2
19	BULL DURHAM	R	F	NM	SP	85	27.3	1.82	26.3
20	CAMEL XX	R	NF	NM	SP	70	25.2	1.86	18.5
21	CAMEL XX	R	F	NM	SP	85	18.7	1.36	20.6
22	CAMEL LIGHTS	L	F	NM	SP	85	10.0	0.90	12.2
23	CAMEL LONG LIGHTS	L	F	NM	SP	100	12.2	0.97	15.8
24	CARLTON	R	F	NM	HP	85	0.1	0.06	0.5
25	CARLTON	R	F	NM	SP	85	0.8	0.13	1.9
26	CARLTON	R	F	M	SP	85	0.5	0.11	1.3
27	CARLTON 100'S	L	F	NM	SP	100	4.0	0.39	6.3
28	CARLTON 100'S	L	F	M	SP	100	3.8	0.36	6.0
29	CHESTERFIELD	R	NF	NM	SP	70	22.7	1.41	16.1
30	CHESTERFIELD	R	NF	NM	SP	85	27.1	1.68	19.8
31	CHESTERFIELD	R	F	NM	SP	85	1.7	1.01	16.5
32	CHESTERFIELD	R	F	NM	SP	101	16.0	1.12	18.5
33	DECADE	R	F	NM	SP	85	4.7	0.41	3.6
34	DECADE	R	F	M	SP	85	4.4	0.41	3.3
35	DECADE	R	F	NM	SP	100	8.6	0.74	7.6
36	DORAL	R	F	NM	SP	85	12.4	0.94	12.4
37	DORAL	R	F	M	SP	85	12.0	0.95	12.4
38	DORAL II	R	F	NM	SP	85	4.7	0.42	4.4
39	DORAL II	R	F	M	SP	85	5.1	0.43	4.8
40	DUMAURIER	R	F	NM	HP	85	14.3	0.97	19.0
41	ENGLISH OVALS	R	NF	NM	HP	70	22.0	1.74	13.2
42	ENGLISH OVALS	R	NF	NM	HP	85	27.8	2.25	15.6
43	EVE	R	F	NM	SP	100	15.1	1.14	17.2
44	EVE	R	F	M	SP	100	14.6	1.06	16.4
45	EVE	R	F	NM	HP	120	13.6	1.05	13.6
46	EVE	R	F	M	HP	120	12.2	0.97	12.4
47	FATIMA	R	NF	NM	SP	85	26.3	1.67	19.5
48	GALAXY	R	F	NM	SP	85	16.1	1.04	18.3
49	GOLDEN LIGHTS	L	F	NM	SP	85	8.2	0.77	9.1
50	GOLDEN LIGHTS	L	F	M	SP	85	7.6	0.68	8.7
51	GOLDEN LIGHTS 100'S	L	F	NM	SP	100	8.4	0.73	9.1
52	GOLDEN LIGHTS 100'S	L	F	M	SP	100	8.1	0.74	9.3
53	HALF & HALF XX	R	F	NM	SP	85	20.3	1.64	18.0
54	HERBERT TAREYTON	R	NF	NM	SP	85	26.8	1.73	20.4
55	HOME RUN	R	NF	NM	SP	70	21.1	1.28	21.1
56	ICEBERG 100'S	R	F	M	SP	100	2.9	0.27	4.9
57	KENT MICRONITE II	R	F	NM	HP	80	12.3	0.96	13.1
58	KENT MICRONITE II	R	F	NM	SP	85	12.5	0.99	13.6
59	KENT III	R	F	NM	SP	85	2.7	0.34	3.6
60	KENT MICRONITE II	R	F	NM	SP	100	14.5	1.14	15.2
61	KENT MICRONITE II	R	F	M	SP	100	14.3	1.21	14.7
62	KENT III	R	F	NM	SP	100	4.3	0.48	7.0
63	KOOL	R	NF	M	SP	70	18.2	1.13	15.5
64	KOOL	R	F	M	HP	80	16.0	1.27	18.3
65	KOOL	R	F	M	SP	85	15.5	1.20	18.6
66	KOOL MILDS	L	F	M	SP	85	11.5	0.89	14.3
67	KOOL SUPER LIGHTS	E	F	M	SP	85	7.2	0.59	8.6
68	KOOL	R	F	M	SP	100	15.1	1.11	19.4
69	KOOL MILDS	L	F	M	SP	100	12.4	1.04	14.6
70	KOOL SUPER LIGHTS	E	F	M	SP	100	8.4	0.68	10.9
71	KOOL INTERNATIONAL	R	F	M	HP	100	8.5	0.68	10.9
72	L & M	R	F	NM	HP	80	15.0	0.97	16.7
73	L & M LIGHTS	L	F	NM	SP	85	7.6	0.68	5.6
74	L & M XX	R	F	NM	SP	100	15.6	1.10	18.3
75	L & M XX	R	F	NM	SP	85	15.3	1.04	17.5
76	L & M LIGHTS	L	F	NM	SP	100	8.0	0.73	6.2
77	L & M	R	F	M	SP	100	16.3	1.14	18.4
78	LARK XX	R	F	NM	SP	85	16.0	1.11	17.4
79	LARK II	R	F	NM	SP	85	7.8	0.63	8.2
80	LARK LIGHTS	L	F	NM	SP	85	7.5	0.60	8.4
81	LARK 100'S XX	R	F	NM	SP	100	17.0	1.19	19.2
82	LARK LIGHTS 100'S	L	F	NM	SP	100	7.5	0.61	8.1
83	LONG JOHNS	R	F	NM	SP	120	16.5	1.33	19.5
84	LONG JOHNS	R	F	M	SP	120	14.3	1.22	18.1
85	LUCKY STRIKE	R	NF	NM	SP	70	22.5	1.32	8.8
86	LUCKY TEN	R	F	NM	SP	85	8.5	0.64	10.9
87	LUCKY 100'S	R	F	NM	SP	100	3.0	0.30	5.3
88	MARLBORO	R	F	NM	HP	80	15.5	1.04	16.4

TABLE 2.6 (cont'd.) Federal Trade Commission Listing of Carbon Monoxide (CO) Rankings for 187 Cigarette Brands

OBS	BRAND	LIGHT	FILTER	MENTHOL	PACK	LENGTH	TAR	NICOTINE	CO
89	MARLBORO	R	F	M	HP	80	14.5	0.96	14.6
90	MARLBORO	R	F	NM	SP	85	16.2	1.09	17.6
91	MARLBORO LIGHTS XX	L	F	NM	SP	85	11.2	0.82	14.2
92	MARLBORO	R	F	M	SP	85	14.1	0.92	15.7
93	MARLBORO	R	F	NM	HP	100	15.6	1.11	17.1
94	MARLBORO	R	F	NM	SP	100	15.9	1.10	17.6
95	MARLBORO LIGHTS	L	F	NM	SP	100	10.6	0.78	14.7
96	MAX	R	F	NM	SP	120	17.5	1.51	18.2
97	MAX	R	F	M	SP	120	17.8	1.52	18.5
98	MERIT XX	R	F	NM	SP	85	7.7	0.57	11.5
99	MERIT XX	R	F	M	SP	85	7.3	0.54	11.4
100	MERIT	R	F	NM	SP	100	9.2	0.72	12.4
101	MERIT	R	F	M	SP	100	9.9	0.75	13.0
102	MONTCLAIR	R	F	M	SP	85	15.6	1.04	17.7
103	MORE	R	F	NM	SP	120	21.8	1.75	25.6
104	MORE	R	F	M	SP	120	21.5	1.78	26.2
105	MULTIFILER	R	F	NM	SP	85	11.2	0.81	11.9
106	MULTIFILER	R	F	M	SP	85	11.8	0.78	12.5
107	NEWPORT	R	F	M	HP	80	15.5	1.19	17.5
108	NEWPORT	R	F	M	SP	85	16.6	1.29	19.4
109	NEWPORT LIGHTS	L	F	M	SP	85	9.7	0.86	11.6
110	NEWPORT	R	F	M	SP	100	19.8	1.57	22.6
111	NOW	R	F	NM	HP	85	1.6	0.19	2.6
112	NOW	R	F	NM	SP	85	2.0	0.25	3.1
113	NOW	R	F	M	HP	85	2.0	0.21	2.9
114	NOW	R	F	M	SP	85	1.8	0.23	2.9
115	OASIS XX	R	F	M	SP	85	15.5	1.07	16.6
116	OLD GOLD STRAIGHTS	R	NF	NM	SP	85	25.2	1.64	19.2
117	OLD GOLD FILTERS	R	F	NM	SP	85	16.5	1.27	19.2
118	OLD GOLD LIGHTS	L	F	NM	SP	85	9.3	0.81	10.9
119	OLD GOLD 100'S	R	F	NM	SP	100	18.5	1.47	20.2
120	PALL MALL	R	NF	NM	SP	85	23.5	1.51	18.6
121	PALL MALL	R	F	NM	SP	85	16.9	1.11	19.5
122	PALL MALL EXTRA LIGHTS	E	F	NM	SP	85	6.5	0.53	7.9
123	PALL MALL	R	F	NM	SP	100	17.0	1.26	18.5
124	PALL MALL LIGHTS XX	L	F	NM	SP	100	11.3	0.89	12.8
125	PALL MALL LIGHTS XX	L	F	M	SP	100	12.1	0.97	13.3
126	PARLIAMENT LIGHTS	L	F	NM	HP	80	8.4	0.61	10.2
127	PARLIAMENT LIGHTS	L	F	NM	SP	85	9.2	0.72	11.3
128	PARLIAMENT LIGHTS 100'S	L	F	NM	SP	100	11.4	0.91	11.6
129	PHILIP MORRIS	R	NF	NM	SP	70	21.6	1.47	13.8
130	PHILIP MORRIS CMDR	R	NF	NM	SP	85	25.4	1.66	16.7
131	PHILIP MORRIS INT'S	R	F	NM	HP	100	17.5	1.21	19.8
132	PHILIP MORRIS INT'S	R	F	M	HP	100	17.0	1.12	19.0
133	PICAYUNE	R	NF	NM	SP	70	22.4	1.36	21.7
134	PIEDMONT	R	NF	NM	SP	70	21.6	1.31	14.8
135	PLAYERS	R	NF	NM	HP	70	23.8	1.89	15.4
136	RALEIGH	R	NF	NM	SP	85	22.7	1.35	19.0
137	RALEIGH	R	F	NM	SP	85	15.2	0.98	19.3
138	RALEIGH LIGHTS	L	F	NM	SP	85	8.5	0.70	11.2
139	RALEIGH	R	F	NM	SP	100	16.0	1.14	20.4
140	RALEIGH LIGHTS 100'S	R	F	NM	SP	100	8.4	0.77	11.7
141	REAL	R	F	NM	SP	85	9.7	0.86	11.7
142	REAL	R	F	M	SP	85	8.7	0.81	10.0
143	ST MORITZ XX	R	F	NM	SP	100	14.6	1.10	15.8
144	ST MORITZ	R	F	M	SP	100	14.1	1.07	15.3
145	SALEM	R	F	M	HP	80	16.2	1.16	18.5
146	SALEM XX	R	F	M	SP	85	16.1	1.15	18.6
147	SALEM LIGHTS XX	L	F	M	SP	85	10.9	0.87	13.9
148	SALEM XX	R	F	M	SP	100	19.0	1.46	20.0
149	SALEM LIGHTS	L	F	M	SP	100	11.2	0.97	12.7
150	SARATOGA	R	F	NM	HP	120	15.4	1.12	18.1
151	SARATOGA	R	F	M	HP	120	14.8	1.06	16.7
152	SILVA THINS	R	F	NM	SP	100	12.7	1.07	11.7
153	SILVA THINS	R	F	M	SP	100	11.3	0.90	11.3
154	SPRING 100'S	R	F	M	SP	100	18.5	1.12	19.1
155	TALL	R	F	NM	SP	120	16.4	1.34	20.6
156	TALL	R	F	M	SP	120	15.1	1.22	18.4
157	TAREYTON	R	F	NM	SP	85	13.8	0.93	17.4
158	TAREYTON LIGHTS XX	L	F	NM	SP	85	7.0	0.60	8.8
159	TAREYTON ULTRA LOW TAR	E	F	M	SP	85	1.2	0.16	1.5
160	TAREYTON	R	F	NM	SP	100	13.6	0.97	18.2
161	TAREYTON LIGHTS XX	L	F	NM	SP	100	7.2	0.64	8.8
162	TEMPO	R	NF	NM	SP	85	6.8	0.55	9.2
163	TRIUMPH	R	F	NM	SP	85	2.8	0.38	3.6
164	TRIUMPH	R	F	M	SP	85	2.6	0.35	3.2
165	TRUE	R	F	NM	SP	85	4.9	0.43	5.9
166	TRUE	R	F	M	SP	85	4.9	0.45	6.1
167	TRUE	R	F	NM	SP	100	7.6	0.59	9.7
168	TRUE	R	F	M	SP	100	7.2	0.57	9.8
169	TWIST	R	F	M	SP	100	15.7	1.24	17.5
170	VANTAGE XX	R	F	NM	SP	85	10.5	0.82	16.6
171	VANTAGE XX	R	F	M	SP	85	10.4	0.81	16.8
172	VANTAGE ULTRA LIGHTS	E	F	NM	SP	85	6.3	0.54	9.5
173	VANTAGE XX	R	F	NM	SP	100	11.1	0.91	16.4
174	VICEROY	R	F	NM	SP	85	14.7	0.99	18.8
175	VICEROY RICH LIGHTS	L	F	NM	SP	85	8.3	0.69	10.8
176	VICEROY	R	F	NM	SP	100	15.4	1.15	19.7
177	VICEROY RICH LIGHTS	L	F	NM	SP	100	8.2	0.71	11.8
178	VIRGINIA SLIMS	R	F	NM	SP	100	14.8	0.99	16.0
179	VIRGINIA SLIMS	R	F	M	SP	100	14.4	0.99	16.1
180	VIRGINIA SLIMS LIGHTS	L	F	NM	HP	100	8.3	0.62	9.7
181	VIRGINIA SLIMS XX	R	F	M	SP	100	8.6	0.67	11.0
182	WINSTON	R	F	NM	HP	80	17.5	1.29	18.5
183	WINSTON XX	R	F	NM	SP	85	19.0	1.42	20.7
184	WINSTON LIGHTS XX	L	F	NM	SP	85	13.3	1.05	15.3
185	WINSTON 100'S XX	R	F	NM	SP	100	18.4	1.38	20.5
186	WINSTON LIGHTS 100'S	L	F	NM	SP	100	12.6	0.98	16.6
187	WINSTON 100'S	R	F	M	SP	100	17.1	1.32	18.1

alphabetically. In addition to carbon monoxide, each brand's tar and nicotine content is listed as well as the following variables:

Light type: Regular (R), Light (L), or Extra light (E)
Filter type: Filter (F) or Nonfilter (NF)
Menthol type: Menthol (M) or Nonmenthol (NM)
Pack Hardness: Hard pack (HP), Soft pack (SP)
Length (in millimeters): 70, 85, 100, or 120

[*Note:* The symbol XX means that the brand formula has been changed since the FTC tests were conducted.]

a. Classify each of the variables in Table 2.6 as either quantitative or qualitative.
b. Consider the following three data sets: the tar content (in milligrams) of the 187 cigarette brands, the nicotine content (in milligrams) of the 187 cigarette brands, and the carbon monoxide content of smoke (in milligrams) of the 187 cigarette brands. Construct relative frequency distributions for each of the three data sets. Compare and interpret the three graphs.

REFERENCES

Alexander, C. "Carter's farewell budget." *Time,* January 26, 1981, 50–51.

Beard, B. H. "The sunflower crop." *Scientific American,* May 1981, 151–161.

Bissell, L., & Jones, R. W., "The alcoholic nurse." *Nursing Outlook,* February 1981.

Book, S. *Statistics: Basic techniques for solving applied problems.* New York: McGraw-Hill, 1977. Chapter 1.

Dun & Bradstreet, Inc., *The business failure record.* 1977, 2–14.

Fowler, L. E. "Hatching success and nest predation in the green sea turtle, *chelonia mydas,* at Tortuguero, Costa Rica." *Ecology,* October 1979, *60,* 946–955.

Friedrich, O. "The robot revolution." *Time,* December 8, 1980, 72–78, 83.

"FTC ranks cigarettes for carbon monoxide." Gainesville *Sun,* May 6, 1981, 10A.

Gray, C. L., Jr., & von Hippel, F. "The fuel economy of light vehicles." *Scientific American,* May 1981, *244,* 48–59.

McClave, J. T., & Benson, P. G. *Statistics for business and economics.* 2d ed. San Francisco: Dellen, 1982.

Miller, H. D. "Projecting the impact of new sentencing law on prison populations." *Policy Sciences,* February 13, 1981, *13,* 51–73.

Miller, L. S. "New York subways: Business (& crises) as usual." *Railway Age,* May 12, 1980, 37–40.

National Soft Drink Association. "NSDA 1977 sales survey of the soft drink industry." 1–16.

"Public knows more about breast cancer: NCI survey." *Nursing Outlook,* February 1981, 71–77.

Steinmetz, S. K. "Elder abuse." *Aging,* January–February, 1981, 6–10.

Tanur, J. M., Mosteller, F., Kruskal, W. H., Link, R. F., Pieters, R. S., & Rising, G. R. *Statistics: A guide to the unknown.* San Francisco: Holden-Day, 1978.

Three

Numerical Methods for Describing Quantitative Data

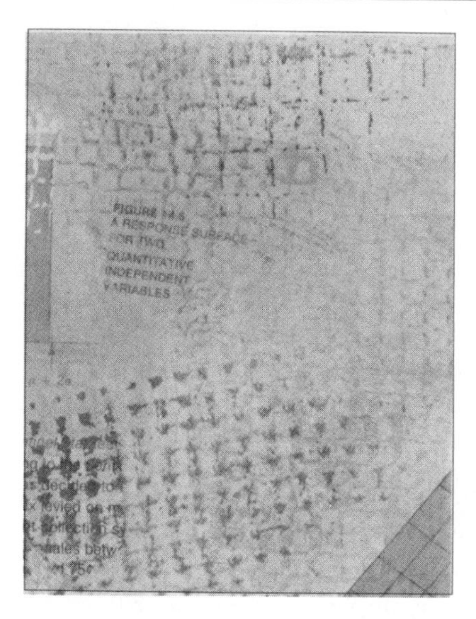

Engineers have a term for unaided human acts of lifting, lowering, pushing, pulling, carrying, or holding and releasing a heavy object—manual materials handling activities (MMHA). Researchers are currently working to develop strength and capacity guidelines for MMHA. The obvious problem is that not every person can safely lift the same heavy object. Suppose you were to determine a guideline for the weight that can be lifted safely for a sample of 75 male college students. How could you describe the capacities of these students with a single number that would characterize the capacity norms of the sample of 75 college students? We will show you in this chapter how numbers (called ***numerical descriptive measures***) can be used to describe the characteristics of a set of measurements, and we will apply it to the MMHA problem in Case Study 3.1.

Contents

3.1 Why We Need Numerical Descriptive Measures

It is probably true that a picture is worth a thousand words, and it is certainly true when you wish to describe a quantitative data set. But sometimes you will want to discuss the major features of a data set and it may not be convenient to produce the relative frequency distribution for the data. When this situation occurs, we seek a few summarizing numbers, called ***numerical descriptive measures,*** that conjure in our minds a picture of the relative frequency distribution.

3.2 Types of Numerical Descriptive Measures

Examine the relative frequency distribution for the 948 starting salaries, Appendix A, that is reproduced in Figure 3.1. If you were allowed to choose two numbers that would help you to construct a mental image of the distribution, which two would you choose?

We think that you would probably choose

1. A number that is located near the "center" of the distribution (see Figure 3.2(a))
2. A number that measures the "spread" of the distribution (see Figure 3.2(b))

A number which would describe the "center" of the distribution would be visually located near the spot where most of the data seem to be concentrated. Consequently, numbers that fulfill this role are called ***measures of central tendency.*** We will define and describe several measures of central tendency for data sets in Section 3.4.

The amount of "spread" in a data set is a measure of the variation in the data. Consequently, numerical descriptive measures that perform this function are called ***measures of variation*** or, sometimes, ***measures of dispersion.*** As you will subsequently see (Section 3.5), there are several ways to measure the variation in a data set.

Measures of central tendency and data variation are not the only numerical descriptive measures for describing data sets. Some are constructed to measure the ***skewness*** of a relative frequency distribution, the tendency of the distribution to tail out to the right (or left). For example, the relative frequency distribution, Figure 3.1, shows that most of the starting salaries were concentrated near $15,000 but that a few graduates had much larger starting salaries. A distribution that tends to spread unusually far to the high side is said to be ***skewed to the right*** or ***positively skewed.*** Similarly, distributions of data are said to be ***skewed to the left*** or ***negatively skewed*** if they tend to spread unusually far to the low side.

We will concentrate in this chapter on measures of central tendency and measures of variation. As you read this material, keep in mind our goal of using a pair of numbers to create a mental image of a relative frequency distribution. Relate each numerical descriptive measure to this objective and verify that it fulfills the role it is intended to play.

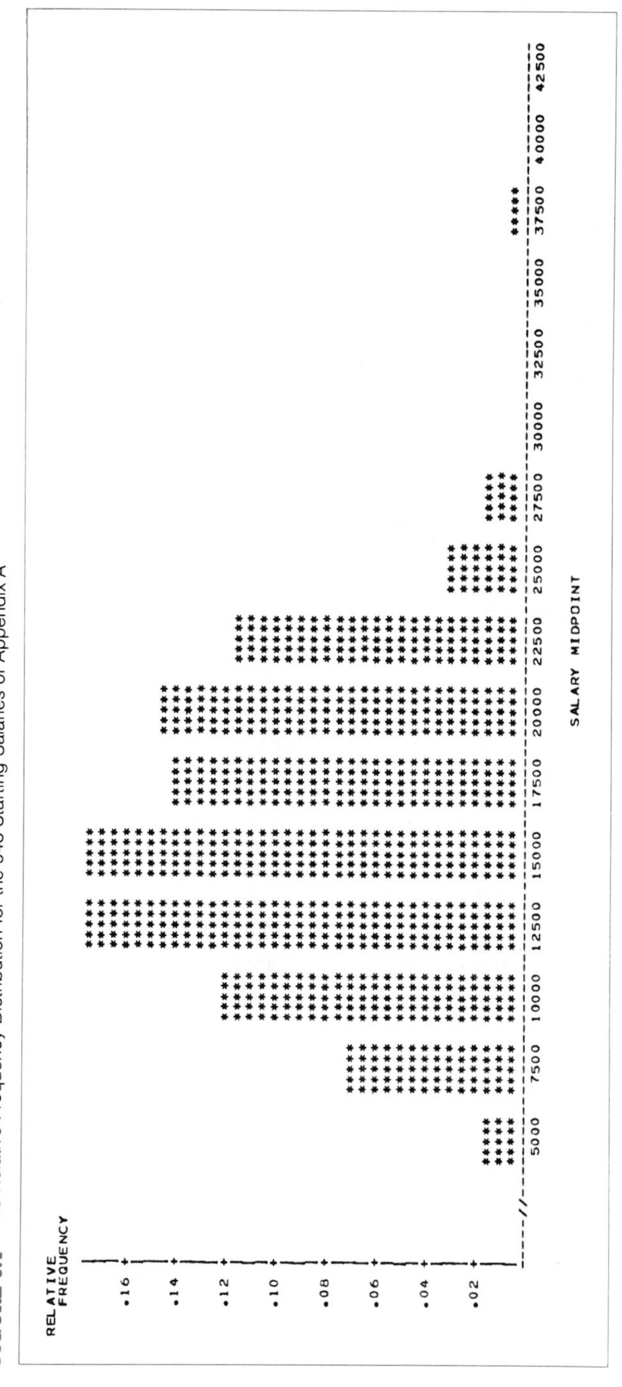

FIGURE 3.1 The Relative Frequency Distribution for the 948 Starting Salaries of Appendix A

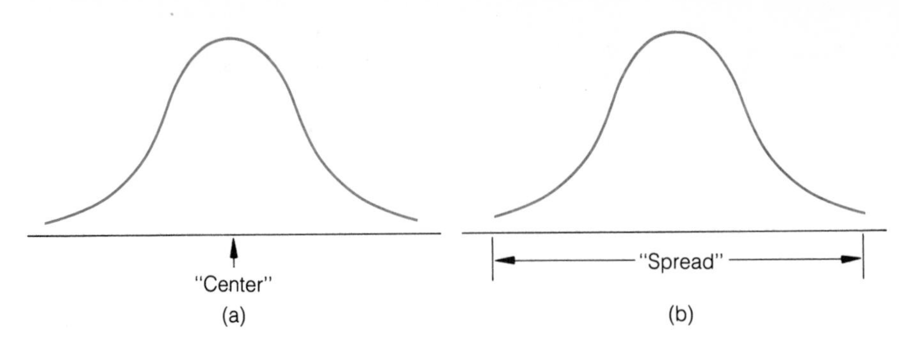

FIGURE 3.2 Numerical Descriptive Measures

3.3 Summation Notation

Explaining how to find specific numerical descriptive measures, particularly some measures of variability, often requires so much verbal explanation that the procedures may become lost in the verbiage. Consequently, we need to express the procedures in terms of formulas.

Suppose that a data set was obtained by observing a quantitative variable, x. For example, x may represent the starting salary of a college graduate. By observing x (starting salary) for the 948 graduates of Appendix A, we obtained the data set consisting of 948 starting salaries.

If we want to represent a particular observation in a data set, say the 26th one, we represent it by the symbol x with a subscript of 26. For example, you can see that the 26th observation in Appendix A is \$14,800. Therefore,

$$x_{26} = \$14,800$$

The complete data set would be represented by the symbols $x_1, x_2, x_3, \ldots, x_{948}$.

Most of the formulas that we shall use require the summation of numbers. For example, we may want to sum the observations in a data set, or we may want to square each observation and sum the squares of all observations in the data set. The sum of the observations in a data set will be represented by the symbol

$$\Sigma x$$

This symbol is read "summation x." The symbol Σ (sigma), the capital letter "S" in the Greek alphabet, is giving you an instruction. It is telling you to sum a set of numbers. The variable to be summed, x, is shown to the right of the Σ symbol.

EXAMPLE 3.1 Suppose that the variable x is used to represent the length of time (in years) for a college student to obtain his or her bachelor's degree. Five graduates are selected and the value of x is recorded for each. These observations are 2, 4, 3, 5, 4.

a. Find Σx. **b.** Find Σx^2.

Solution **a.** The symbol Σx tells you to sum the x-values in the data set. Therefore,

$$\Sigma x = 2 + 4 + 3 + 5 + 4 = 18$$

b. The symbol Σx^2 tells you to sum the squares of the x-values in the data set. Therefore,

$$\begin{aligned}\Sigma x^2 &= (2)^2 + (4)^2 + (3)^2 + (5)^2 + (4)^2 \\ &= 4 + 16 + 9 + 25 + 16 = 70\end{aligned}$$

EXAMPLE 3.2 Refer to Example 3.1.

a. Find $\Sigma(x-4)$. **b.** Find $\Sigma(x-4)^2$.

Solution **a.** The symbol $\Sigma(x-4)$ tells you to subtract 4 from each x-value and then sum. Therefore,

$$\begin{aligned}\Sigma(x-4) &= (2-4) + (4-4) + (3-4) + (5-4) + (4-4) \\ &= (-2) + 0 + (-1) + 1 + 0 = -2\end{aligned}$$

b. The symbol $\Sigma(x-4)^2$ tells you to subtract 4 from each x-value in the data set, square these differences, and then sum them as follows:

$$\begin{aligned}\Sigma(x-4)^2 &= (2-4)^2 + (4-4)^2 + (3-4)^2 + (5-4)^2 + (4-4)^2 \\ &= (-2)^2 + (0)^2 + (-1)^2 + (1)^2 + (0)^2 \\ &= 4 + 0 + 1 + 1 + 0 = 6\end{aligned}$$

MEANING OF SUMMATION NOTATION Σx

Sum observations on the variable that appears to the right of the summation (i.e., the sigma) symbol.

EXERCISES

3.1 A data set contains the observations 5, 1, 3, 2, 1. Find:

a. Σx **b.** Σx^2 **c.** $\Sigma(x-1)$ **d.** $\Sigma(x-1)^2$

3.2 Suppose that a data set contains the observations 3, 8, 4, 5, 3, 4, 6. Find:

a. Σx **b.** Σx^2 **c.** $\Sigma(x-5)^2$ **d.** $\Sigma(x-2)^2$

3.3 Refer to Exercise 3.1. Find:

a. $\Sigma x^2 - \dfrac{(\Sigma x)^2}{5}$ **b.** $\Sigma(x-2)^2$

3.4 Refer to Exercise 3.2. Find:

a. $\Sigma x^2 - \dfrac{(\Sigma x)^2}{7}$ **b.** $\Sigma(x-5)^2$

3.5 A data set contains the observations 6, 0, −2, −1, 3. Find:

a. Σx **b.** Σx^2 **c.** $\Sigma x^2 - \dfrac{(\Sigma x)^2}{5}$

3.4 Measures of Central Tendency

The word ***center,*** as applied to a relative frequency distribution, is not a well-defined term. In our minds, we know vaguely what we mean: a number somewhere near the "middle" of the distribution, a single number that tends to typify the data set. The measures of central tendency that we define often generate different numbers for the same data set but all will satisfy our general objective. If we visually imagine a hump-shaped relative frequency distribution, all measures of central tendency will fall near the middle of the hump.

The most common measure of the central tendency of a data set, one that is familiar to you, is the ***arithmetic mean*** of the data. The arithmetic average, or arithmetic mean, is defined as follows:

DEFINITION 3.1

The ***arithmetic mean*** of a set of n observations, $x_1, x_2, \ldots, x_n$, is denoted by the symbol $\bar{x}$, and is computed as:

$$\bar{x} = \frac{\text{Sum of the } x\text{-values}}{\text{Number of observations}} = \frac{\Sigma x}{n}$$

(the symbol $\bar{x}$ is read "x-bar")

EXAMPLE 3.3 Find the mean for the data set, 5, 1, 6, 2, 4.

Solution You can see that the data set contains $n = 5$ observations. Therefore,

$$\bar{x} = \frac{\Sigma x}{n} = \frac{5+1+6+2+4}{5} = \frac{18}{5} = 3.6$$

EXAMPLE 3.4 Find the mean for the 948 starting salaries of Appendix A. Locate it on the relative frequency distribution, Figure 3.1. Does the mean fall near the center of the distribution?

Solution We summed the starting salary observations in Appendix A by computer and divided this sum by the number of observations, 948, to obtain

$$\bar{x} = \$15{,}909$$

This mean or average starting salary should be located near the center of the relative frequency distribution for the 948 starting salaries, Figure 3.1. If you examine Figure 3.1, you will see that the mean $\bar{x}$ does indeed fall near the center of the mound-shaped portion of the distribution. If we did not have Figure 3.1 available, we could reconstruct the distribution in our minds as a mound-shaped figure centered in the vicinity of $\bar{x} = \$15{,}909$.

A second measure of central tendency for a data set is the ***median.*** The ***median*** for a data set is a number chosen so that half of the observations are less than the median and half are larger. Since the areas of the bars used to construct the relative frequency distribution are proportional to the numbers of observations falling within the classes, it follows that the median is a value of x that divides the area of the relative frequency distribution into two equal portions. Half of the area will lie to the left of the median (see Figure 3.3) and half will lie to the right. For example, the median for the 948 starting salaries, Figure 3.1, is $15,300. You can see from Figure 3.1 that this starting salary divides the data into two sets of equal size. Half of the 948 starting salaries will be less than $15,300; half will be larger.

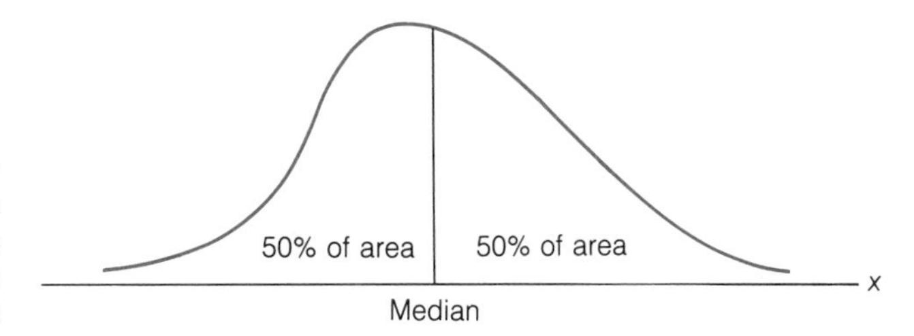

FIGURE 3.3 The Median Divides the Area of a Relative Frequency Distribution into Two Equal Portions

EXAMPLE 3.5 Find the median for the data set, 7, 4, 3, 5, 3.

Solution We first arrange the data in increasing (or decreasing) order:

3, 3, 4, 5, 7

Since we have an *odd number* of measurements, the choice for the median is easy; we will choose 4. Half of the measurements are less than 4 and half are greater than 4.

EXAMPLE 3.6 Suppose that you have an *even number* of measurements in the data set, say 5, 7, 3, 1, 4, 6. Find the median.

Solution If we arrange the data in increasing order, we obtain

1, 3, 4, 5, 6, 7

You can see that there are now many choices for the median. Any number between 4 and 5 will divide the data set into two groups of three each. There are many ways to choose this number, but the simplest is to choose the median as the point halfway between the two middle numbers when the data are arranged in order. Thus,

$$\text{Median} = 4.5$$

DEFINITION 3.2

The ***median*** of the n observations in a data set is defined as follows:

If n is odd: the middle observation when the data are arranged in order.
If n is even: the number halfway between the two middle observations when the data are arranged in order.

A third measure of central tendency for a data set is the ***mode.*** The ***mode*** is the value of x that occurs with greatest frequency (see Figure 3.4). If the data have been grouped into classes, we will define the mode as the center of the class with the largest class frequency (or relative frequency). For example, the modal starting salary for the data of Appendix A is \$15,000. (Although the vertical bars above the intervals with midpoints at \$12,500 and \$15,000 have the same height in Figure 3.1, the interval with a midpoint at \$15,000 actually contains two more starting salaries.)

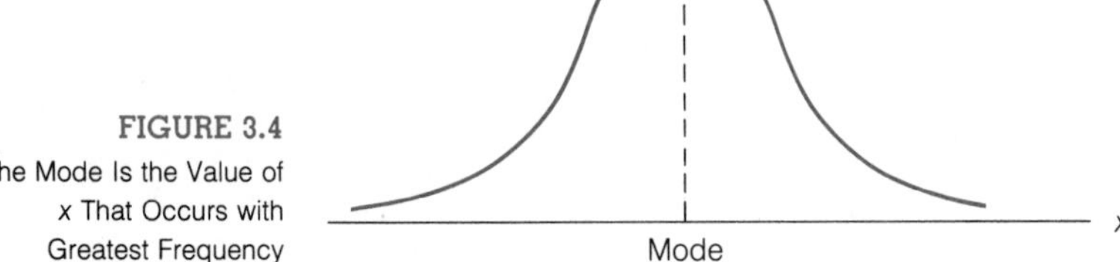

FIGURE 3.4
The Mode Is the Value of x That Occurs with Greatest Frequency

DEFINITION 3.3

The ***mode*** of a data set is the value of x that occurs with greatest frequency.

The mean, median, and mode are shown (Figure 3.5) on the graph of the relative frequency distribution for the 948 starting salaries, Appendix A. Which is the best measure of central tendency for this distribution? The answer is that it depends upon the descriptive information that you desire. If the distribution figure were made of plywood and balanced on a line, that line would coincide with the mean. The median is a point that equally divides the area of the distribution, and the mode is located beneath the point at which the highest frequency occurs. If your notion of a typical or "central" starting salary is one that is larger than half of the starting salaries and less than the remainder, then you will prefer the median to the mean or mode.

The mean is sensitive to very large or very small observations. Consequently, the mean will shift toward the direction of skewness. The median is often preferred as a measure of central tendency because it is insensitive to skewness. If the relative frequency of occurrence of values of x can be viewed as a measure of employer perception of a college graduate's first-year value (for example, the greatest frequency of starting salaries occurred in the \$13,750 to \$16,250 class), then the mode might be the preferred measure of central tendency. You may regard these differences in the mean, median, and mode to be rather minor when you examine Figure 3.5. Notice that there is little difference in the numerical values of these three measures of central tendency and that all three accomplish their objective. Knowing the value of any one of the three would help you mentally locate the "center" of a relative frequency distribution.

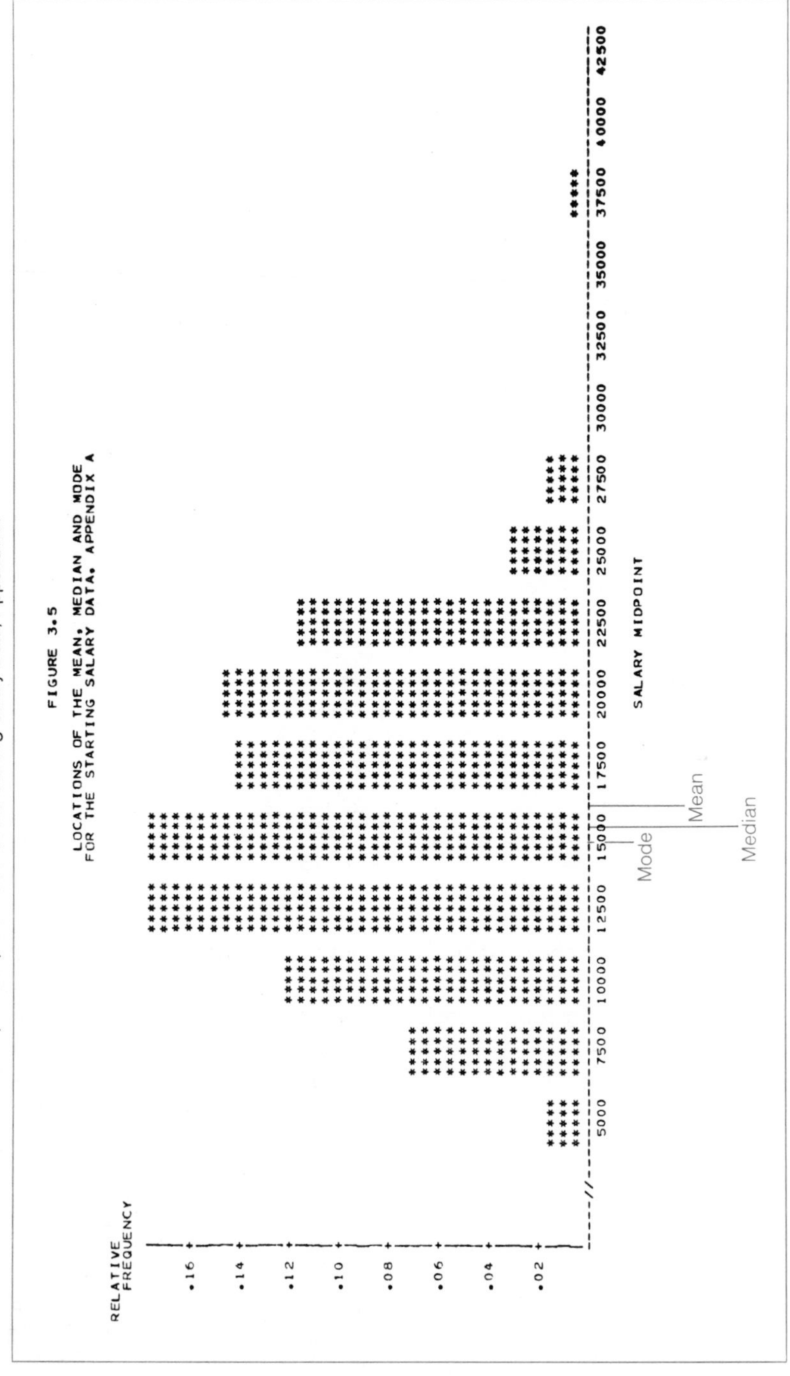

FIGURE 3.5 Locations of the Mean, Median, and Mode for the Starting Salary Data, Appendix A

EXERCISES

3.6 Find the mean and median for the data set consisting of the five measurements, 3, 9, 0, 7, 4.

3.7 Find the mean and median for the sample of $n = 6$ measurements, 7, 3, 4, 1, 5, 6.

3.8 Find the mean and median for the $n = 50$ starting salaries of Table 2.3 (page 31). Locate the mean and median on the relative frequency distribution for the data set (see Figure 2.10). Notice that they fall near the center of the distribution.

3.9 Find the ***modal class*** (the class in which the mode occurs) for the relative frequency distribution, Figure 2.10. Take the mode to be the midpoint of this class interval. Compare your answer with the mean and median obtained in Exercise 3.8.

3.10 Suppose that a distribution of data is skewed to the right. Would you expect the mean of this data set to be larger or smaller than the median? See if your answer agrees with the results of Exercise 3.8.

3.11 Appendix B contains the starting salaries for University of Florida graduates in five specific colleges. These data were extracted from Appendix A. The mean starting salaries (to the nearest dollar) for the five data sets are shown in the accompanying table. Use these measures of central tendency to construct a mental picture of the relative locations of the relative frequency distributions for the five data sets.

COLLEGE	MEAN STARTING SALARY
Business Administration	$15,225
Education	10,212
Engineering	20,178
Liberal Arts	12,460
Sciences	16,293

3.12 A psychologist has developed a new technique intended to improve rote memory of college students. To test the method against other standard methods, 20 college students are selected and each is taught the new technique. The students are then asked to memorize a list of 100 word phrases using the technique. The following are the numbers of word phrases memorized correctly by the students.

68	90	93	83	84	79	80
66	71	72	86	98	90	75
83	64	80	83	87	91	

a. Construct a relative frequency distribution for the data.

b. Compute the mean, median, and mode (modal class) for the above data set and locate them on the relative frequency distribution. Do these measures of central tendency appear to locate the center of the distribution of data?

3.5 Measures of Data Variation

Just as measures of central tendency locate the "center" of a relative frequency distribution, measures of variability measure its "spread." Examine the relative frequency distribution for the 948 starting salaries, Figure 3.5, and think how you might describe its spread. The first thought that probably comes to mind is the ***range.***

DEFINITION 3.4

The ***range*** of a quantitative data set is equal to the difference between the largest and the smallest measurements in the set.

EXAMPLE 3.7 Find the range for the data set, 3, 7, 2, 1, 8.

Solution The smallest and largest members of the data set are 1 and 8, respectively. Therefore,

$$\text{Range} = \text{Largest measurement} - \text{Smallest measurement}$$
$$= 8 - 1 = 7$$

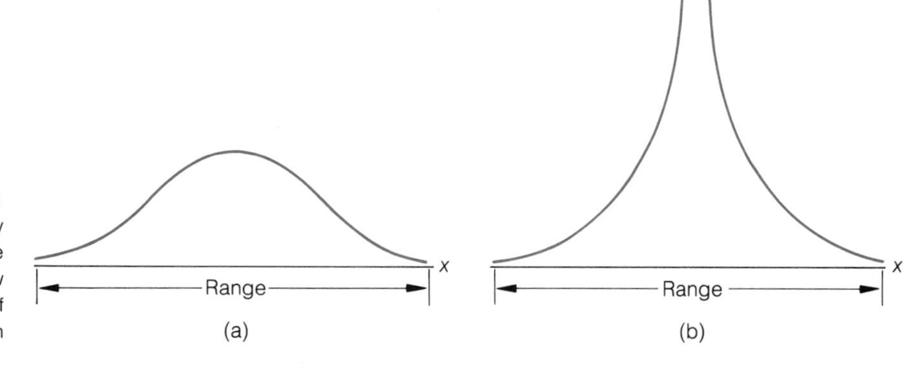

FIGURE 3.6 Two Relative Frequency Distributions That Have Equal Ranges, but Show Differing Amounts of Data Variation

The range of a data set is easy to acquire, but it is an insensitive measure of variation and is not very informative. To demonstrate insensitivity, examine Figure 3.6. Both relative frequency distributions possess the *same range,* but it is clear that the relative frequency distribution, Figure 3.6(b), indicates much less data variation than the distribution, Figure 3.6(a). Most of the observations, Figure 3.6(b), lie close to the mean. In contrast, most of the observations, Figure 3.6(a), deviate substantially from the center of the distribution. Since the ranges for the two distributions are equal, it is clear that the range is a fairly insensitive measure of data variation. It was unable to detect the differences in data variation for the data sets represented in Figure 3.6.

To demonstrate the fact that the range is not very informative, examine the relative frequency distribution, Figure 3.5, and note that most of the data fall between $6,250 and $28,750, i.e., visually, we see the range as approximately $22,500. But if you examine the set of all 948 starting salaries, Appendix A, you will find that the smallest observation is $4,900, the largest is $42,000, and the range is really $37,100. In other words, the range for the data set quantifies the spread of the extreme largest and smallest members of the data set. *What would be more useful is a range within which most of the data would lie.* This, along with the mean or median, would help us construct a mental image of the relative frequency distribution, Figure 3.5.

The magical measure that will perform this chore for us is known as the ***standard deviation*** of a data set. The standard deviation of a data set is based on how much the observations deviate from their mean. Notice that most of the observations, Figure 3.6(b), deviate very little from the mean of the distribution. In contrast, most of the observations, Figure 3.6(a), deviate substantially from the mean of that distribution.

The deviation between an observation x and the mean $\bar{x}$ of a data set is the difference,

$$x - \bar{x}$$

If a quantitative data set contains n observations, the ***variance*** of the data set is equal to the sum of the squares of the deviations from the mean of all n observations, divided by $(n - 1)$. That is, the variance, denoted by the symbol s^2, is

$$s^2 = \frac{\Sigma (x - \bar{x})^2}{n - 1}$$

The standard deviation, s, is the square root of this quantity. Before we give practical significance to the standard deviation, we will show you how it is computed for a small data set.

DEFINITION 3.5

The ***variance,*** s^2, of a set of n measurements is equal to the sum of squares of deviations of the measurements about their mean, divided by $(n - 1)$, i.e.,

$$s^2 = \frac{\Sigma (x - \bar{x})^2}{n - 1}$$

DEFINITION 3.6

The ***standard deviation*** of a set of n measurements is equal to the square root of the variance, i.e.,

$$s = \sqrt{\frac{\Sigma (x - \bar{x})^2}{n - 1}}$$

EXAMPLE 3.8 Find the standard deviation for the data set, 3, 7, 2, 1, 8.

Solution The five observations are listed in the first column of Table 3.1. You can see that $\Sigma x = 21$ and, therefore,

$$\bar{x} = \frac{\Sigma x}{n} = \frac{21}{5} = 4.2$$

This value of $\bar{x}$, 4.2, is subtracted from each observation to determine how much each observation deviates from the mean. These deviations are shown in the second column of Table 3.1. A *negative* deviation means that the observation fell *below* the mean; a *positive* deviation indicates that the observation fell *above* the mean. *Notice that the sum of the deviations equals 0. This will be true for all data sets.*

TABLE 3.1
Data and Computation Table

	OBSERVATION x	$x - \bar{x}$	$(x - \bar{x})^2$
	3	−1.2	1.44
	7	2.8	7.84
	2	−2.2	4.84
	1	−3.2	10.24
	8	3.8	14.44
TOTALS	21	0	38.8

The squares of the deviations are shown in Column 3 of Table 3.1. The total at the bottom of the column gives the sum of squares of deviations,

$$\Sigma (x - \bar{x})^2 = 38.8$$

Then the *variance* is

$$s^2 = \frac{\Sigma (x - \bar{x})^2}{n - 1} = \frac{38.8}{4} = 9.7$$

and the *standard deviation* is

$$s = \sqrt{s^2} = \sqrt{9.7} = 3.1$$

The procedure illustrated in Example 3.8 for calculating a standard deviation is tedious and often leads to rounding errors in finding the sum of squares of deviations, $\Sigma (x - \bar{x})^2$. A shortcut procedure for calculating the sum of squares of deviations is illustrated in the following example.

EXAMPLE 3.9 Use the ***shortcut procedure*** to calculate the sum of squares of deviations, $\Sigma (x - \bar{x})^2$, for the data set, Example 3.8.

Solution The shortcut procedure provides an easy way to calculate $\Sigma (x - \bar{x})^2$. Instead of calculating the deviation of each measurement from the mean, we calculate the squares of the observations, as shown in Table 3.2.

TABLE 3.2
Table for Calculating a Standard Deviation: The Shortcut Procedure

	OBSERVATION x	x^2
	3	9
	7	49
	2	4
	1	1
	8	64
TOTALS	21	127

Then it can be shown (proof omitted) that $\Sigma(x-\bar{x})^2$ is equal to

$$\Sigma(x-\bar{x})^2 = \Sigma x^2 - \frac{(\Sigma x)^2}{n}$$

Substituting the sum of squares Σx^2 and the sum Σx of the observations into this formula, we obtain

$$\Sigma(x-\bar{x})^2 = \Sigma x^2 - \frac{(\Sigma x)^2}{n} = 127 - \frac{(21)^2}{5}$$
$$= 127 - 88.2 = 38.8$$

This is exactly the same total that you obtained for the sum of squares of deviations in Table 3.1.

SHORTCUT PROCEDURE FOR CALCULATING $\Sigma(x-\bar{x})^2$

$$\Sigma(x-\bar{x})^2 = \Sigma x^2 - \frac{(\Sigma x)^2}{n}$$

Now that you know how to calculate a standard deviation, we will demonstrate with examples how it can be used to measure the spread or variation of a relative frequency distribution.

EXAMPLE 3.10 The mean and standard deviation of the 948 starting salaries of Figure 3.5 were calculated by computer and found to be

$\bar{x} = \$15{,}909$

$s = \$5{,}353$

Form an interval by measuring one standard deviation on either side of the mean, i.e., $\bar{x} \pm s$. Also, form the intervals $\bar{x} \pm 2s$ and $\bar{x} \pm 3s$. These three intervals are shown on the relative frequency distribution for the data, Figure 3.7. Find the proportions of the total number (948) of measurements falling within these intervals.

Solution It is too tedious to check by hand each of the starting salaries in Appendix A to determine whether they fall within the three intervals, so we did it by computer. The proportions of the total number of starting salaries falling within the three intervals are

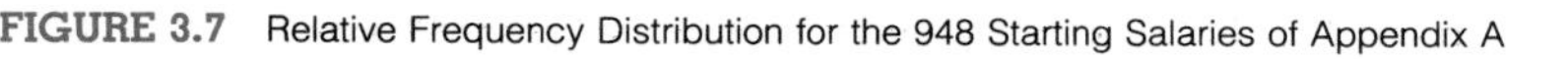

FIGURE 3.7 Relative Frequency Distribution for the 948 Starting Salaries of Appendix A

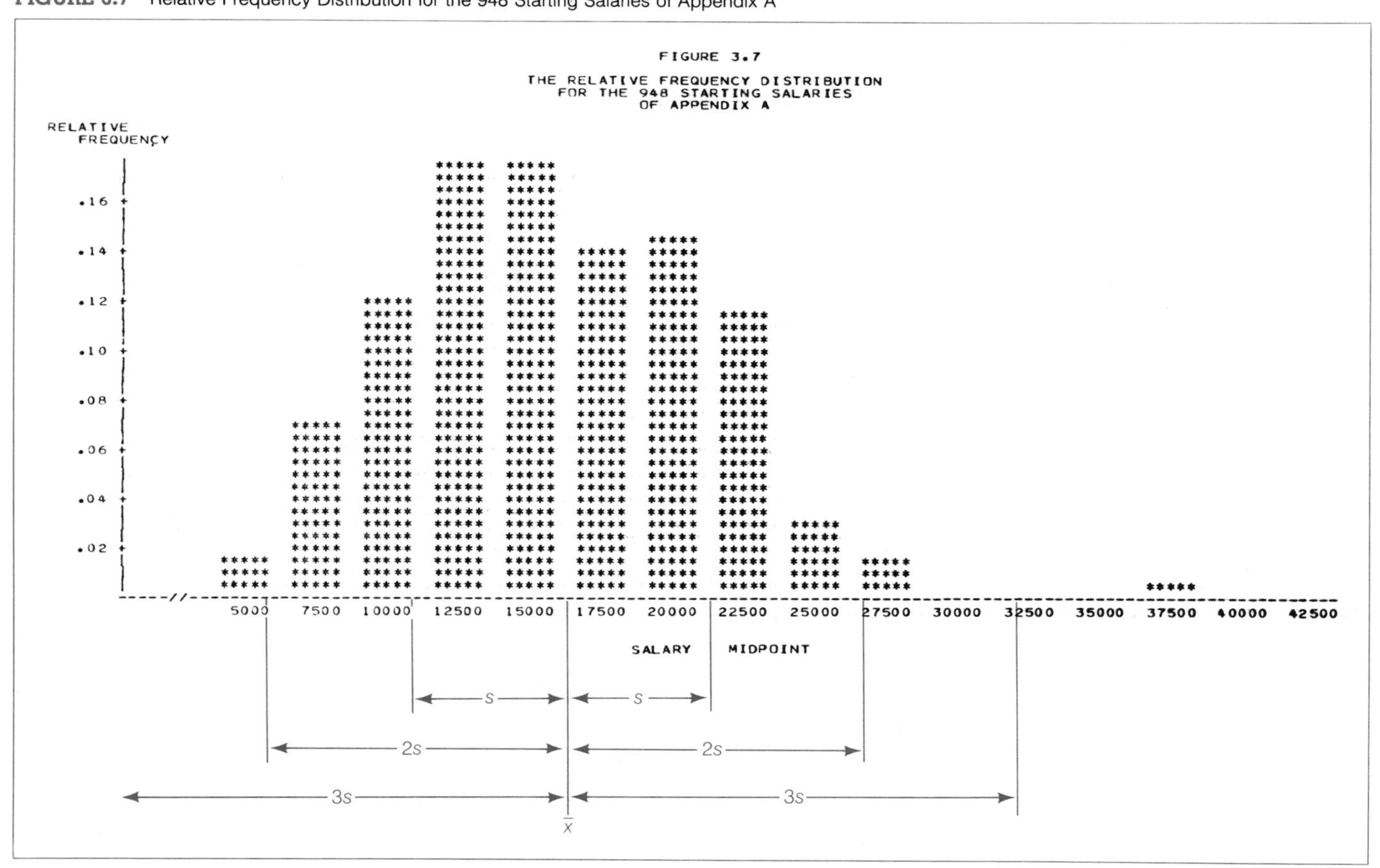

shown in Table 3.3. You can see that the proportions of the total area under the relative frequency distribution, Figure 3.7, that lie over the three intervals agree with these proportions.

TABLE 3.3 Proportions of the Total Number of Starting Salaries in Intervals $\bar{x} \pm s$, $\bar{x} \pm 2s$, and $\bar{x} \pm 3s$

INTERVAL	PROPORTION IN INTERVAL
$\bar{x} \pm s$ or ($10,556, $21,262)	0.71
$\bar{x} \pm 2s$ or ($5,203, $26,615)	0.98
$\bar{x} \pm 3s$ or (−$150, $31,968)	0.99

Will the proportions of the total number of observations falling within the intervals $\bar{x} \pm s$, $\bar{x} \pm 2s$, and $\bar{x} \pm 3s$ remain fairly stable for most distributions of data? To examine this possibility, consider the following example:

EXAMPLE 3.11 Calculate the mean and standard deviation of the following data sets (Appendix C):

a. The 500 customer checkout times at supermarket A
b. The 500 customer checkout times at supermarket B

Solution Because of the large amount of data involved, we computed the means and standard deviations on a computer. They are shown in Table 3.4. The relative frequency distributions for the two data sets are shown in Figures 3.8 and 3.9, respectively.

TABLE 3.4 Means and Standard Deviations for Two Data Sets of Appendix C

DATA SET	MEAN $\bar{x}$	STANDARD DEVIATION s
Checkout times at supermarket A	104.4 seconds	48.2 seconds
Checkout times at supermarket B	50.1 seconds	49.1 seconds

The means and standard deviations of Table 3.4 were used to calculate the intervals $\bar{x} \pm s$, $\bar{x} \pm 2s$, and $\bar{x} \pm 3s$ for each data set, and we obtained a computer count of the number and proportion of the total number of observations falling within each interval. These proportions are presented in Tables 3.5(a) and (b).

TABLE 3.5 Proportions of the Total Number of Observations Falling within $\bar{x} \pm s$, $\bar{x} \pm 2s$, and $\bar{x} \pm 3s$

(a) Checkout Times at Supermarket A

INTERVAL	PROPORTION IN INTERVAL
$\bar{x} \pm s$ or (56.2, 152.6)	0.72
$\bar{x} \pm 2s$ or (8.0, 200.8)	0.96
$\bar{x} \pm 3s$ or (−40.2, 249.0)	0.99

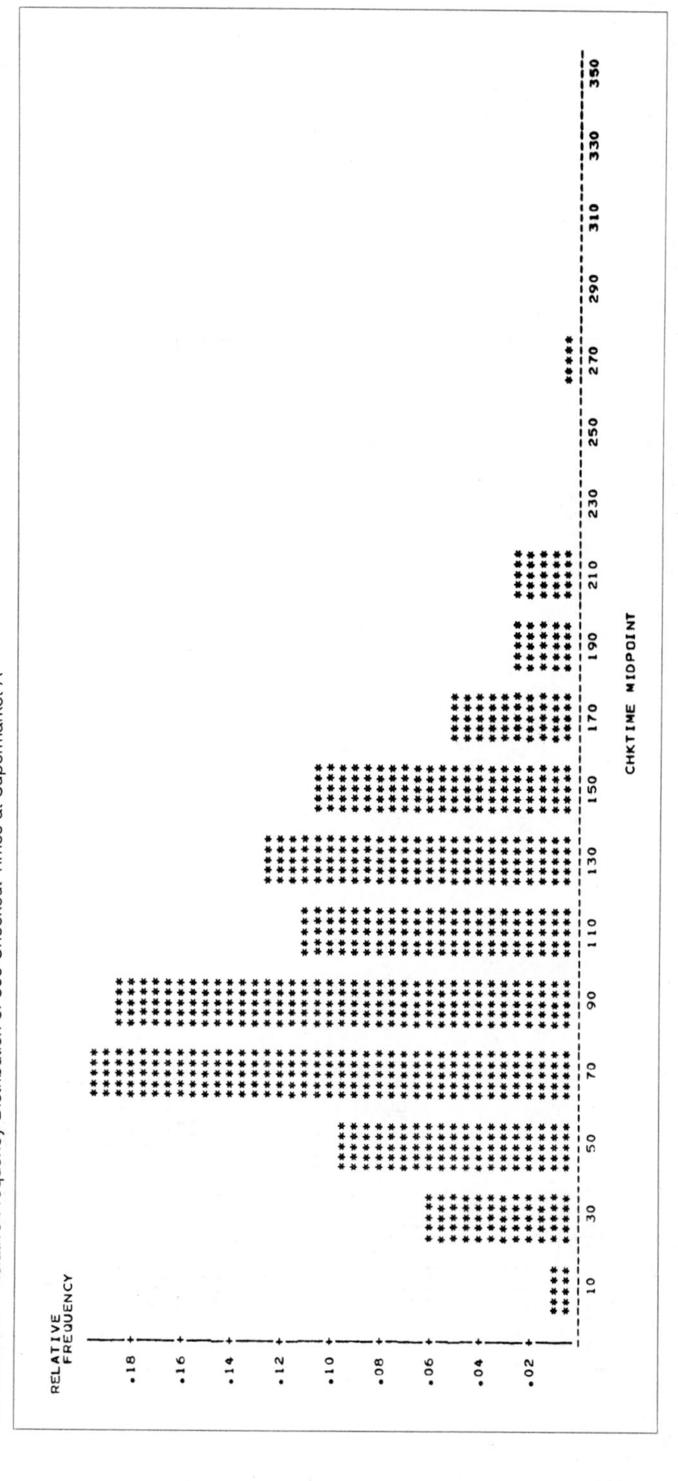

FIGURE 3.8 Relative Frequency Distribution of 500 Checkout Times at Supermarket A

FIGURE 3.9 Relative Frequency Distribution of 500 Checkout Times at Supermarket B

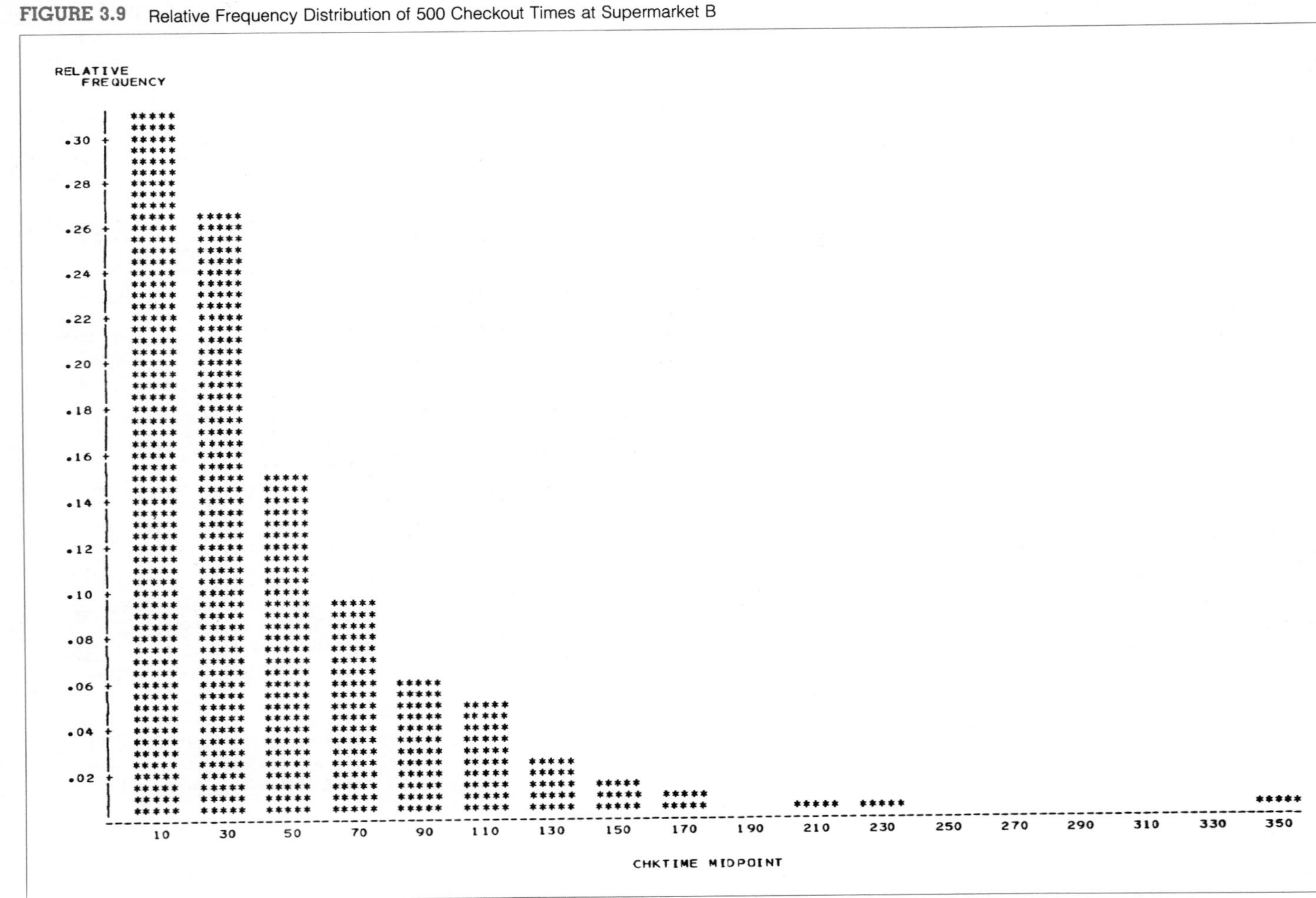

(b) Checkout Times at Supermarket B

INTERVAL	PROPORTION IN INTERVAL
$\bar{x} \pm s$ or (1.0, 99.2)	0.86
$\bar{x} \pm 2s$ or (−48.1, 148.3)	0.96
$\bar{x} \pm 3s$ or (−92.2, 197.4)	0.98

Tables 3.3 and 3.5(a) and (b) demonstrate a property that is common to many data sets. The percentage of observations that lie within one standard deviation of the mean $\bar{x}$, i.e., in the interval $(\bar{x} \pm s)$, is fairly large and variable, usually from 60% to 80% of the total number, but the percentage can reach 90% or more for highly skewed distributions of data. The percentage within two standard deviations of $\bar{x}$, i.e., in the interval $(\bar{x} \pm 2s)$, is close to 95% but, again, this percentage will be larger for highly skewed sets of data. Finally, the percentage of observations within three standard deviations of $\bar{x}$, i.e., in the interval $(\bar{x} \pm 3s)$, is almost 100%, meaning that almost all of the observations in a data set will fall within this interval. This property, which seems to hold for most data sets which contain at least 20 observations, is called the ***Empirical Rule.*** The Empirical Rule provides a very good rule of thumb for forming a mental image of a distribution of data when you know the mean and standard deviation of the data set. Calculate the intervals $\bar{x} \pm s$, $\bar{x} \pm 2s$, and $\bar{x} \pm 3s$ and then picture the observations grouped as follows:

THE EMPIRICAL RULE

If a relative frequency distribution of data has mean $\bar{x}$ and standard deviation s, then the proportions of the total number of observations falling within the intervals $\bar{x} \pm s$, $\bar{x} \pm 2s$, and $\bar{x} \pm 3s$ are as follows:

$\bar{x} \pm s$: Usually 60% to approximately 80%. The percentage will be larger (near 90%) for highly skewed distributions. The percentage will be near 70% for distributions that are mound-shaped and nearly symmetric.

$\bar{x} \pm 2s$: Close to 95%

$\bar{x} \pm 3s$: Near 100%

If you examine the relative frequency distribution of starting salaries, Figure 3.7, you will see that most of the starting salaries are concentrated in a mound-shaped distribution lying between $7,500 and $25,000. The distribution then skews to the right, and even contains two starting salaries larger than $40,000 and several in the range of $26,000 to $40,000 (see the original data set, Appendix A). These very large starting salaries create a moderate degree of skewness in the distribution which inflates the value of the standard deviation, s. For this reason, the percentages of observations falling within the intervals $\bar{x} \pm s$, $\bar{x} \pm 2s$, and $\bar{x} \pm 3s$ for the distribution,

Figure 3.7, will tend to be on the high side of the range of values given by the Empirical Rule. Notice also that the relative frequency distribution of customer checkout times for supermarket B of Appendix C is highly skewed to the right (see Figure 3.9). Consequently, the percentage of observations falling in the interval $\bar{x} \pm s$ is fairly large (86%).

EXERCISES

3.13 Find the range, variance, and standard deviation for the data set, 3, 9, 0, 7, 4. (Use the shortcut procedure to calculate $\Sigma(x-\bar{x})^2$.)

3.14 Find the range, variance, and standard deviation for the $n = 6$ measurements, 7, 3, 4, 1, 5, 6.

3.15 Find the variance and standard deviation for the $n = 25$ measurements, 2, 1, 7, 6, 5, 3, 8, 5, 2, 4, 5, 6, 3, 4, 4, 6, 9, 4, 3, 4, 5, 5, 7, 3, 5.

3.16 Refer to the data, Exercise 3.15. Construct the intervals $\bar{x} \pm s$, $\bar{x} \pm 2s$, and $\bar{x} \pm 3s$. Count the number of observations falling within each interval and find the corresponding proportions. Compare your results to the Empirical Rule.

3.17 Use the information, Exercise 3.16, to construct a sketch of the relative frequency distribution for the data set, Exercise 3.15. Then construct a relative frequency distribution for the data using class intervals .5 to 1.5, 1.5 to 2.5, . . . , 8.5 to 9.5. Compare your sketch with the relative frequency distribution for the data set, Exercise 3.16.

3.18 The National Advisory Committee on Criminal Justice Standards and Goals compiles yearly reports on the cost and resource implications of correctional standards related to halfway houses. The main purpose of these reports is to provide state and local decision makers with cost information on the many kinds of activities carried out by halfway houses. One variable included in the report is Yearly per bed rental costs. Given below are the yearly per bed rental costs (in dollars) for a random sample of 35 halfway houses:

417	282	280	303	400	76	643	480	472
317	264	384	205	257	136	250	100	732
750	402	422	373	325	408	345	749	313
791	196	891	283	186	693	137	52	

Compute $\bar{x}$ and s for this sample, then find the intervals $\bar{x} \pm s$, $\bar{x} \pm 2s$, and $\bar{x} \pm 3s$. Count the number of yearly per bed rental costs in each interval and construct a table similar to Table 3.3. Compare your results with the Empirical Rule. Do you agree that the values of $\bar{x}$ and s, in conjunction with the Empirical Rule, provide a reasonably good description of the data set?

3.19 Considering the climate, is it economically feasible to start an orange grove in northern California? If the temperature falls below 32°F, costly oil-burning smudge pots must be lit to keep the orange trees from freezing. Suppose a prospective grower randomly selects 20 years since 1900 and obtains the total number of days per year that the temperature fell below 32°F. The data are given on the next page.

15	14	14	13	9	10	15
17	20	25	18	28	13	12
16	11	13	12	16	6	

a. Compute $\bar{x}$ and s for this sample.
b. Construct the intervals $\bar{x} \pm s$, $\bar{x} \pm 2s$, and $\bar{x} \pm 3s$.
c. Does the Empirical Rule provide a good description of the sample data?

3.20 The Trail Making Test (TMT) is frequently used in neuropsychological assessment to provide a quick estimate of brain damage in humans. Subjects taking the TMT are asked to perform a certain task as quickly as possible. Like other tests of speed, the TMT is sensitive to the effects of age—an older person normally takes longer to complete the task. In order to investigate the neuropsychological deficits in alcoholics, 50 problem drinkers (25 under the age of forty and 25 forty years or older) were given the TMT and their performance scores observed (the higher the score, the more extensive the brain damage). The results are reported below.

	ALCOHOLICS UNDER AGE 40	ALCOHOLICS 40 OR OLDER
Mean performance score	39.6	49.7
Standard deviation	19.7	19.1

a. Use the above information, in conjunction with the Empirical Rule, to sketch your mental images of the relative frequency distributions of TMT performance scores for the two groups of alcoholics.
b. Estimate the fraction of alcoholics under age 40 who score between 19.9 and 59.3 on the TMT.
c. Approximately what percentage of alcoholics aged 40 or older score between 11.5 and 68.8 on the TMT?

3.21 Worker productivity is often gauged by the amount that a company invests in capital equipment divided by the number of workers at the company. Recent studies show that office productivity lags far behind the productivity figures for industry and agriculture. The means and standard deviations of Company investment per worker in capital equipment for the three types of workers as reported by a major consulting firm are shown in the accompanying table.

	OFFICE WORKER	INDUSTRIAL WORKER	FARM WORKER
Mean productivity	\$2,000	\$25,000	\$35,000
Standard deviation	500	7,000	8,000

a. Use the above information, in conjunction with the Empirical Rule, to sketch your mental images of the three relative frequency distributions. Construct them on the same graph so that you can see how they appear relative to each other.

b. Estimate the proportion of companies which invest between $11,000 and $39,000 in capital equipment per industrial worker.

c. Estimate the proportion of companies which invest between $27,000 and $43,000 in capital equipment per farm worker.

3.6 Measures of Relative Standing

You may want to describe the relative position of a particular measurement in a data set. For example, suppose that a college graduate in the data set, Appendix A, had a starting salary of $25,100. You might want to know whether this is a relatively low or high starting salary, etc. What percentage of the starting salaries were less than $25,100; what percentage were larger? Descriptive measures that locate the relative position of a measurement, in relation to the other measurements, are called ***measures of relative standing.*** One that expresses this position in terms of a percentage is called a ***percentile*** for the data set.

DEFINITION 3.7

Let $x_1, x_2, \ldots, x_n$ be a set of n measurements arranged in increasing (or decreasing) order. The ***pth percentile*** is a number x such that $p\%$ of the measurements fall below the pth percentile and $(100 - p)\%$ fall above it.

The starting salary, $25,100, falls at the 96.5 percentile of the starting salary data, Appendix A. This tells you that 96.5% of the starting salaries were less than $25,100 and $(100 - 96.5)\% = 3.5\%$ were greater.

Another measure of relative standing is the ***z-score*** for a measurement. For example, suppose that you were told that $25,100 lay 1.72 standard deviations above the mean for the 948 starting salaries of Appendix A. Knowing that most of the starting salaries will be less than 2 standard deviations from the mean and almost all will be within 3, you would have a good idea of the relative standing of the $25,100 starting salary.

The distance that a measurement, x, lies above or below the mean, $\bar{x}$, of a data set, measured in units of the standard deviation, s, is called the ***z-score*** for the measurement. Negative z-scores indicate that the observation lies to the left of the mean; positive z-scores indicate that the observation lies to the right of the mean.

DEFINITION 3.8

The ***z-score*** for a measurement, x, is

$$z = \frac{x - \bar{x}}{s}$$

EXAMPLE 3.12 We have noted that the mean and standard deviation for the 948 starting salaries, Appendix A, are $\bar{x} = \$15{,}909$ and $s = \$5{,}353$. Use these values to find the z-score for a starting salary of $25,100.

Solution Substituting the values of x, $\bar{x}$, and s into the formula for z, we obtain

$$z = \frac{x - \bar{x}}{s} = \frac{\$25{,}100 - \$15{,}909}{\$5{,}353}$$

$$= 1.72$$

Since the z-score is positive, we conclude that the $25,100 starting salary lies a distance of 1.72 standard deviations above (to the right of) the mean of $15,909.

EXERCISES

3.22 The 70th percentile for the 948 starting salaries, Appendix A, is $19,000. Interpret this value.

3.23 Refer to the 50 starting salaries, Table 2.3 (page 31). The mean and standard deviation are $\bar{x} = \$15{,}400$ and $s = \$5{,}363$. Find the z-score for a college graduate with a starting salary of:

a. $8,700 **b.** $21,300

What is the interpretation of negative z-scores?

3.24 The relative frequency distribution of the daily single-room lodging rates for a sample of 68 motels and hotels in Las Vegas, Nevada, is summarized by the following numerical descriptive measures:

$$\bar{x} = \$31.25$$

$$s = \$6.50$$

$$\text{Median} = \$28.75$$

a. Find the 50th percentile for the 68 single-room lodging rates.

b. A Las Vegas motel, selected at random from the 68, advertises its daily single-room lodging rate as $23.97. Find the z-score for this motel rate and interpret its value.

c. If the median for a set of data has a z-score which is positive, the relative frequency distribution is skewed to the left; a negative z-score implies that the relative frequency distribution is skewed to the right. Find the z-score for the median Las Vegas single-room lodging rate. Interpret this value.

3.25 A research cardiologist is interested in the age when adult males suffer their first heart attack. The cardiologist randomly sampled the medical records of 70 male coronary patients and obtained the following summary statistics on age at the time of the first heart attack:

$$\bar{x} = 54.9 \text{ years}$$

$$s = 9.8 \text{ years}$$

a. Use the Empirical Rule to sketch the approximate relative frequency distribution for the age at the time of the first heart attack for the 70 male coronary patients.

b. One of the 70 male coronary patients experienced his first heart attack at the age of 38. Locate this observation on your sketch and compute its *z*-score.

c. Refer to your answer to part b. Would you expect many of the 70 male coronary patients to have *z*-scores below the *z*-score for this particular patient? Explain.

3.26 A biologist investigating the parent-young conflict in herring gulls recorded the feeding rates (the number of feedings per hour) for two groups of gulls: those parents with only 1 chick to feed and those with 2 or 3 chicks to feed. The results are summarized below.

	BROOD SIZE (NUMBER OF CHICKS)	
	1	2 or 3
Mean feeding rate	.18	.31
Standard deviation	.15	.13

a. A particular pair of parent gulls fed the 3 chicks in their brood at a rate of .44 time per hour. Find the *z*-score for this feeding rate and interpret its value.

b. Would you expect to observe a feeding rate of .44 for parent gulls with a brood size of only 1 chick? Explain.

3.27 An electric company serving a mid-sized central Florida city reported that the number of kilowatt-hours (KWH) used by 1,200 apartment dwellers during the month of September had a mean of 630 and a standard deviation of 80.

a. September utility statements showed that the 97th percentile for the set of 1,200 electric meter readings was 782 KWH. Interpret this value.

b. The September utility bill for one of the 1,200 apartments was selected at random from the company's files. The number of KWH used by this apartment was recorded as 874. Find the *z*-score for this electrical usage value.

c. Refer to your answer to part b. Assuming that the mean and standard deviation reported by the company were correct, would you expect an apartment with a September kilowatt usage as high as 874 KWH? [*Hint:* Apply the Empirical Rule.] Does this suggest that the mean and/or standard deviation of the 1,200 kilowatt usages may have been incorrectly reported? Explain.

3.7 Numerical Descriptive Measures for Populations

Remember, when you analyze a data set, you are doing so for a reason. Presumably, you want to use the information in the data set to infer the nature of some larger set of data, a population. For example, we might want to know something about the complete set of June 1980–March 1981 bachelor's degree graduates of the University of Florida. If we take the starting salary for each graduate to be the financial compensation that the graduate *would have received* during the year following graduation (if the graduate had secured a job), then the conceptual set of starting salaries for all June 1980–March 1981 University of Florida bachelor's degree graduates would be the population of interest to us. However, many of these graduates *did not secure a job* immediately after graduation and, consequently, we can never obtain the starting salaries for all graduates in the population. We do know that the entire population of starting salaries possesses a relative frequency distribution (the exact form of which is unknown to us), and we will want to infer the nature of this distribution based on the sample of 948 starting salaries contained in Appendix A. It is natural that we would want to use the descriptive measures of this sample to infer the nature of the population relative frequency distribution.

The numerical descriptive measures that characterize the relative frequency distribution for a population are called ***parameters.*** Since we will often use the numerical descriptive measures of a sample to estimate the corresponding unknown descriptive measures of the population, we will need to make a distinction between the numerical descriptive measure symbols for the population and for the sample.

In our previous discussion, we used the symbols $\bar{x}$ and s to denote the mean and standard deviation, respectively, of a sample of n observations. Similarly, we will use the symbol μ (mu), the Greek letter "m," to denote the mean of a population, and the symbol σ (sigma), the Greek letter "s," to denote the standard deviation of a population. As you will subsequently see, we will use the sample mean $\bar{x}$ to estimate the population mean μ, and the sample standard deviation s to estimate the population standard deviation σ. In doing so, we will be using the sample to help us infer the nature of the population relative frequency distribution.

SAMPLE AND POPULATION NUMERICAL DESCRIPTIVE MEASURES

Sample mean: $\bar{x}$

Population mean: μ

Sample standard deviation: s

Population standard deviation: σ

Sample z-score: $z = \dfrac{x - \bar{x}}{s}$

Population z-score: $z = \dfrac{x - \mu}{\sigma}$

3.8 Calculating a Mean and Standard Deviation from Grouped Data (Optional)

In Sections 3.4 and 3.5 we gave formulas for computing the mean and standard deviation of a data set. However, these formulas apply only to ***raw*** data sets, i.e., those in which the value of each of the individual observations in the data set is known. If your data have already been grouped into classes of equal class width and arranged in a frequency table, you must use an alternative method to compute the mean and standard deviation.

EXAMPLE 3.13 Refer to Example 2.9 (page 31). Calculate the mean and standard deviation for the starting salary data, Table 2.3, using the grouping shown in the frequency table, Table 2.4.

Solution Since the data of Table 2.3 are raw, i.e., the starting salaries for each of the 50 college graduates selected from Appendix A are given, we could compute the sample mean starting salary and sample standard deviation of starting salaries directly, using the formulas of Sections 3.4 and 3.5. For the purposes of illustration, however, we will assume that we have access only to the grouped data of Table 2.4. The formulas for calculating $\bar{x}$, s^2, and s from grouped data are given in the box.

FORMULAS FOR CALCULATING A MEAN AND STANDARD DEVIATION FROM GROUPED DATA

x_i = Midpoint of the ith class

f_i = Frequency of the ith class

k = Number of classes

n = Total number of observations in the data set

$$\bar{x} = \frac{(x_1 f_1 + x_2 f_2 + x_3 f_3 + \cdots + x_k f_k)}{n} = \frac{\Sigma x_i f_i}{n}$$

$$s^2 = \frac{(x_1^2 f_1 + x_2^2 f_2 + x_3^2 f_3 + \cdots + x_k^2 f_k) - \dfrac{(\Sigma x_i f_i)^2}{n}}{n-1} = \frac{\Sigma x_i^2 f_i - \dfrac{(\Sigma x_i f_i)^2}{n}}{n-1}$$

$$s = \sqrt{s^2}$$

The eleven class intervals, midpoints, and frequencies of Table 2.4 are reproduced in Table 3.6. Substituting the class midpoints and frequencies into the formulas, we obtain

$$\bar{x} = \frac{\Sigma x_i f_i}{n} = \frac{(7{,}050)(4) + (10{,}050)(8) + (13{,}050)(11) + \cdots + (37{,}050)(1)}{50}$$

$$= \frac{766{,}500}{50} = 15{,}330$$

$$s^2 = \frac{\Sigma x_i^2 f_i - \dfrac{(\Sigma x_i f_i)^2}{n}}{n-1}$$

$$= \frac{\{(7{,}050)^2(4) + (10{,}050)^2(8) + \cdots + (37{,}050)^2(1)\} - \dfrac{(\Sigma x_i f_i)^2}{50}}{50-1}$$

TABLE 3.6
Class Intervals, Midpoints, and Frequencies for the 50 Starting Salaries, Table 2.4

CLASS	CLASS INTERVAL	CLASS MIDPOINT x_i	CLASS FREQUENCY f_i
1	5,550— 8,550	7,050	4
2	8,550—11,550	10,050	8
3	11,550—14,550	13,050	11
4	14,550—17,550	16,050	14
5	17,550—20,550	19,050	5
6	20,550—23,550	22,050	6
7	23,550—26,550	25,050	1
8	26,550—29,550	28,050	0
9	29,550—32,550	31,050	0
10	32,550—35,550	34,050	0
11	35,550—38,550	37,050	1
			$n = 50$

Since we found $\Sigma x_i f_i = 766{,}500$ when calculating $\bar{x}$, we have

$$s^2 = \frac{\{(7{,}050)^2(4) + (10{,}050)^2(8) + \cdots + (37{,}050)^2(1)\} - \dfrac{(766{,}500)^2}{50}}{49}$$

$$= \frac{(13{,}218{,}525{,}000) - (11{,}750{,}445{,}000)}{49} = 29{,}960{,}816$$

and

$$s = \sqrt{29{,}960{,}816} = 5{,}473.6$$

Thus, using the grouped data method, we have $\bar{x} = \$15{,}330$ and $s = \$5{,}473.6$.

The values of $\bar{x}$, s^2, and s based on the formulas for grouped data will usually not agree exactly with those obtained using the raw or ungrouped data. Applying the formulas of Sections 3.4 and 3.5 to the raw data of Table 2.3, we obtain $\bar{x} = 15{,}400$ and $s = 5{,}363.2$. These values are different from those computed above because, in the grouped data method, we have substituted the value of the class midpoint for each value of x, Starting salary, in a class interval. Only when every value of x in each class is equal to its respective class midpoint (which is rarely the case) will the formulas for grouped and for ungrouped data give identical values of $\bar{x}$, s^2, and s. Therefore, the formulas for grouped data are approximations to these numerical descriptive measures.

EXERCISES

3.28 Refer to the yearly per bed rental cost data of Exercise 3.18.

a. Using the six class intervals, 51.5—191.5, 191.5—331.5, . . ., 751.5—891.5, construct a frequency table similar to Table 3.6.

b. Compute $\bar{x}$ and s from the grouped data formulas. Compare these values to the values of $\bar{x}$ and s you obtained in Exercise 3.18 using the raw data.

3.29 Refer to Exercise 2.15.

a. Calculate the mean and standard deviation for the sample of 35 response rates, using the raw data.

b. Repeat part a, but use the grouped data method. Use the class intervals that you formed in Exercise 2.15.

3.30 Each year *Billboard* magazine conducts a Disco Data poll, querying disco clubs across the country about equipment update expenditures, attendance records, and promotional tactics. The 1980 Disco Data poll included 3,500 discos and produced the following information on weekly attendance. Assuming that the information in the table pertains to one randomly selected week, compute the mean weekly attendance $\bar{x}$ and the standard deviation s for the 3,500 discos. [Note that the weekly attendance class intervals are not of equal width. However, this will not affect the procedure for computing $\bar{x}$ and s—follow the steps outlined in the box.]

WEEKLY ATTENDANCE	NUMBER OF DISCOS
0—1500	805
1500—2500	1610
2500—3500	630
3500—5000	280
5000—7500	175
	3500

Source: *Billboard,* June 7, 1980, © 1980 by Billboard Publications, Inc. Reprinted by permission.

3.31 Many of the top producers in the liquor industry saw little growth in the consumption of their brand during 1979. Information on the percent change in sales volume from 1978 to 1979 for the 35 highest ranked brands of liquor is given in the table. Compute $\bar{x}$, s^2, and s for the percent change in sales volume data, using the grouping given in the table.

PERCENT CHANGE IN SALES VOLUME, 1978 TO 1979	NUMBER OF BRANDS
−8.05 to −3.55	8
−3.55 to +0.95	3
+0.95 to +5.45	16
+5.45 to +9.95	2
+9.95 to +14.45	4
+14.45 to +18.95	1
+18.95 to +23.45	0
+23.45 to +27.95	1
	35

Source: *Advertising Age,* March 31, 1980. Copyright 1980 by Crain Communications Inc. Reprinted with permission.

3.9 Summary

Numerical descriptive measures enable us to construct a mental image of the relative frequency distribution for a data set. The two most important types of numerical descriptive measures are those that measure central tendency and data variation.

Three numerical descriptive measures are used to locate the "center" of a relative frequency distribution: the mean, median, and mode. Each conveys a special piece of information. In a sense, the mean is the balancing point for the data. The median divides the data; half of the observations will be less than the median, and half will be larger. The mode is the value of x that occurs with greatest frequency. It is the value of x that locates the point where the relative frequency distribution achieves its maximum relative frequency.

The range and the standard deviation measure the spread of a relative frequency distribution. Particularly, we can obtain a very good notion of the way data are distributed by constructing the intervals $\bar{x} \pm s$, $\bar{x} \pm 2s$, and $\bar{x} \pm 3s$ and referring to the Empirical Rule. The percentages of the total number of observations falling within these intervals will be approximately as shown in the accompanying table.

INTERVAL	PERCENTAGE
$\bar{x} \pm s$	60% to 80%
$\bar{x} \pm 2s$	95%
$\bar{x} \pm 3s$	Almost 100%

KEY WORDS

Numerical descriptive measures
Measures of central tendency
Measures of data variation or spread
Measures of relative standing
Mean
Median
Mode
Skewness
Range
Variance
Standard deviation
Empirical Rule
Percentile
z-Score
Parameters
Raw data*
Grouped data*

*From the optional section.

KEY SYMBOLS

For a sample:
Mean: $\bar{x}$
Variance: s^2
Standard deviation: s
z-Score: $z = \dfrac{x - \bar{x}}{s}$

For a population:
Mean: μ
Variance: σ^2
Standard deviation: σ
z-Score: $z = \dfrac{x - \mu}{\sigma}$

SUPPLEMENTARY EXERCISES

[*Note:* Starred (*) exercises refer to the optional section in this chapter.]

3.32 Compute Σx, Σx^2, and $(\Sigma x)^2$ for each of the following data sets:
a. 7, 8, 4, −2, 12, 8
b. 123, 247, 0, 100
c. −3, 4, −2, 0, −3, −2, −4
d. 17, 17, 20, 23, 12

3.33 Compute the mean, median, and mode for each of the data sets in Exercise 3.32.

3.34 Compute the range, s^2, and s for each of the data sets in Exercise 3.32.

3.35 Calculate $\bar{x}$, s^2, and s for each of the following data sets:
a. $\Sigma x^2 = 13.3$, $\Sigma x = 7.6$, $n = 10$
b. $\Sigma x^2 = 863$, $\Sigma x = 112$, $n = 27$
c. $\Sigma x^2 = 45$, $\Sigma x = 8$, $n = 4$

3.36 For each of the following data sets, compute $\bar{x}$, s^2, and s:
a. 14, 7, 0, 0, 9, 0
b. −8, −6, 10, 16
c. −4, −2, −2, 1, 1, 8, 6, 4, 6
d. 1,242, 1,793, 485, 480

3.37 Give a realistic example in your field of interest for which the best measure of central tendency of a quantitative data set is provided by:
a. The mean **b.** The median **c.** The mode

3.38 What is the best measure of the variability of a quantitative data set? Why?

3.39 Give two different methods of measuring the relative standing of an observation within a set of measurements. How are they different? How are they alike?

3.40 In an attempt to reduce the revenue lost due to passengers who have reserved a seat but fail to show, most major airlines frequently overbook their flights. *Overbooking* is the practice of selling more ticket reservations for a flight than there are seats available on the plane. Thus, at times, more passengers with reservations show up for a flight than there are seats available, and some must wait until the next available flight before departing (they have been *bumped*). The following data represent the number of passengers who were bumped from a Chicago to Atlanta flight due to overbooking on 10 randomly selected days:

3, 0, 0, 1, 4, 2, 1, 1, 0, 2

a. Determine the mean, median, mode, range, variance, and standard deviation for this data set.
b. Which of the numerical descriptive measures in part a are measures of variability, and which are measures of central tendency?

3.41 The thesis submitted by a student as a requirement for a master's degree in psychology consisted of 30 typewritten pages. The number of typographical errors on each page was recorded as given below. Compute $\bar{x}$, s^2, and s for this data set.

0	0	3	6	11	5
1	0	5	0	4	0
3	1	2	0	3	1
2	0	2	0	2	1
4	1	1	3	2	3

3.42 A recently hired coach of distance runners was interested in knowing how many miles America's top distance runners usually run in a week. The coach surveyed 15 of the best distance runners with the following results (in miles):

120	95	110	95	70
90	80	100	125	75
85	100	115	130	90

a. Construct a relative frequency distribution for the data set.
b. Compute $\bar{x}$ and s.
c. Calculate the intervals $\bar{x} \pm s$, $\bar{x} \pm 2s$, and $\bar{x} \pm 3s$ and locate these intervals on the graph of the relative frequency distribution.
d. Does the Empirical Rule adequately describe the distribution of number of miles run per week by America's top distance runners?

3.43 An important factor in an oil company's decision on whether to continue drilling in a certain region is the productivity of the oil wells currently in use in the region. A major oil company now operates one oil well in a flatland region of Texas. The decision on whether to drill for more wells in the region will be based upon the daily production of the oil well for the 25 days that it has been in operation. These daily productions (in hundreds of barrels) are given below. Compute $\bar{x}$, s^2, and s for this data set.

82	91	43	75	88	28	57	74	76
44	55	83	94	93	37	77	86	68
93	42	58	78	49	63	63		

3.44 What is the relationship between ants and plants? Do ants protect certain plants from insect seed predators? A naturalist, wishing to investigate the phenomenon of ants protecting plants, sampled 100 flower heads of a certain sunflower plant which attracts ants. (Each flower head was on a different sunflower plant.) The flower heads were divided into two groups of 25 each. However, ants were prevented from reaching the flower heads of one group (this was accomplished by painting ant repellant around the flower stalks of the 25 plants). After a specified period of time, the number of insect seed predators on each flower head was counted. The results are summarized below. Sketch your mental images of the two relative frequency distributions. (The Empirical Rule will help you to do this.) Does it appear that ants protect sunflower plants from insect seed predators?

	PLANTS WITH ANTS	PLANTS WITHOUT ANTS
Mean number of predators per flower head	2.9	7.6
Standard deviation	2.4	4.4

3.45 The start of the 1980–81 television ratings race among the three major TV networks was delayed due to a strike by the Guild of Screen and Television Actors in September 1980. Items under negotiation with television producers included higher minimum salaries and more fringe benefits for the actors. At the time of the strike, salaries were structured so that the majority of TV actors earned from $30,000 to $100,000 per year, while a few "big-name" actors earned millions of dollars annually.

a. Based on this salary information, discuss possible skewness in the pre-strike distribution of TV actors' salaries.

b. Which measure of central tendency of TV actors' salaries—the mean or the median—should the actors use in an attempt to convince TV producers that they (the actors) are deserving of higher salaries and more fringe benefits? Explain.

c. Which measure of central tendency of TV actors' salaries should the producers quote in order to rebuff the actors' demands? Explain.

3.46 In 1981, Lucky Strike brand cigarettes fell in the 93rd percentile of the nicotine content (as determined by the Federal Trade Commission) distribution of the country's top selling cigarette brands. Describe Lucky Strike's nicotine content ranking in relation to the other 186 cigarette brands.

3.47 *Muck* is a rich, highly organic type of soil which serves as a growth medium for most vegetation in the Florida Everglades. Because of the high concentration of organic material, muck can be destroyed by drought, fire, and windstorms. In April and May of 1980, south Florida experienced its worst drought in more than a decade, and the Everglades lost a considerable amount of muck. In order to determine the amount of loss, members of the Florida Fish and Game Commission marked 40 plots with stakes at various locations in the Everglades and measured the depth of the muck (in inches) at each stake. The data are given below.

27	32	35	57	23	32	18
26	33	45	30	38	31	35
23	40	7	15	27	30	29
41	19	26	16	39	30	30
35	22	40	33	15	21	37
27	24	28	26	36		

a. Compute the mean of the data set.
b. Find the median of the data set.
c. Find the mode of the data set.

3.48 Refer to Exercise 3.47. Find the range, variance, and standard deviation of the data set. (Use the shortcut procedure to calculate $\Sigma(x-\bar{x})^2$.)

3.49 Refer to Exercises 3.47 and 3.48. Construct the intervals $\bar{x} \pm s$, $\bar{x} \pm 2s$, and $\bar{x} \pm 3s$. Count the number of observations falling within each interval and find the corresponding proportions. Compare the results to the Empirical Rule.

3.50 Refer to Exercise 3.47. Find the 10th percentile for the 40 muck depths.

3.51 Market-demand studies indicate that the typical wine-purchasing household has a higher income, fewer members, and more education than average. A random sample of 4,500 wine-purchasing households showed incomes with an average of $16,039 and a standard deviation of $5,010; a sample of 7,000 nonpurchasing households showed incomes with an average of $13,126 and a standard deviation of $3,102.

a. Roughly sketch on a piece of graph paper the relative frequency distributions for the purchasing and nonpurchasing household incomes.
b. What is the approximate proportion of wine-purchasing households in the sample with incomes between $6,019 and $26,059?
c. What is the approximate proportion of nonpurchasing households in the sample with incomes larger than $16,228 or smaller than $10,024?

3.52 Five applicants for real estate appraisal training, two from Oregon and three from Florida, have submitted their scores on different real estate appraisal aptitude exams. The two from Oregon had taken the Pacific Appraisers' Aptitude Test (PACAT), while the three from Florida had taken the Southeastern Appraisers' Aptitude Test (SEAT). Scores on the PACAT have a mean of 50 and a standard deviation of 10, while scores on the SEAT have a mean of 120 and a standard deviation of 15. The applicants and their scores are listed in the table.

APPLICANT	EXAM TAKEN	SCORE
#1	PACAT	60
#2	PACAT	45
#3	SEAT	130
#4	SEAT	115
#5	SEAT	110

a. Find the respective z-scores for each of the applicants.
b. If the five applicants are ranked from highest to lowest entirely on the basis of who had the best appraisal aptitude test score *in relation to the type of exam taken* (highest score receiving rank 1, and so forth), which applicant would have the highest rank? The lowest? [*Hint:* Use your results from part a.]

3.53 Research shows that couples who do not become parents until their mid or late 30s increase the risk of their babies' becoming senile in old age (*Family Weekly*, April 26, 1981). A study of 80 elderly patients with atrophy of the prefrontal brain areas—the most prevalent cause of senility in the U.S.—indicated that the average age of the patients' mothers at the time of their births was 36.5 years, 10 years older than the average for first-time parents. Assume that the standard deviation of the 80 ages was 7.2 years.

a. Use the Empirical Rule to describe the relative frequency distribution of the ages of the 80 elderly patients' mothers at the time of the patients' births.

b. Approximately what proportion of the 80 patients had mothers who were between 22.1 and 50.9 years at the time of the patients' births?

c. The median age of the 80 patients' mothers at the time of their births was 35.5 years. What proportion of the 80 mothers were younger than 35.5 years at the time of the patients' births?

d. One of the 80 senile patients is selected at random. Would you expect the age of this patient's mother at the time of the patient's birth to be 20 years or less? Explain.

3.54 Refer to the starting salaries of bachelor's degree graduates in the five colleges of Appendix B. The means and standard deviations of the starting salaries of the five data sets are shown in the table.

COLLEGE	MEAN STARTING SALARY	STANDARD DEVIATION
Business Administration	$15,225	$5,172
Education	10,212	2,997
Engineering	20,178	3,428
Liberal Arts	12,460	3,939
Sciences	16,293	6,019

a. Use the means and standard deviations of the starting salaries, in conjunction with the Empirical Rule, to sketch your mental images of the five relative frequency distributions. Construct them on the same graph so that you can see how they appear relative to each other.

b. Construct tables similar to Table 3.3 for the starting salary data for each of the five colleges. Do the proportions agree with the Empirical Rule? [*Note:* The relative frequency distributions for these five data sets are shown in Figures 3.10 to 3.14 (pp. 85–89), respectively.]

3.55 An all-news cable television network claims that the distribution of the number of homes reached nightly by their prime-time (8:00 PM) newscast has a mean of 2.7 million homes and a standard deviation of .35 million homes. Assume that the claim is true and also that the distribution is mound-shaped.

a. Approximately what proportion of the prime-time newscasts would you expect to reach between 2.0 and 3.4 million homes?

b. Approximately what proportion of the prime-time newscasts would you expect to reach between 1.65 and 3.75 million homes?

FIGURE 3.10 June 1980–March 1981 Bachelor's Degree Graduates' Starting Salaries (College of Business Administration)

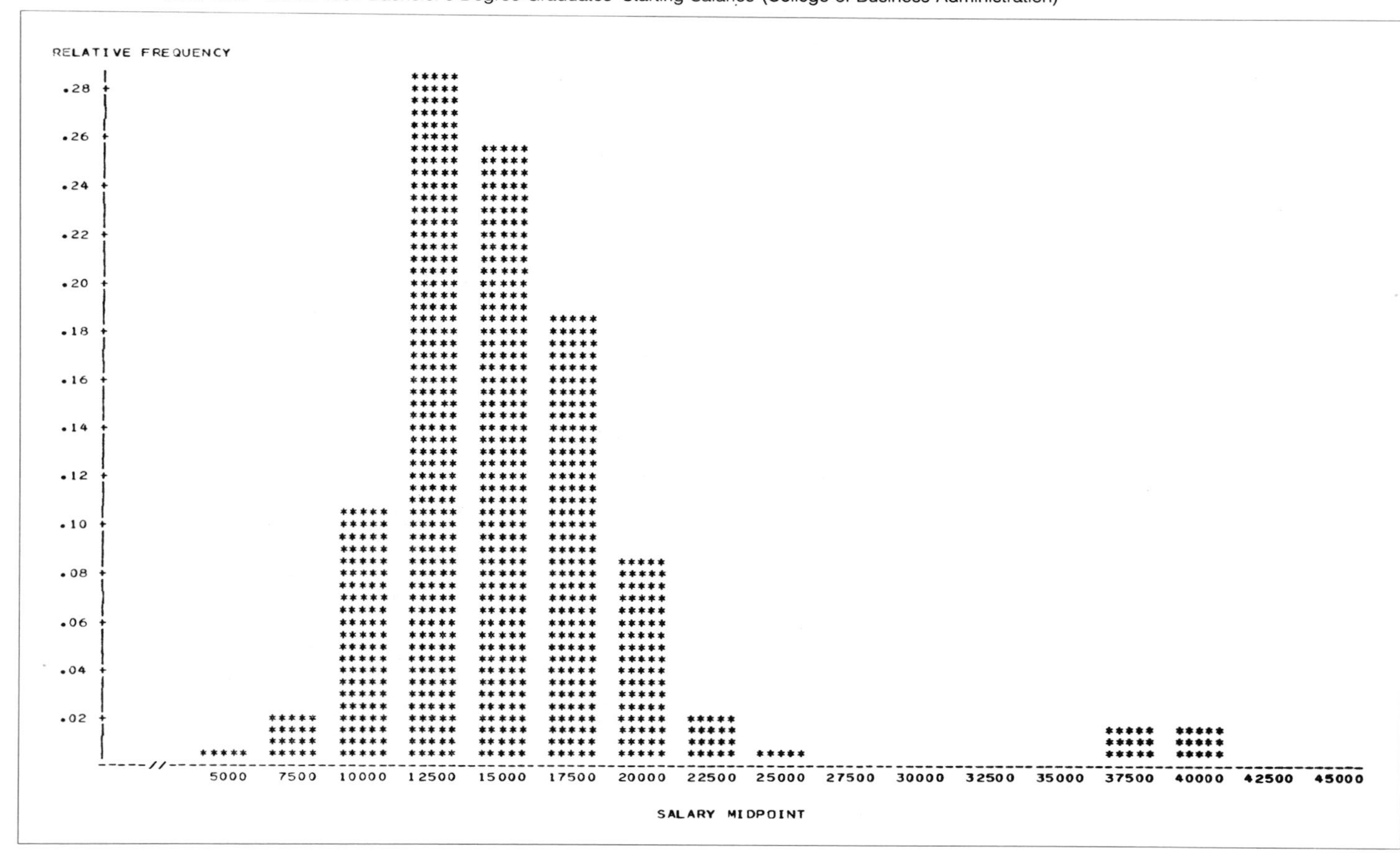

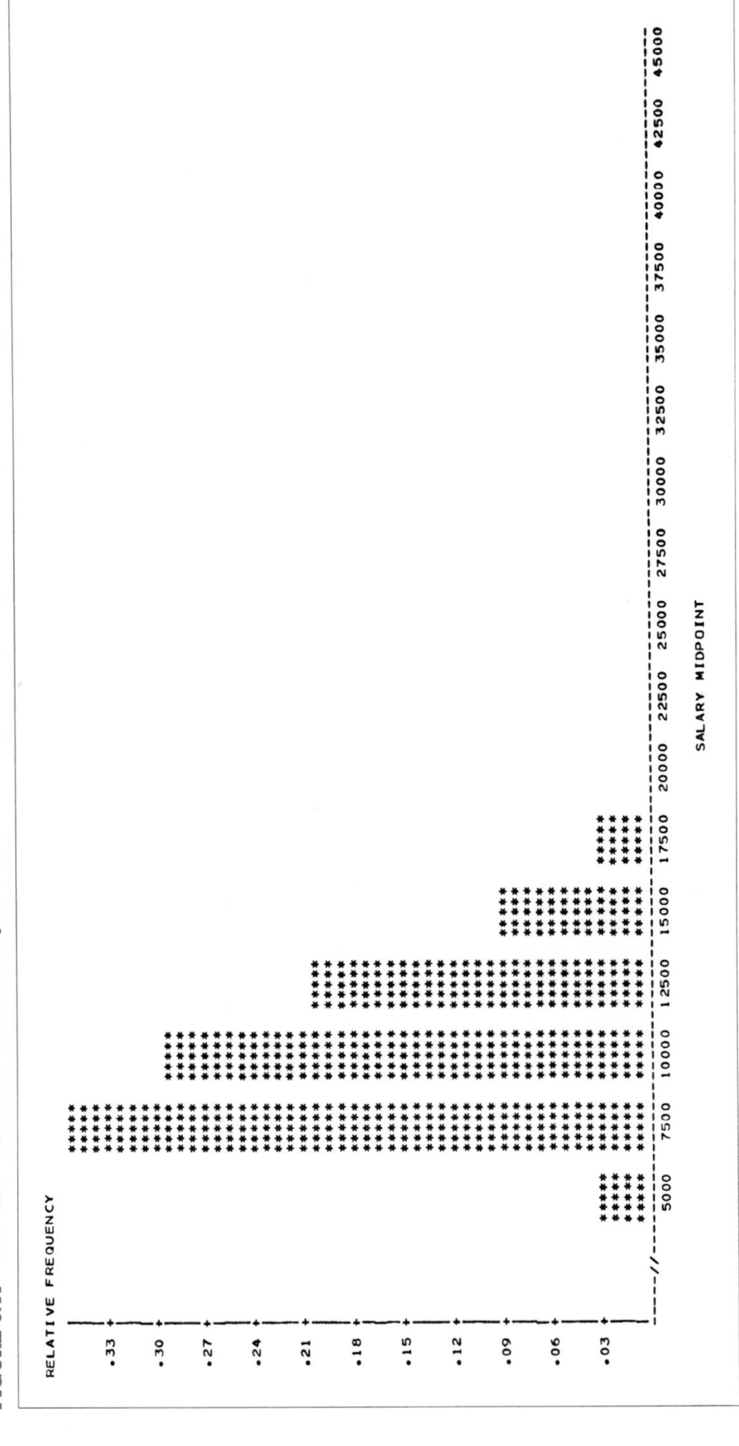

FIGURE 3.11 June 1980–March 1981 Bachelor's Degree Graduates' Starting Salaries (College of Education)

FIGURE 3.12 June 1980–March 1981 Bachelor's Degree Graduates' Starting Salaries (College of Engineering)

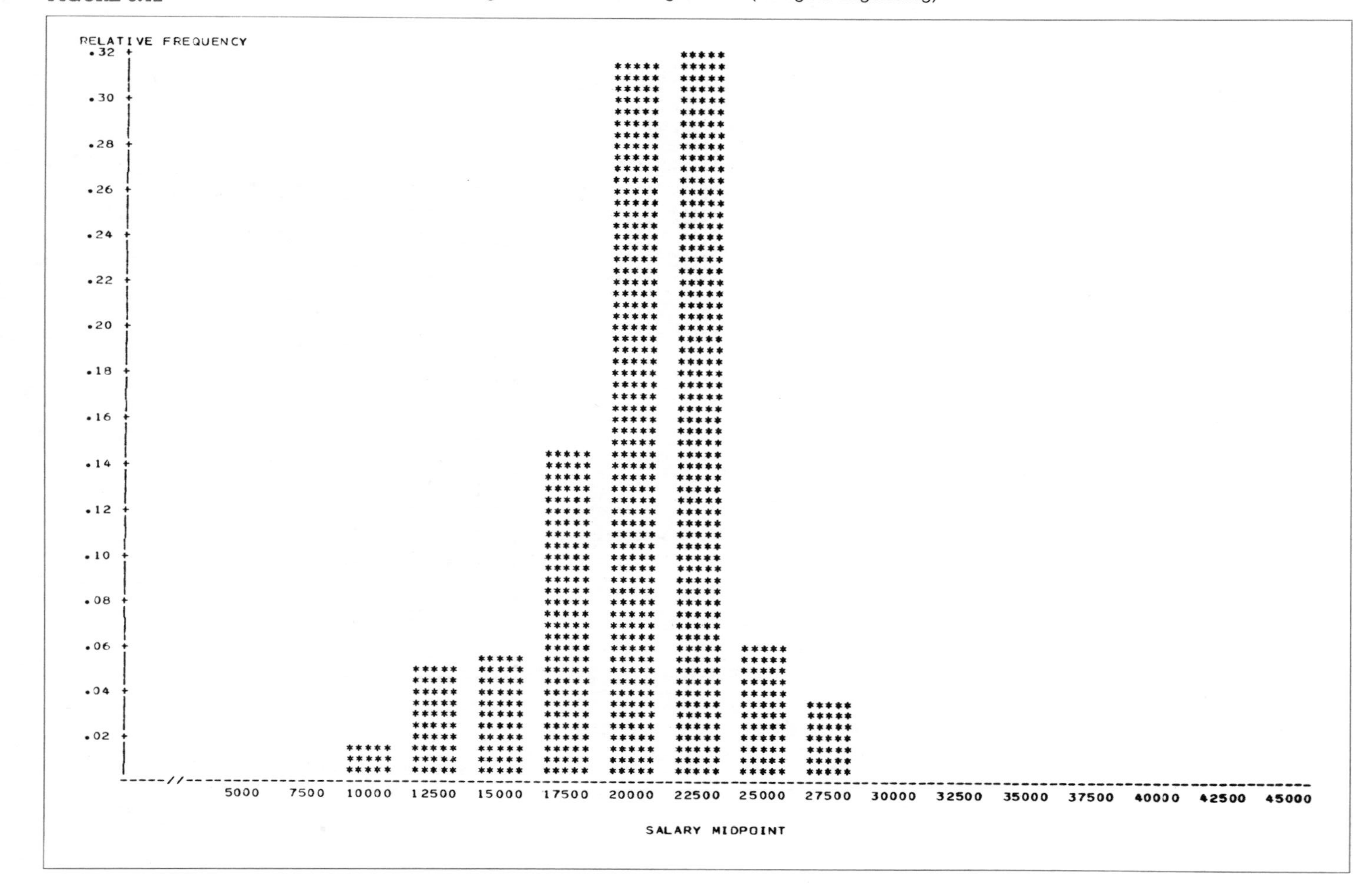

FIGURE 3.13 June 1980–March 1981 Bachelor's Degree Graduates' Starting Salaries (College of Liberal Arts)

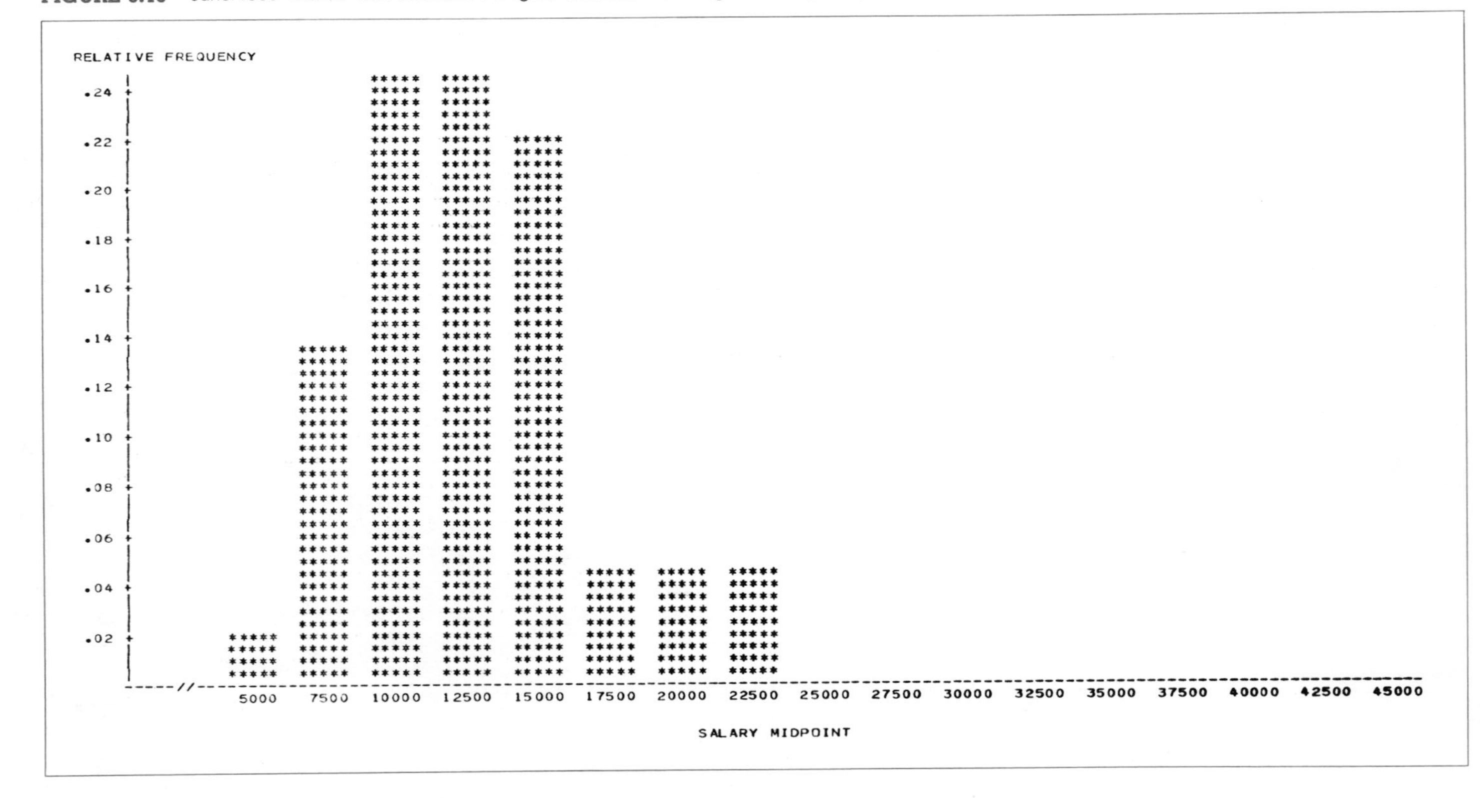

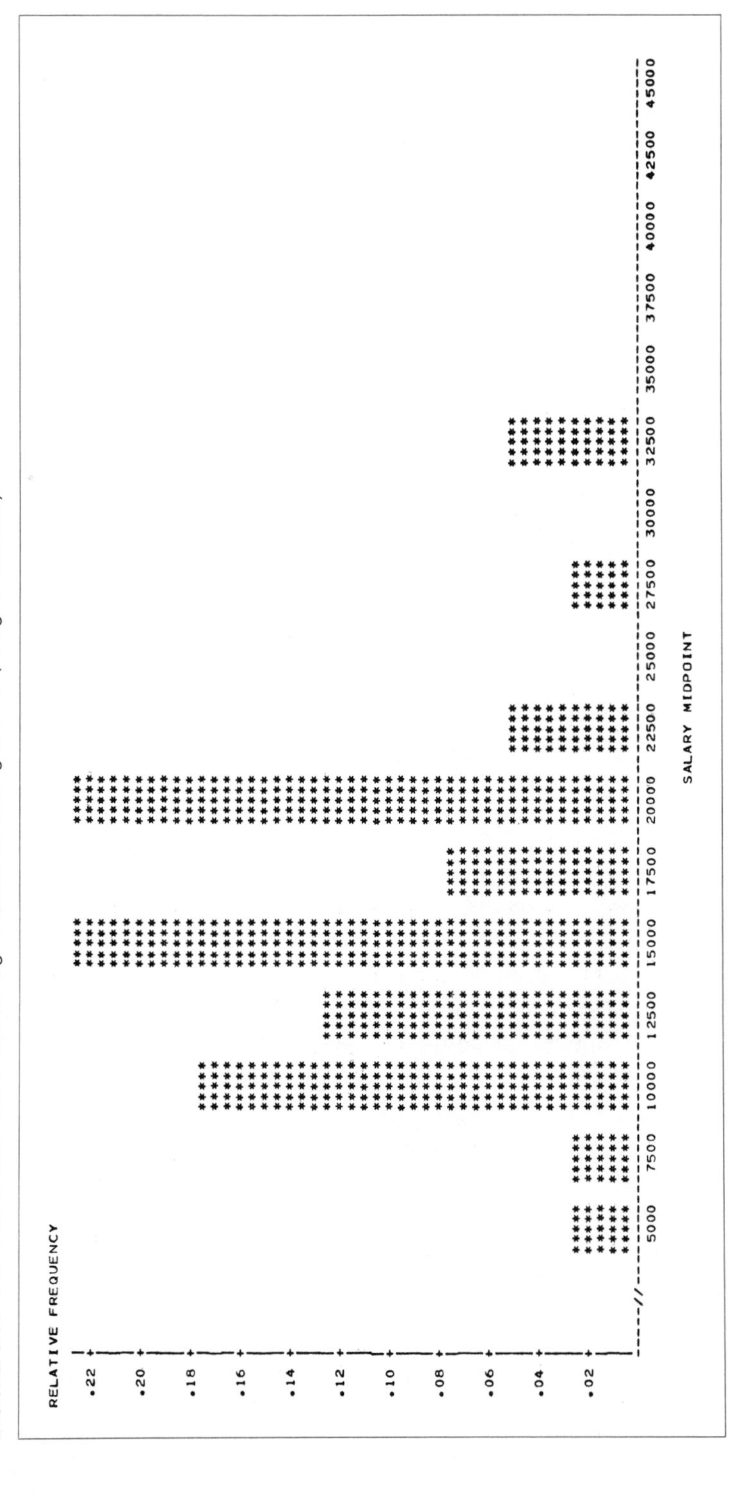

FIGURE 3.14 June 1980–March 1981 Bachelor's Degree Graduates' Starting Salaries (College of Sciences)

3.56 Refer to Exercise 3.55. A potential advertiser questions the validity of the cable network's claim concerning the mean number of homes reached nightly by the prime-time newscast. Specifically, the advertiser believes the claimed mean of 2.7 million homes is too high. In an attempt to support this belief, the advertiser hires a rating service that uses the same rating system as the cable network. Suppose that the rating service found that, for one randomly selected night, the prime-time newscast was viewed by 1.5 million homes.

a. Compute the z-score for the observation, 1.5 million homes, assuming that the information provided by the cable network is correct. Interpret its value.

b. On the basis of the rater's findings and your knowledge of the Empirical Rule, do you believe that the cable network has overstated the mean number of homes reached by their prime-time newscast? Explain.

***3.57** An important part of human resource planning by an organization's personnel department is an analysis of labor force trends. In order to assess the supply of labor during the next decade, a company obtained U.S. labor force projections, by age group, for 1990. The data are presented in grouped form in the accompanying table.

AGE GROUP	PROJECTED NUMBER IN THE LABOR FORCE IN 1990 (THOUSANDS)
15—24	22,139
25—34	32,301
35—44	28,532
45—54	18,733
55—64	11,218
	112,923

Source: U.S. Department of Labor, *Employment and Training Report of the President,* 1977.

a. Compute the approximate mean projected age of persons in the U.S. labor force in 1990.

b. Compute the approximate standard deviation of the projected ages of persons in the U.S. labor force in 1990.

***3.58** The time patients spend waiting in a physician's office before they receive health care services, i.e., *patient waiting time,* plays an important role in the efficient operation of the physician's practice (see Case Study 7.2). The number of minutes (after their appointment times) that each of a sample of 55 patients had to wait in a dentist's office before being served are shown in the table on page 91.

a. Calculate approximate values for the mean waiting time and standard deviation of waiting times from this sample.

b. Form the interval $\bar{x} \pm 2s$. Using the frequency table, determine the number of patients in the sample with waiting times in this interval.

c. Compare the results of part b with the Empirical Rule.

WAITING TIME (MINUTES)	NUMBER OF PATIENTS
0— 3.5	12
3.5— 7.0	11
7.0—10.5	7
10.5—14.0	15
14.0—17.5	6
17.5—21.0	3
21.0—24.5	1
	55

3.59 Arizona's housing market has recently become one of the nation's strongest. This is due to the steady stream of new residents fleeing harsh northern winters, along with former Californians searching for a relatively less-populated area in which to live. In a random sample of 40 new homes sold in Arizona this year, the average cost was $83,350 and the standard deviation was $7,650.

a. Describe the relative frequency distribution of the costs of the 40 Arizona homes.
b. Estimate the fraction of the 40 Arizona homes which cost between $75,700 and $91,000.
c. One of the homes in the sample, located in Flagstaff, Arizona, had a cost of $73,787. Compute the z-score for the cost of this home.
d. If an Arizona home were chosen at random from the sample of 40, is it more likely that the cost of the home would be greater than $73,787 or less than $73,787? Refer to your answer to part c.

3.60 A chemical manufacturing plant, under investigation by the Environmental Protection Agency (EPA) for possible violation of air pollution standards, reports that the distribution of the amount of dangerous chemicals in the plant's air emissions is approximately mound-shaped, with a mean of 5 parts per million (ppm) and a standard deviation of 1.5 ppm.

a. Use the Empirical Rule to describe the distribution of the amount of dangerous chemicals in the plant's air emissions.
b. The EPA sets a limit of 10 ppm on dangerous chemicals in plant air emissions. Plants found exceeding this standard more than once per year are required to install expensive air pollution control devices. Based upon your description of the air emissions distribution of part a, do you believe that the chemical plant is in violation of air quality standards? Explain.

3.61 Most people living in metropolitan areas receive impressions of what is happening in their area primarily through their major newspapers. A study was conducted to determine whether the *Uniform Crime Report,* compiled by the United States Federal Bureau of Investigation (FBI), and the daily newspaper gave consistent information about the trend and distribution of crime in a metropolitan area.

An attention score, based on the amount of space devoted to a story, was calculated for each paper's coverage of murders, assaults, robberies, etc. Suppose the murder attention scores of metropolitan newspapers across the country in 1981 had a mean of 62 and a standard deviation of 7.

a. Use the Empirical Rule to determine the approximate proportion of metropolitan newspapers which had a murder attention score between 41 and 83.

b. The *Atlanta Constitution* had a 1981 murder attention score of 76. Describe the *Constitution's* rank among all metropolitan newspapers.

*__3.62__ The accompanying table gives the distribution of the grade point average (GPA) of 200 randomly selected University of Florida freshmen.

GPA	NUMBER OF FRESHMEN
1.35—1.55	6
1.55—1.75	12
1.75—1.95	15
1.95—2.15	43
2.15—2.35	53
2.35—2.55	27
2.55—2.75	18
2.75—2.95	9
2.95—3.15	12
3.15—3.35	4
3.35—3.55	1
	200

a. Find the modal class of the grouped data.

b. Compute the approximate mean and standard deviation of the data set, based on the grouping shown in the table.

c. Estimate the percentage of University of Florida freshmen with GPAs within 2 standard deviations of the mean GPA.

3.63 The accompanying data set represents the sales volume (in shares) of the 15 most active stocks on the New York Stock Exchange for Tuesday, February 10, 1981.

Prime Cm	790,300	Texaco Inc.	391,000	TW Corp	326,100
Exxon	666,300	Cont. Air Lines	385,500	Twc Fox	304,400
Sony Corp	523,900	US Air	375,700	Gillette Co.	303,600
Boeing	420,900	Whirlpool	341,000	K Mart	303,000
IBM	392,600	AT&T	329,500	Standard Oil	301,400

a. Compute $\overline{x}$, s^2, and s for the sample.

b. Compute the median and the range.

c. Find the *z*-score for IBM's sales volume. Interpret this value.

d. Find the stock which falls in the 80th percentile of the sales volume distribution of the 15 most active stocks.

e. Compute the intervals $\bar{x} \pm s$, $\bar{x} \pm 2s$, and $\bar{x} \pm 3s$. Count the number of stock sales volumes which fall within each interval and compare the interval percentages to those of the Empirical Rule.

3.64 Suppose that the average loss payment per insurance claim by a sample of 1,000 motorcycle owners is \$615 and the standard deviation is \$85. Assume that the distribution of loss payment per claim is approximately mound-shaped.

a. Form a mental image of the distribution of loss payments for this sample. Sketch your image on a piece of graph paper.
b. Estimate the percentage of the loss payments in the interval \$445 to \$785.
c. Would you expect to observe a loss payment in the sample as large as \$900?

***3.65** Refer to Exercise 3.47. Construct a frequency table for the muck depths, using 10 class intervals. Compute $\bar{x}$, s^2, and s using the grouped data method, and compare these values to your previous results (Exercises 3.47 and 3.48).

3.66 Refer to Exercise 2.21.

a. Calculate $\bar{x}$, s^2, and s for the EPA estimated highway mileage data.
b. Use the numerical measures of part a to describe the relative frequency distribution of the data.
c. Compute a measure of the relative standing of Pontiac Firebird's estimated mpg of 20.

3.67 A Las Vegas tourist hotel has recently experienced problems with people who reserve rooms for a nonholiday weekend but fail to show up. These "no shows" cost the hotel thousands of dollars in lost revenue. An examination of hotel records for a random selection of 35 nonholiday weekends indicated that the number of empty rooms (per weekend) due to "no shows" had a mean of 8.5 and a standard deviation of 2.0.

a. Describe the distribution of the number of empty rooms due to "no shows" on nonholiday weekends at the Las Vegas hotel.
b. Suppose that for an upcoming nonholiday weekend, after all the rooms were reserved, 4 additional reservations were accepted by the hotel. Management thus expects that there will be enough "no shows" (in fact, 4 or more) so that the hotel can honor all of its reservations. Based upon your description of the distribution in part a, do you think it is very likely that the hotel will be able to honor all of its reservations? Explain.

CASE STUDY 3.1 How Much Weight Can You Lift—And Still Avoid Serious Injury?

Most typical Americans either have lifted or know someone who has lifted (or at least attempted to lift) by themselves a heavy load (e.g., a refrigerator, a wheelbarrow filled with bricks, the dining room table) on one day and who, as a result, was forced to remain in bed the following day with chronic back pain or a slipped disk. An unusually large number of injuries, both at work and at home, arise from the handling (or mishandling) of heavy materials, and most of these result from slipping and falling, dropping of the load, or strain caused by lifting the load. Engineers have a term for unaided human acts of lifting, lowering, pushing, pulling, carrying, or holding and releasing an object—manual materials handling activities (MMHA).

M. M. Ayoub et al. (1980) have attempted to develop strength and capacity guidelines for MMHA.* These include norms for lifting and lowering, pushing and pulling, and carrying. The authors point out that a clear distinction between strength and capacity must be made. "Strength implies what a person can do in a single attempt, whereas capacity implies what a person can do for an extended period of time. Lifting strength, for example, determines the amount that can be lifted at infrequent intervals."

Ayoub et al. arrive at these norms through the current "state of the art," i.e., by combining the results of other researchers. Much of the successful research in determining norms for MMHA relies on the psychophysical methodology. In a study of MMHA using psychophysics, subjects are required to adjust one of the task variables (e.g., weight of the load) according to their own perception of muscular effort or force. For example, "Switzer (1962) instructed 75 male college students to find reasonable weights that could be lifted without excessive strain or discomfort for a single lift. Subjects varied the weight lifted by adding or subtracting 2.26-, 4.53-, and 9.06-[kilogram] bags of lead shot." In another study, Kroemer (1974) had 73 male subjects maintain (what they perceived to be) a maximum push force steadily over a 5-second period while standing in various postures on a slippery floor. (An interesting result of this experiment was that body support, not body weight or body size, was most useful in predicting force output. "A physically strong subject," conclude Ayoub et al., "may be able to exert only weak push or pull forces due to lack of body support.") A third example of the psychophysical approach [Snook (1974)] involved both male and female workers who determined their own work load and the walking speed with which they carried the load.

In Table 3.7 we present a portion of the recommendations of Ayoub et al. for the lifting capacities of males and females based on the psychophysical technique. The table gives means and standard deviations of the maximum weight (in kilograms) of a box 30 centimeters wide that can be safely lifted from the floor to knuckle height at two different lift rates: 1 lift per minute and 4 lifts per minute.

TABLE 3.7 Mean and Standard Deviation of the Maximum Recommended Weight of Lift (in Kilograms)

SEX	LIFTS/MINUTE	MEAN	STANDARD DEVIATION
Male	1	30.25	8.56
	4	23.83	6.70
Female	1	19.79	3.11
	4	15.82	3.23

a. Roughly sketch the relative frequency distribution of maximum recommended weight of lift for each of the four sex/lifts-per-minute combinations. The Empirical Rule will help you do this.

b. Construct the interval $\bar{x} \pm 2s$ for each of the four data sets and give the approximate proportion of measurements which fall within the interval.

*M. M. Ayoub, A. Mital, G. M. Bakken, S. S. Asfour, & N. J. Bethea. "Development of strength capacity norms for manual materials handling activities: The state of the art," *Human Factors,* June 1980, *22,* 271–283. Copyright 1980 by the Human Factors Society, Inc., and reproduced by permission.

c. Assuming the MMHA recommendations of Ayoub et al. are reasonable, would you expect that a randomly selected male could safely lift a box (30 centimeters wide) weighing 25 kilograms from the floor to knuckle height at a rate of 4 lifts per minute? A randomly selected female? Explain.

[*Note:* Ayoub et al. warn that, "At the present time there is no general agreement regarding the strength of capacity limits for various MMHA for males and females. . . . The recommendations made (in the table) are based on the current state of the art. More work is needed before final recommendations can be made."]

CASE STUDY 3.2 Consumer Complaints: Due to Chance or Specific Causes?

The degree of sensitization on the part of a firm to the needs and wants of its consumers is frequently an important factor in determining the firm's overall success. Jean Namias* presents a procedure for achieving such sensitivity. This procedure uses the rate of consumer complaints about a product to determine when and when not to conduct a search for specific causes of consumer complaints. This rate may change or vary merely as a result of chance or fate, or it may be due to some specific cause, such as a decline in the quality of the product. Concerning the former, Namias writes:

> In any operation or production process, variability in the output or product will occur, and no two operational results may be expected to be exactly alike. Complete constancy of consumer rates of complaint is not possible, for the vagaries of fate and chance operate even within the most rigid framework of quality or operation control.

Namias has determined that the complaint rate of a product (e.g., the number of customer complaints per 10,000 units sold) has a relative frequency distribution which is approximately mound-shaped. This leads to a ***Decision Rule*** with which to determine when the observed variation in the rate is due to chance and when it is due to specific causes.

DECISION RULE

If the observed rate is 2 standard deviations or less away from the mean rate of complaint, it is attributed to chance. If the observed rate is farther than 2 standard deviations above the mean rate, it is attributed to a specific problem in the production or distribution of the product.

The reasoning is that if there are no problems with the production and distribution of the product, 95% of the time the rate of complaint should be within 2 standard deviations of the mean rate. If the production and distribution process were operating normally, it would be very unlikely for a rate higher than 2 standard deviations above the mean to occur. Instead, it is more likely that the high complaint

*J. Namias, "A method to detect specific causes of consumer complaints." *Journal of Marketing Research,* August 1964, 63–68.

rate is caused by abnormal operation of the production and/or distribution process, i.e., something specific is wrong with the process.

Namias recommends searching for the cause (or causes) only if the observed variation in the rate of complaints is determined by the rule to be the result of a specific cause (or causes). The degree of variability due to chance must be tolerated. Namias says:

> As long as the results exhibit chance variability, the causes are common, and there is no need to attempt to improve the product by making specific changes. Indeed, this may only create more variability, not less, and may inject trouble where none existed, with waste of time and money. . . . On the other hand, time and money are again wasted through failure to recognize specific conditions when they arise. It is therefore economical to look for a specific cause when there is more variability than is expected on the basis of chance alone.

Namias collected data from the records of a beverage company for a 2-week period to demonstrate the effectiveness of the rule. Consumer complaints primarily concerned chipped bottles that looked dangerous. For one of the firm's brands, the complaint rate was determined to have a mean of 26.01 per 10,000 bottles sold and a standard deviation of 11.28. The complaint rate observed during the 2 weeks under study was 93.12 complaints per 10,000 bottles sold.

a. Compute the *z*-score for the observed rate of 93.12.
b. Make a general interpretation of its value.
c. Use the Namias Decision Rule to determine whether the observed rate is due to chance or whether it is due to some specific cause. (In actuality, a search for a possible problem in the bottling process led to a discovery of rough handling of the bottled beverage in the warehouse by newly hired workers. As a result, a training program for new workers was instituted.)

CASE STUDY 3.3
Estimating the Mean Weight of Short-Leaf Pine Trees—Grouped Data Method

Consider the problem of estimating the true mean weight of short-leaf pine trees in the 40-acre tract of land located in western Arkansas (Case Study 1.2, page 13). How can we use the sample information given in Table 1.1 to obtain this estimate? From our discussion in this chapter, we need to compute the sample mean, i.e., the mean weight of the sample of 117 trees in the 20 ⅕-acre plots which were "cruised." However, recall that the actual weight of each tree had to be estimated, based upon the diameter of the tree at chest height and the number of 16-foot logs the tree was capable of producing. This was best accomplished by grouping the trees according to diameter and assigning the same estimated weight to each tree in a particular group. For example, the 38 trees which had chest height diameters measured at 10 inches had all been assigned the estimated weight of 580 pounds, while the 34 trees with 11-inch diameters had all been assigned a weight of 750 pounds (see Table 1.1). In one sense, we can think of these weights as representing the midpoints of various weight class intervals. The weight of 580 pounds could represent the midpoint of the class interval 579.5—580.5, 750 pounds the midpoint of the class interval 749.5—750.5, etc. Notice that there are numerous class intervals (e.g., 580.5—581.5, 581.5—582.5, . . ., 748.5—749.5) which have a class frequency of 0.

a. Use the formulas given in Section 3.8 to compute the approximate sample mean weight $\bar{x}$ and the sample standard deviation s from the grouped data of Table 1.1.

b. Suppose that the tree weights were given in raw data form and that the estimated weights in Table 1.1 were the actual weights of the trees. If you were to compute $\bar{x}$ and s from the raw data, how would these values compare to those computed in part a?

REFERENCES

Ayoub, M. M., Mital, A., Bakken, G. M., Asfour, S. S., & Bethea, N. J. "Development of strength and capacity norms for manual materials handling activities: The state of the art." *Human Factors,* June 1980, *22,* 271-283.

Maxwell, J. C., Jr. "Higher prices driving consumers from liquor." *Advertising Age,* March 31, 1980, 64.

McClave J. T., & Dietrich, F. H. *Statistics*. 2d ed. San Francisco: Dellen, 1982. Chapter 2.

Mendenhall, W. *Introduction to probability and statistics.* 5th ed. North Scituate, Mass.: Duxbury, 1979. Chapter 3.

Namias, J. "A method to detect specific causes of consumer complaints." *Journal of Marketing Research,* August 1964, 63–68.

Peterson, S. "Disco-data general operation." *Billboard,* June 7, 1980, D-4, D-8.

Four Probability

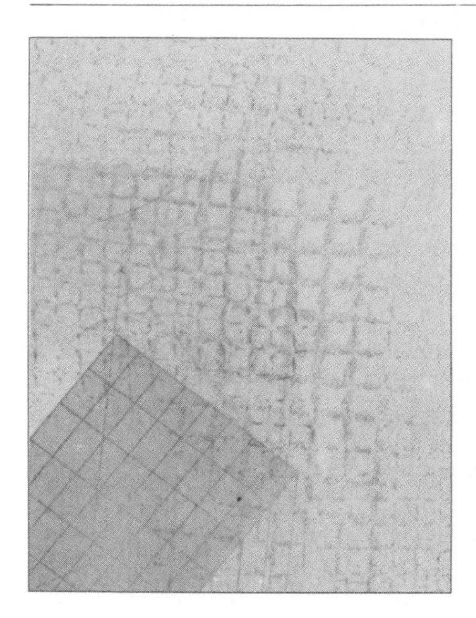

How would you like to strike it rich with a $10 stake (Case Study 4.3), or obtain a winning edge in blackjack (Case Study 4.4)? And what do either of these ventures have to do with statistics? The answer is ***uncertainty.*** The return in real dollars on most investments cannot be predicted with certainty. Neither can we be certain that an inference about a population, based on the partial information contained in a sample, will be correct. In this chapter we shall learn how probability can be used to measure uncertainty, and we shall take a brief glimpse at its role in assessing the reliability of statistical inferences.

Contents

4.1 Blackjack, Investing, and Statistics

If you play blackjack, a popular gambling game, whether you win in any one game is an outcome that is very uncertain. Similarly, investing in bonds, stock, a new business, etc., is a venture whose success is, again, subject to uncertainty. (In fact, some would argue that investing is a form of educated gambling, one in which knowledge, experience, and good judgment can improve the odds of winning.*)

Much like blackjack and investing, inferences based on sample data are also subject to uncertainty. A sample rarely tells a perfectly accurate story about the population from which it was selected. There is always a margin of error (as the pollsters tell us) when sample data are used to estimate the proportion of people in favor of a particular political candidate, some consumer product, or some political or social issue. There is always uncertainty about how far the sample estimate will depart from the true population proportion of affirmative answers that you are attempting to estimate. Consequently, a measure of the amount of uncertainty associated with an estimate (which we called ***the reliability of an inference*** in Chapter 1) plays a major role in statistical inference.

How do we measure the uncertainty associated with events? Anyone who has observed a daily newscast can answer that question. The answer is ***probability.*** Thus, it may be reported that the probability of rain on a given day is 20%. Such a statement acknowledges that it is uncertain whether it will rain on the given day and indicates that the forecaster measures the likelihood of its occurrence as 20%.

In this chapter we will examine the meaning of probability and develop some properties of probability that will be useful in our study of statistics. We will do so by building our discussion around a game of chance, namely blackjack, and then we will apply our knowledge to the solution of some practical problems.

4.2 The Game of Blackjack

Blackjack (or Twenty-One as it is sometimes called) is a game played by two or more players using an ordinary 52-card bridge deck. Each card is assigned a value. Cards numbered from 2 to 10 are assigned the values shown on the card. For example, a 7 of spades has a value of 7; a 3 of hearts has a value of 3. Face cards (kings, queens, and jacks) are each valued at 10, and an ace can be assigned a value of either 1 or 11, a decision at the discretion of the player holding the card.

The essence of the game is that you are dealt two cards and you may request more. The objective is to acquire cards whose total value is as near as possible to, but not exceeding, 21. If the sum of the initial two cards is equal to 21, you have drawn a "blackjack," in most circumstances a winning hand. Each player bets and plays against one of the players who is designated as the "dealer." You win the bet if the total value of your hand exceeds the total value of the dealer's hand. If the total

*One can also utilize sample data (experience) to improve the chances of winning at blackjack. The *Wall Street Journal* (October 24, 1980) reports that the Treasury Hotel Casino in Las Vegas offers seminars on a "surefire method for beating the house" in blackjack.

value of the cards in your hand exceeds 21, you automatically lose; if you hold 5 cards and their total value is 21 or less, you automatically win; if you tie the dealer, you lose. Some typical initial draws are shown in Figure 4.1.

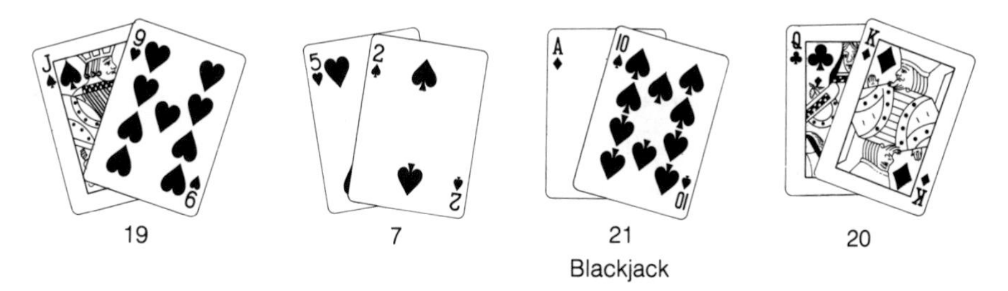

FIGURE 4.1
Some Typical Two-Card Draws and Their Values

The game is played as follows: The dealer distributes two cards, face down, to each player and two cards to himself, one face up and one face down. You see the total value of your hand and, seeing ***one*** of the dealer's two cards, you have some information on the total value of the cards in the dealer's hand. The next move belongs to you. You can decide to face the dealer with the total value of the two cards contained in your initial draw, or you can elect to draw one or more new cards, hoping to push the total value of your hand nearer to, but not exceeding, 21. If additional draws lead to a total exceeding 21, you lose. If not, presumably at some draw, you decide to face the dealer with the total value you then hold in your hand.

After you and the other players finish playing, the dealer must decide whether to face you with the total value contained in his initial pair of cards (if the initial pair is a blackjack, the dealer automatically wins) or whether to draw additional cards in an attempt to increase the total value of his hand.* In doing so, he risks the possibility of drawing a card whose value will push the total value of his hand over 21. If this were to occur, the dealer would lose to all players still in the game.

When the dealer concludes his draw, he displays his cards face up. If the total value of your cards exceeds the total value of the dealer's cards, you win. If not, you lose. Naturally, you hope to draw an ace and a 10 or a face card (blackjack) on the initial draw. This draw is an automatic win for you unless the dealer matches it.

Now that you have a basic understanding of the game of blackjack, we are ready to use this game of chance to illustrate some important and useful probability concepts.

4.3 Experiments and Events

In the language employed in a study of probability, the word ***experiment*** possesses a very broad meaning. In this language, an experiment is a process of making an observation. For example, dealing a pair of cards in the game of blackjack could be

*Different variations of the game are played in casinos throughout the world. For example, "blackjack" is an automatic win for a player in Las Vegas. Also according to Las Vegas rules, the dealer has no decision on a "hit." If the total value of the dealer's cards is 16 or less, he must "hit" (draw another card). If his hand has a total value of 17 through 21, he must "stick" (play the cards in his hand).

viewed as an experiment. Counting the number of U.S. citizens living in a particular state or county is an experiment. Similarly, recording a voter's opinion on an important political issue is an experiment. Observing the fraction of insects killed by a new insecticide is an experiment. Note that most experiments result in outcomes which cannot be predicted with certainty in advance.

DEFINITION 4.1

The process of making an observation is called an ***experiment.***

EXAMPLE 4.1 Consider the following experiment: You are dealt a pair of cards in the game of blackjack. List some possible outcomes of this experiment that cannot be predicted with certainty in advance.

Solution Some possible outcomes of this experiment that cannot be predicted with certainty in advance are:

a. You draw an ace of hearts and a 3 of clubs.
b. You draw an 8 of diamonds and a 9 of hearts.
c. You draw a blackjack—an ace and a card whose value is 10 (a total value of 21).
d. You do not draw a blackjack.

EXAMPLE 4.2 Consider the following experiment: Five hundred urban residents with children are selected from a large number of urban residents with children to determine the proportion who favor a busing plan which would, in theory, racially balance public schools in the city. The response of each urban resident is recorded. List some possible outcomes of this experiment that cannot be predicted with certainty in advance.

Solution Some of the very large number of outcomes of this experiment are as follows:

a. Exactly 387 of the 500 urban residents favor the busing plan.
b. Exactly 388 favor the busing plan.
c. A particular resident, the Jones family, favors the busing plan.

Clearly, we could define many other outcomes of this experiment that cannot be predicted in advance.

In the language of probability theory, outcomes of experiments are called ***events.*** One particular property of events can be seen in Examples 4.1 and 4.2. Two events are said to be ***mutually exclusive*** if, when one occurs, the other cannot occur. To illustrate, the events listed under parts c and d of Example 4.1 are mutually exclusive. You cannot conduct an experiment and "draw a blackjack" (the event listed under part c) and at the same time "not draw a blackjack" (the event listed under part d). If one of these two events occurs when an experiment is conducted, the other event cannot have occurred. Therefore, we say that they are mutually exclusive events.

DEFINITION 4.2

Outcomes of experiments are called ***events.*** [*Note:* In order to simplify our discussion, we will use italic capital letters, A, B, C, . . . , to denote specific events.]

DEFINITION 4.3

Two events are said to be ***mutually exclusive*** if, when one of the two events occurs in an experiment, the other cannot occur.

EXAMPLE 4.3 Refer to Example 4.2 and define the following events:

A: Exactly 387 of the 500 urban residents favor the busing plan.
B: Exactly 388 favor the busing plan.
C: A particular resident, the Jones family, favors the busing plan.

State whether the pairs of events, A and B, A and C, B and C, are mutually exclusive.

Solution

a. Events A and B are mutually exclusive because if you have observed precisely 387 residents who favor the busing plan, then you could not, at the same time, have observed precisely 388.
b. Events A and C are ***not*** mutually exclusive because the Jones family may be one of the residents among the 387 residents in event A who favor busing. Therefore, it is possible for both events A and C to occur simultaneously.
c. Events B and C are not mutually exclusive for the same reason given in part b.

EXAMPLE 4.4 Suppose that an experiment consists of two players (you and the dealer) being dealt the initial two cards in a game of blackjack. Define the following events:

A: You draw a blackjack (21).
B: The dealer draws a blackjack.

Are A and B mutually exclusive events?

Solution A and B are not mutually exclusive because both could occur when the cards are dealt. That is, both you and the dealer could each draw a blackjack (21). If both events A and B occurred, you would be in a tie with the dealer and the dealer would win. (At some casinos, the play is a "standoff" and the bet is canceled.)

As you will subsequently learn, mutually exclusive events play an important role in calculating the probability of an event. The concept of the probability of an event is the topic of our next section.

EXERCISES

4.1 Consider the following experiment: You enroll in three upper-level courses this term and, at the end of the school term, you record your grade (passing or failing) for each. We will define the following events:

A: You receive a passing grade in all three courses.
B: You receive a passing grade in course #1.
C: Your receive a passing grade in course #2.

Explain whether the pairs of events, *A* and *B*, *A* and *C*, *B* and *C*, are mutually exclusive.

4.2 Consider the following experiment: The flight director of the U.S. space shuttle Columbia monitors and records the heartbeat pattern of Columbia's commander while the spacecraft is attempting landing maneuvers. We will define the following events:

A: The commander's heartbeat rises to 135 beats per minute.
B: The commander's heartbeat stays below 120 beats per minute.
C: The commander's heartbeat exceeds his normal heartbeat, namely 60 beats per minute.

Explain whether the pairs of events, *A* and *B*, *A* and *C*, *B* and *C*, are mutually exclusive.

4.3 A child psychologist is interested in the ability of 5-year-old children to distinguish between imaginative and nonfiction stories. Two 5-year-old children are selected from a kindergarten class and are given a test to determine if they can distinguish imagination from reality. Consider the following events:

A: The first child can distinguish imagination from reality.
B: Neither child can distinguish imagination from reality.
C: The second child can distinguish imagination from reality, but the first child cannot.

Explain whether the pairs of events, *A* and *B*, *A* and *C*, *B* and *C*, are mutually exclusive.

4.4 The loan officer of a bank performs the following experiment: Observe the size of a loan request and the financial characteristics (as they relate to credit risk) of the applicant. If the experiment is conducted once (the loan officer observes the characteristics described above for a single loan application), some of the events he or she might observe are

A: The loan request exceeds $100,000.
B: The loan request exceeds $60,000.
C: The applicant's net worth is $185,000.

Explain whether the pairs of events, *A* and *B*, *A* and *C*, *B* and *C*, are mutually exclusive.

4.5 A housewife is asked to rank three brands of laundry detergent, Wisk, Era, and Shout, according to her preference. Consider the following experiment: Observe the rankings (1st, 2nd, and 3rd) of the three brands.

a. Give two of the many different events which could be observed.

b. Are the two events you listed in part a mutually exclusive? Explain. If not, define a pair of mutually exclusive events.

4.4 The Probability of an Event

The outcome for each of the experiments described in the preceding sections was shrouded in uncertainty; i.e., prior to conducting the experiment, we could not be certain whether a particular event would occur. This uncertainty is measured by the ***probability*** of the event.

EXAMPLE 4.5 Suppose that we define the following experiment: Toss a coin and observe whether the upside of the coin is a head or a tail. If an event H is defined by

H: Observe a head

what do we mean when we say that the probability of H, denoted by $P(H)$, is equal to $\frac{1}{2}$?

Solution We mean that, in a very long series of tosses, we believe that approximately half would result in a head. Therefore, the number, $\frac{1}{2}$, measures the likelihood of observing a head on a single toss of the coin.

Stating that the probability of observing a head is $P(H) = \frac{1}{2}$ does not mean that exactly half of a number of tosses will result in heads. For example, we do not expect to observe exactly 1 head in 2 tosses of a coin or exactly 5 heads in 10 tosses of a coin. Rather, we would expect the proportion of heads to vary in a random manner and to approach closer and closer the probability of a head, $P(H) = \frac{1}{2}$, as the number of tosses increases. This property can be seen in the graphs, Figure 4.2.

Figure 4.2(a) on page 106 shows the proportion of heads observed after $n = 25, 50, 75, 100, 125, \ldots, 1{,}450, 1{,}475$, and 1,500 repetitions of a coin-tossing experiment simulated by the author. The number of tosses is marked along the horizontal axis of the graph, and the corresponding proportions of heads are plotted on the vertical axis above the values of n. We have connected the points by line segments to emphasize the fact that the proportion of heads moves closer and closer to .5 as n gets larger (as you move to the right on the graph).

The results of a similar experiment ($n = 10, 11, 12, \ldots, 99, 100$) are shown in Figure 4.2(b) on page 107. You can see that the proportions of heads for a given value of n may differ from Figures 4.2(a) to 4.2(b), but the proportion of heads, for both experiments, moves closer and closer to $\frac{1}{2}$ as n gets larger (as you move to the right along the horizontal axes).

We conducted a similar experiment to determine the approximate probability of drawing a blackjack when two cards are dealt from a well-mixed standard deck of

FIGURE 4.2 The Proportion of Heads in n Tosses of a Coin

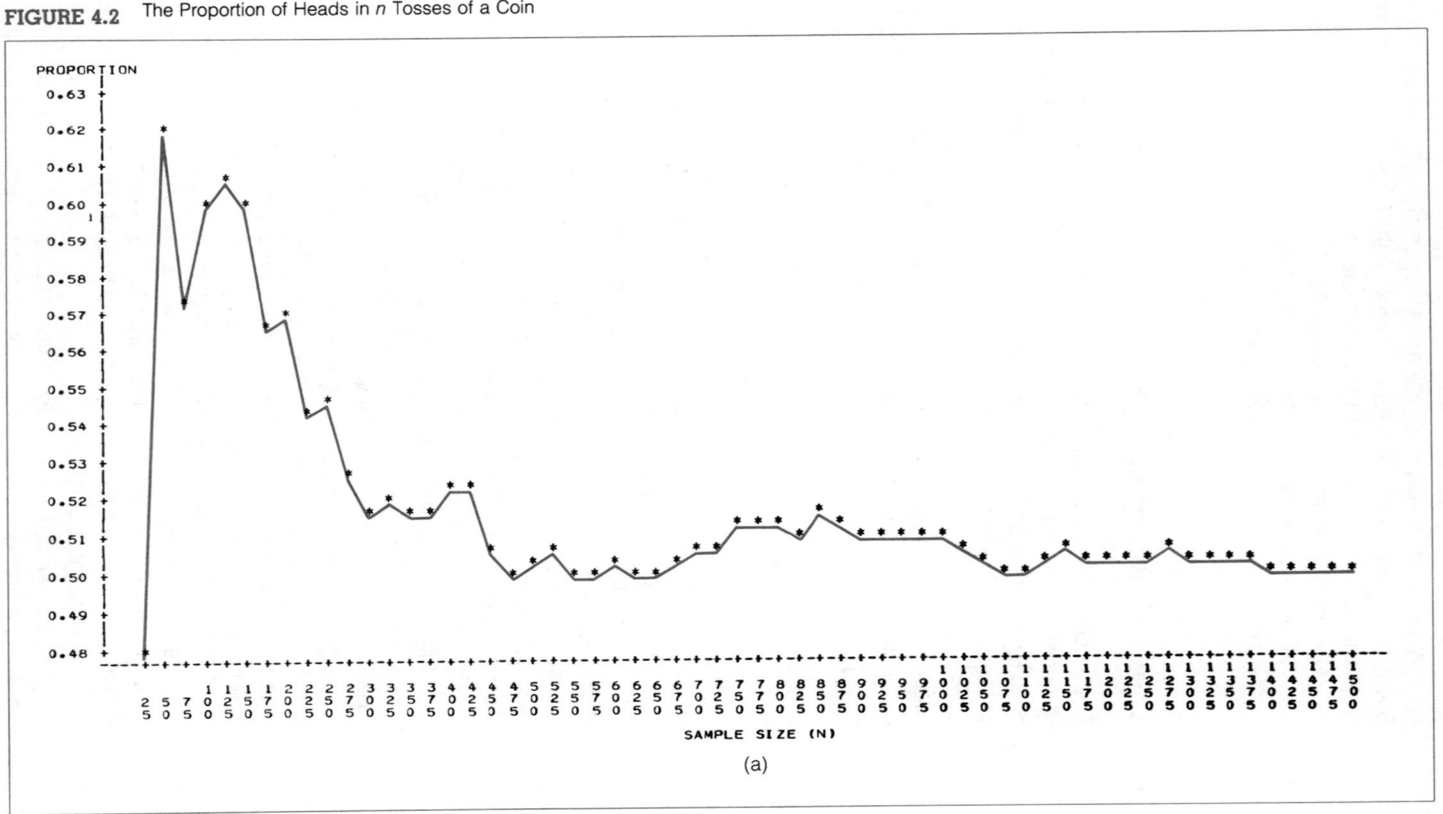

(a)

FIGURE 4.2 (cont'd.) The Proportion of Heads in n Tosses of a Coin

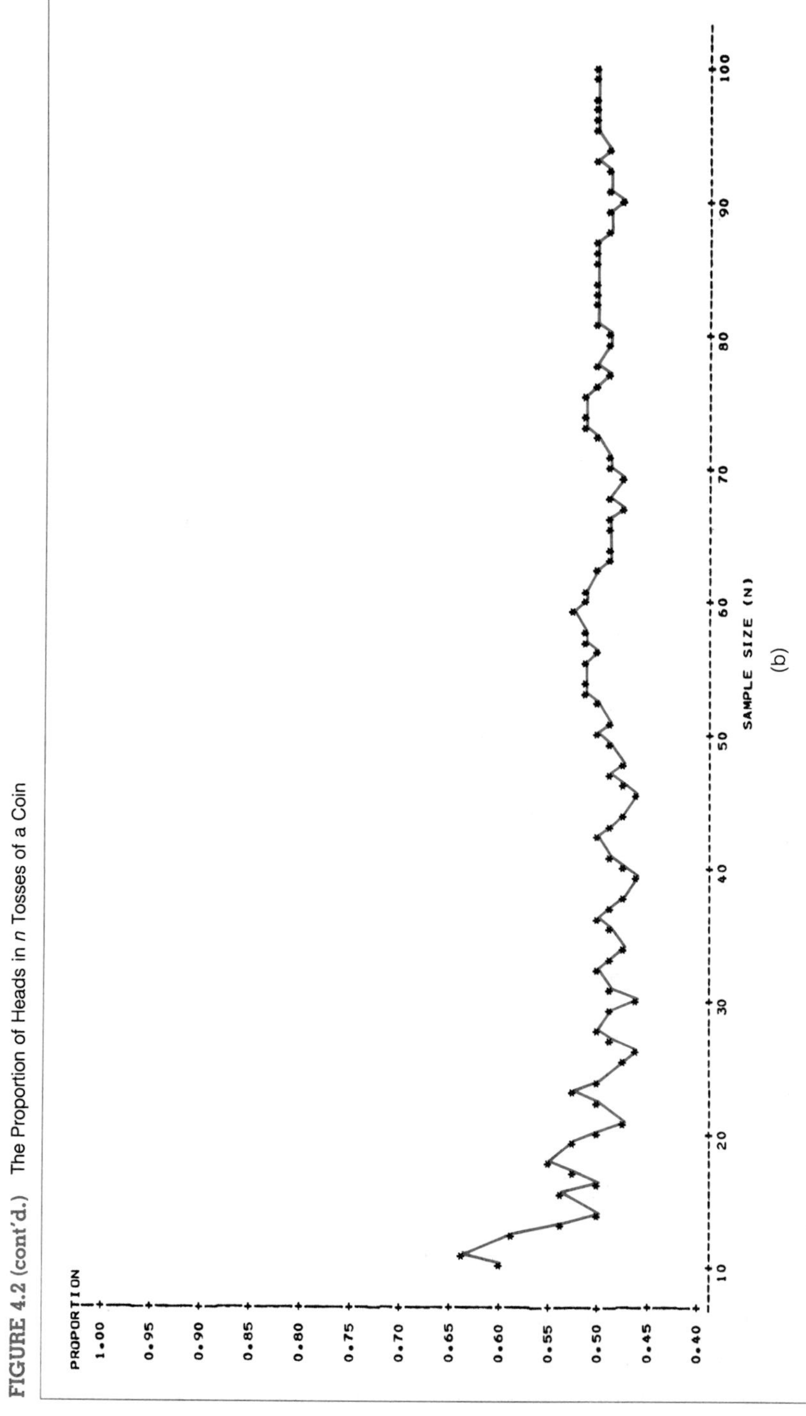

(b)

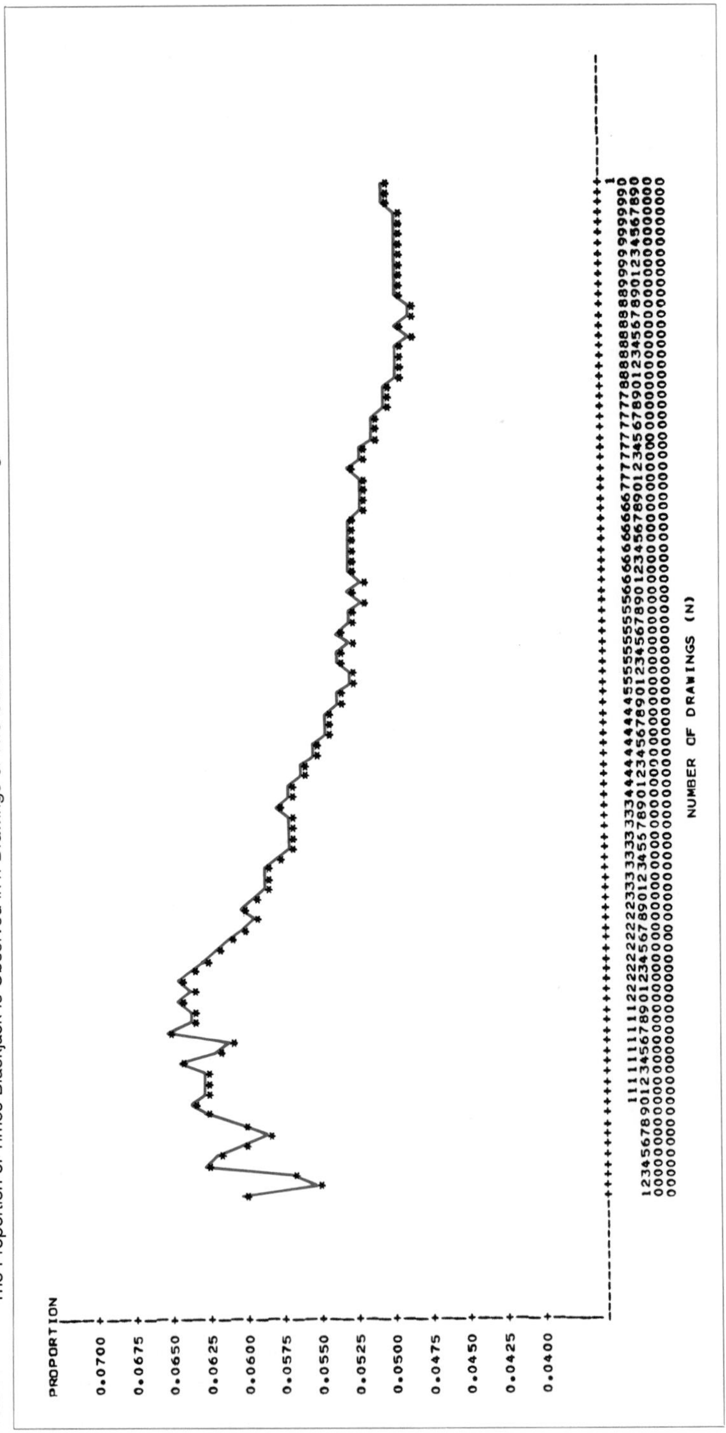

FIGURE 4.3 The Proportion of Times Blackjack Is Observed in n Drawings of Two Cards from a Well-Mixed Bridge Deck

bridge cards. Figure 4.3 on page 108 shows the proportion of times a blackjack is observed when the deal is repeated $n = 100, 200, 300, 400, \ldots, 9{,}800, 9{,}900$, and 10,000 times. The proportions are plotted in Figure 4.3 for each value of n. Notice that the proportion of times a blackjack is observed moves closer and closer to 0.048 as n gets larger (as you move to the right along the horizontal axis). We will show you how to calculate the exact probability of drawing a blackjack, 0.04826546, in Section 4.6.

Although most people think of the probability of an event as the proportion of times the event occurs in a very long series of trials, some experiments can never be repeated. For example, if you invest $50,000 in starting a new business, the probability that your business will survive five years has some unknown value that you will never be able to evaluate by repetitive experiments. The probability of this event occurring is a number that has some value, but it is unknown to us. The best that we could do, in estimating its value, would be to attempt to determine the proportion of similar businesses that survived five years and take this as an approximation to the desired probability. In spite of the fact that we may not be able to conduct repetitive experiments, the relative frequency definition for probability appeals to our intuition.

DEFINITION 4.4

The ***probability of an event A,*** denoted by $P(A)$, is the proportion of times that A is observed when the experiment is repeated a very large number of times.

EXERCISES

4.6 Consider the following experiment: Thirteen cards are dealt face up from a standard, well-mixed, 52-card bridge deck and their suits (hearts, spades, clubs, and diamonds) are observed. Define the event A as follows:

A: 4 hearts, 4 spades, 3 clubs, and 2 diamonds are observed.

It can be shown that the probability of A is .01796, i.e., $P(A) = .01796$. What do we mean by the statement, "$P(A) = .01796$"?

4.7 A power plant which discharges its waste into a nearby gulf performed the following experiment each day for a period of one year (365 days): A water sample is selected from an area near the plant's discharge, analyzed for the presence of PCB (a dangerous chemical), and the amount of PCB (in parts per million) in the sample is recorded. Suppose that the power plant observed that the amount of PCB in the water samples exceeded government pollution standards on two of the days. Find the approximate probability of A, where A is the event that, on any randomly selected day, the amount of PCB in the water sample will exceed the standard.

4.8 The U.S. government estimates that unemployment in the labor force will reach 8% by year-end 1981 (Board of Economics, *Time,* January 19, 1981). That is, of all those who are considered to be eligible workers by year-end 1981, 8% will be unemployed. Suppose that we select one eligible worker at random from the labor force at the end of 1981 and determine that person's employment status. Based on the government estimate, what is the probability that we will find this person unemployed?

4.9 Artificial skin made from cowhide, shark cartilage, and plastic has been used successfully by doctors at Massachusetts General Hospital to replace human skin destroyed by burns (Gainesville *Sun,* April 24, 1981).

a. A total of ten burn patients were treated with the new skin and three of the ten "probably would have died" without it, reported the chief surgeon. Based on these figures, what is the approximate probability that the new artificial skin will be responsible for saving a burn patient's life?

b. Doctors also report that the new skin does not increase the burn patient's chance of getting a fatal infection. (Infections are a common cause of death among burn patients.) Interpret this statement.

4.5 The Additive Probability Rule for Mutually Exclusive Events

If two events, A and B, are mutually exclusive, the probability that ***either*** A ***or*** B occurs is equal to the sum of their probabilities. We will illustrate with an example.

EXAMPLE 4.6 Consider the following experiment: You toss two coins and observe the upper faces of the two coins. What is the probability that you toss exactly one head?

Solution The experiment can result in one of four mutually exclusive events. One possibility is that you will observe a head on coin #1, call it H_1, and a head on coin #2, H_2. We could denote this event as H_1H_2. Similarly, we could observe a head on coin #1 and a tail on coin #2, call this H_1T_2. The other two possible outcomes are a tail on coin #1 and a head on coin #2, T_1H_2, or tails on both coins, T_1T_2. These four events are shown diagrammatically in Figure 4.4.

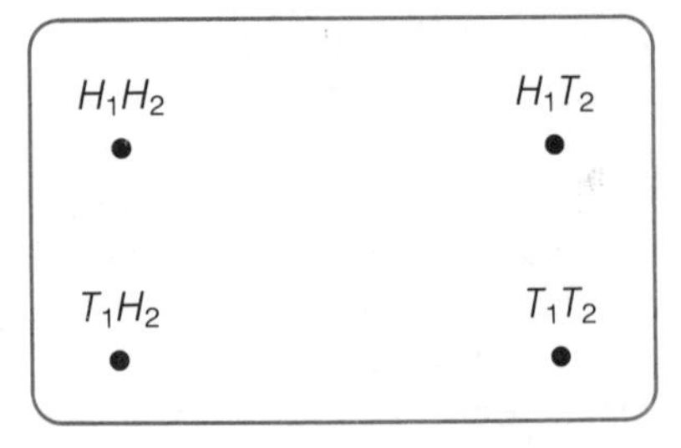

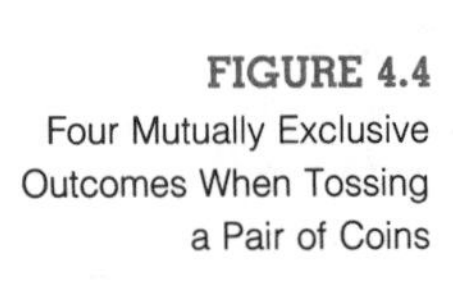

FIGURE 4.4 Four Mutually Exclusive Outcomes When Tossing a Pair of Coins

Since the chance of tossing either a head or a tail on each coin is ½, we would expect each of the four mutually exclusive outcomes of Figure 4.4 to occur with approximately equal relative frequency, ¼, if the coin-tossing experiment were repeated over and over again a large number of times. Since you will observe exactly one head *only if T_1H_2 occurs or if H_1T_2 occurs,* and these events are mutually exclusive, either one or the other of these events will occur $(1/4) + (1/4) = 1/2$ of the time. Therefore, the probability of observing exactly one head in a toss of a pair of coins is equal to the probability of observing ***either*** T_1H_2 ***or*** H_1T_2, which is $1/2$. You can verify this result experimentally, using the procedure employed in Section 4.4.

PROBABILITY RULE #1

The Additive Rule for Mutually Exclusive Events If two events, A and B, are mutually exclusive, the probability that ***either*** A ***or*** B occurs is equal to the sum of their respective probabilities; i.e.,

$$P(\text{either } A \text{ or } B \text{ occurs}) = P(A) + P(B)$$

You can now see why the concept of mutually exclusive events is important. We will illustrate with another example.

EXAMPLE 4.7 Consider the following experiment: A pair of dice is tossed and the number of dots on the upper faces of the dice are observed. Find the probability that the sum of the two numbers is equal to 7 (a winning number in the game of craps).

Solution Mark the dice so that they are identified as die #1 and die #2. Then there are $6 \times 6 = 36$ distinctly different ways that the dice could fall. You could observe a 1 on die #1 and a 1 on die #2; a 1 on die #1 and a 2 on die #2; a 1 on die #1 and a 3 on die #2, etc. In other words, you can pair the 6 values (shown on the six sides) of die #1 with the 6 values of die #2 in $6 \times 6 = 36$ mutually exclusive ways. These combinations are shown in Figure 4.5.

Die #2 \ Die #1	1	2	3	4	5	6
1	2	3	4	5	6	7
2	3	4	5	6	7	8
3	4	5	6	7	8	9
4	5	6	7	8	9	10
5	6	7	8	9	10	11
6	7	8	9	10	11	12

FIGURE 4.5 The Sum of the Dots for the 36 Mutually Exclusive Outcomes in the Tossing of a Pair of Dice

Since there are 36 possible ways the dice could fall and, since these ways should occur with equal frequency, the probability of observing any one of the 36 events listed above is $1/36$. Then to find the probability of tossing a 7, we only need to add the probabilities of those events corresponding to a sum on the dice equal to 7.

If we denote the event that you observe a 6 on die #1 and a 1 on die #2 as (6, 1), etc., then as shown in Figure 4.5, you will toss a 7 if you observe a (6, 1), (5, 2), (4, 3), (3, 4), (2, 5), or (1, 6). Therefore, the probability of tossing a 7 is

$$\begin{aligned} P(7) &= P[(6,1) \text{ or } (5,2) \text{ or } (4,3) \text{ or } (3,4) \text{ or } (2,5) \text{ or } (1,6)] \\ &= P(6,1) + P(5,2) + P(4,3) + P(3,4) + P(2,5) + P(1,6) \\ &= \frac{1}{36} + \frac{1}{36} + \frac{1}{36} + \frac{1}{36} + \frac{1}{36} + \frac{1}{36} = \frac{6}{36} = \frac{1}{6} \end{aligned}$$

Example 4.7 suggests a modification of Probability Rule #1 when an experiment can result in one and only one of a number of equally likely (equiprobable) mutually exclusive events.

PROBABILITY RULE #2

The Probability Rule for an Experiment That Results in One of a Number of Equally Likely Mutually Exclusive Events Suppose that an experiment can result in one and only one of M equally likely mutually exclusive events and that m of these events result in event A. Then the probability of event A is

$$P(A) = \frac{m}{M}$$

EXAMPLE 4.8 Two high school tennis players are randomly selected to play in a championship doubles match from among four players who appear to have equal tennis skills. If two are really better than the others, what is the probability that at least one of these two will be selected?

Solution Identify the four tennis players as C_1, C_2, C_3, and C_4, and let C_3 and C_4 be the two best players. Then the six distinctly different and mutually exclusive ways that the two players may be selected from the four are

$$C_1C_2 \quad C_1C_3 \quad C_1C_4 \quad C_2C_3 \quad C_2C_4 \quad C_3C_4$$

Since the players were randomly chosen, we would expect the likelihood that any one pair would be chosen would be the same as for any other pair. Then, since $M = 6$ and since $m = 5$ of the pairs result in the choice of either C_3 or C_4 or both, the probability of drawing at least one of these two players is

$$P(\text{at least one of } C_3 \text{ or } C_4) = \frac{m}{M} = \frac{5}{6}$$

Examples 4.6, 4.7, and 4.8 identify the properties of the probabilities of all events, as shown in the box.

PROPERTIES OF PROBABILITIES

1. The probability of an event always assumes a value between 0 and 1.
2. If two events A and B are mutually exclusive, then the probability that ***either*** *A* ***or*** *B* occurs is equal to $P(A) + P(B)$.
3. If we list all possible mutually exclusive events associated with an experiment, then the sum of their probabilities will always equal 1.

Before concluding this section, we will comment on two important mutually exclusive events and their probabilities. Consider, once again, the game of blackjack and define the following two events:

A: The first two cards dealt result in a blackjack.
$\overline{A}$: The first two cards dealt do not result in a blackjack.

Thus $\overline{A}$ is the event that A ***does not*** occur. You can see that A and $\overline{A}$ are mutually exclusive events and, further, that if A occurs p% of the time in a long series of trials, then $\overline{A}$ will occur $(100 - p)$% of the time. In other words,

$$P(A) + P(\overline{A}) = 1$$

DEFINITION 4.5

The ***complement*** of an event A, denoted by the symbol $\overline{A}$, is the event that A ***does not*** occur.

Complementary events are important because sometimes it is difficult to find the probability of an event A but easy to find the probability of its complement, $\overline{A}$. In this case, we can find $P(A)$ using the relationship

$$P(A) = 1 - P(\overline{A})$$

EXERCISES

4.10 Refer to Examples 4.6, 4.7, and 4.8 and verify that property 3 (shown in the box that lists the properties of probabilities) holds for each of these examples.

4.11 Suppose that an experiment involves tossing two coins and observing the upper faces of the coins. Find the probabilities of:

a. A: Observing exactly two heads
b. B: Observing exactly two tails

c. *C:* Observing at least one head
d. Describe the complement of event *A* and find its probability.

4.12 Two dice are tossed and the upper faces of the dice are observed. Use Figure 4.5 to find the probability that the sum shown on the dice is equal to:
a. 12 b. 5 c. 11

4.13 Refer to the dice-tossing experiment, Exercise 4.12. Find the approximate probability of tossing a total of 7 by conducting an experiment similar to the experiments illustrated in Figure 4.2: Toss a pair of dice a large number of times and record the proportion of times a 7 is observed. Compare your value with the exact probability, $1/6$.

4.14 An English literature professor hands out a list of four essay questions, two of which will appear on the final examination for the course. One of the students taking the course is pressed for time and can prepare for only two of the four questions. If the professor chooses the two questions at random from the four, what is the probability that the student will be prepared for both questions?

4.6 Random Sampling and the Probability of Blackjack

Most samples are selected from populations that contain a finite number of experimental units. For example, two tennis players were selected from among a total of four tennis players in Example 4.8. Similarly, polls conducted by the Gallup, Harris, and other polling organizations select a sample of opinions from a population that contains a finite number of people. If the sample is selected in such a way that every different sample of size *n* has an equal probability of being selected, the procedure is called ***random sampling*** and the sample is called a (simple) ***random sample*** of size *n*.

DEFINITION 4.6

A ***random sample*** of *n* experimental units is one that is selected in such a way that every different sample of size *n* has an equal probability of selection.

Dealing two cards from a well-mixed bridge deck is equivalent to selecting a sample of 2 from a population containing 52 experimental units. Therefore, finding the probability of drawing a blackjack is just a generalization of Example 4.8.

EXAMPLE 4.9 Find the probability of drawing a blackjack.

Solution Since every possible pair of cards is as likely to be drawn as any other, we will use Probability Rule #2 to find the probability of drawing a blackjack.

1. The first step is to find *M*, the number of distinctly different samples of 2 cards that can be selected from a deck of 52. We will use the formula contained in the

box (proof omitted) which gives the number of ways for selecting n objects from a total of N.

COMBINATORIAL RULE FOR DETERMINING THE NUMBER OF DIFFERENT SAMPLES THAT CAN BE SELECTED FROM A POPULATION

The number of different samples of n objects that can be selected from among a total of N is

$$C_n^N = \frac{N!}{n!(N-n)!}$$

where

$$N! = N(N-1)(N-2)(N-3)\ldots(3)(2)(1)$$
$$n! = n(n-1)(n-2)(n-3)\ldots(3)(2)(1)$$
$$0! = 1$$

For this problem, we are selecting $n = 2$ cards from $N = 52$. Therefore, the total number of different pairs of cards that could be dealt is

$$M = C_n^N = \frac{N!}{n!(N-n)!} = \frac{52!}{2!50!} = \frac{(52)(51)(50)\ldots(3)(2)(1)}{[(2)(1)][(50)(49)(48)\ldots(3)(2)(1)]}$$

or

$$M = \frac{(52)(51)}{2} = 1{,}326$$

2. The second step in solving the problem is to find the number m of samples that result in a blackjack (21). We can find m by considering the pairs of cards that could produce a total of 21. We will first list the cards valued 10 that could be paired with the ace of hearts. These are

K ♡,	Q ♡,	J ♡,	10 ♡	The 16 cards that could be matched with the ace of hearts to produce blackjack
K ◇,	Q ◇,	J ◇,	10 ◇	
K ♧,	Q ♧,	J ♧,	10 ♧	
K ♤,	Q ♤,	J ♤,	10 ♤	

You can see that the ace of hearts could be paired with any of 16 different cards, each valued 10, to produce a blackjack. Likewise, the same 16 cards could be paired with either the ace of diamonds, the ace of spades, or the ace of clubs. Therefore, there are $m = (4)(16) = 64$ pairs of cards that could result in a blackjack.

3. In step 3, we apply Probability Rule #2. We found in step 1 that $M = 1{,}326$ and in step 2 that $m = 64$. Therefore, the probability of drawing a blackjack is

$$P(\text{blackjack}) = \frac{m}{M} = \frac{64}{1{,}326} = .04826546$$

or, approximately 1 chance in 20.

EXAMPLE 4.10 A case of wine contains 12 bottles of which two are improperly sealed and therefore spoiled. If two bottles are selected from the case, what is the probability that both are spoiled?

Solution **Step 1.** We are sampling $n = 2$ bottles from a total of $N = 12$. Therefore, there are

$$M = C_n^N = C_2^{12} = \frac{12!}{2!10!} = \frac{(12)(11)(10)\ldots(3)(2)(1)}{[(2)(1)][(10)(9)\ldots(3)(2)(1)]}$$
$$= 66$$

different pairs of bottles that could be selected.

Step 2. There is only $m = 1$ way of selecting two spoiled bottles from the case since the case contains only two bottles which are spoiled.

Step 3. Therefore, the probability of drawing the two spoiled bottles of wine is

$$P(\text{draw two spoiled bottles}) = \frac{m}{M} = \frac{1}{66}$$

EXAMPLE 4.11 Consider the following problem in statistical inference: Regard the 12 wine bottles in the case in Example 4.10 as a population, and suppose you now think that the case contains at most 2 spoiled bottles. You sample 2 bottles from the 12 and observe that both are spoiled. What would you infer about the population (i.e., the case) of 12 bottles?

Solution You have a theory that the population contains at most 2 spoiled bottles, and you have sampled 2 bottles to see if the sample data disagree with your theory. If your theory is true and, at worst, the population contains 2 spoiled bottles, the chance of observing a sample of 2 spoiled out of 2 sampled is very small; i.e., only $1/66$ or 0.0152. Naturally, the probability of drawing 2 spoiled bottles in a sample of 2 is much higher if the case contains a larger number of spoiled bottles. For example, if 6 are spoiled, the probability of drawing 2 spoiled out of 2 is (proof omitted) $15/66 = .2273$. Consequently, we can reach one of two conclusions.

1. Our theory is correct. The number of spoiled bottles in the case does not exceed 2. The fact that we drew 2 spoiled in a sample of 2 was an unlucky and rare event.
2. Our theory is incorrect. The case contains more than 2 spoiled bottles, a situation that makes the observed sample more probable.

Which of these two conclusions would you choose? We think that you will choose the second one. In doing so, you have used the probability of an observed sample to infer the nature of a population. If you choose the first one, you are implying that the observed sample is an unlucky and rare event, with only a $1/66$ chance of occurring. The fact that the event did occur leads you to believe that its probability of occurring was much higher, i.e., that the case contains more than 2 spoiled bottles.

EXERCISES **4.15** In Example 4.8, we found that one could choose six different samples of 2 tennis players from a group of 4. Use the Combinatorial Rule to arrive at this result.

4.16 How many different samples of size $n = 3$ can be selected from a population containing $N = 5$ elements?

a. Use the Combinatorial Rule to acquire the answer.

b. Suppose that the $N = 5$ elements are denoted as A, B, C, D, E. List the different samples (for example, one sample would be A, B, C; another would be A, B, E, etc.).

4.17 A National Merit Scholarship award committee recently claimed that each of five grant applications received equal consideration in awarding two grants and that, in fact, the recipients were randomly selected from among the five. Three of the applicants were from a majority group and two were from a minority group. Suppose that both grants were awarded to members of a majority group.

a. What is the probability of this event occurring, if, in fact, the committee's claim is true?

b. Is the probability, part a, inconsistent with the committee's claim that the selection was at random?

4.18 Entomologists are often interested in studying the effect of chemical attractants (pheromones) on insects. One common technique is to release several insects equidistant from the pheromone being studied and from a control substance. If the pheromone has an effect, more insects will travel toward it rather than toward the control. Otherwise, the insects are equally likely to travel in either direction. Suppose that five insects are released. If we are interested in the number of insects which travel toward the pheromone, how many outcomes are possible? [*Hint:* If the five insects are denoted as *A*, *B*, *C*, *D*, and *E*, one possible outcome is two insects, *A* and *B*, travel toward the pheromone. Another possible outcome is four insects, *A*, *B*, *C*, and *D*, travel toward the pheromone.]

4.19 Refer to Exercise 4.18. Suppose that the pheromone under study has no effect and, therefore, it is equally likely that an insect will move toward either the pheromone or the control, i.e., the possible outcomes are equiprobable. Find the probability that both insect *A* and insect *E* travel toward the pheromone.

4.20 Two differently styled blue jeans, *A* and *B*, are being considered for mass production by a large clothing manufacturer. A marketing experiment is to be run in a local department store to determine if the public has a preference for one of the two styles. Each of four randomly selected customers will be shown a pair of style *A* blue jeans and a pair of style *B* blue jeans, and asked to voice their preference. The four responses are then observed.

a. List the possible outcomes for this experiment.

b. If the public has no preference for one style over the other, then the events you listed in part a are equiprobable. Assuming equally likely events, what is the probability that all four customers will prefer style *A*?

c. Suppose that the experiment is run and, in fact, all four customers voice their preference for style *A*. Use your answer to part b to infer whether the public has a preference for one of the two styles of blue jeans.

4.7 Conditional Probability and Independence

The *Wall Street Journal* (October 24, 1980) reports that the Treasury Hotel Casino in Las Vegas offers seminars on a "surefire method for beating the house" in blackjack. The system that will "beat the house" is based on the observation of cards dealt by the dealer.

To give a simple example, we found the probability of drawing a blackjack from a well-mixed deck of cards to be equal to 0.04826546. But suppose that the deck has not been newly shuffled, and that we have already seen 10 cards fall on the table: 3 aces, 4 tens, and 3 cards with values less than 10. Now what is the probability that you will draw blackjack in the next pair of cards dealt to you? Intuitively, you will realize that the probability will be less than when drawing from a complete deck because now the deck contains only 1 ace. In fact, given that 10 cards have been removed from the deck, including 3 aces and 4 tens, the probability of drawing blackjack on the next pair of cards is only $^{12}/_{861}$ or 0.0139373.

The probability of drawing blackjack from a mixed complete deck is called the ***unconditional probability*** of that event. In contrast, the probability of drawing a blackjack, ***given that you know some other event has already occurred,*** is called the ***conditional probability*** of the event.

DEFINITION 4.7

The probability of an event A, given that an event B has occurred, is called the ***conditional probability of A given B*** and is denoted by the symbol

$P(A|B)$

[*Note:* The vertical bar between A and B is read "given."]

EXAMPLE 4.12 A box contains three fuses, one good and two defective. Two fuses are drawn in sequence, first one and then the other.

a. What is the probability that the second fuse is defective?

b. What is the probability that the second fuse is defective if you know, for certain, that the first fuse drawn is defective?

Solution **a.** We will denote the good fuse by the letter G and the two defective fuses as D_1 and D_2. If the fuses are drawn at random from the box, the six possible orders of selection are

G, D_1 G, D_2 D_1, G D_1, D_2 D_2, G D_2, D_1

Step 1. Since these six mutually exclusive events are equally likely and comprise all possible outcomes of the draw, we have $M = 6$.

Step 2. Next, we must find the number of selections in which a defective is selected in the second draw. You can see from the listed draws that $m = 4$.

Step 3. Using Probability Rule #2, the unconditional probability of obtaining a defective fuse on the second draw is

$$P(\text{defective fuse on the 2nd draw}) = \frac{m}{M} = \frac{4}{6} = \frac{2}{3}$$

b. The probability of observing a defective on the second draw, given that you have observed a defective on the first draw, is a conditional probability, $P(A \mid B)$, where

A: Observe a defective fuse on the second draw.
B: Observe a defective fuse on the first draw.

If the first fuse drawn from the box is defective, then the box now contains only two fuses, one defective and one nondefective. This means that you have a fifty-fifty chance of drawing a defective fuse on the second draw, given that a defective fuse has already been drawn. That is,

$$P(A \mid B) = \frac{1}{2}$$

The probability obtained in part a, the unconditional probability of event *A*, was equal to ⅔. Clearly, the probability has changed when we know that event *B* has occurred.

EXAMPLE 4.13 A balanced coin is tossed 10 times, resulting in 10 tails. If the coin is tossed one more time, what is the probability of observing a head?

Solution We are asked to find the conditional probability of event *A*, given that event *B* has occurred, where

A: The 11th toss results in a head.
B: The first ten tosses resulted in 10 heads.

Intuitively, it seems reasonable to expect the probability of observing a head on the 11th toss (given that the ten previous tosses resulted in heads) to be greater than ½, but such is not the case. If the coin is truly balanced and is tossed in an unbiased manner, the probability of observing a head on the 11th toss is still ½. (This has been verified both theoretically and experimentally.) Therefore, this is a case where the conditional probability of an event *A* is equal to the unconditional probability of *A*.

Example 4.13 illustrates an important relationship that exists between some pairs of events. If the probability of one event does not depend upon whether a second event has occurred, then the events are said to be ***independent.***

The notion of independence is particularly important when we want to find the probability that ***both*** of two events will occur. When the events are independent, the probability that both events will occur is equal to the product of their unconditional probabilities.

DEFINITION 4.8

Two events A and B are said to be ***independent*** if

$$P(A \mid B) = P(A)$$

or if

$$P(B \mid A) = P(B)$$

[*Note:* If one of these equalities is true, then the other will also be true.]

PROBABILITY RULE #3

The Probability that Both of Two Independent Events A and B Occur If two events, A and B, are independent, the probability that ***both*** A ***and*** B occur is equal to the product of their respective unconditional probabilities; i.e.,

$$P(\text{both } A \text{ and } B \text{ occur}) = P(A)P(B)$$

EXAMPLE 4.14 Find the probability of observing two heads in two tosses of a balanced coin.

Solution Define the following events:

A: Observe a head on the first toss.
B: Observe a head on the second toss.

Since we know that events A and B are independent and that $P(A) = P(B) = 1/2$, the probability that we observe two heads, i.e., both events A and B, is

$$\begin{aligned} P(\text{observe two heads}) &= P(A)P(B) \\ &= \left(\frac{1}{2}\right)\left(\frac{1}{2}\right) = \frac{1}{4} \end{aligned}$$

You can see that this answer agrees with our reasoning in Example 4.6.

EXAMPLE 4.15 Experience has shown that a manufacturing operation produces, on the average, only 1 defective unit in 10. These are removed from the production line, repaired, and returned to the warehouse. Suppose that during a given period of time you observe 5 defective units emerging from the production line in sequence.

a. If prior history has shown that defective units usually emerge randomly from the production line, what is the probability of observing a sequence of 5 consecutive defective units?

b. If the event in part a really occurred, what would you conclude about the process?

Solution **a.** If the defectives really occur randomly, then whether any one unit is defective should be independent of whether the others are defective. Second, the unconditional probability that any one unit is defective is known to be 1/10. We will define the following events:

D_1: The first unit is defective.
D_2: The second unit is defective.
. .
. .
. .
D_5: The fifth unit is defective.

Then,

$$P(D_1) = P(D_2) = P(D_3) = P(D_4) = P(D_5) = \frac{1}{10}$$

and the probability that all 5 are defective is

$$\begin{aligned} P(\text{all 5 are defective}) &= P(D_1)P(D_2) \ldots P(D_5) \\ &= \left(\frac{1}{10}\right)\left(\frac{1}{10}\right)\left(\frac{1}{10}\right)\left(\frac{1}{10}\right)\left(\frac{1}{10}\right) \\ &= \frac{1}{100{,}000} \end{aligned}$$

b. We do not need a knowledge of probability to know that something must be wrong with the production line. Intuition would tell us that observing 5 defectives in sequence is highly improbable (given past history), and we would immediately infer that past history no longer describes the condition of the process. In fact, we would infer that something is disturbing the stability of the process.

EXERCISES **4.21** Find the probability of throwing a pair of 4's when tossing a pair of dice. Compare your answer with the answer that you would obtain from Figure 4.5.

4.22 Recently, the National Aeronautics and Space Administration (NASA) purchased a new solar-powered battery guaranteed to have a failure rate of only 1 in 20. A new system to be used in a space vehicle operates on one of these batteries. To increase the reliability of the system, NASA installed three batteries, each designed to operate if the preceding batteries in the chain fail. If the system is operated in a practical situation, what is the probability that all three batteries would fail?

4.23 Responding to complaints of spoiled milk being served to customers of a late-night diner, the Food and Drug Administration (FDA) has sent an official government inspector to the diner. Suppose that 10 of the 50 bottles of milk the diner currently has on hand contain spoiled milk. The FDA official randomly selects 5 bottles from the 50 (drawn in sequence) for inspection and testing.

a. What is the probability that the first bottle selected contains spoiled milk?
b. If the first bottle selected does not contain spoiled milk, what is the probability that the second bottle selected will contain spoiled milk?

c. Suppose that none of the first four bottles selected contain spoiled milk. What is the probability that the last (fifth) bottle selected will contain spoiled milk?

4.24 A pharmaceutical company produces a drug which is designed to reduce high blood pressure. It advertises that only 2% of the patients who take the drug will experience adverse side effects. Four patients who have taken the drug for a trial period are randomly and independently selected.

a. Assuming the pharmaceutical company's claim is true, find the probability that all four of the patients experience adverse side effects from the drug.
b. Suppose that the event, part a, actually occurs. What would you infer about the pharmaceutical company's claim?

4.25 The game of craps is played with two dice. A player throws both dice, winning unconditionally if he produces either of the outcomes 7 or 11 (the sum of the numbers showing on the two dice), which are designated as "naturals." If the player casts the outcomes 2, 3, or 12—referred to as "craps"—he loses unconditionally.

a. Find the probability of a player throwing a "natural."
b. Find the probability of a player throwing "craps."
c. Suppose that a "hot" player has thrown five "naturals" in a row. What is the probability that the player throws a "natural" on his next toss?
d. Suppose that a "cold" player has thrown five "craps" in a row. What is the probability that the player throws "craps" on his next toss?

4.8 The Additive and Multiplicative Laws of Probability (Optional)

Probability Rule #3 (Section 4.7) gave a formula for finding the probability that both events A and B occur for the special case where A and B were independent events. We can also give a formula, called the ***Multiplicative Law of Probability,*** that applies in general; i.e., regardless of whether A and B are independent events. Although this law is not needed for a study of the remaining material in this text, it (along with the ***Additive Law of Probability*** which we will subsequently present) is needed to complete an introductory coverage of probability.

THE MULTIPLICATIVE LAW OF PROBABILITY

The probability that ***both*** of two events, A and B, occur is

$$P(\text{both } A \text{ and } B \text{ occur}) = P(A)P(B \mid A)$$
$$= P(B)P(A \mid B)$$

EXAMPLE 4.16 Refer to Example 4.12 where we selected two fuses from a box that contained three, two of which were defective. Use the Multiplicative Law of Probability to find the probability that you first draw defective fuse D_1 and then draw D_2.

Solution Define the following events:

A: The second draw results in D_2.

B: The first draw results in D_1.

The probability of event B is $P(B) = 1/3$. Also, from Example 4.12, the conditional probability, $P(A|B) = 1/2$. Then, the probability that both events A and B occur is

$$P(\text{both } A \text{ and } B \text{ occur}) = P(B)P(A|B) = \left(\frac{1}{3}\right)\left(\frac{1}{2}\right) = \frac{1}{6}$$

You can verify this result by rereading Example 4.12.

An additive probability rule, Probability Rule #1, was given for the event that either A or B occurs, but it applied only to the case where A and B were mutually exclusive events. A rule that applies in general is given by the ***Additive Law of Probability.***

THE ADDITIVE LAW OF PROBABILITY

The probability that ***either*** an event A ***or*** an event B or ***both*** occur is

$$P(\textit{either } A \text{ or } B \text{ or } \textit{both} \text{ occur}) = P(A) + P(B) - P(\text{both } A \text{ and } B \text{ occur})$$

EXAMPLE 4.17 Suppose that an experiment consists of tossing a pair of coins and observing the upper faces. Define the following events:

A: Observe at least one head.

B: Observe at least one tail.

Use the Additive Law of Probability to find the probability of observing either A or B or both.

Solution We know the answer to this question before we start because the probability of observing at least one head or at least one tail is 1; i.e., the event is a certainty. To obtain this answer using the Additive Law of Probability, we could use the method of Example 4.6 to find

$$P(A) = P(\text{at least one head}) = \frac{3}{4}$$

$$P(B) = P(\text{at least one tail}) = \frac{3}{4}$$

The event "both A and B occur"—observing at least one head and at least one tail—is the event that you observe exactly one head and exactly one tail. We found this probability in Example 4.6 to be $1/2$. Therefore,

$$P(\text{either } A \text{ or } B \text{ or both occur}) = P(A) + P(B) - P(\text{both } A \text{ and } B \text{ occur})$$

$$= \frac{3}{4} + \frac{3}{4} - \frac{1}{2} = 1$$

This answer confirms what we already knew, that the probability of the event is equal to 1.

EXAMPLE 4.18 Psychologists tend to believe that there is a relationship between aggressiveness and order of birth. To test this belief, a psychologist randomly chose 1,000 elementary school children and administered to each a test designed to measure the student's aggressiveness. Each student was then classified according to aggressiveness (aggressive or unaggressive) and according to order of birth (firstborn, secondborn, or other). The percentages of students falling in the six categories are shown in Table 4.1.

TABLE 4.1 Results of the Aggressiveness Test

		ORDER OF BIRTH			TOTALS
		Firstborn	Secondborn	Other	
AGGRESSIVENESS	Aggressive	6	10	8	24
	Unaggressive	12	33	31	76
TOTALS		18	43	39	100

Suppose we use the percentages contained in the cells of the table to give the approximate probabilities that a single elementary school student would fall in the respective categories. We will define the following events:

A: The student is aggressive.
B: The student was firstborn.

a. Find the probability that both A and B occur.
b. Find the conditional probability that A will occur given that B has occurred.
c. Find the probability that A will not occur.
d. Find the probability that either A or B or both occur.

Solution **a.** We can see from the table that 6% of the students were classified as both aggressive (A) and firstborn (B). Therefore,

$$P(\text{both } A \text{ and } B \text{ occur}) = .06$$

b. Again, examining Table 4.1, we find that 24% of all the students were classified as aggressive and 18% were the firstborn in their family. Therefore,

$$P(A) = .24 \quad \text{and} \quad P(B) = .18$$

To find $P(A|B)$, we substitute the answer to part a and the value of $P(B)$ into the formula for the Multiplicative Law of Probability. Thus,

$$P(\text{both } A \text{ and } B \text{ occur}) = P(B)P(A|B)$$

or

$$.06 = (.18)P(A|B)$$

Solving for $P(A|B)$ yields

$$P(A|B) = \frac{.06}{.18} = .333$$

c. The event that A does not occur is the complement of A, denoted by the symbol $\bar{A}$. Since A is the event that a student is aggressive, $\bar{A}$ is the event that a student is unaggressive. Recall that $P(A)$ and $P(\bar{A})$ bear a special relationship to each other; i.e.,

$$P(A) + P(\bar{A}) = 1$$

From part b, we have $P(A) = .24$. Therefore,

$$P(\bar{A}) = 1 - .24 = .76$$

which can be verified by examining Table 4.1.

d. The probability that either A or B or both occur is given by the Additive Law of Probability. From parts a and b we know that

$$P(A) = .24, \qquad P(B) = .18$$

and

$$P(\text{both } A \text{ and } B \text{ occur}) = .06$$

Then,

$$\begin{aligned} P(\text{either } A \text{ or } B \text{ or both occur}) &= P(A) + P(B) - P(\text{both } A \text{ and } B \text{ occur}) \\ &= .24 + .18 - .06 = .36 \end{aligned}$$

EXERCISES

4.26 Experience has shown that 1% of all novels published in the United States make the "Best Sellers" list and, of these, 50% are dramatized on television, in the theater, or at the movies. Define the following events:

A: A novel makes the "Best Sellers" list.
B: A novel is dramatized on television, in the theater, or at the movies.

a. Give the unconditional probability of event A.
b. Give the probability of event B, given that event A has occurred.
c. Find the probability that a novel will make the "Best Sellers" list and also be dramatized on television, in the theater, or at the movies.

4.27 A recent survey of the American public showed that a majority believed that when they retired their retirement income (from Social Security, company retirement plans, etc.) would be inadequate. A breakdown of the percentages in each category is shown in the table.

		PRIMARY TYPE OF RETIREMENT SUPPORT				TOTALS
		Social Security	Job Pensions	Personal Savings	Other	
BELIEVE SUPPORT WILL BE	Adequate	16	9	11	1	37
	Inadequate	41	12	4	6	63
TOTALS		57	21	15	7	100

Assume that the percentage of the total number of people who fall in a given cell of the table gives the approximate probability that a person selected at random will fall in that cell category. Define the following events:

A: A person believes that his or her retirement income will be inadequate.
B: The major source of retirement income will be a Social Security pension.
C: The major source of retirement income will be a job pension.

Find:

a. $P(A)$
b. $P(B)$
c. $P(C)$
d. $P(\text{both } A \text{ and } B)$
e. $P(\text{both } B \text{ and } C)$
f. $P(\text{either } A \text{ or } B \text{ or both})$
g. $P(B|A)$

4.28 Your broker has set up a deal which involves two separate investments, London gold and American silver. You are informed that the gold investment has a .95 chance of being successful and the silver investment has a .80 chance of being successful. However, the deal is worked so that if either the gold or silver investment is a success, then your overall investment will also be a success. If the success of the gold investment is independent of the success of the silver investment, find the probability that your overall investment will be a success. [*Hint:* Use the Additive Law of Probability and Probability Rule #3.]

4.29 The family-oriented "400 Club" has a membership of 400 people and operates facilities that include an outdoor Olympic-sized swimming pool and indoor racquetball courts. Before the decision on whether to build a new, smaller indoor pool and additional racquetball courts was made, the club manager surveyed members to determine the percentages who regularly use each facility. The survey showed that 60% of the members regularly use the swimming pool and, of these, 25% also regularly use the racquetball courts. Consider the following events:

A: A randomly selected club member regularly uses the swimming pool.
B: A randomly selected club member regularly uses the racquetball courts.

a. Define the event, (both *A* and *B*), in the words of the problem.
b. Find $P(\text{both } A \text{ and } B)$.
c. Define the event, (either *A* or *B* or both), in the words of the problem.
d. If the unconditional probability of event *B* is .45, find $P(\text{either } A \text{ or } B \text{ or both})$. (Use your answer to part b.)

4.30 Do children understand television commercials? In order to answer this question, a major TV advertiser conducted a survey which included 1,000 randomly selected children of various ages. The percentages of children in each category are given in the table.

	CHILDREN'S AGE (YEARS) 4–6	7–9	10–12	TOTALS
Do not understand	41.0	7.3	2.0	50.3
Understand	8.1	18.5	23.1	49.7
TOTALS	49.1	25.8	25.1	100.0

Suppose a child is selected at random from the 1,000 surveyed and assume the percentage of children who fall in a cell of the table gives the approximate probability that the randomly selected child will fall in that cell category.

a. Given that the child does not understand TV commercials, what is the probability that the child is 7–9 years old?
b. What is the probability that a 4–6 year old child who understands TV commercials is selected?
c. What is the probability that a child who understands TV commercials or a child at least 7 years old is selected?
d. What is the probability that a 4–6 year old child is selected?

4.9 Summary

In this chapter we introduced the notion of experiments whose outcomes could not be predicted with certainty in advance. The uncertainty associated with these outcomes (events) was measured by their probabilities—the relative frequencies of their occurrence in a very large number of repetitions of the experiment.

We presented three rules for finding the probabilities of events. Rules 1 and 2 enabled us to find the probability that either one or the other of two events would occur when the events were mutually exclusive (Probability Rule #1), and when all possible outcomes of the experiment were both mutually exclusive and equiprobable (Probability Rule #2). Rule 3 provides a formula for finding the probability that both of two events will occur when the two events are independent. In the optional section of this chapter we gave probability rules which apply in general—the Multiplicative Law of Probability and the Additive Law of Probability. (These probability rules are summarized in the box.)

Finally, we suggested in several of the examples how probability plays a role in statistical inference. We drew a sample from a population and then, based on the probability of observing the sample under various assumptions about the population, we made a decision concerning the nature of the sampled population.

We will not be using our probability rules to solve probability problems in the succeeding chapters because the sample probabilities that we need are too difficult to obtain (that is, their calculation is beyond the scope of this text). Nevertheless, the basic concepts of probability covered in this chapter will be of considerable benefit in understanding how probability plays a role in the inferential methods that follow.

PROBABILITY RULES

1. ***Additive Rule:*** If two events A and B are mutually exclusive then

   $P(\text{either } A \text{ or } B \text{ occurs}) = P(A) + P(B)$

   In general,*

   $P(\text{either } A \text{ or } B \text{ or both occur}) = P(A) + P(B) - P(\text{both } A \text{ and } B \text{ occur})$

2. ***Modified Additive Rule for Equally Likely Mutually Exclusive Events:*** If an experiment results in one and only one of M equally likely mutually exclusive events of which m of these result in an event A, then

   $$P(A) = \frac{m}{M}$$

3. ***Multiplicative Rule:*** If two events, A and B, are independent then

   $P(\text{both } A \text{ and } B \text{ occur}) = P(A)P(B)$

   In general,*

   $P(\text{both } A \text{ and } B \text{ occur}) = P(A)P(B \mid A) = P(B)P(A \mid B)$

Other Helpful Rules:

4. ***Rule of Complements:*** $P(A) = 1 - P(\bar{A})$
5. ***Combinatorial Rule:*** The number of different samples of n objects that can be selected from a total of N is

   $$C_n^N = \frac{N!}{n!(N-n)!}$$

KEY WORDS

Uncertainty
Experiment
Event
Probability
Mutually exclusive events
Complementary events
Conditional probability
Independent events

SUPPLEMENTARY EXERCISES

[Starred (*) exercises refer to the optional section in this chapter.]

4.31 To find the probability that a randomly selected United States citizen was born in a given state, Virginia for example, an introductory statistics student divides the number of favorable outcomes, one, by the total number of states, 50. Thus, the student reports that the probability that John Q. Citizen was born in Virginia is $1/50$.

*These rules are from the optional section in this chapter.

Explain why this probability is incorrect. If you have access to the complete 1980 census, how could the correct probability be obtained?

4.32 There are very few (if any) tests for pregnancy that are 100% accurate. Sometimes a test may indicate that a woman is pregnant even though she really is not. This is known as a false positive result. Similarly, a false negative result occurs when the test indicates that the woman is not pregnant even though she really is. Suppose a woman submits to a certain pregnancy test. Define the events A, B, and C as follows:

A: The woman is really pregnant.
B: The test gives a false positive result.
C: The test gives a false negative result.

Which pair(s) of events are mutually exclusive?

4.33 Rexford Manor is an apartment complex which rents furnished and unfurnished one-, two-, and three-bedroom units. Currently, the apartment complex has only one unit available for rent. Define the events A, B, and C as follows:

A: The unit is an unfurnished apartment.
B: The unit is a one-bedroom, furnished apartment.
C: The unit is a two- or three-bedroom apartment.

Explain whether the pairs of events, A and B, B and C, A and C, are mutually exclusive.

4.34 The positions of president and vice-president of a large labor union are vacant. In the upcoming election, union members are to choose from among four men to fill the positions: Anson (A), Bostock (B), Carrithers (C), and Dennison (D). The highest vote-getter is awarded the presidency and the runner-up is awarded the vice-presidency.

a. List all possible outcomes of the election.
b. Assuming that the events of part a are equiprobable, what is the probability that Bostock is elected president?
c. Assuming that the events of part a are equiprobable, what is the probability that Dennison is elected to one of the two positions?

4.35 As part of an advertising campaign to attract new listeners, an AM radio station conducts a cash jackpot drawing each weekday morning. Every morning a name is chosen at random from the local phone directory (residential phones only) and that person is called for a chance to win money. A housewife, whose last name begins with a "Y," calls the radio station and inquires about her chance of being selected. She is informed that since there are 26 letters in the alphabet, the probability of choosing a last name beginning with "Y" is $1/26$. Is this probability correct? Explain.

4.36 The crocodile is being slowly driven toward extinction both by man's greed for its hide and by cannibalism of its own young. Suppose that a zoo keeper wishes to acquire two male crocodiles for breeding purposes. Four male crocodiles are available from a crocodile ranch in Africa; however, unknown to the zoo keeper, one of the crocodiles has eaten its own young after its mate gave birth and will have a tendency to do so again. If the zoo keeper makes his selections at random, what is the probability that the keeper will acquire the crocodile with cannibalistic tendencies? [*Hint:* Denote the cannibalistic crocodile by C and the three other crocodiles by O_1, O_2, and O_3.]

4.37 A study of the urban mass transportation habits of a city's workers revealed the following: Fifteen percent of the city workers regularly drive their own car to work. Of those who do drive their own car to work, 80% would gladly switch to public mass transportation if it were available. Forty percent of the city workers live more than three miles from the center of the city.

Suppose that one city worker is chosen at random. Define the events A, B, and C as follows:

- A: The person regularly drives their own car to work.
- B: The person would gladly switch to public mass transportation if it were available.
- C: The person lives within three miles of the center of the city.

Find:

a. $P(A)$ **b.** $P(B|A)$ **c.** $P(C)$

d. Explain whether the pairs of events, A and B, A and C, B and C, are mutually exclusive.

***4.38** Each year, American income earners must file a tax report with the Internal Revenue Service (IRS). The IRS has simplified the tax-filing process for many income earners by developing a short income tax form (Form 1040A). Workers earning less than $20,000 if single ($40,000 if filing a joint return) may be eligible to file the short form. Workers earning over this amount are required to file the long income tax form (Form 1040) and may also choose to itemize tax deductions. The accompanying table shows the number of correctly and incorrectly filed tax forms for a random sample of 500 forms examined by the IRS last year. Suppose that one filed tax form is selected at random from the 500 and examined. Define the events A, B, C, and D as follows:

- A: The tax form was filed incorrectly.
- B: The tax form was a short form (1040A).
- C: The tax form was a long form (1040).
- D: The tax form was a long form (1040) with no itemized deductions.

	TAX FORM			TOTALS
	Short Form (1040A)	**Long Form (1040) No Itemized Deductions**	**Long Form (1040) Itemized Deductions**	
Incorrect	10	61	19	90
Correct	115	206	89	410
TOTALS	125	267	108	500

Find:

a. $P(\bar{A})$ b. $P(B)$ c. $P(C)$
d. $P(\text{either } A \text{ or } B)$ e. $P(\text{either } B \text{ or } C)$ f. $P(\text{both } A \text{ and } D)$
g. $P(\text{both } B \text{ and } D)$ h. $P(A|B)$ i. $P(C|A)$
j. Are A and D independent events? Explain.
k. Are B and C mutually exclusive events? Explain.

4.39 The New York State Bureau of Fisheries reported that acid rain and snowfall—originating from oxides of nitrogen and sulfur deposited in the atmosphere from industrial burning of coal and from automobile exhaust—had killed off all fish and many plants in 50% of the high elevation lakes in the Adirondack Mountains.* Of all 2,800 lakes in the entire American Adirondack area (including the high elevation lakes), at least 10% had been found to contain no fish.

a. If one of the high elevation lakes in the Adirondack Mountains is randomly selected and tested, what is the probability that it contains fish?
b. If one of the 2,800 lakes in the entire American Adirondack area is randomly selected and tested, what is the probability that it will be found to contain no fish?

4.40 Despite penicillin and other antibiotics, bacterial pneumonia still kills thousands of Americans every year. A new antipneumonia vaccine, called Pneumovax, is designed especially for elderly or debilitated patients, who are usually the most vulnerable to bacterial pneumonia. Suppose the probability of an elderly or debilitated person being exposed to these bacteria is .45. If exposed, the probability that an elderly or debilitated person inoculated with the vaccine acquires pneumonia is only .10. What is the probability that an elderly or debilitated person inoculated with the vaccine does acquire bacterial pneumonia?

4.41 A daily newspaper operates with two high-speed printing presses (presses #1 and #2). The manufacturer of these high-speed presses claims that, when operating properly, the machines shut down for repairs on only 1% of the operating days. Suppose that the presses operate independently, i.e., the chance of one press breaking down is in no way influenced by the current operating condition of the other. One operating day is randomly selected and the performance of the presses observed.

a. What is the probability that press #1 will be shut down for repairs?
b. What is the probability that press #2 will not need to be shut down for repairs?
c. What is the probability that both presses will be shut down for repairs?
d. Suppose that both presses actually do need to be shut down for repairs during the operating day. Based on this observation, what would you infer about the printing press manufacturer's claim?

4.42 One of the most popular card games among Americans is the game of poker. Each player in the game is dealt 5 cards from a standard 52-card bridge deck. The player with the best (as defined by the rules of the game) 5-card hand is declared

---

**Facts on File*, February 20, 1981.

the winner. Use the Combinatorial Rule to determine the number of different 5-card poker hands which can be dealt from a 52-card deck.

4.43 Advertisers often hire television, movie, and sports personalities to endorse their products. Suppose that a controversial sports personality is being considered to endorse a new product developed by a consumer-oriented company. As a result of this proposed advertising campaign, market strategists have determined that only one of three possible events would occur. These events and their predicted probabilities of occurring are given below.

		Probability
A:	The majority of consumers identify with the personality, and the advertising campaign is a huge success—product sales increase tremendously.	.30
B:	The majority of consumers do not identify with the personality but continue to buy the product, resulting in a moderate increase in sales.	.60
C:	The majority of consumers are antagonized by the personality and refuse to buy the product, which causes an immediate decrease in sales.	.10

Suppose that the company decides to use the controversial sports personality to endorse their product. Based on the market strategists' predictions, find:

a. The probability that the advertising campaign will increase sales tremendously

b. The probability that the advertising campaign will not cause an immediate decrease in sales

4.44 Researchers at a major oil company have developed a new oil-drilling device. Six drilling sites are being considered as potential testing sites for the new device, each in a different state—Arizona, California, Louisiana, New Mexico, Oklahoma, and Texas. Due to the high cost of experimenting with this new device, only three testing sites will be used. These three sites will be chosen randomly from the six states.

a. Use the Combinatorial Rule to determine the number of different three-state selections which the oil company can make.

b. List the different outcomes of part a.

c. What is the probability that both Arizona and Texas are selected for testing the new device?

d. What is the probability that California or Louisiana or both are selected for testing the new device?

e. Given that New Mexico is selected, what is the probability that Oklahoma is also selected?

***4.45** Recently, teenage unemployment hit a four-year high of 19% (i.e., 19% of all teenagers who seek work remain unemployed). Congress is considering a bill designed to provide incentive for employers to hire disadvantaged youths.* The bill,

---

**Wall Street Journal*, February 9, 1981.

if passed, would allow employers to pay teenage workers (16 to 19 years old) less than the minimum wage for six months. The bill is supported by the National Restaurant Association, which reports that 16% of all working teenagers are employed by the food service industry. Suppose that a teenager is selected at random from all teenagers who are seeking or have obtained work.

a. What is the probability that the teenager is employed?

b. What is the probability that the teenager works in the food service industry?

4.46 Refer to Exercise 4.43. A group of 50 potential television advertisers was asked to identify the television broadcaster who they believe the majority of consumers identify with and would thus be ideally suited to endorse the product they sell. The results of the survey are given in the table. Consider the broadcaster identified by an advertiser randomly selected from the group of 50 surveyed.

BROADCASTER	NUMBER OF VOTES
Dan Rather, CBS News	12
Vin Scully, CBS Sports	9
Max Robinson, ABC News	7
Jim McKay, ABC Sports	6
Dick Enberg, NBC Sports	5
Tom Brokaw, NBC News	4
Bryant Gumbel, NBC Sports	3
Mike Wallace, CBS News	3
Barbara Walters, ABC News	1
	50

a. What is the probability that the advertiser favors a sportscaster?

b. What is the probability that the advertiser favors a broadcaster from CBS?

c. What is the probability that the advertiser favors a newscaster or an NBC broadcaster or both?

d. What is the probability that the advertiser favors a sportscaster from ABC?

e. Given that a newscaster is selected, what is the probability that the newscaster is from CBS?

4.47 A survey of 100 members of Congress produced the breakdown given in the table according to political affiliation and position on continuation of the grain embargo against the Soviet Union. Suppose that one congressional member is randomly chosen from the group of 100 surveyed.

	REPUBLICAN	DEMOCRAT	INDEPENDENT	TOTALS
Favor grain embargo	22	31	7	60
Oppose grain embargo	18	12	10	40
TOTALS	40	43	17	100

a. What is the probability that the member is a Republican?
b. What is the probability that the member is a Democrat and opposes the grain embargo?
c. What is the probability that the member is an Independent or favors the grain embargo or both?
d. Given that the congressional member is a Republican, what is the probability that the member opposes the grain embargo?

4.48 Recently, the stock market has gained publicity through the "Witches of Wall Street." These "witches" are people who predict whether a stock will go up or down using Tarot cards, astrological readings, or other supernatural means. One such "witch" claims that she can correctly predict a daily increase or decrease in price for any stock 60% of the time. In order to test the "witch," a broker selects three stocks at random from the New York Stock Exchange and asks the "witch" to predict the next day's increase or decrease in price for each. Assume that the "witch's" three predictions are independently made.
a. What is the probability that the "witch" correctly predicts an increase or decrease in price for all three stocks? (Assume that the "witch's" claim is true.)
b. What is the probability that the "witch" makes an incorrect prediction for all three stocks? (Assume that the "witch's" claim is true.)
c. Suppose that you observe three incorrect predictions by the "witch." Considering your answer to part b, do you believe that the "witch's" claim is true? Explain.

4.49 In order to compare a new brand of peanut butter, Brand X, to the two best-selling brands, Brand Y and Brand Z, a taste test is performed. Two expert tasters are presented specimens of each of the three different brands and asked to select the brand with the preferred taste. Referring to the two selections, (X, Z) denotes the event that the first taster selects Brand X and the second taster selects Brand Z. Note that the same brand may be selected by both expert tasters.
a. List the different ways in which the tasters may select the brands.
b. If the tasters randomly select the brands, what is the probability that Brand X is selected by at least one of the tasters?
c. If the tasters randomly select the brands, what is the probability that the same brand is selected by both tasters?

4.50 Five $1 bills, four $5 bills, and one $20 bill are placed in a box and shuffled. You are permitted to reach blindly into the box and pull out one bill. If your selection is a $1 bill or a $20 bill, the game ends and you keep the bill. If you select a $5 bill, you may replace it and try again.
a. What is the probability that you choose a $1 bill on the first attempt?
b. Given that your first selection was a $5 bill and you have opted to play again, what is the probability that you will select the $20 bill on the second attempt?

***4.51** A county welfare agency employs twelve welfare workers who interview prospective food stamp recipients. The welfare supervisor suspects that two of the twelve have been giving illegal deductions to applicants, although the identity of the two is unknown. Suppose that the supervisor examines the forms completed by two randomly selected workers to audit for illegal deductions.

a. What is the probability that the first worker chosen has been giving illegal deductions?

b. Given that the first worker chosen has been giving illegal deductions, what is the probability that the second worker chosen has also been giving illegal deductions?

c. What is the probability that both of the workers chosen have been giving illegal deductions? [*Hint:* Use the Multiplicative Law of Probability and your answers to parts a and b.]

d. Suppose the supervisor observes that both workers chosen have been giving illegal deductions. If in fact only two of the twelve workers have been giving illegal deductions, is the observed event considered rare? (Use your answer to part c.) Is there evidence that more than two workers are giving illegal deductions?

4.52 Food labelled as "low calorie" is required by law to contain no more than 40 calories per serving. The Food and Drug Administration (FDA) suspects a company of marketing illegally labelled cans of "low calorie" chocolate pudding, i.e., the cans contain more than 40 calories per serving even though they are labelled as "low calorie." From a supermarket shelf containing ten cans of "low calorie" chocolate pudding, the FDA randomly selects three for inspection. Unknown to the FDA, exactly seven of the ten cans contain more than 40 calories per serving.

a. In how many different ways can the FDA randomly select three of the ten cans for inspection? [*Hint:* Apply the Combinatorial Rule.]

b. What is the probability that the FDA observes all three cans to be legally labelled, i.e., the cans contain no more than 40 calories per serving?

c. What is the probability that the FDA observes at least one of the three cans to be illegally labelled? [*Hint:* The complement of the event "at least one of three cans is illegal" is the event "all three cans are legal."]

d. If the event of part b actually occurred, would you consider it to be a rare event, given the condition that seven of the ten cans are illegally labelled?

***4.53** The Coast Guard operates a high-speed motor boat for emergency purposes. Two motors have been installed in the speed boat, but the motors do not operate simultaneously. The second motor acts as a backup to the first motor, and operates only when the first motor fails to start. The probability that the first motor fails to start is .20. If in fact the first fails to start, then the probability that the second motor also fails to start is .30.

a. What is the probability that the Coast Guard's speed boat will be unable to respond to an emergency? [*Hint:* The boat will be unable to respond to an emergency if both motors fail to start. Use the Multiplicative Law of Probability.]

b. The reliability of the Coast Guard's emergency speed boat is the probability that the boat is able to respond to an emergency. Find the reliability of the emergency speed boat. [*Hint:* Use your answer to part a and the notion of complementary events.]

4.54 The U.S. Census Bureau hopes to achieve at least 90% accuracy with the long form of the census. That is, the form is designed so that at least 90% of all those who receive the long form will complete the entire form correctly. To check the accuracy, a random sample of three households that received the long form in the last census is selected, and a follow-up interview is conducted to determine the accuracy of the completed forms. Define the events A, B, and C as follows:

> A: All three households complete the form correctly.
> B: Households #1 and #2 complete the form correctly, but household #3 completes the form incorrectly.
> C: All three households complete the form incorrectly.

Assuming a 90% accuracy rate, find:

a. $P(A)$ **b.** $P(B)$ **c.** $P(C)$
d. $P(\text{either } A \text{ or } B)$ **e.** $P(\text{both } B \text{ and } C)$ **f.** $P(A|B)$
g. If event C is observed, do you think that the U.S. Census Bureau has achieved at least 90% accuracy with the long form of the census? Explain.

4.55 There are two job openings for registered nurses at a general hospital. Out of the five applicants for the jobs, one is a male nurse. Since the applicants are equally qualified, the openings will be randomly filled.

a. List the different ways in which the two nursing positions can be filled. Use the notation F_1, F_2, F_3, F_4 for the four female applicants and M for the male applicant. [*Hint:* Apply the Combinatorial Rule to determine the number of outcomes in your list.]
b. What is the probability that the male nurse is hired?

4.56 A panel studying emergency evacuation plans for Florida's Gulf Coast in the event of a hurricane has determined that it would take between 14 and 17 hours to evacuate people living in low-lying land with the probabilities shown in the table.

EVENT	TIME TO EVACUATE	PROBABILITY
A	14 hours	.27
B	15 hours	.42
C	16 hours	.18
D	17 hours	.13
		1.00

a. What is the probability that the residents of low-lying coastal areas will take at least 16 hours to evacuate their homes? [*Hint:* Find $P(\text{either } C \text{ or } D)$.]
b. What is the probability that residents of low-lying coastal areas will take less than 17 hours to evacuate their homes? [*Hint:* Find $P(\text{either } A \text{ or } B \text{ or } C)$ or $1 - P(D)$.]
c. Weather forecasters say they cannot accurately predict a hurricane landfall more than 14 hours in advance. If the civil engineering department of the Gulf Coast waits until the 14-hour warning before beginning evacuation, what is the

probability that all residents of low-lying areas are evacuated safely (i.e., before the hurricane hits the Gulf Coast)?

***4.57** A psychologist believes that the helping behavior of a person may be related to the person's age. To test her belief, the following experiment was performed. Subjects were presented with a staged situation where help was obviously needed, but not specifically requested, by the experimenter. The time elapsed before the subject offered assistance was recorded. The results of the helping behavior experiment are reported in the table. (The entries in the table are percentages.)

		TIME (IN SECONDS)			
		Less than 30	**30–60**	**60–90**	**Over 90**
AGE (IN YEARS)	**Under 30**	5	10	13	5
	30–50	20	8	15	1
	Over 50	7	6	10	0

Suppose that one of the subjects who participated in the experiment is randomly selected.

a. What is the probability that the subject is over 50 years of age?

b. What is the probability that the subject is under 30 years of age and had an elapsed time of 60–90 seconds before offering assistance?

c. What is the probability that the subject had an elapsed time of 90 seconds or less before offering assistance?

d. Given that the subject had an elapsed time of more than 90 seconds before offering assistance, what is the probability that the subject is under 30 years of age?

e. What is the probability that the subject is over 50 years of age or had an elapsed time of less than 30 seconds before offering assistance?

4.58 Refer to Exercise 4.25. In the two-dice game of craps, a player wins if he throws a "natural" (a 7 or 11) and loses if he throws "craps" (a 2, 3, or 12). However, if the sum of the two dice is 4, 5, 6, 8, 9, or 10 (each of these is known as a "point"), the player continues throwing the dice until the same outcome (point) is repeated (in which case the player wins), or the outcome 7 occurs (in which case the player loses). For example, if a player's first toss results in a 6, the player continues to toss the dice until a 6 or 7 occurs. If a 6 occurs first, the player wins. If a 7 occurs first, the player loses.

a. What is the probability that a player throws a "point" on his first toss? [*Hint:* Find *P*(either 4 or 5 or 6 or 8 or 9 or 10).]

b. If a player throws a "point" of 6 on his first toss, what is the probability that the player wins the game on his next toss?

c. If a player throws a "point" of 6 on his first toss, what is the probability that the player loses the game on his next toss?

4.59 Refer to Exercises 4.25 and 4.58. From the information provided by these exercises, there are basically three events which result in a win for the craps player:

A: The player throws a 7 on his first toss.
B: The player throws an 11 on his first toss.
C: The player throws a "point" on his first toss, and throws the same "point" on a subsequent toss before throwing a 7.

Since the events *A*, *B*, and *C* form pairs of mutually exclusive events, the probability that the player wins the game, i.e., the probability of "making a pass," is simply $P(A) + P(B) + P(C)$. It can be shown (proof omitted) that the probability of a player "making a pass" is .493.

a. Interpret this win probability.

b. In most casinos, betting that a player "makes a pass" pays off at "even odds," i.e., for every $1 bet, you win $1 if the player "makes a pass." Considering the .493 probability of winning, do you think that the even payoff odds are "fair," that is, if you repeatedly bet on a player to "make a pass," would you expect to win as much money as you lose? [*Hint:* When the payoff odds for a winning bet are "even," the game is deemed "fair" if the probability of winning the bet is .50.]

4.60 Roadside breath-testing surveys conducted throughout the U.S. indicate that 6% of the people driving at night have a blood alcohol concentration (BAC) of at least .10%, the legal definition of driving while intoxicated (DWI) in most states. Thus, the probability of a police patrol officer detecting a person DWI from a random stop at night is only .06. However, D. H. Harris (1980) has found that certain visual clues can aid the patrol officer in discriminating between DWI and driving while sober (DWS). These visual clues serve to increase the DWI detection rate above the chance probability of .06. As a result, Harris developed the Drunk Driver Detection Guide for patrol officers. The guide, a portion of which is shown below, gives the percentage of nighttime drivers stopped for a traffic violation (based on various characteristics which serve as visual clues for the patrol officer) with a BAC of at least .10%.

Suppose a patrol officer has stopped a nighttime driver for a traffic violation.

VISUAL CLUE	PERCENTAGE OF NIGHTTIME DRIVERS STOPPED FOR TRAFFIC VIOLATION (BASED ON VISUAL CLUE) WHO HAVE BAC OF AT LEAST .10%
Stopping (without cause) in traffic lane	70%
Following too closely	60%
Slow response to traffic signals	50%
Headlights off (at night)	50%
Weaving	45%
Accelerating or decelerating rapidly	45%
Braking erratically	35%
Driving into opposing or crossing traffic	30%

Source: *Human Factors*, December 1980, pp. 725–732. Copyright 1980 by the Human Factors Society, Inc., and reproduced by permission.

a. Given that the driver has been charged with following too closely behind another vehicle, what is the probability that he or she is DWI?
b. Given that the driver has been charged with stopping (without cause) in a traffic lane, what is the probability that he or she is DWI?
c. Consider the following events:

A: A nighttime driver has a BAC of at least .10%.
B: A nighttime driver is driving with headlights off.
C: A nighttime driver is weaving.

Special adjustments must be made to the Drunk Driver Detection Guide when two or more visual clues are detected. If two clues are detected, add 5% to the larger of the two corresponding percentages to obtain the conditional probability of DWI; if three or more clues are detected, add 10% to the largest of the corresponding percentages. Use this information to find the conditional probability of event *A*, given that both event *B* and event *C* have occurred.

CASE STUDY 4.1 The Illegal Numbers Game

Forbes magazine (October 27, 1980) reports on the illegal "numbers game" which it describes as the poor man's alternative to the gambling playgrounds of Atlantic City. The numbers game, one of organized crime's largest sources of revenue, operates in the following manner: A player selects a three-digit number, such as 987, 243, etc., and then places a bet with an agent, who might be a shopkeeper, an office worker, newsstand operator, etc. These bets are picked up by a "runner" (who receives a commission of between 10% and 25%) who passes the money on to the numbers "bank" which finances the operation and pays off the winners.

In New York, the winning number is usually based on "the handle," the total dollar amount bet on each of the third, fifth, and seventh races at one of the local racetracks. The winning number is obtained by taking the last digit of the handle for each of the three races. Bets can be as small as 50 cents or a dollar, and the payoff to a winner is 500 to 1. For example, if you bet $1 and win, your return is $499 ($500, less the dollar placed with the agent).

Suppose that a person played the lottery and selected the number 139.

a. What is the probability that the winning first digit will be a 1?
b. What is the probability that the winning second digit will be a 3? The winning third digit will be a 9?
c. What is the probability that a person selecting one three-digit number will win?
d. Considering your answer to part c, do you think the payoff rate is reasonable for the player?

CASE STUDY 4.2 The Iranian Hostage Rescue Mission

According to an article in the Orlando *Sentinel Star* (April 29, 1980), a House of Representatives Subcommittee Chairman stated on April 28, 1980, that his panel would investigate "last week's attempt by American commandos to rescue the U.S. hostages in Iran." According to the report, Representative Samuel S. Stratton, Democrat, N.Y., said that the purpose of the inquiry by the House Armed Services Investigations Subcommittee would be to find out why three out of eight helicopters in the mission failed. It was estimated, prior to embarking on the mission, that at least six helicopters were needed to provide a reasonable probability of success.

Specifically, Stratton said, "The failure rate of three out of eight doesn't match the record. The President said this was practiced 20 times and the whole thing ran perfectly. We need to find out what the problem was and how it can be prevented in the future." Representative Stratton went on to say that he was not sure if the investigation will be public. (If Representative Stratton's investigation was public, and if he found a reasonable answer to the probabilistic inconsistency described above, it has received little attention by the press.)

A complete examination of the inconsistency between an observed failure rate of 3 helicopters out of 8 during the week of April 21 and a zero failure rate for 8 helicopters, each flown on 20 missions, is beyond the scope of this chapter, but we do have the tools to make some comments on the inconsistency noted by Representative Stratton, i.e., to see whether a "failure rate of 3 out of 8 doesn't match the record."

Suppose that the helicopters were very reliable (which would disagree with a failure rate of 3 out of 8 helicopters—or even 1 out of 8). For example, suppose that the probability that a single helicopter would fail on a single flight was only 1 chance in 100, i.e., $P(\text{failure of 1 helicopter}) = 1/100$, and that this probability is the same for all 8 helicopters.

a. Assume that the helicopters were serviced between flights and that the failure on any one flight is independent of failure on any other flight. What is the probability that a single helicopter would be able to successfully complete two flights? [*Hint:* If $P(\text{failure on a single flight}) = p = 1/100$, then $P(\text{success on a single flight}) = 1 - p = 1 - 1/100 = 99/100$. Then use Probability Law #3.]

b. Refer to part a. What is the probability that the helicopter would be able to successfully complete 3 flights?

c. 160 flights? [*Note:* You will need a pocket calculator to find this probability.]

d. You will note from part c that the probability of successfully completing 160 flights is very small, assuming $p = P(\text{failure on a single flight})$ is as small as .01. Does this result suggest that $P(\text{success on a single flight})$ is larger or smaller than .99?

e. Since the probability of failure p on a single flight is a *very* small number, is it reasonable to assume that, on the day of the rescue attempt, you would observe 3 failures among the 8 helicopters? Does this suggest that the value of p on the day of the rescue mission differed from the value that existed during the training flights? Explain.

CASE STUDY 4.3 Oil Leases: Striking It Rich with a $10 Stake

How would you like to be an oil baron? Jeffrey Zaslow, in an article* entitled "Striking It Rich with a $10 Stake," explains how you might be able to do it.

> Since 1960, parcels of land that may contain oil have been placed in a lottery with the winner receiving leasing rights for 10 years. Any U.S. citizen older than 21 can play by paying a $10 filing fee to the Bureau of Land Management. If you win—and if an oil

*Orlando *Sentinel Star,* October 12, 1980.

company is interested in drilling on the parcel you obtain—a large sum of money could be tossed your way.

The parcels, most of which are located in Wyoming or New Mexico, are auctioned each month by the Bureau of Land Management. Only one entry is allowed per person or per company, and the winner is permitted to lease the land from the U.S. government for $1 per acre per year for a period of up to 10 years. Not all parcels contain oil but, if you are a lucky winner, some oil company will want to buy your lease. Zaslow mentions four particular winners who sold their leases to oil companies: "An army veteran won $63,000. A real-estate broker won $128,000. A shopkeeper in Texas won $166,000. And an Iowa newspaper circulation manager won $265,000."

Since not many Americans participate in the lottery, and because you have the same chance of winning as Exxon or Phillips Petroleum, this lottery provides the "little guy" with a real opportunity to strike it rich. Zaslow states that, "According to Chuck Wheeler, Department of Interior paralegal specialist, the probability of finding oil on the good parcels—the ones commanding $35,000 to $200,000 on initial sale—is usually a 1-in-20 shot." You can get some guidance on which parcels are the "good ones" by seeking the services of one of the many companies that not only provide this information but will also, for a fee, file the subscriber cards with the Bureau of Land Management.

As you might suspect, some problems associated with the lottery have emerged. Some of the service companies are knowledgeable and others are not. In addition, there have been some suspected cases of "hanky-panky" in the conduct of the lottery and, for several months in 1980, the lottery was suspended. One case reported by Zaslow involved a player who won three times in one month. The three parcels had 1,836, 1,365, and 495 entries, respectively. In this particular case, an Interior Department audit stated that "federal workers did a poor job of shaking the drum before the drawing."

a. Find the probability that a player would win on three parcels involving 1,836, 1,365, and 495 entries.
b. Is this probability consistent with the Interior Department's explanation of this particular event?
c. Based on your knowledge of probability and "rare events," would you make the same inference as that made by the auditor? Explain.

CASE STUDY 4.4 The Blackjack Victory of the Seven Samurai

Since we commenced this chapter with a discussion of the game of blackjack, it is only appropriate that we terminate with the thoughtful consideration of a problem faced by gambling houses. Is card counting by blackjack players a threat to the house? And, if so, what can be done about it? To stimulate your thinking on the subject, we offer the following article from the November 18, 1980, issue of the *Wall Street Journal*.

> Card counters—the bane of American casino operators—have turned up in Macao. And the house has been screaming for a new deal.
>
> Seven touring gamblers from the U.S. and Austria, dubbed the Seven Samurai, until this past weekend were winning big at Macao's blackjack tables by memorizing

cards to get a better jump on the odds. They hinted their net winnings were running $50,000 a week.

"We're just using our heads," said Gabriel Tirado, a 29-year-old recording studio operator from San Francisco, while his friends nodded in agreement.

"This is nonsense," fumed Stanley Ho, managing director of Macao Tourism & Amusement Co., which owns four casinos and has a monopoly on gambling in this Portuguese territory that is tucked under China about 40 miles from Hong Kong. "In all casinos in the world these people aren't welcome," Mr. Ho said.

While the seven didn't admit it, observers say the gamblers were winning by using a variation of a counting system.

In blackjack, players try to beat the dealer by accumulating cards that total as close to 21 as possible without exceeding it. If the player's count is higher than the dealer's, the player wins.

In card counting, the player memorizes the cards dealt from the pack to determine when the remaining pile is rich in picture cards, which are counted as 10, or in low numbers.

Casinos in Nevada and New Jersey bar counters because they can seize as much as a 1% to 1½% advantage over the house. But Macao's laws prevent Mr. Ho from barring anyone.

Mr. Ho has been trying to get the law changed and has altered some of the blackjack rules, including ordering the casino to shuffle the cards after each hand to frustrate the counters. But that switch in procedure worked only briefly before the counters were back to winning again.

The youngest of the crew is 22-year-old Harold Zima, a post-graduate mathematics student from Vienna. Besides Mr. Tirado, the others are Americans who say they are "involved in real estate."

Four members of the group were traveling in Europe when they met Mr. Zima at a casino in Baden-Baden, West Germany. At the casino's blackjack tables their common interest became apparent. The five, along with two other friends, decided to play Asia's casinos because "it is commonly known that Asian casinos are easier to play in," Mr. Tirado said.

About seven weeks ago, the group began arriving separately in Macao. They pretended they didn't know one another. Casino employees spotted several of the counters immediately. But the house didn't know they were working together. Eventually the size of their bets gave them all away. "We put a lot of money in action," said Joseph Maly, one of the seven. "And we would do that for eight to 10 hours at a time."

The casino encouraged them to leave and even threatened them, the gamblers said. Finally, it was hotel space—or the lack of it—that forced the seven out. The players had to leave Macao this past weekend because all the rooms were booked in advance for an annual auto race.

Whether the players will be allowed back is anybody's gamble—and, on this bet, the odds favor the house.

REFERENCES

Baker, T. "Reality takes the wheel." *Forbes,* October 27, 1980, 133–134.

Epstein, R. A. *The theory of gambling and statistical logic.* Revised ed. New York: Academic Press, 1977.

Feller, W. *An introduction to probability theory and its applications.* 3rd ed. Vol. I. New York: Wiley, 1968. Chapters 1, 3, 4, and 5.

Harris, D. H. "Visual detection of driving while intoxicated." *Human Factors,* December 1980, *22,* 725–732.

McClave, J. T., & Dietrich, F. H. *Statistics*. 2d ed. San Francisco: Dellen, 1982. Chapter 3.

Mendenhall, W., Scheaffer, R., & Wackerly, D. *Mathematical statistics with applications*. 2d ed. Boston: Duxbury, 1981. Chapter 2.

Mosteller, F., Rourke, R., & Thomas, G. *Probability with statistical applications*. 2d ed. Reading, Mass.: Addison-Wesley, 1970.

Parzen, E. *Modern probability theory and its applications*. New York: Wiley, 1960. Chapters 1 and 2.

Spaeth, A. "Blackjack victory of 'Seven Samurai'." *Wall Street Journal,* November 18, 1980.

Zaslow, J. "Striking it rich with a $10.00 stake." Orlando *Sentinel Star,* October 12, 1980.

Five

Opinion Polls and the Binomial Probability Distribution

Was the ill-fated helicopter attempt to rescue the American hostages held by Iran (described in Case Study 4.2) doomed from the start, or was there a reasonable chance for the mission's success? The Department of Defense claimed that the 8 helicopters used in the attempt each flew 20 practice missions without malfunction. Consider the number x of the 8 helicopters which malfunction during a mission. What value of x should we expect to observe, and what values of x would be considered unusual or rare events, in view of the pre-raid training record? In this chapter we will learn how to answer questions of this type, and will specifically address the questions concerning the Iran rescue mission in Case Study 5.2.

Contents

5.1 Random Variables

In a practical setting, an experiment (as defined in Chapter 4) involves selecting a sample of data consisting of one or more observations on some variable. For example, we might survey 1,000 physicians concerning their preferences for aspirin and record x, the number preferring a particular brand. Or we might randomly select a single supermarket customer from Appendix C and record his or her total checkout time, x. Since we can never know with certainty the exact value that we will observe when we record x for one experimental unit, we call x a ***random variable.***

DEFINITION 5.1

A ***random variable*** is a variable which, when observed, assumes numerical values which are associated with events.

The two random variables described above are examples of two different types of random variables, ***discrete*** and ***continuous.*** The number x of physicians in a sample of 1,000 who prefer a particular brand of aspirin is said to be a discrete random variable because it can assume only a countable number of values, 0, 1, 2, 3, . . . , 999, 1,000. In contrast, the checkout time of a supermarket customer could theoretically assume any one of an infinite number of values, any value from 0 seconds upwards. Of course, in practice we record checkout time to the nearest second but, *in theory,* the checkout time of a supermarket customer could assume any value, say 137.21471 seconds.

A good way to distinguish between discrete and continuous random variables is to imagine the values that they may assume as points on a line. Discrete random variables may assume any one of a countable number (say 10, 21, 100, etc.) of values corresponding to points on a line. In contrast, a continuous random variable can assume ***any*** value corresponding to the points in one or more intervals on a line. For example, theoretically, the checkout time of a supermarket customer could be represented by any of the infinitely large number of points on some portion of the positive half of a line.

DEFINITION 5.2

A ***discrete random variable*** is one that can assume only a countable number of values.

DEFINITION 5.3

A ***continuous random variable*** can assume any value in one or more intervals on a line.

5.2 Probability Models for Populations

We learned in Chapter 4 that we make inferences based on the probability of observing a particular sample outcome. Since we never know the *exact* probability of some event, we must construct probability models for the values assumed by random variables. For example, if we toss a die, we assume that the values 1, 2, 3, 4, 5, and 6 represent equiprobable events, i.e., $P(1) = P(2) = P(3) = \cdots = P(6) = 1/6$. In doing so, we have constructed a probabilistic model for the relative frequency distribution for the number of dots x that we would observe if we tossed the die thousands and thousands of times and recorded x for each toss. It is unlikely that a perfectly balanced die exists, but most dice would produce relative frequencies very close to $1/6$; and a relative frequency distribution of the results of a large number of tosses would appear as shown in Figure 5.1.

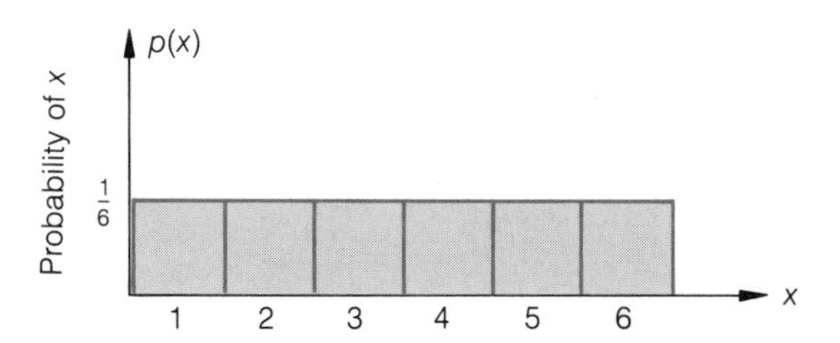

FIGURE 5.1 The Probability Distribution for x, the Number of Dots Observed [*Note:* All the Probability Is Concentrated at the Midpoint of the Interval beneath Each Rectangle.]

Figure 5.1, which gives the relative frequency for each value of x in a very large number of tosses of a die, is called the ***probability distribution for the discrete random variable x.***

> **DEFINITION 5.4**
>
> The ***probability distribution for a discrete random variable x*** is a table, graph, or formula that gives the probability of observing each value of x. We shall denote the probability of x by the symbol $p(x)$.

EXAMPLE 5.1 Consider the following sampling situation: Draw a random sample of $n = 5$ physicians from a very large number, say 10,000, and record the number x of physicians who favor aspirin brand X. Suppose that 2,000 of the physicians actually prefer brand X. Replace the 5 physicians in the population and randomly draw a new sample of 5 physicians. Record the value of x again. Repeat this process over and over again 100,000 times and construct a relative frequency distribution for the 100,000 values of x.

Solution We simulated (on a computer) the drawing of 100,000 samples of 5 physicians from 10,000. Table 5.1 gives the possible values of x (0, 1, 2, 3, 4, and 5), the frequency (the number of times we observed a particular value of x), and the relative frequency (frequency/100,000) for each of these values. The relative frequency distribution is shown in Figure 5.2.

TABLE 5.1
Relative Frequencies for 100,000 Observations on x, the Number of Physicians in a Sample of 5 Who Prefer Aspirin Brand X

x	FREQUENCY	RELATIVE FREQUENCY	$p(x)$
0	32,891	.32891	.32768
1	40,929	.40929	.40960
2	20,473	.20473	.20480
3	5,104	.05104	.05120
4	599	.00599	.00640
5	4	.00004	.00032

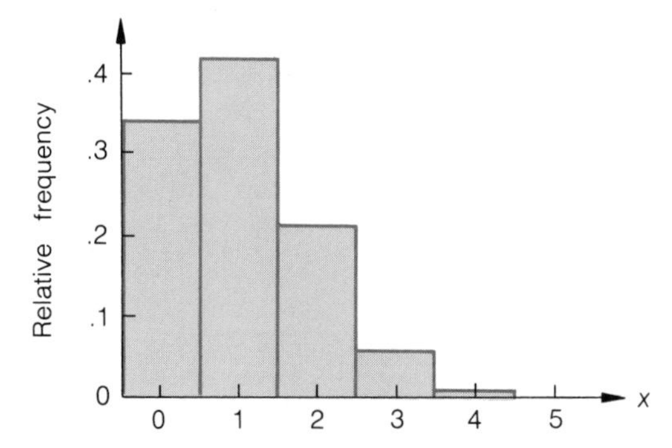

FIGURE 5.2
Relative Frequency Distribution for x

Figure 5.2 provides a very good approximation to the probability distribution for x, the number of physicians in a sample of 5 who prefer brand X (assuming that 20% of the physicians in the population prefer brand X).

In the next section, we will show you how to find a good model for this probability distribution. As you will subsequently learn, this model will give the probabilities, $p(x)$, shown in Table 5.1.

5.3 The Binomial Probability Distribution

Consumer preference and opinion polls are conducted so frequently in political, psychological, sociological, medical, and business situations that it is useful for us to know the probability distribution of the number x in a random sample of n experimental units (people) that prefer some specific proposition. This probability distribution, known as a ***binomial probability distribution,*** is applicable when the sample size n is small relative to the number N of experimental units in the population.

EXAMPLE 5.2 Use the binomial probability distribution (shown in the box on page 149) to calculate the probabilities for the sampling experiment, Example 5.1.

Solution For this sample survey, $n = 5$ and p, the proportion of physicians in the population preferring aspirin brand X, is .2. We will substitute these values of n and p and each value of x into the formula for $p(x)$,

$$p(x) = C_x^n p^x q^{n-x} = \frac{n!}{x!(n-x)!} p^x q^{n-x}$$

THE BINOMIAL PROBABILITY DISTRIBUTION

$$p(x) = C_x^n p^x q^{n-x} = \frac{n!}{x!(n-x)!} p^x q^{n-x}$$

where n = sample size

x = number in the sample who favor the proposition

p = proportion in the population that favor the proposition

$q = 1 - p$

Assumption: The sample size n is small relative to the number N of elements in the population, say, n/N smaller than $1/20$.

Then, remembering that $0! = 1$ and $q = 1 - p$,

$$P(x = 0) = p(0) = C_0^5 (.2)^0(.8)^5$$
$$= \frac{5!}{0!5!}(.2)^0(.8)^5 = (1)(1)(.32768)$$
$$= .32768$$

Similarly,

$$P(x = 1) = p(1) = C_1^5 (.2)^1(.8)^4 = \frac{5!}{1!4!}(.2)^1(.8)^4 = .40960$$

$$P(x = 2) = p(2) = C_2^5 (.2)^2(.8)^3 = \frac{5!}{2!3!}(.2)^2(.8)^3 = .20480$$

$$P(x = 3) = p(3) = C_3^5 (.2)^3(.8)^2 = \frac{5!}{3!2!}(.2)^3(.8)^2 = .05120$$

$$P(x = 4) = p(4) = C_4^5 (.2)^4(.8)^1 = \frac{5!}{4!1!}(.2)^4(.8)^1 = .00640$$

$$P(x = 5) = p(5) = C_5^5 (.2)^5(.8)^0 = \frac{5!}{5!0!}(.2)^5(.8)^0 = .00032$$

Example 5.2 illustrates an important property of discrete probability distributions: the sum of the probabilities for all values of x equals 1. Thus,

$$p(0) + p(1) + p(2) + \cdots + p(5) = .32768 + .40960 + .20480 + \cdots + .00032 = 1$$

PROPERTY #1 FOR ALL DISCRETE PROBABILITY DISTRIBUTIONS

The sum of the probabilities $p(x)$, $x = 0, 1, 2, \ldots, n, \ldots$ for all values of x is always equal to 1, i.e.,

$$\Sigma p(x) = 1$$

EXAMPLE 5.3 Refer to Example 5.2 and find the probability that 3 or more physicians in the sample prefer brand X.

Solution The values that a random variable x can assume are always mutually exclusive events, i.e., you could not observe $x = 2$ and, at the same time, observe $x = 3$. Therefore, the event "$x = 3$ or more" (the event that $x = 3$ *or* $x = 4$ *or* $x = 5$) can be found using Probability Rule #2 (Chapter 4). Thus,

$$P(x = 3 \text{ or } 4 \text{ or } 5) = P(x = 3) + P(x = 4) + P(x = 5)$$
$$= p(3) + p(4) + p(5)$$

Substituting the probabilities found in Example 5.2, we obtain

$$P(x = 3 \text{ or } 4 \text{ or } 5) = .0512 + .0064 + .00032 = .05792$$

Example 5.3 illustrates a second important property of discrete probability distributions: The probability that $x = a$ or b is equal to $p(a) + p(b)$.

PROPERTY #2 FOR ALL DISCRETE PROBABILITY DISTRIBUTIONS

The probability that $x = a$ ***or*** $x = b$ is equal to $p(a) + p(b)$, i.e.,

$$P(x = a \text{ or } x = b) = p(a) + p(b)$$

We have defined a binomial probability distribution in the context of opinion polling, but the distribution possesses much wider applications—any that satisfy a set of five conditions characterizing a ***binomial experiment.*** These conditions are listed in the box.

CONDITIONS REQUIRED FOR A BINOMIAL EXPERIMENT

1. A sample of n experimental units is selected from a population.
2. Each experimental unit possesses one of two characteristics. We conventionally call the characteristic of interest a "success" and the other a "failure."
3. The probability that a single experimental unit possesses the "success" characteristic is equal to p. This probability is the same for all experimental units.
4. The outcome for any one experimental unit is independent of the outcome for any other experimental unit (i.e., the draws are independent).
5. The random variable x counts the number of "successes" in a sample of size n.

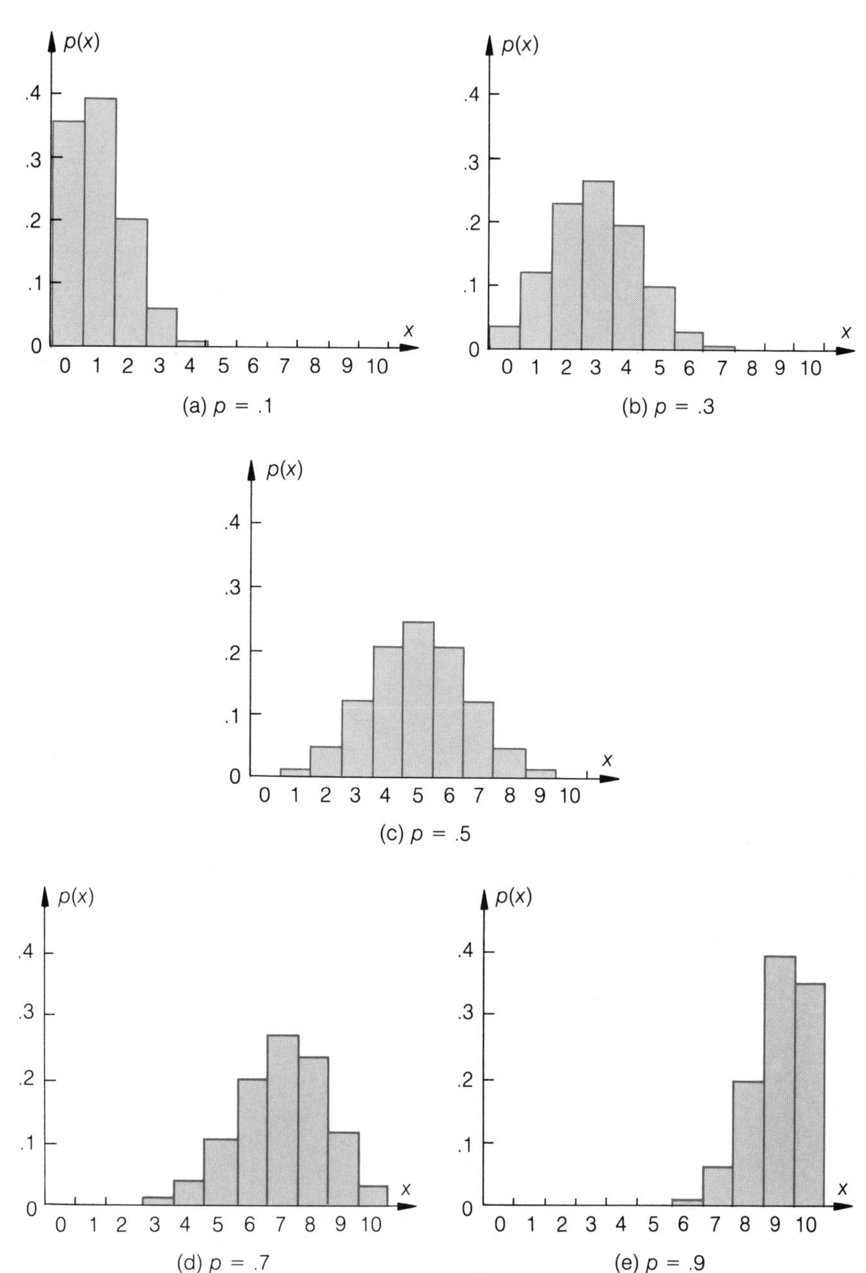

FIGURE 5.3
Binomial Probability Distributions for $n = 10$, $p = .1, .3, .5, .7, .9$

The binomial probability distributions for a sample of $n = 10$ and

a. $p = .1$ **b.** $p = .3$ **c.** $p = .5$ **d.** $p = .7$ **e.** $p = .9$

are shown in Figure 5.3. Note that the probability distribution is skewed to the right for small values of p, skewed to the left for large values of p, and is symmetric for $p = .5$.

As you work the exercises in this chapter, you will encounter several nonsurvey applications of the binomial experiment. Check each to convince yourself that the five characteristics of a binomial experiment are satisfied.

EXERCISES

5.1 A coin is tossed 10 times and the number of heads is recorded. To a reasonable degree of approximation, is this a binomial experiment? Check to see whether each of the five identifying characteristics of a binomial experiment is satisfied.

5.2 Four coins are selected without replacement from a group of five pennies and five dimes. Let x equal the number of pennies in the sample of four coins. To a reasonable degree of approximation, is this a binomial experiment? Check to see whether each of the five identifying characteristics of a binomial experiment is satisfied.

5.3 A parachutist selects two parachutes from among a group of five. Unknown to the parachutist, two of the five are defective and will fail to open. Let x equal the number of defective parachutes in the parachutist's sample of two. To a reasonable degree of approximation, is this a binomial experiment? Check to see whether each of the five identifying characteristics of a binomial experiment is satisfied.

5.4 Suppose that the parachutist, Exercise 5.3, had chosen two parachutes from among a group of 1,000, of which 400 are defective and will fail to open. To a reasonable degree of approximation, is this a binomial experiment? Explain.

5.5 Suppose that you have invested $10 in each of three oil lease lotteries of the type described in Case Study 4.3. Let x equal the number of lotteries in which you will be the winner. To a reasonable degree of approximation, is this a binomial experiment? Explain.

5.6 A random sample of 1,200 people is selected, and each person is asked to choose the news magazine that he or she prefers. Let x equal the number who select the *U.S. News and World Report.* To a reasonable degree of approximation, is this a binomial experiment? Explain.

5.7 A random sample of 100 persons was selected to make a visual comparison of two floor waxes, A and B. Two hundred 9-inch squares of floor tile were used for the experiment and were randomly divided into two groups of 100 tiles each. The first group received a coat of floor polish A, the second group received a coat of floor polish B, and the tiles were marked so that they could be identified only by the person conducting the experiment. The tiles were then assigned, one of each type, to each of the 100 evaluators. Each evaluator chose the tile that he or she preferred, and the experimenter tabulated the number of persons preferring each of the two waxes. Suppose that x is the number of persons preferring floor wax A.

a. Is x a binomial random variable? [*Hint:* Check to see if each of the five characteristics of a binomial experiment is satisfied.]

b. Suppose that the majority of persons inspecting pairs of tiles receiving polishes A and B prefer polish A. What does this imply about the value of p?

c. Suppose that the two waxes are, in fact, identical. What does this imply about the value of p?

5.8 Use the formula for the binomial probability distribution to find the probabilities for $n = 4$, $p = .5$, and $x = 0, 1, 2, 3$, and 4.

5.9 Repeat the instructions of Exercise 5.8 for $n = 4$ and $p = .2$.

5.10 Construct graphs similar to Figure 5.3 of the two probability distributions, Exercises 5.8 and 5.9.

5.11 Refer to Exercise 5.8 and find the probability that:

a. x is less than 2.

b. x is less than or equal to 2.

c. Locate the probabilities, parts a and b, on the graph that you constructed in Exercise 5.10.

d. Verify that (except for rounding) the sum of the probabilities for $x = 0, 1, 2, 3$, and 4 equals 1.

5.12 Refer to Exercise 5.9 and find the probability that:

a. x is less than 2.

b. x is equal to 2 or more.

c. How are the events, parts a and b, related?

d. What relationship must the probabilities, parts a and b, satisfy?

5.13 Records of the Federal Bureau of Investigation (FBI) show that 20% of all convenience stores in a certain state reported one or more armed robberies last year. Thus, the probability that a randomly selected convenience store reported an armed robbery last year is .2. If the FBI contacts the owners of three convenience stores in the state, what is the probability that:

a. All three stores reported an armed robbery last year? (Define x as the number of convenience stores in the sample that reported one or more armed robberies last year.)

b. Exactly 1 store reported an armed robbery last year?

c. At most 1 store reported an armed robbery last year? [*Hint:* The event, "at most 1," is equivalent to "$x = 0$ or $x = 1$".]

5.14 In Exercise 5.13, a "success" was defined as the event that a convenience store reported an armed robbery last year. In contrast, suppose we define a "success" as the event that a convenience store reported no armed robberies last year.

a. What is the value of p for this binomial experiment?

b. Define x in terms of the problem.

c. What values of x correspond to the event that all 3 convenience stores contacted by the FBI reported an armed robbery last year?

d. What value of x corresponds to the event that exactly 1 out of 3 convenience stores reported an armed robbery last year?

e. What values of x correspond to the event that at most 1 out of 3 convenience stores reported an armed robbery last year?

5.15 In Exercise 5.4, we selected two parachutes from among a group of 1,000, of which 400 were defective. Let x equal the number of parachutes in a sample of 2 that open.

a. What is meant by a "success" in the context of this problem?
b. What is the value of p?
c. Find the approximate probability that $x = 1$.

5.16 Refer to Exercise 5.15. Let x equal the number of parachutes in a sample of 2 that fail to open.

a. What is meant by a "success" in the context of this problem?
b. What is the value of p?
c. Find the approximate probability that $x = 1$.

5.4 Tables of the Binomial Probability Distribution

As you can see from the preceding section, the calculation of binomial probabilities is often very tedious. In practice, we would refer to one of the many tables that give the computed values of $p(x)$ for a wide range of values of n and p. To aid you in using these tables, we have included binomial probability tables for $n = 5, 6, 7, 8, 9, 10, 15, 20, 25$. These are listed in Table 1, Appendix E. References to more extensive sets of tables are listed at the end of the chapter.

The tabulated binomial probability distribution for $n = 5$ (Table 1, Appendix E) is reproduced in Table 5.2.

Examining Table 5.2, you note that the values of p are given in the top row of the table, and the values that x can assume, $x = 0, 1, 2, 3, 4, 5$, are shown in the first column. To find the binomial probability for some value of p, say $p = .2$, move across the top row of the table to $p = .2$. The values of $p(x)$ appear in the column beneath this value of p. We will illustrate with an example.

EXAMPLE 5.4 Find the probability that $x = 1$ for a binomial random variable with $n = 5$ and $p = .2$.

Solution Move down the column beneath $p = .2$ until you reach the row corresponding to $x = 1$. The value of $p(1)$ is given as .4096. This value is shaded in Table 5.2. Compare this and the tabulated values of $p(x)$ for $x = 0, 2, 3, 4$, and 5 with the values that we computed (using the formula for $p(x)$) in Example 5.2.

TABLE 5.2 A Reproduction of a Portion of Table 1, Appendix E: The Binomial Probability Distribution, $n = 5$

	P													
X	0.01	0.05	0.1	0.2	0.3	0.4	0.5	0.6	0.7	0.8	0.9	0.95	0.99	X
0	.9510	.7738	.5905	.3277	.1681	.0778	.0313	.0102	.0024	.0003	.0000	.0000	.0000	0
1	.0480	.2036	.3280	.4096	.3601	.2592	.1563	.0768	.0283	.0064	.0005	.0000	.0000	1
2	.0010	.0214	.0729	.2048	.3087	.3456	.3125	.2304	.1323	.0512	.0081	.0011	.0000	2
3	.0000	.0011	.0081	.0512	.1323	.2304	.3125	.3456	.3087	.2048	.0729	.0214	.0010	3
4	.0000	.0000	.0004	.0064	.0283	.0768	.1563	.2592	.3601	.4096	.3280	.2036	.0480	4
5	.0000	.0000	.0000	.0003	.0024	.0102	.0313	.0778	.1681	.3277	.5905	.7738	.9510	5

EXERCISES

5.17 Refer to the binomial probability table for $n = 6$, Table 1, Appendix E.

a. Find $p(2)$ when $p = .30$.
b. Find $p(3)$ when $p = .50$.
c. Find $p(5)$ when $p = .10$.

5.18 Refer to the binomial probability distribution for $n = 10$, Table 1, Appendix E.

a. Find $p(0)$ when $p = .1$.
b. Find $p(1)$ when $p = .1$.
c. Find $p(2)$ when $p = .1$.
d. Compare these tabulated values with the values shown on the graph, Figure 5.3.

5.19 Refer to the binomial probability distribution for $n = 10$, Table 1, Appendix E.

a. Find $p(3)$ when $p = .5$.
b. Find $p(4)$ when $p = .5$.
c. Find $p(5)$ when $p = .5$.
d. Compare your answers, parts a, b, c, with the values shown on the graph, Figure 5.3.

5.20 Construct graphs of the binomial probability distributions for:

a. $n = 6, p = .2$
b. $n = 6, p = .5$
c. $n = 6, p = .8$

5.21 Construct graphs of the binomial probability distributions for:

a. $n = 15, p = .1$
b. $n = 15, p = .5$
c. $n = 15, p = .9$

5.5 Cumulative Binomial Probability Tables

Sometimes we will want to compare an observed value of x obtained from an opinion poll (or some other binomial experiment) with some theory or claim associated with the sampled population. Particularly, we will want to see if the observed value of x represents a "rare event," assuming that the claim is true.

EXAMPLE 5.5 A manufacturer of photographic flash cubes claims that 95% of all flash cubes that it produces will function properly. Suppose that you purchase a box of ten of these flash cubes and find that only six flash. Is this sample outcome highly improbable (a "rare event"), if in fact the manufacturer's claim is true?

Solution If p really does equal .95 (or some larger value), then observing a small number, x, of flash cubes that function properly would represent a rare event. Since we observed $x = 6$, we want to know the probability of observing a value of $x = 6$ or some other value of x even more contradictory to the manufacturer's claim, i.e., we want to find the probability that $x = 0$ or $x = 1$ or $x = 2 \ldots$ or $x = 6$. Using the additive rule for values of $p(x)$, we obtain

$$P(x = 0 \text{ or } x = 1 \text{ or } x = 2 \ldots \text{ or } x = 6) = p(0) + p(1) + p(2) + p(3) + p(4) + p(5) + p(6)$$

when $n = 10$ and $p = .95$. Referring to Table 1, Appendix E, for these values of $p(x)$ and substituting, we find

$$\begin{aligned} P(x = 0 \text{ or } x = 1 \text{ or } x = 2 \ldots \text{ or } x = 6) \\ &= .0000 + .0000 + .0000 + .0000 + .0000 + .0001 + .0010 \\ &= .0011 \end{aligned}$$

This small probability tells us that observing as few as 6 good flash cubes in a pack of 10 is indeed a rare event, if in fact the manufacturer's claim is true. Such a sample result suggests either that the manufacturer's claim is false or that the 10 flash cubes in the box do not represent a random sample from the manufacturer's total production. Perhaps they came from a particular production line that was temporarily malfunctioning.

Solving practical problems of the type illustrated in Example 5.5 requires us to sum values of $p(x)$. Partial sums of the values of $p(x)$, called ***cumulative probabilities,*** are given for $n = 5, 6, 7, 8, 9, 10, 15, 20,$ and 25 in Table 2, Appendix E.

A reproduction of the cumulative binomial probability table for $n = 5$ is shown in Table 5.3.

EXAMPLE 5.6 Consider a binomial experiment with $n = 5$ and $p = .2$. Find the sum of the binomial probabilities $p(x)$ for $x = 0, 1, 2$; i.e., find $P(x = 0 \text{ or } x = 1 \text{ or } x = 2)$.

Solution In general. to find the sum of the binomial probabilities $p(x)$ for $x = 0, 1, 2, \ldots, a$ from Table 2, Appendix E, search for the tabled entry corresponding to the row $x = a$ under the appropriate column for p. The cumulative sum of probabilities for this example is given in the column corresponding to $p = .2$ and the row corresponding to $x = 2$. Therefore,

$$P(x = 0 \text{ or } x = 1 \text{ or } x = 2) = p(0) + p(1) + p(2) = .9421$$

This value is shaded in Table 5.3 below.

TABLE 5.3 A Reproduction of a Portion of Table 2, Appendix E: The Cumulative Binomial Probability Distribution, $n = 5$

x	P 0.01	0.05	0.1	0.2	0.3	0.4	0.5	0.6	0.7	0.8	0.9	0.95	0.99	x
0	.9510	.7738	.5905	.3277	.1681	.0778	.0313	.0102	.0024	.0003	.0000	.0000	.0000	0
1	.9990	.9774	.9185	.7373	.5282	.3370	.1875	.0870	.0308	.0067	.0005	.0000	.0000	1
2	1.0000	.9988	.9914	.9421	.8369	.6826	.5000	.3174	.1631	.0579	.0086	.0012	.0000	2
3	1.0000	1.0000	.9995	.9933	.9692	.9130	.8125	.6630	.4718	.2627	.0815	.0226	.0010	3
4	1.0000	1.0000	1.0000	.9997	.9976	.9898	.9687	.9222	.8319	.6723	.4095	.2262	.0490	4

EXAMPLE 5.7 If you toss a balanced coin 10 times, what is the probability that you will observe 8 or more heads?

Solution First, suppose that we define a success to be a head. Then, the probability of a success is .5, and

$$P(x = 8 \text{ or more}) = P(x = 8 \text{ or } x = 9 \text{ or } x = 10) = p(8) + p(9) + p(10)$$

Remember that Table 2, Appendix E, gives cumulative sums, and that the sum of the values of $p(x)$ over all values of x is equal to 1, i.e., $\Sigma p(x) = 1$. Therefore,

$$P(x = 8 \text{ or } x = 9 \text{ or } x = 10) = p(8) + p(9) + p(10) = 1 - [p(0) + p(1) + \cdots + p(7)]$$

where the partial sum $p(0) + p(1) + \cdots + p(7)$ is represented by the symbol

$$\sum_{x=0}^{7} p(x)$$

The next step is to turn to the cumulative binomial probability table in Table 2, Appendix E, for $n = 10$ and find

$$\sum_{x=0}^{7} p(x) = p(0) + p(1) + \cdots + p(7)$$

The tabulated value in the $p = .5$ column and the row corresponding to $x = 7$ is .9453. Then, the probability of tossing 8 or more heads in 10 tosses of a balanced coin is

$$P(x = 8 \text{ or } x = 9 \text{ or } x = 10) = 1 - \sum_{x=0}^{7} p(x) = 1 - .9453 = .0547$$

EXAMPLE 5.8 Find the probability of tossing 8 or more heads, Example 5.7, by defining a success as observing a tail.

Solution If a success is observing a tail, then the probability of a success is still $p = .5$, and x is the number of tails in 10 tosses of the coin. The next step is to define the event "observe 8 or more heads" in terms of x, the number of tails. This event will occur if the number of tails is 0, 1, or 2. Therefore,

$$P(\text{observe 8 or more heads}) = P(x = 0 \text{ or } x = 1 \text{ or } x = 2) = \sum_{x=0}^{2} p(x)$$

We can read this cumulative sum directly from Table 2, Appendix E, in the table corresponding to $n = 10$. Looking in the column corresponding to $p = .5$ and the row corresponding to $x = 2$, we read

$$P(\text{observe 8 or more heads}) = P(x = 0 \text{ or } x = 1 \text{ or } x = 2) = \sum_{x=0}^{2} p(x) = .0547$$

This is exactly the same answer as was obtained in Example 5.7.

EXERCISES

5.22 Refer to the cumulative binomial probability table for $n = 5$, Table 2, Appendix E.

a. Find $\sum_{x=0}^{2} p(x) = p(0) + p(1) + p(2)$ when $p = .3$.

b. Find $\sum_{x=0}^{4} p(x)$ when $p = .3$.

c. Find $\sum_{x=0}^{5} p(x)$ when $p = .3$.

5.23 Refer to the cumulative binomial probability table for $n = 10$ and $p = .4$, Table 2, Appendix E.

a. Find the probability that x is less than or equal to 8.
b. Find the probability that x is less than 8.
c. Find the probability that x is larger than 8.

5.24 If x is a binomial random variable with $n = 10$ and $p = .1$, use Table 2, Appendix E, to:

a. Find the probability that x is less than or equal to 2.
b. Locate the probability rectangles in Figure 5.2 that correspond to the probability, part a.
c. Find the probability that x is greater than 2.

5.25 If you flip a balanced coin 5 times, what is the probability that:

a. You will toss 3 or fewer heads?
b. You will toss more than 3 heads?
c. You will toss at most 1 head?
d. You will toss fewer than 3 heads?

5.26 Executives in the chemical industry claim that only 5% of all chemical plants discharge more than the EPA's suggested maximum amount of toxic waste into the air and water. Suppose that the EPA randomly samples 20 of the very large number of U.S. chemical plants for inspection. If in fact the executives' claim is true, what is the probability that the number x of plants in violation of the EPA's standard is:

a. Less than 1
b. Less than or equal to 1
c. Less than 2
d. More than 1
e. What would you infer about the executives' claim if the observed value of x is 3? Explain.

5.27 According to a criminal lawyer, 60% of all Americans chosen for jury duty favor capital punishment. To check this claim, a random sample of 25 prospective jurors is selected, and x, the number favoring capital punishment, is recorded. If the lawyer is correct in his assertion, what is the probability that:

a. x is less than 20?
b. x is less than 9?
c. x is more than 9?
d. What would you think about the lawyer's assertion if the observed value of x is less than 9? Explain.

5.28 As we suggested in Chapter 4, investing may be viewed as a form of educated gambling where knowledge and experience will change your odds (chance) of winning. Suppose that the probability of your winning in a single venture is .6 and that you repeat the process for a total of $n = 20$ independent ventures. What is the probability that:

a. You will win on at least half of the ventures (i.e., $x = 10$ or more)?
b. You will win on at least 60% of the ventures (i.e., $x = 12$ or more)?
c. You will win on at least 80% of the ventures?

5.6 The Mean and Standard Deviation for a Binomial Probability Distribution

The samples collected in most sample surveys (or other binomial experiments) are usually quite large. For example, the Gallup and Harris survey results reported in the news media are usually based on samples from $n = 1{,}000$ to $n = 2{,}000$ people. Since we would not wish to calculate $p(x)$ for values of n this large, we need an easy way to describe the probability distribution for x to help us decide whether an observed value of x represents a rare event. To do this, we need to know the mean and standard deviation for the distribution. Then we can describe it using the Empirical Rule of Chapter 3.

In general, the probability distribution for a discrete random variable is a theoretical frequency distribution for a population. Consequently, we can describe it by finding its mean μ and standard deviation σ. These quantities are found using the following definitions:

DEFINITION 5.5

The ***mean*** μ (or ***expected value***) of a discrete random variable x is equal to the sum of the products of each value of x and the corresponding value of $p(x)$, i.e.,

$$\mu = \Sigma x p(x)$$

DEFINITION 5.6

The ***variance*** σ^2 of a discrete random variable x is equal to the sum of the products of $(x - \mu)^2$ and the corresponding value of $p(x)$, i.e.,

$$\sigma^2 = \Sigma (x - \mu)^2 p(x)$$

DEFINITION 5.7

The ***standard deviation*** σ of a random variable x is equal to the positive square root of the variance.

EXAMPLE 5.9 The formulas given in Definitions 5.5, 5.6, and 5.7 provide numerical measures (μ and σ) that will describe the location and spread of a discrete probability distribution. We shall demonstrate this by finding the mean and standard deviation for the binomial probability distribution with $n = 5$, $p = .2$. The graph of this probability distribution is shown in Figure 5.4. Find the mean μ and standard deviation σ for this distribution.

Solution By Definition 5.5, the mean, μ, is given by

$$\mu = \Sigma x p(x)$$

The values of $p(x)$ are given in Table 1, Appendix E, for $n = 5$ and $p = .2$. Since x can take values, $x = 0, 1, 2, \ldots, 5$, we have:

$$\mu = 0p(0) + 1p(1) + 2p(2) + \cdots + 5p(5)$$
$$= 0(.32768) + 1(.4096) + 2(.2048) + 3(.0512) + 4(.0064) + 5(.00032)$$
$$= 1.0$$

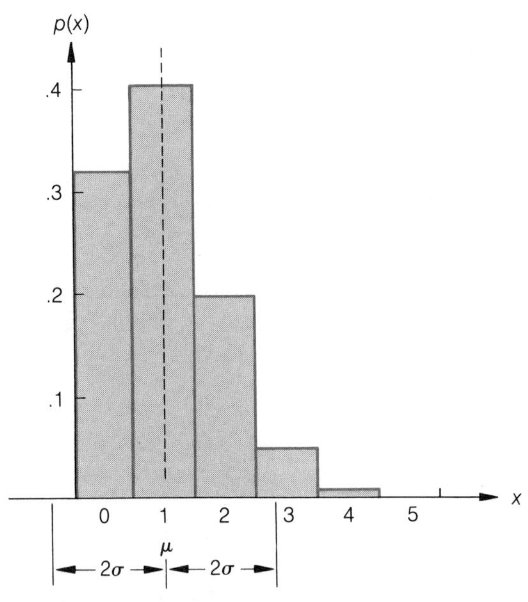

FIGURE 5.4
The Binomial Probability Distribution
$n = 5$, $p = .2$

Similarly, by Definition 5.6,

$$\sigma^2 = \Sigma (x - \mu)^2 p(x) \qquad \text{where} \quad \mu = 1.0$$
$$= (0 - 1.0)^2 p(0) + (1 - 1.0)^2 p(1) + (2 - 1.0)^2 p(2) + \cdots + (5 - 1.0)^2 p(5)$$
$$= (0 - 1.0)^2(.32768) + (1 - 1.0)^2(.4096) + (2 - 1.0)^2(.2048) + (3 - 1.0)^2(.0512) + (4 - 1.0)^2(.0064) + (5 - 1.0)^2(.00032)$$
$$= .80$$

and by Definition 5.7, the standard deviation, σ, is given by

$$\sigma = \sqrt{\sigma^2} = \sqrt{.80} = .894$$

EXAMPLE 5.10 Locate the interval $\mu \pm 2\sigma$ on the graph of the binomial probability distribution for $n = 5$, $p = .2$. Confirm that most of the (theoretical) population falls within this interval.

Solution Figure 5.4 gives the graph of the binomial probability distribution when $n = 5$, $p = .2$ (the actual probabilities are given in the table corresponding to $n = 5$, Table 1, Appendix E). Recall that $\mu = 1.0$ and $\sigma = .894$. Then,

$$\mu - 2\sigma = 1.0 - 2(.894) = -.788$$
$$\mu + 2\sigma = 1.0 + 2(.894) = 2.788$$

The interval $-.788$ to 2.788, shown in Figure 5.4, includes the values of $x = 0$, $x = 1$, and $x = 2$. Thus, the probability (relative frequency) that a population value falls within this interval is

$$\begin{aligned} p(0) + p(1) + p(2) &= .3277 + .4096 + .2048 \\ &= .9421 \end{aligned}$$

This certainly agrees with the Empirical Rule, which states that approximately 95% of the data will lie within 2σ of the mean μ.

You can show (proof omitted), using the formulas in Definitions 5.5, 5.6, and 5.7, that the mean, variance, and standard deviation for a binomial probability distribution are as follows:

MEAN, VARIANCE, AND STANDARD DEVIATION FOR A BINOMIAL PROBABILITY DISTRIBUTION

$$\mu = np$$

$$\sigma^2 = npq$$

$$\sigma = \sqrt{npq}$$

where n = Sample size

p = Probability of a success on a single trial

= Proportion of experimental units in a large population that are "successes"

$q = 1 - p$

EXAMPLE 5.11 Although one primary objective is to describe binomial probability distributions based on large samples (i.e., when n is large), we can see how well μ and σ characterize the binomial probability distributions shown in the graphs, Figure 5.3. Find μ and σ for a binomial probability distribution with $n = 10$, $p = .1$, and find the theoretical proportion of the population which lies within the interval $\mu \pm 2\sigma$. Does this result agree with the Empirical Rule, which states that approximately 95% of the measurements in the distribution should lie within this interval?

Solution Using the formulas for μ and σ, we obtain

$$\mu = np = (10)(.1) = 1$$

$$\sigma = \sqrt{npq} = \sqrt{(10)(.1)(.9)} = .949$$

Then,

$$\mu - 2\sigma = 1 - 2(.949) = -.898$$

$$\mu + 2\sigma = 1 + 2(.949) = 2.898$$

The values of x in the interval, $-.898$ to 2.898, are $x = 0$, 1, and 2. To find the sum of the probabilities in the interval $\mu \pm 2\sigma$ (i.e., $-.898$ to 2.898), we need to find $p(0) + p(1) + p(2)$. The easy way to find this partial sum is to refer to the cumulative

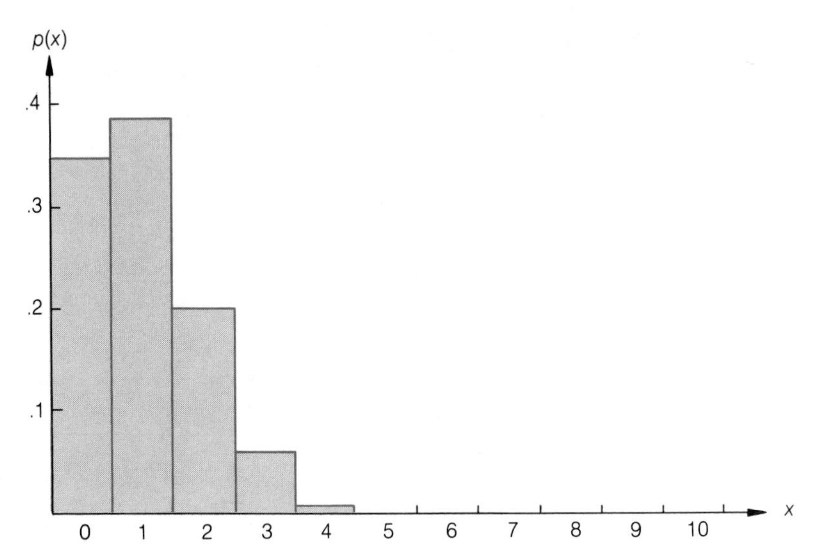

FIGURE 5.5
The Binomial Probability Distribution
$n = 10$, $p = .1$

binomial probability table for $n = 10$ and $p = .1$. This gives

$$\sum_{x=0}^{2} p(x) = p(0) + p(1) + p(2) = .9298$$

You can see from this value, .9298, that the sum of the probabilities for values of x in the interval $\mu \pm 2\sigma$ agrees very closely with the Empirical Rule. You can see this graphically in Figure 5.5.

EXAMPLE 5.12 Describe a binomial probability distribution for $n = 1{,}000$ and $p = .4$.

Solution We will use the formulas to find μ and σ:

$$\mu = np = (1{,}000)(.4) = 400$$
$$\sigma = \sqrt{npq} = \sqrt{1{,}000(.4)(.6)} = \sqrt{240} = 15.49$$
$$\mu - 2\sigma = 400 - 2(15.49) = 369.02$$
$$\mu + 2\sigma = 400 + 2(15.49) = 430.98$$

Therefore, we envision a distribution that centers about $\mu = 400$ with (from the Empirical Rule) approximately 95% of the distribution lying between 369.02 and 430.98. Since x can assume only integer values, this means that approximately 95% of the probability lies within the interval $x = 370$ to $x = 430$.

EXAMPLE 5.13 Refer to Example 5.12. Is it likely that x could assume a value as large as 450?

Solution Not likely! We know that most (approximately 95%) of the distribution lies in the interval $x = 370$ to $x = 430$. A value of $x = 450$ lies outside this interval. In fact, the z-score (discussed in Chapter 3) corresponding to $x = 450$ is

$$z = \frac{x - \mu}{\sigma} = \frac{450 - 400}{15.49} = 3.23$$

Therefore, $x = 450$ lies 3.23 standard deviations above $\mu = 400$. The probability of observing a value of x this far above the mean μ is very small (almost negligible).

EXAMPLE 5.14 One hundred thousand early votes have been counted in a national election. If candidate Anderson has only 49,000 of the votes, does this imply that his opponent will win?

Solution We cannot answer this question unless we can be fairly certain that the 100,000 early votes represent a random sample from the entire voting population. If we are willing to make this assumption, and if in fact Anderson will receive 50% of all the votes, then $p = .5$ and $q = .5$, and

$$\mu = np = (100{,}000)(.5) = 50{,}000$$
$$\sigma = \sqrt{npq} = \sqrt{(100{,}000)(.5)(.5)} = 158.1$$

Then, if Anderson really has 50% of *all* votes (i.e., if $p = .5$), we would expect the number of voters favoring Anderson to fall in the interval

$$\mu - 2\sigma = 50{,}000 - 2(158.1) = 49{,}683.8$$

to

$$\mu + 2\sigma = 50{,}000 + 2(158.1) = 50{,}316.2$$

If Anderson has only $x = 49{,}000$ votes in a sample of 100,000, you can see that this value of x is highly improbable (assuming $p = .5$). It lies a long way outside the interval $\mu \pm 2\sigma$. The z-score for this value of x is

$$z = \frac{x - \mu}{\sigma} = \frac{49{,}000 - 50{,}000}{158.1} = -6.33$$

If Anderson is to win (i.e., if $p = .5$ or larger), the probability that the number, x, of voters in a sample of 100,000 favoring him is as small as 49,000 is almost 0. Therefore, we are inclined to believe (based on the sample value, $x = 49{,}000$) that the proportion of voters favoring Anderson is much less than $p = .5$.

EXERCISES

5.29 Calculate μ and σ for a binomial probability distribution with

a. $n = 15, p = .1$ **b.** $n = 15, p = .5$

5.30 Since a probability distribution is a theoretical model for a population relative frequency distribution, what proportion of the total probability would you expect to lie within the interval $\mu \pm 2\sigma$? [*Hint:* Use the Empirical Rule, Chapter 3.]

5.31 Refer to Exercise 5.29. Locate the interval $\mu \pm 2\sigma$ on the graphs of the probability distributions, Exercise 5.21, parts a and b. For each graph, find the probability that x lies within the interval $\mu \pm 2\sigma$. Do these probabilities agree with your answer to Exercise 5.30?

5.32 Find the mean and standard deviation for the following binomial probability distributions:

a. $n = 100, p = .99$ **b.** $n = 100, p = .8$ **c.** $n = 100, p = .5$
d. $n = 100, p = .2$ **e.** $n = 100, p = .01$

5.33 Use the values of μ and σ calculated in Exercise 5.32 to construct rough sketches of the five binomial probability distributions.

5.34 Find the means and standard deviations for the following binomial probability distributions:

a. $n = 900$, $p = .99$
b. $n = 900$, $p = .8$
c. $n = 900$, $p = .5$
d. $n = 900$, $p = .2$
e. $n = 900$, $p = .01$

5.35 Use the values of μ and σ calculated in Exercise 5.34 to construct rough sketches of the five binomial probability distributions.

5.36 A physician claims that only 10% of all American adults suffer from high blood pressure. The American Medical Association (AMA) conducted a study involving 1,200 randomly selected American adults.

a. What is the expected number of adults in the sample who suffer from high blood pressure, if in fact the physician's claim is true?
b. Within what limits would the AMA expect x, the number of sampled adults who suffer from high blood pressure, to fall? [*Hint:* Find the interval $\mu \pm 2\sigma$.]
c. Suppose that the number, x, in the sample of 1,200 who suffer from high blood pressure is equal to 151. If the physician's claim is true, would this represent a rare event? Would you doubt the physician's claim? Explain.

5.37 Many people perceive themselves as "middle class," but the term is difficult to define. According to the *Wall Street Journal* (February 10, 1981), New York's Mayor Koch earns $60,000 per year and "definitely" considers himself middle class. John Shannon of the Advisory Commission on Intergovernmental Relations, in defining "middle class," states that "it's that group of damned irritated people who receive no direct federal aid but aren't rich enough to work any sophisticated tax angles." Joseph Minarik of the Brookings Institution argues that the middle class live on income derived from labor (as opposed to dividend, other investment income, or government support), and that 98% of all American households received more than 90% of their income from labor in 1977. Suppose we accept the theory that approximately 98% of us view ourselves as middle class, and that a random sample of 2,000 adults is asked whether they perceive themselves in that category or not. Let x be the number of people in the sample who consider themselves middle class.

a. What is the expected number in the sample who will consider themselves to be middle class?
b. What is the standard deviation of x?
c. Within what limits would you expect x to fall if in fact 98% of all people view themselves as middle class?
d. Suppose that the number in the sample of 2,000 who consider themselves as middle class is 1,850. Does this contradict the theory that 98% of all people view themselves as middle class? Explain.

5.38 Refer to Exercise 5.37. Suppose that each member of your class is asked to write on a slip of paper whether or not they regard themselves as middle class. Let x equal the number in the class who consider themselves middle class.

a. Find the mean and standard deviation of x. Assume that 98% of all Americans consider themselves middle class.
b. Find the interval $\mu \pm 2\sigma$.
c. Is the value of x observed for your class consistent with the 98% theory? Explain. [*Note:* Regardless of the outcome of your survey, remember that your class is not a random sample from the population about which the inference is being made.]

5.39 Refer to Exercise 5.37. According to the *Wall Street Journal,* the median income of unemployed heads of households in 1979 was approximately $14,200. You randomly select 1,000 unemployed heads of households.
a. What proportion would you expect to have incomes exceeding $14,200? [*Hint:* If the median is known to be $14,200, then the probability that an unemployed head of household has an income exceeding $14,200 is .5.]
b. What are the mean and standard deviation of the number x who have incomes exceeding $14,200 in the sample of 1,000?
c. Within what range would you expect x to lie? [*Hint:* Use the Empirical Rule.]

5.7 Summary

This chapter utilizes a very practical sampling problem, a public opinion poll, as a vehicle for discussing discrete random variables and the binomial probability distribution. We are first led to a discussion of random variables, particularly discrete random variables, which are those that can assume a countable number of values. A complete list (or graph or formula) that gives the probabilities associated with each value of a random variable x is called its probability distribution.

The number, x, of people (consumers, voters, etc.) in a random sample of n people from a large population who favor some proposition follows a binomial probability distribution. Other random variables that possess the same probability distribution are those that satisfy the five conditions that describe a binomial experiment. The formula for a binomial probability distribution (along with the binomial probability tables) enables us to calculate probabilities about x. Particularly, in a practical situation, we can determine whether an observed sample value of x is a highly improbable event.

When the sample size n is large, it is difficult to calculate the values of $p(x)$ but it is easy to describe the binomial probability distribution by finding the mean μ and standard deviation σ of the distribution. Then we can use the Empirical Rule to describe $p(x)$ and to identify values of x that are improbable.

Keep in mind that this chapter is written about a very practical sampling situation, opinion polling, and that the binomial probability distribution gives the probability of observing a specific number x of people who favor some proposition. As you will subsequently see, we will use these probabilities to make an inference about the proportion p of people in the population who favor the proposition.

KEY WORDS

Random variable
Discrete random variable
Continuous random variable
Binomial random variable
Binomial probability distribution
Binomial experiment
Cumulative binomial probabilities
Rare event

SUPPLEMENTARY EXERCISES

5.40 Use the formula for the binomial probability distribution to find the probabilities for $n = 3$, $p = .1$, $x = 0, 1, 2, 3$. Graph this probability distribution.

5.41 Repeat Exercise 5.40 for $n = 3$, $p = .3$.

5.42 Repeat Exercise 5.40 for $n = 3$, $p = .5$.

5.43 Repeat Exercise 5.40 for $n = 3$, $p = .7$.

5.44 Repeat Exercise 5.40 for $n = 3$, $p = .9$.

5.45 A coin is tossed 8 times and the number x of heads is recorded. Find the probability that:

a. $x = 4$ **b.** x is larger than 4 **c.** x is less than 2

5.46 A Gallup poll, conducted after the Washington *Post* admitted that one of its reporters had fabricated a 1980 Pulitzer Prize–winning story of an 8-year-old heroin addict, suggests that many Americans are skeptical about the information they obtain from the news media (Gainesville *Sun,* April 27, 1981). The poll, commissioned by *Newsweek* magazine, revealed that only 5% of those responding believe all news reports are true. Suppose you randomly sample 9 individuals, ask their opinion on the credibility of the news media, and find that 2 believe all news reports are true.

a. Find the probability of observing 2 or more individuals in the sample who believe all news reports are true if in fact the Gallup poll figure is correct.

b. Would the observed sample result be regarded as a rare event?

c. Would your answer to part b cause you to doubt the Gallup poll figure?

d. The Gallup pollsters reported that their result has a margin of error of "plus or minus 4 percentage points." That is, the true proportion of Americans who believe all news reports are true could be as low as 1% or as high as 9%. (We will learn how to construct such an interval in Chapter 8.) Answer parts a, b, and c under the assumption that the true proportion of Americans who believe all news reports are true is $p = .09$. (Obtain an approximate value for the probability of part a using $p = .10$ in the binomial tables, Appendix E.)

5.47 The Internal Revenue Service (IRS) claims that the probability an income tax return will be audited (reviewed for possible illegal deductions) is only one chance in 10 (i.e., $p = .1$). The IRS has completed a check on last year's tax returns for four of your acquaintances and two of the four have been audited.

a. What is the probability of observing 2 or more audited returns in a sample of 4 if, in fact, p is as small as .1?

b. Do you doubt the IRS's claim that the probability it will audit an income tax return is only .1? Explain.

c. Can you think of a reason why you might observe as many as 2 audited returns in a sample of 4 even if p really is equal to .1?

5.48 According to a Federal Bureau of Investigation (FBI) report, 40% of all assaults on federal workers in 1980 were directed at IRS employees (*Family Weekly*, April 26, 1981). Suppose the FBI records for a random sample of 50 federal workers who were assaulted in 1980 is selected, and the number x of workers employed by the IRS is recorded. Is x a binomial random variable? If so, what are the values of n and p?

5.49 A bottling company randomly samples one bottle of a popular drink from the production line every 6 minutes and conducts a test of its contents. Each sampled bottle is then rated as acceptable or unacceptable. At the end of two hours, 20 bottles have been tested. Is x, the number of unacceptable bottles in the sample of 20, a binomial random variable? Justify your answer.

5.50 A group of 20 college graduates contains 10 highly motivated persons, as determined by a company psychologist. Suppose that a personnel director selects 10 persons from among this group of 20 for employment. Let x equal the number of highly motivated persons included in the personnel director's selection. Is this a binomial experiment? Explain.

5.51 In a long series of free-throws, a basketball player has a record of 80% successes. At the conclusion of an important game, the player is fouled and given two free-throws. With one successful free-throw out of two, his team wins the game.

a. Do you think that the probability of a basket on the first free-throw is equal to .8?

b. Is the outcome on the second free-throw likely to be independent of the outcome on the first free-throw?

c. Considering your answers to parts a and b, is it likely that the two free-throws constitute a binomial experiment?

5.52 An international survey involving thousands of past Olympic participants indicated that an overwhelming majority—70%—believe that athletes should be barred from Olympic competition if they have been treated with anabolic steroids during the month prior to competition. In a random sample of 10 former Olympic participants, let x be the number who believe that athletes should be barred from Olympic competition if they have been treated with anabolic steroids during the month prior to competition. What is the probability that:

a. x is equal to 10? **b.** x is at least 3? **c.** x is at most 1?

5.53 Only 3½ out of every 1,000 childbirths result in identical twins. That is, the probability that a childbirth will result in identical twins is $p = .0035$. If a random sample of 20 childbirths is selected, what is the probability of observing at least 1 childbirth in the sample which results in identical twins?

5.54 Find the probabilities associated with a binomial probability distribution with $n = 6$ and $p = .5$. Graph $p(x)$.

5.55 Refer to Exercise 5.54.

a. Find μ and σ for the binomial probability distribution.

b. Construct the interval $\mu \pm 2\sigma$ and find the probability that x will fall within this interval. How does this result compare with the Empirical Rule?

5.56 In recent years, the use of the telephone as a data collection instrument for public opinion polls has been steadily increasing. However, one of the major factors bearing on the extent to which the telephone will become an acceptable data collection tool in the future is the refusal rate, i.e., the percentage of the eligible subjects actually contacted who refuse to take part in the poll. Suppose that past records indicate a refusal rate of 20% in a large city. If 25 prospective survey subjects are contacted by telephone, what is the probability that:

a. 5 or more refuse to take part in the poll?

b. 8 or more refuse to take part in the poll?

c. Fewer than 3 refuse to take part in the poll?

5.57 A series of experiments was conducted to determine whether a certain chemical, used in permanent hair dyes, is linked with cancer. When the chemical was ingested in large doses by experimental rats, data showed that the substance produced thyroid tumors in 24% of the female rats. If each of a group of 150 female rats is given a large dose of the chemical:

a. What is the expected number of female rats that will develop thyroid tumors? (Assume that the true proportion of female rats that develop thyroid tumors after receiving a large dose of the chemical is in fact .24.)

b. Let x be the number of female rats in the sample of 150 that develop thyroid tumors. What is the standard deviation of x?

c. Within what limits would you expect x to fall? [*Hint:* Use the Empirical Rule.]

d. What would you infer if the observed number x of female rats that develop thyroid tumors is less than 22?

5.58 A 1980 *Wall Street Journal*–Gallup survey (*Wall Street Journal,* November 21, 1980) of 282 corporate chief executives of the nation's largest firms found that 59% believed that their managers were more likely to fire an incompetent worker than they were a few years ago (see Case Study 14.1).

a. If you were to randomly select five corporate chief executives from this group and obtain their opinions on this question, where x is the number in the sample who believe their managers are more likely to fire incompetent workers, would x be (to a reasonable degree of approximation) a binomial random variable? Explain.

b. What is the probability that all five executives would believe that their managers are more likely to fire incompetent workers?

c. Fewer than 5?

d. More than 3?

5.59 Refer to Exercise 5.58. The 282 firms were selected from among 1,300 large corporations listed by *Fortune* magazine. Suppose that the actual proportion of chief executives in the entire population of 1,300 companies who believe that their managers are more likely to fire incompetent workers is equal to p. Would this survey of 282 firms from a total of 1,300 be, to a reasonable degree of approximation, a binomial experiment? Explain.

CASE STUDY 5.1 Signalling the Rise or Fall of the Stock Market: Corporate Insider Theory

Is the stock market going up or down in the immediate future? If you plan to invest, you may want to observe the behavior of those in the know—the corporate insiders. Insiders are officers, directors, or investors who hold 10% or more of a company's stock and, according to the Orlando *Sentinel Star,** whose buying behavior may signal a pending rise or fall in stock prices. The theory is that when insiders buy, they do so to make money: they expect their company's earnings to rise. When they sell, they see a drop in future earnings and their selling signals foul weather for their company.

According to one theory, an analysis of corporate insider stock transactions can be used to signal the future behavior of the stock market. If the proportion p of insider transactions which are buys increases from one time period (say one month) to another, the increase suggests a future rise in stock prices. If the proportion decreases, it portends a future decline in stock prices.

Let p_1 denote the probability that an insider transaction during the current month is a buy, and let p_2 denote the probability of the same event during the next month. Thus, according to the "insider theory," if $p_2 > p_1$ then the market is likely to rise, and if $p_2 < p_1$ then the market is likely to fall.

Suppose that n insider trades are observed next month and the number x of buys is recorded.

a. If conditions motivating the stock market's behavior are identical to last month's, within what limits would you expect x to fall? [*Hint:* Use the Empirical Rule, Chapter 3.]

b. How could you use the information obtained in part a to provide a signal, an indicator of a potential market rise or fall? Explain.

CASE STUDY 5.2 The Probability of a Successful Helicopter Flight during the Iran Rescue Mission

In Case Study 4.2, we discussed the ill-fated April 28, 1980, helicopter attempt to rescue the American hostages held by Iran. Particularly, we noted that the government (presumably, the Department of Defense) claimed that the 8 helicopters each flew 20 missions without malfunction, a total of 160 missions. On the day of the rescue mission, however, 3 of the 8 helicopters that embarked on the mission failed.

---

*Orlando *Sentinel Star,* November 12, 1980.

Some members of the public, including Congressional Representative Samuel S. Stratton, thought that the data collected on the day of the rescue attempt disagreed with the Department of Defense's claim. In fact, it seemed extremely doubtful that helicopters capable of flying 160 successful missions out of a total of 160 could produce 3 malfunctions in a sample of $n = 8$ flights.

Use the data to show that the probability that a single helicopter would complete a successful flight was probably much less during the rescue attempt than during the pre-raid training period. Explain why this might have been the case.

CASE STUDY 5.3 Expected Gains in the Oil Lease Lottery

In Case Study 4.3, we introduced you to the oil lease lottery conducted by the Bureau of Land Management. Given various probabilities of winning, it is interesting to compute the expected gain associated with a single $10 entry.

For example, it is estimated that the probability of winning on a parcel worth $25,000 to $150,000 is approximately $1/5{,}000$.

a. Use the values presented above to calculate the expected gain associated with a single $10 entry.

b. Suppose that you enter the lottery for three parcels of land. If the probability that you will win $25,000 on a single entry is $1/5{,}000$, what is the expected gain for the $30 investment in the three parcels? [*Hint:* The gain can assume one of four values depending upon whether x, the number of wins, is 0, 1, 2, or 3.]

REFERENCES

Allis, S. "Era of middle-class has arrived but it's hard to say who's in it." *Wall Street Journal,* February 10, 1981.

McClave, J. T., & Dietrich, F. H. *Statistics.* 2d ed. San Francisco: Dellen, 1982. Chapter 4.

Mendenhall, W. *Introduction to probability and statistics.* 5th ed. North Scituate, Mass.: Duxbury, 1979. Chapters 5 and 6.

Mosteller, F., Rourke, R., & Thomas, G. *Probability with statistical applications.* 2d ed. Reading, Mass.: Addison-Wesley, 1970.

Tables of the binomial probability distribution. Department of Commerce, National Bureau of Standards, Applied Mathematics Series, Vol. 6, 1952.

Zuwaylif, F. H. *General applied statistics.* 3d ed. Reading, Mass.: Addison-Wesley, 1979. Chapter 4.

Six

The Normal Distribution

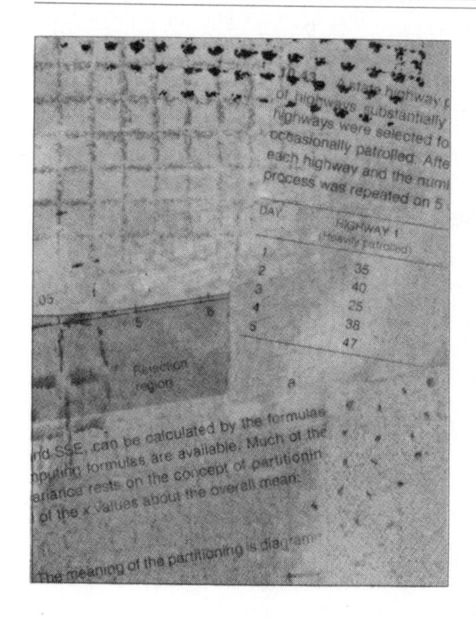

According to one theory of stock price behavior, publicized and popularized by B. G. Malkiel's book, *A Random Walk Down Wall Street* (Case Study 6.2), stock prices actually change (walk) upward and downward in a random manner and produce a relative frequency distribution of changes in price that possesses a familiar bell-shaped curve known as a ***normal distribution.*** Does the real world of stock price changes agree with this theory? For reasons that you will subsequently learn, many random variables possess normal relative frequency distributions. Consequently, we shall study the characteristics of normal distributions in this chapter and, particularly, learn how to identify improbable or rare events. This will help us decide whether the real world of stock prices disagrees with the random walk theory of stock price changes.

Contents

6.1 Probability Models for Continuous Random Variables

Suppose that you want to predict the length of time of your wait in a dentist's office, the sale price of a home, the annual amount of rainfall in your city, or an evening's winnings at blackjack. If you do, you will need to know something about continuous random variables.

You will recall that continuous random variables are those that, at least in theory, can assume any of the infinitely large number of values contained in an interval. Thus, we might envision a population of patient waiting times in a dentist's office, the sale prices of houses, the annual amounts of rainfall in your city since 1900, or the gains (or losses) of many evenings of blackjack. Since we shall want to make inferences about a population based on the measurements contained in a sample, we shall need to know the probability that the sample values (or sample statistics) assume specific values.

For example, suppose that all University of Florida bachelor's degree graduates during the period June 1980–March 1981 had secured employment at the time of graduation, instead of only the 948 graduates recorded in Appendix A. Then the resulting set of starting salaries would be the population that characterizes the financial compensations of all bachelor's degree graduates of the university. What is the probability that a single graduate selected at random had a starting salary of less than $20,000? To answer this question, we need to know the proportion of all graduates who had starting salaries less than $20,000. This proportion is given by the shaded area under the population relative frequency distribution that lies to the left of $20,000 in Figure 6.1. For example, if this area is equal to seven-tenths of the total area, then the probability that a randomly selected graduate had a starting salary of less than $20,000 is .7.

The problem, of course, is that not all of the June 1980–March 1981 graduates secured employment at the time of graduation and responded to the employment questionnaire. Hence, we do not know the exact shape of the population relative frequency distribution, i.e., we can talk about Figure 6.1 but its exact shape is unknown. Then, as in the case of a coin-tossing experiment, we postulate a model, i.e., we select a smooth curve (similar to the one shown in Figure 6.1) as a ***model*** for

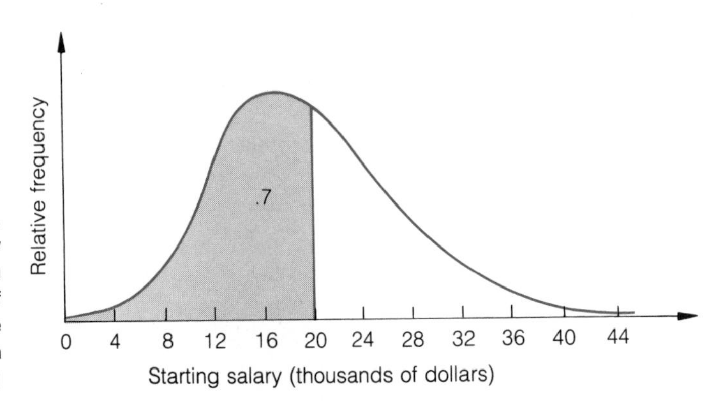

FIGURE 6.1 Relative Frequency Distribution of Starting Salaries of University of Florida Bachelor's Degree Graduates in June 1980–March 1981

the population relative frequency distribution. To find the probability that a particular observation, say a starting salary, will fall in a particular interval, we use the model and find the area under the curve that falls over that interval.

Of course, in order for this approximate probability to be realistic, we need to be fairly certain that the smooth curve, the model, and the population relative frequency distribution are very similar. In Chapter 7 we shall show why we believe that the models we use are good approximations to reality.

In the following section, we shall introduce one of the most important and useful models for population relative frequency distributions and show how it can be used to find probabilities associated with specific sample observations.

6.2 The Normal Distribution

One of the most useful models for population relative frequency distributions is known as the ***normal distribution.*** A graph of the normal distribution, often called the ***normal curve,*** is shown in Figure 6.2.

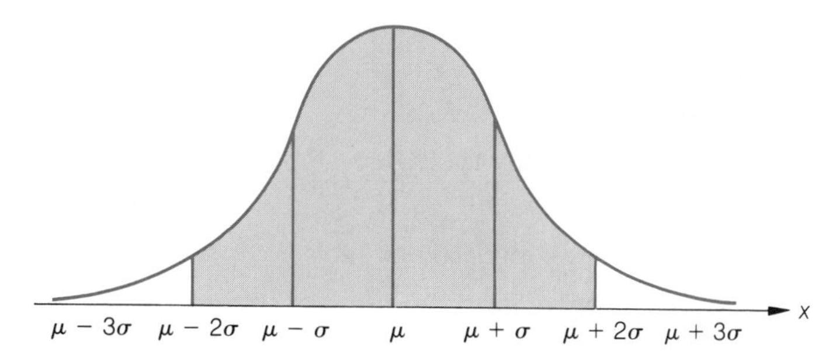

FIGURE 6.2
The Normal Curve

You can see that the mound-shaped normal curve is symmetric about its mean μ. Furthermore, approximately 68% of the area under a normal curve lies within the interval, $\mu \pm \sigma$ (see Figure 6.2). Approximately 95% of the area lies within the interval $\mu \pm 2\sigma$ (shaded in Figure 6.2), and almost all (99.7%) lies within the interval $\mu \pm 3\sigma$. Note that the mound-shaped normal curve agrees with the Empirical Rule of Section 3.5. This is because the Empirical Rule is based on consideration of a normal distribution.

Remember that areas under the normal curve have a probabilistic interpretation. Thus, if a population of measurements has approximately a normal distribution, then the probability that a randomly selected observation falls within the interval $\mu \pm 2\sigma$ is approximately .95.

The areas under the normal curve have been computed and they appear in Table 3 of Appendix E. Since the normal curve is symmetric, we need give areas on only one side of the mean. Consequently, the entries in Table 3 are areas between the mean and a point x to the right of the mean. This area is shaded in Figure 6.3.

Since the values of μ and σ will vary from one normal distribution to another, the easiest way to express a distance from the mean is in terms of a z-score (see

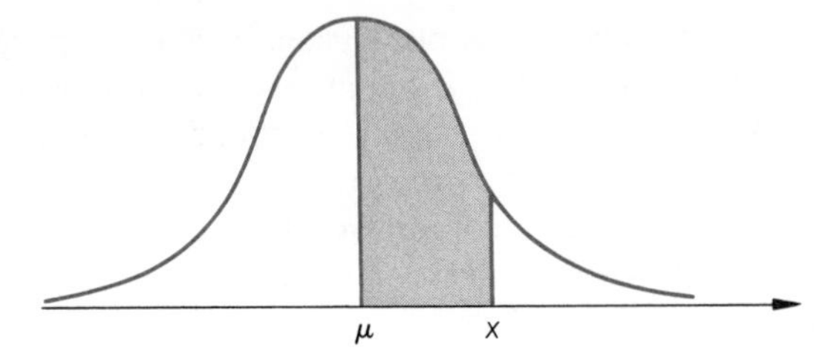

FIGURE 6.3
Tabulated Area Given in Table 3, Appendix E

Section 3.6), the number of standard deviations between the mean and the point x. Thus,

$$z = \frac{x - \mu}{\sigma}$$

is the distance between x and μ expressed in units of σ.

EXAMPLE 6.1 Suppose that a population has a relative frequency distribution with mean $\mu = 500$ and standard deviation $\sigma = 100$. Give the z-score corresponding to $x = 650$.

Solution You can see that $x = 650$ lies 150 units above $\mu = 500$. This distance, expressed in units of $\sigma (\sigma = 100)$, is 1.5. We can get this answer directly by substituting x, μ, and σ into the formula for z. Thus,

$$z = \frac{x - \mu}{\sigma} = \frac{150 - 100}{100} = \frac{150}{100} = 1.5$$

A partial reproduction of Table 3, Appendix E, is shown in Table 6.1. The entries in the complete table give the areas to the right of the mean for distances $z = 0.00$ to $z = 3.09$.

EXAMPLE 6.2 Find the area under a normal curve between the mean and a point $z = 1.26$ standard deviations to the right of the mean.

Solution To locate the proper entry, proceed down the left (z) column of the table to the row corresponding to $z = 1.2$. Then move across the top of the table to the column headed .06. The intersection of the .06 column with the 1.2 row contains the desired area, .3962. This area is shaded in Figure 6.4.

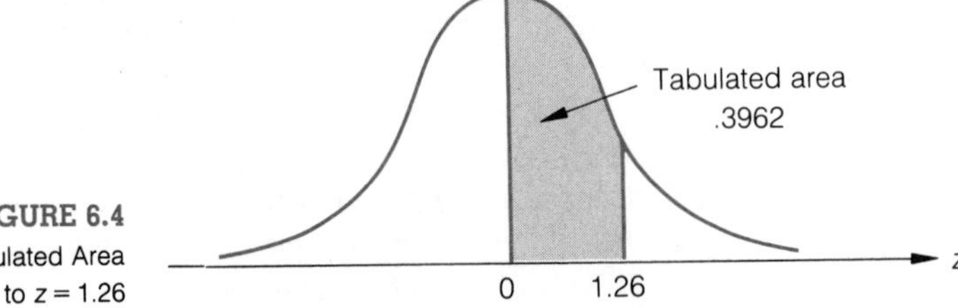

FIGURE 6.4
The Tabulated Area Corresponding to $z = 1.26$

TABLE 6.1 Reproduction of Part of Table 3, Appendix E

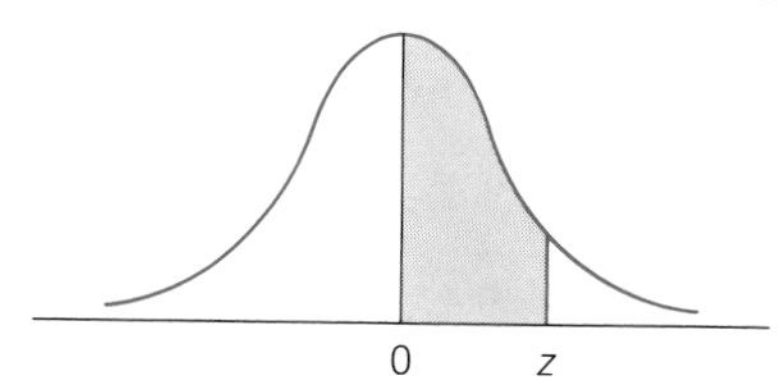

z	.00	.01	.02	.03	.04	.05	.06	.07	.08	.09
0.0	.0000	.0040	.0080	.0120	.0160	.0199	.0239	.0279	.0319	.0359
0.1	.0398	.0438	.0478	.0517	.0557	.0596	.0636	.0675	.0714	.0753
0.2	.0793	.0832	.0871	.0910	.0948	.0987	.1026	.1064	.1103	.1141
0.3	.1179	.1217	.1255	.1293	.1331	.1368	.1406	.1443	.1480	.1517
0.4	.1554	.1591	.1628	.1664	.1700	.1736	.1772	.1808	.1844	.1879
0.5	.1915	.1950	.1985	.2019	.2054	.2088	.2123	.2157	.2190	.2224
0.6	.2257	.2291	.2324	.2357	.2389	.2422	.2454	.2486	.2517	.2549
0.7	.2580	.2611	.2642	.2673	.2704	.2734	.2764	.2794	.2823	.2852
0.8	.2881	.2910	.2939	.2967	.2995	.3023	.3051	.3078	.3106	.3133
0.9	.3159	.3186	.3212	.3238	.3264	.3289	.3315	.3340	.3365	.3389
1.0	.3413	.3438	.3461	.3485	.3508	.3531	.3554	.3577	.3599	.3621
1.1	.3643	.3665	.3686	.3708	.3729	.3749	.3770	.3790	.3810	.3830
1.2	.3849	.3869	.3888	.3907	.3925	.3944	.3962	.3980	.3997	.4015
1.3	.4032	.4049	.4066	.4082	.4099	.4115	.4131	.4147	.4162	.4177
1.4	.4192	.4207	.4222	.4236	.4251	.4265	.4279	.4292	.4306	.4319
1.5	.4332	.4345	.4357	.4370	.4382	.4394	.4406	.4418	.4429	.4441

The normal distribution of the z statistic (as shown in Figure 6.4) is called the ***standard normal distribution.*** The mean of a standard normal distribution is 0 ($z = 0$ when $x = \mu$); the standard deviation is equal to 1. Since the mean is 0, z-values to the right of the mean are positive; those to the left are negative.

EXAMPLE 6.3 Find the area beneath a normal curve between the mean and the point, $z = -1.26$.

Solution The best way to solve a problem of this sort is to draw a sketch of the distribution (see Figure 6.5). Since $z = -1.26$ is negative, we know that it lies to the left of the mean, and the area that we seek will be the shaded area shown.

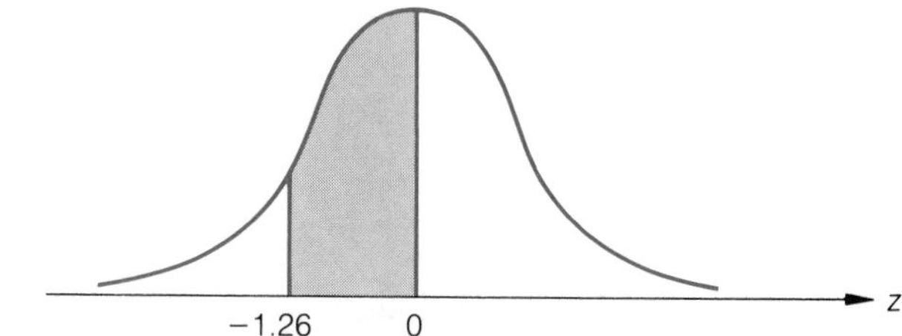

FIGURE 6.5 Standard Normal Distribution: Example 6.3

Since the normal curve is symmetric, the area between the mean 0 and $z = -1.26$ is exactly the same as the area between the mean 0 and $z = +1.26$. We found this area in Example 6.2 to be .3962. Therefore, the area between $z = -1.26$ and $z = 0$ is .3962.

EXAMPLE 6.4 Find the probability that a normally distributed random variable will lie within $z = 2$ standard deviations of its mean.

Solution The probability that we seek is the shaded area shown in Figure 6.6. Since the area between the mean 0 and $z = 2.0$ is exactly the same as the area between the mean and $z = -2.0$, we need find only the area between the mean and $z = 2$ standard deviations to the right of the mean. This area is given in Table 3, Appendix E, as .4772. Therefore, the probability P that a normally distributed random variable will lie within two standard deviations of its mean is

$$P = 2(.4772) = .9544$$

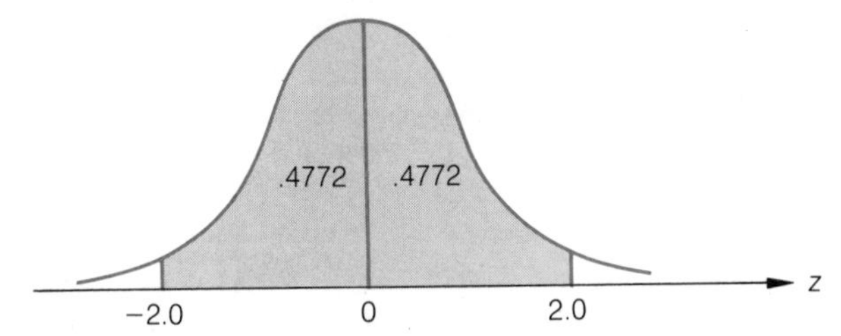

FIGURE 6.6 Standard Normal Distribution: Example 6.4

EXAMPLE 6.5 Find the probability that a normally distributed random variable x will lie more than $z = 2$ standard deviations above its mean.

Solution The probability we seek is the shaded area shown in Figure 6.7. The total area under a standard normal curve is 1; half the area lies to the left of the mean, half to the right. Consequently, the probability P that x will lie more than 2 standard deviations above the mean is equal to .5 less the area A, i.e.,

$$P = .5 - A$$

The area A corresponding to $z = 2.0$ is .4772. Therefore,

$$P = .5 - .4772 = .0228$$

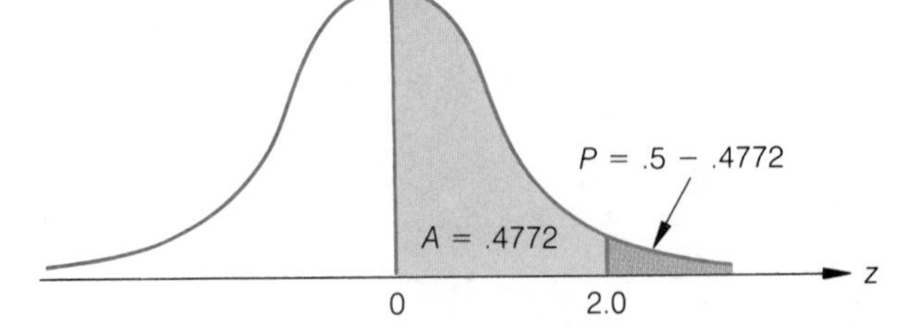

FIGURE 6.7 Standard Normal Distribution: Example 6.5

EXAMPLE 6.6 Find the area under the normal curve between $z = 1.2$ and $z = 1.6$.

Solution The area A that we seek lies to the right of the mean because both z-values are positive. It will appear as the shaded area shown in Figure 6.8. Let A_1 represent the area between $z = 0$ and $z = 1.2$, and A_2 represent the area between $z = 0$ and $z = 1.6$. Then the area A that we desire is

$$A = A_2 - A_1$$

From Table 3, Appendix E, we obtain:

$$A_1 = .3849$$
$$A_2 = .4452$$

Then

$$\begin{aligned} A &= A_2 - A_1 \\ &= .4452 - .3849 \\ &= .0603 \end{aligned}$$

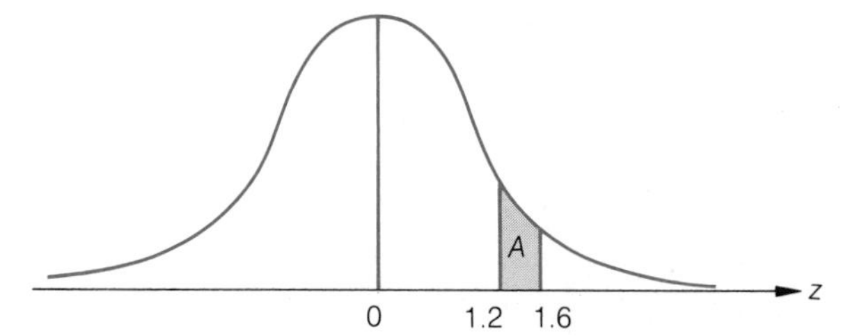

FIGURE 6.8 Standard Normal Distribution: Example 6.6

EXAMPLE 6.7 Find the value of z such that the area to the right of z is .10.

Solution The z-value that we seek appears as shown in Figure 6.9. Note that we show an area to the right of z equal to .10. Since the total area to the right of the mean $z = 0$ is equal to .5, the area between the mean 0 and the unknown z-value is $.5 - .1 = .4$ (as shown in the figure). Consequently, to find z, we must look in Table 3, Appendix E, for the z-value that corresponds to an area equal to .4.

The area .4000 does not appear in Table 3. The closest values are .3997, corresponding to $z = 1.28$, and .4015, corresponding to $z = 1.29$. Since the area .3997 is closer to .4000 than is .4015, we will choose $z = 1.28$ as our answer.

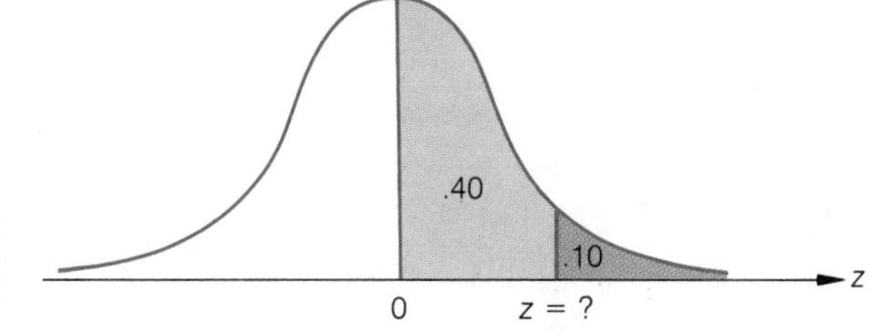

FIGURE 6.9 Standard Normal Distribution: Example 6.7

EXAMPLE 6.8 Recent research has linked excessive consumption of salt to hypertension (high blood pressure). Yet, after sugar, salt is America's leading food additive, both in factory-processed foods and home cooking. The average amount of salt consumed per day by an American is 15 grams (15,000 milligrams). (The actual physiological minimum daily requirement for salt is only 220 milligrams.) Suppose also that the amount of salt intake per day is approximately normally distributed with a standard deviation of 5 grams. What proportion of all Americans consume more than 23.75 grams of salt per day?

Solution The proportion P of Americans who consume more than $x = 23.75$ grams of salt per day is shown (shaded) in Figure 6.10(a). The next step is to find the z-value corresponding to $x = 23.75$. Substituting $x = 23.75$, $\mu = 15$, and $\sigma = 5$ into the formula for z, we obtain

$$z = \frac{x - \mu}{\sigma} = \frac{23.75 - 15}{5} = \frac{8.75}{5} = 1.75$$

Then, the area P to the right of $x = 23.75$ in Figure 6.10(a) is the same as the area to the right of $z = 1.75$ in Figure 6.10(b). Since half (.5) of the area under the z-distribution lies to the right of $z = 0$, it follows that

$$P = .5 - A$$

where A is the area shown in Figure 6.10(b), an area that corresponds to $z = 1.75$. This area, given in Table 3, is $A = .4599$. Then

$$\begin{aligned} P &= .5 - A \\ &= .5 - .4599 \\ &= .0401 \end{aligned}$$

Therefore, 4.01% of all Americans consume more than 23.75 grams of salt per day.

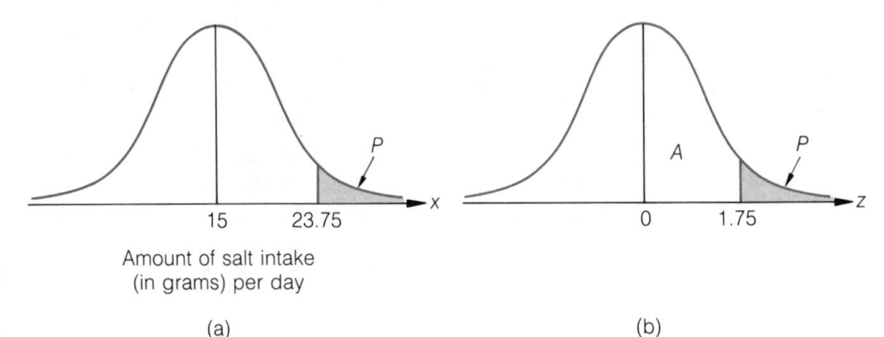

FIGURE 6.10 Normal Curve Sketches for Example 6.8

EXAMPLE 6.9 Refer to Example 6.8. Physicians recommend that those Americans who desire to reach a level of salt intake at which hypertension is less likely to occur should consume less than 1 gram of salt per day. What is the probability that an American will consume less than 1 gram of salt per day?

Solution The probability P that an American will consume less than 1 gram of salt per day is the shaded area in Figure 6.11(a). The z-value corresponding to $x = 1$ (shown in Figure 6.11(b)) is

$$z = \frac{x - \mu}{\sigma} = \frac{1 - 15}{5} = \frac{-14}{5} = -2.80$$

Since the area to the left of $z = 0$ is equal to .5, the probability that x is less than or equal to 1 (the shaded area in Figure 6.11(b)) is

$$P = .5 - A$$

where A is the tabulated area corresponding to $z = -2.80$. This value, given in Table 3, Appendix E, is $A = .4974$. Then, the probability that an American will consume no more than 1 gram of salt per day is

$$P = .5 - A = .5 - .4974 = .0026$$

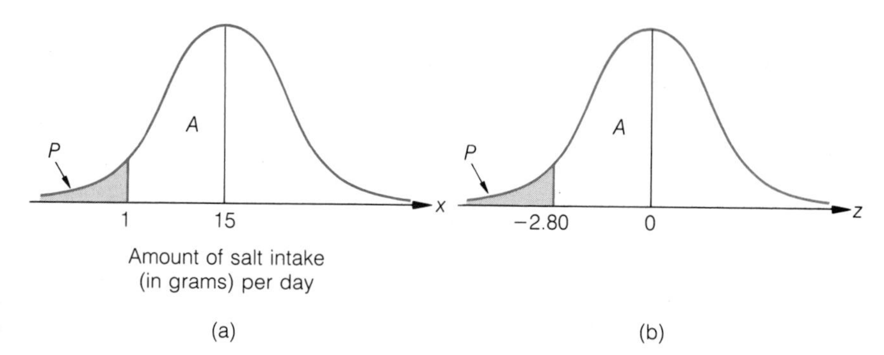

FIGURE 6.11
Normal Curve Sketches for Example 6.9

EXAMPLE 6.10 Recently, the federal government passed a law which enables banks to offer interest-on-checking-account programs. This interest-checking plan allows customers to earn 5¼% interest on the money that is normally kept in checking accounts. (Previously, interest could be earned only on savings account deposits.)

A national bank, serving the metropolitan area of Atlanta, Georgia, determined that the interest earned by its customers after one year of the interest-checking plan was approximately normally distributed with a mean of \$270 and a standard deviation of \$110. What proportion of the bank's customers earned between \$138 and \$300 on the interest-checking plan last year?

Solution The proportion P of the bank's customers who earned between $x = 138$ dollars and $x = 300$ dollars is the shaded area in Figure 6.12(a). Before we can compute this area, we need to determine the z-values which correspond to $x = 138$ and $x = 300$. Substituting $\mu = 270$ and $\sigma = 110$ into the formula for z, we compute the z-value for $x = 138$ as

$$z = \frac{x - \mu}{\sigma} = \frac{138 - 270}{110} = \frac{-132}{110} = -1.2$$

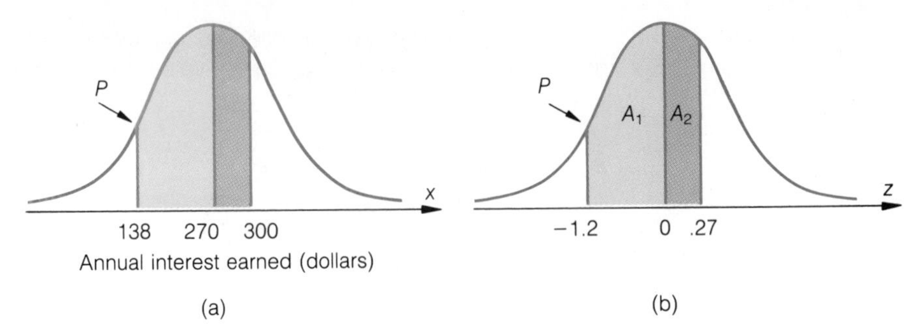

FIGURE 6.12
Normal Curve Areas for Example 6.10

The corresponding z-value for $x = 300$ (rounded to the nearest hundredth) is

$$z = \frac{x - \mu}{\sigma} = \frac{300 - 270}{110} = \frac{30}{110} = .27$$

Figure 6.12(b) shows these z-values along with P. From this figure, we see that

$$P = A_1 + A_2$$

where A_1 is the area corresponding to $z = -1.2$, and A_2 is the area corresponding to $z = .27$. These values, given in Table 3, Appendix E, are $A_1 = .3849$ and $A_2 = .1064$. Thus,

$$P = A_1 + A_2 = .3849 + .1064 = .4913$$

So 49.13% of the bank's customers earned between \$138 and \$300 interest from the interest-checking plan last year.

EXAMPLE 6.11 Refer to Example 6.10. Only 10% of the bank's customers earned more than x dollars of annual interest on checking last year. Find x.

Solution In this example, we are not requested to compute a probability or area under the normal relative frequency distribution, but instead we are asked to find a particular value of the normal random variable x, the annual interest on checking. We need to determine the annual checking interest that divides the top 10% of the annual checking interests from the remainder, that is, the value of the normal random variable x, say x_0, for which the area under the normal curve to the right of x_0 is $P = .10$. We call P the ***tail probability associated with x_0.*** This value x_0, along with its tail probability, is shown in Figure 6.13(a).

Notice that we have placed x_0 in the upper tail of the normal curve, i.e., to the right of the mean of 270. Our choice of the location of x_0 is not arbitrary, but depends upon the value of P. In order for the area under the curve to the right of x_0 to be $P = .10$, x_0 must lie above 270. To see this more clearly, try placing x_0 below (to the left of) 270. Now (mentally) shade in the corresponding tail probability P, i.e., the area to the right of x_0. You can see that this shaded area cannot possibly equal .10—it is too large (.50 or greater). Thus x_0 must lie to the right of 270. The first step, then, in solving problems of this type is to determine the location of x_0 in relation to the mean μ. The corresponding z-value for x_0, say z_0, is shown in Figure 6.13(b). Note that for $\mu = 270$ and $\sigma = 110$, we have the relation

$$z_0 = \frac{x_0 - \mu}{\sigma} = \frac{x_0 - 270}{110}$$

or the equivalent relation

$$x_0 = \mu + z_0(\sigma) = 270 + z_0(\sigma)$$

You can see from this relationship that if we can determine z_0, then we will be able to compute x_0. Thus, our next step is to find z_0.

Recall how Table 3, Appendix E, is constructed. For each particular value of the standard normal random variable z, the table gives the area under the curve between 0 and z. In order to find z_0 then, we must first determine the corresponding area between 0 and z_0. From Figure 6.13(b), it is easy to see that the area between 0 and z_0, say A, is the difference between .5 and the tail probability, i.e.,

$$A = .5 - P = .5 - .1 = .4$$

We now know that the z-value which we seek corresponds to the area $A = .4000$ in the table.

The next step is to locate $A = .4000$ in Table 3, Appendix E. Searching among the areas given in the body of the table, we see that the value closest to .4000 is .3997. The z-value corresponding to the area .3997 is 1.28. Thus, we take

$$z_0 = 1.28$$

The final step is to use this value of z_0 to determine x_0. Since $z_0 = 1.28$ is the corresponding z-value for x_0, then by definition we say that x_0 falls a distance of 1.28 standard deviations away from $\mu = 270$. Further, from Figure 6.13(a), x_0 falls a distance of 1.28 standard deviations above (to the right of) $\mu = 270$. For $\mu = 270$ and $\sigma = 110$, we compute x_0 as follows:

$$\begin{aligned} x_0 &= \mu + z_0(\sigma) \\ &= 270 + (1.28)(110) \\ &= 270 + 140.8 = 410.8 \end{aligned}$$

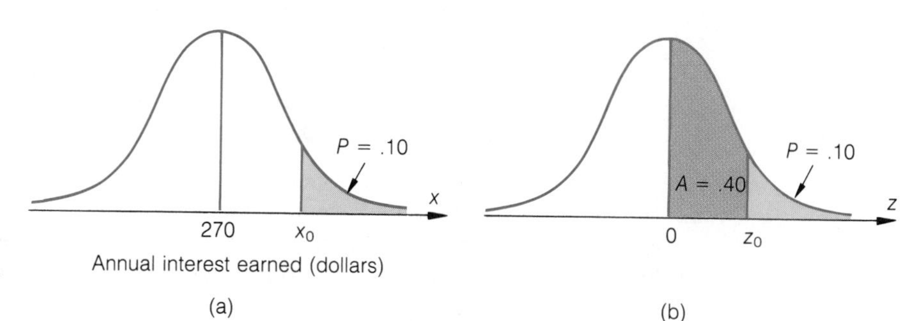

FIGURE 6.13
Normal Curve Areas for Example 6.11

Thus, our answer is \$410.80; that is, only 10% of the bank's customers earned over \$410.80 in interest on checking last year.

To help in solving problems of this type throughout the remainder of the chapter, we outline in the box the steps leading to the solutions.

THE DETERMINATION OF A PARTICULAR VALUE OF THE NORMAL RANDOM VARIABLE GIVEN AN ASSOCIATED TAIL PROBABILITY

1. Graph the relative frequency distribution of the normal random variable x. Shade the tail probability P and locate the corresponding value x_0 on the graph. Remember that x_0 will be to the right or left of the mean μ depending upon the value of P.
2. Graph the corresponding relative frequency distribution of the standard normal random variable z. Locate the z-value corresponding to x_0, say z_0, and shade the area corresponding to P.
3. Compute the area A associated with z_0 as follows:

   $$A = .5 - P$$

4. Use the area A, i.e., the area between 0 and z_0, to find z_0 in Table 3, Appendix E. (If you cannot find the exact value of A in the table, use the closest value.) Note that z_0 will be negative if you place x_0 to the left of the mean in step 1.
5. Compute x_0 as follows:

   $$x_0 = \mu + z_0(\sigma)$$

EXAMPLE 6.12 Many experts believe that the answer to the current energy crisis may lie in the field of synthetic fuels—the conversion of coal, shale oil, and tar sands to liquid fuel. The federal government is currently planning for synthetic fuel plants around the country to produce an average of 50,000 barrels of fuel per day by the turn of the century.

Let us suppose that by the year 2000 the daily production of a particular synthetic fuel plant will have a mean of 50,000 barrels and a standard deviation of 8,000 barrels. Further, let us assume that the daily production at the plant is approximately normally distributed. The federal government has allocated \$81 billion to be used to aid synthetic fuel producers in meeting the projected demand. However, this aid, in the form of loan and price guarantees, will not be provided to any synthetic fuel plant which produces less than a specified number of barrels on 33% of the days. On 33% of the days, this particular plant's production is less than what amount?

Solution We are asked to find the number of barrels which the plant's daily production falls below on 33% of the days. That is, we must determine the value x_0 of the normal random variable x, daily production (in barrels), for which the area under the curve to the left of this value x_0 is $P = .33$. We will find x_0 by following the steps given in the box.

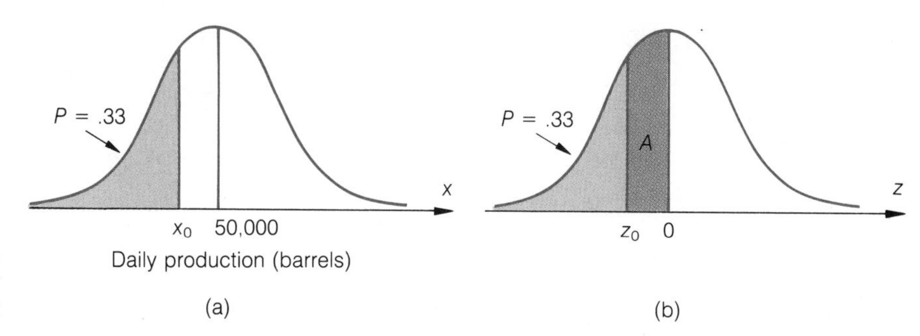

FIGURE 6.14
Normal Curve Areas for Example 6.12

The first two steps are completed by drawing figures similar to those shown in Figure 6.14. In Figure 6.14(a), x_0 is shown in relation to $\mu = 50{,}000$ and the tail probability $P = .33$ (the shaded area). Notice that x_0 must lie to the left of 50,000. You can justify this by recognizing that if x_0 is located to the right of 50,000, the corresponding tail probability P (the area under the curve to the left of x_0) cannot possibly equal .33. Location of the corresponding z-value for x_0 is shown in Figure 6.14(b), as is $P = .33$. The third step is to compute the area A (see Figure 6.14(b)) which is associated with z_0 in Table 3, Appendix E. For this example, we have

$$A = .5 - P = .5 - .33 = .17$$

Searching for $A = .1700$ in the appropriate portion of the table, we see that the corresponding z-value is .44. Since z_0 lies to the left of 0, it is negative; thus, we have $z_0 = -.44$. This completes step 4. Our interpretation of this value is that the number we seek, x_0, falls a distance of .44 standard deviation below (to the left of) the mean $\mu = 50{,}000$.

Finally, we compute x_0 as follows:

$$\begin{aligned} x_0 &= \mu + z_0(\sigma) \\ &= 50{,}000 + (-.44)(8{,}000) \\ &= 50{,}000 - 3{,}520 = 46{,}480 \end{aligned}$$

We conclude then, that the synthetic fuel plant's daily production will fall below 46,480 barrels on 33% of the days. By checking 46,480 against the figure specified by the federal government, we could determine whether the plant is eligible for loan and price guarantees.

The preceding examples should help you to understand the use of the table of areas under the normal curve. The practical applications of this information to inference making will become apparent in the following chapters.

EXERCISES

6.1 Find the area under the standard normal curve:

a. Between $z = 0$ and $z = 1.2$

b. Between $z = 0$ and $z = 1.49$

c. Between $z = -.48$ and $z = 0$

d. Between $z = -1.37$ and $z = 0$

e. For values of z larger than 1.33

Show the respective area and the corresponding z-values on a sketch of the normal curve for each part of the exercise.

6.2 Find the area under the standard normal curve:

a. Between $z = 0$ and $z = 1.96$ **b.** Between $z = -1.96$ and $z = 0$
c. Between $z = -1.96$ and $z = 1.96$ **d.** For values of z larger than .55
e. For values of z less than -1.24

Show the areas and corresponding values of z on a sketch of the normal curve for each part of the exercise.

6.3 Find the area under the standard normal curve:

a. For values of z less than -1.32 **b.** For values of z larger than 0
c. For values of z less than 1.35 **d.** For values of z larger than -1.32
e. For values of z between $z = -1.33$ and $z = 1.33$

Show the areas and corresponding values of z on a sketch of the normal curve for each part of the exercise.

6.4 Find the area under the standard normal curve:

a. Between $z = 1.21$ and $z = 1.94$ **b.** For values of z larger than 2.33
c. For values of z less than -2.33 **d.** Between $z = -1.50$ and $z = 1.79$

Show the areas and corresponding values of z on a sketch of the normal curve for each part of the exercise.

6.5 Find the z-value (to two decimal places) that corresponds to a tabulated area, Table 3, Appendix E, equal to:

a. .1000 **b.** .3200 **c.** .4000 **d.** .4500 **e.** .4750

Show the area and corresponding value of z on a sketch of the normal curve for each part of the exercise.

6.6 Find the z-value (to two decimal places) that corresponds to a tabulated area, Table 3, Appendix E, equal to:

a. .3700 **b.** .4900 **c.** .2500 **d.** .3000 **e.** .3413

Show the area and corresponding value of z on a sketch of the normal curve for each part of the exercise.

6.7 Find the value of z (to two decimal places), Table 3, Appendix E, that cuts off an area in the upper tail of the standard normal curve equal to:

a. .025 **b.** .05 **c.** .005 **d.** .01 **e.** .10

Show the area and corresponding value of z on a sketch of the normal curve for each part of the exercise.

6.8 Suppose that a normal random variable x has mean $\mu = 20.0$ and standard deviation $\sigma = 4.0$. Find the z-score corresponding to:

a. $x = 23.0$ **b.** $x = 16.0$ **c.** $x = 13.5$ **d.** $x = 28.0$ **e.** $x = 12.0$

Locate x and μ on a sketch of the normal curve. Check to make sure that the sign and magnitude of your z-score agree with your sketch.

6.9 Find a value z_0 such that the probability that z is larger than z_0 is:

a. .2266 **b.** .0721 **c.** .0344 **d.** .3520 **e.** .1611

Show the area and the value of z_0 on a sketch of the normal curve for each part of the exercise.

6.10 Find the approximate value for z_0 such that the probability that z is larger than z_0 is:

a. $P = .10$ **b.** $P = .15$ **c.** $P = .20$ **d.** $P = .25$

Locate z_0 and the corresponding probability P on a sketch of the normal curve for each part of the exercise.

6.11 Find the approximate value for z_0 such that the probability that z is less than z_0 is:

a. $P = .10$ **b.** $P = .15$ **c.** $P = .30$ **d.** $P = .50$

Locate z_0 and the corresponding probability P on a sketch of the normal curve for each part of the exercise.

6.12 Weight reduction clinics for both men and women are fast becoming major business ventures. For advertising purposes, the proprietor of one such clinic measured his patients' performance on his "weight loss through diet and exercise" program. He found that the weight losses of his clients follow an approximate normal distribution with a mean of 10 pounds and a standard deviation of 5 pounds. Find the probability that a new patient on the diet and exercise program will:

a. Lose at least 12 pounds
b. Gain at least 2 pounds [*Hint:* Treat a weight "gain" as a negative loss.]
c. Lose between 6 and 15 pounds

Locate μ and the desired probability on a sketch of the normal curve for each part of the exercise.

6.13 One measure of the water quality of a lake is its dissolved oxygen (DO) content. A marine biologist investigating a large northern lake found that the DO content at a certain time and place in the lake is approximately normally distributed with a mean of 7.8 units and a standard deviation of .4 unit.

a. What percentage of time will water samples from the lake at the designated time and place have a DO content less than 7.25 units?
b. What percentage of time will water samples from the lake at the designated time and place have a DO content greater than 6.55 units?
c. What percentage of time will water samples from the lake at the designated time and place have a DO content between 8.0 and 8.25 units?

Locate μ and the desired probability on a sketch of the normal curve for each part of the exercise.

6.14 Suppose that the distribution of monthly rents for 2-bedroom apartments in a region is approximately normally distributed with mean $\mu = \$390$ and standard deviation $\sigma = \$35$.

a. What proportion of the apartments rent for more than \$450?
b. What proportion of the apartments rent for less than \$350?
c. What proportion of the apartments rent for less than \$400?

Locate μ and the desired probability on a sketch of the normal curve for each part of the exercise.

6.15 Bond ratings and bond interest rates depend upon the current level of the cost of borrowing and the credit ratings of the respective cities. Returns on a class of municipal tax-free bonds currently are normally distributed with a mean of 9% and a standard deviation of .4%. A city's bonds are rated within this class. Find the probability that the city will have to pay:

a. More than 9% **b.** More than 9.5% **c.** Less than 8.5%

Locate μ and the desired probability on a sketch of the normal curve for each part of the exercise.

6.16 In order to graduate, all high school students in Florida must demonstrate their competence in mathematics by scoring at least 70 (out of a possible total of 100) on a mathematics achievement test. From published accounts, it is known that the mean score on this test is 78 and the standard deviation is 11.5. Assuming that the test scores are approximately normally distributed, what percentage of all high school students in Florida score at least 70 on the mathematics achievement test? Locate μ and the desired probability on a sketch of the normal curve.

6.17 Refer to Exercise 6.16. What percentage of all high school students in Florida score below 50 on the mathematics achievement test? Locate μ and the desired probability on a sketch of the normal curve.

6.18 Experience has shown that the proportion of packaged seeds of a particular variety of cucumber which fail to germinate is best described by a normal distribution with a mean of .075 and a standard deviation of .025. In a randomly selected package of cucumber seeds, what is the probability that fewer than .06 of the seeds fail to germinate? Show the pertinent quantities on a sketch of the normal curve.

6.19 A television cable company receives numerous phone calls throughout the day from customers reporting service troubles and from would-be subscribers to the cable network. Most of these callers are put "on hold" until a company operator is free to help them. The company has determined that the length of time that a caller is on hold is normally distributed with a mean of 3.1 minutes and a standard deviation of .9 minute. Company experts have decided that if as many as 5% of the callers are put on hold for 4.8 minutes or longer, more operators should be hired. What proportion of the company's callers are put on hold for at least 4.8 minutes? Should the company hire more operators? Show the pertinent quantities on a sketch of the normal curve.

6.20 Refer to Exercise 6.19. At this company, 5% of the callers are put on hold for a period longer than what length of time? Show the pertinent quantities on a sketch of the normal curve.

6.21 Refer to Exercise 6.14. The monthly rents of 20% of the 2-bedroom apartments in the region fall below what value? Show the pertinent quantities on a sketch of the normal curve.

6.3 Summary

This chapter introduced an important probability model for continuous random variables—the normal distribution. Since the normal distribution is a model for a population relative frequency distribution, it follows that the total area under the curve is equal to 1, and areas under the curve correspond to the probabilities of drawing observations that fall within particular intervals.

You will learn in Chapter 7 why we expect many population relative frequency distributions to be approximately normally distributed. You will also begin to develop, in Chapter 7 and following chapters, an understanding of how areas under the normal curve will be used to assess the uncertainty associated with sample inferences.

KEY WORDS

Probabilistic model
Normal distribution
Normal curve
z-Score: $z = \dfrac{x - \mu}{\sigma}$
Standard normal distribution

SUPPLEMENTARY EXERCISES

6.22 Find the area under the normal curve:

a. Between $z = .73$ and $z = 1.55$
b. Between $z = -1.44$ and $z = -.49$
c. Between $z = -1.03$ and $z = 2.00$
d. For values of z less than 1.59
e. For values of z greater than $-.77$

Show the respective areas and corresponding z-values on sketches of the normal curve.

6.23 Find the approximate value for z_0 such that the probability that z is larger than z_0 is:

a. .5
b. .12
c. .45
d. .25

Locate z_0 and the corresponding probability P on a sketch of the normal curve for each part of the exercise.

6.24 Find the approximate value for z_0 such that the probability that z is less than z_0 is:

a. .05
b. .025
c. .16
d. .22

Locate z_0 and the corresponding probability P on a sketch of the normal curve for each part of the exercise.

6.25 Teleconferences, electronic mail, and word processors are among the tools that can reduce the length of business meetings. A recent survey indicated that the percent reduction x in time spent by business professionals in meetings due to automated office equipment is approximately normally distributed with mean equal to 15% and standard deviation equal to 4%.

a. What proportion of all business professionals with access to automated office equipment have reduced their time in meetings by more than 22%?
b. What proportion of all business professionals with access to automated office equipment have reduced their time in meetings by 10% or less?

Locate μ and the desired probability on a sketch of the normal curve for each part of the exercise.

6.26 Suppose that, among sedentary young adult females (i.e., those females aged 20–40 who spend less than 5 hours per week in active exercise), the distribution of systolic blood pressure has an approximate normal distribution with $\mu = 125$ and $\sigma = 15$. If a sedentary young adult female is randomly selected and her systolic blood pressure checked, find:

a. The probability that the female will have a systolic blood pressure of 90 or less.
b. The probability that the female will have a systolic blood pressure between 140 and 155.
c. Would you expect the female to have a systolic blood pressure below 80? Explain.

For each part of the exercise, show the pertinent quantities on a sketch of the normal curve.

6.27 A food processor packages instant orange juice in small jars. The weights of the filled jars are approximately normally distributed with a mean of 10.82 ounces and a standard deviation of .30 ounce.

a. Find the probability that a randomly selected jar of instant orange juice will exceed 10.2 ounces in weight.
b. Suppose that the Food and Drug Administration sets the minimum weight of the jars at 10 ounces. Jars with weights below the allowable minimum must be removed from the supermarket shelf. What proportion of the jars should we expect to be removed from the supermarket shelf?
c. Two percent of the packaged jars are below what weight?

For each part of the exercise, show the pertinent quantities on a sketch of the normal curve.

6.28 The average life of a certain steel-belted radial tire is advertised as 60,000 miles. Assume that the life of the tires is normally distributed with a standard deviation of 2,500 miles. (The "life" of a tire is defined as the number of miles the tire is driven before blowing out.)

a. Find the probability that a randomly selected steel-belted radial tire will have a life of 61,800 miles or less.
b. Find the probability that a randomly selected steel-belted radial tire will have a life between 62,000 miles and 66,000 miles.
c. In order to avoid a tire blowout, the company manufacturing the tires will warn purchasers to replace each tire after it has been used for a given number of miles. What should the replacement time (in miles) be so that only 1% of the tires will blow out?

For each part of the exercise, show the pertinent quantities on a sketch of the normal curve.

6.29 The difference between the actual and the scheduled arrival time for your local commuter train is normally distributed with a mean of 5 minutes (i.e., on the average, it is 5 minutes late) and a standard deviation of 11 minutes. On a given day, what is the probability that:

a. The train will be late?
b. The train will be early?
c. The train will be more than 5 minutes late?
d. The train will be at least 10 minutes late?

Locate μ and the desired probability on a sketch of the normal curve for each part of the exercise.

6.30 Refer to Exercise 6.29. The management of the commuter line would like to adjust the scheduled arrival time so that the commuter line appears to be operating more efficiently. To do this, they would like to adjust the scheduled time of arrival so that only 10% of the trains would arrive late, i.e., later than the scheduled time. How many minutes should they add to the current scheduled time in order to accomplish this goal? Show the pertinent quantities on a sketch of the normal curve.

6.31 Blending feeders are used to break up tobacco that has been aged in tightly packed hogsheads. One cigarette manufacturer determined that the time between breakdowns for each of its blending feeders is best represented by a normal distribution with a mean of 100 hours and a standard deviation of 35 hours. Suppose that a particular feeder was just repaired and put back into service.

a. What is the probability that the feeder will not break down for at least 50 more hours?
b. What is the probability that the feeder will break down within the next 100 hours?

Locate μ and the desired probability on a sketch of the normal curve for each part of the exercise.

6.32 A team of University of Utah medical researchers believe they will soon be able to successfully implant an artificial heart in humans (*Time,* June 1, 1981). Although no mechanical difficulties are expected, there is a chance of internal infection, a problem that has occurred with implants in animals. Experiments show that calves implanted with artificial hearts can live an average of 80 days. Suppose that the distribution of the number of days that a calf implanted with an artificial heart can live is approximately normally distributed with a standard deviation of 25 days.

a. What is the probability that a randomly selected calf implanted with an artificial heart will live longer than 120 days?
b. Twenty-five percent of all calves implanted with an artificial heart live longer than N days. Find the value of N.

Show the pertinent quantities on a sketch of the normal curve for each part of the exercise.

6.33 The length of time required for a rodent to escape a noxious experimental situation was found to have a normal distribution with a mean of 6.3 minutes and a standard deviation of 1.5 minutes.

a. What is the probability that a randomly selected rodent will take longer than 7 minutes to escape the noxious experimental situation?

b. Some rodents may become so disoriented in the experimental situation that they will not escape in a reasonable amount of time. Thus, the experimenter will be forced to terminate the experiment before the rodent escapes. When should each trial be terminated if the experimenter wishes to allow sufficient time for 90% of the rodents to escape?

Show the pertinent quantities on a sketch of the normal curve for each part of the exercise.

6.34 The Taylor Manifest Anxiety Scale (TMAS) is a widely used test in psychological research. The TMAS scores of mental patients suffering from manic depression are approximately normally distributed with a mean of 47.6 and a standard deviation of 10.3. One of these patients is randomly selected and this patient's TMAS score is noted. Within what limits would you expect the TMAS score to fall? Explain.

CASE STUDY 6.1 Comparing Reality with the Normal Curve

You may wish to compare a relative frequency distribution of data with a normal curve to decide whether the normal curve provides an adequate model for the population relative frequency distribution. In this case study we shall graph the normal curve for various values of μ and σ and then apply this knowledge in Case Study 6.2.

The mathematical equation of a standard normal curve is

$$f(z) = \frac{e^{-z^2/2}}{\sqrt{2\pi}}$$

where

$$e = 2.7183$$

$$\pi = 3.1416$$

and

$$z = \frac{x - \mu}{\sigma}$$

The values of $f(z)$ for $z = 0$, .4, .8, 1.2, 1.6, 2.0, 2.4, and 2.8 are shown in the accompanying table. We do not show values of $f(z)$ for negative values of z because the normal distribution is symmetric about the mean, $z = 0$. Therefore, $f(-z) = f(z)$. For example, $f(-.4) = f(.4)$.

z	$f(z)$
0	.3989
.4	.3683
.8	.2897
1.2	.1942
1.6	.1109
2.0	.0540
2.4	.0224
2.8	.0079

a. Graph the standard normal curve $f(z)$, i.e., plot the points corresponding to z and $f(z)$ for $z = -2.8, -2.4, \ldots, -.4, 0, .4, \ldots, 2.4, 2.8$. Then sketch a smooth curve through the points.

To graph any other normal curve, say $f(x)$, which has a mean μ and standard deviation σ, you find the x-value corresponding to a z-value and then read $f(x) = f(z)$ from the table. For example, if $\mu = 10$ and $\sigma = 2$, then the x-value corresponding to $z = .4$ is found by solving the equation

$$z = \frac{x - \mu}{\sigma}$$

Substituting for z, μ, and σ yields

$$.4 = \frac{x - 10}{2} \quad \text{or} \quad x = 10.8$$

The value of $f(x)$ corresponding to $x = 10.8$ would be the same as $f(z)$ for $z = .4$, i.e., $f(x) = .3683$. To plot this normal curve, you would repeat this process for the z-values given in the table, plot the points x and $f(x)$, and sketch a smooth curve through these points.

b. Graph the normal curve with $\mu = 10$, $\sigma = 2$. Then sketch a similar curve with $\mu = 10$, $\sigma = 4$.

CASE STUDY 6.2 A Random Walk Down Wall Street

In his interesting and readable book, *A Random Walk Down Wall Street,** Burton G. Malkiel devotes Chapter 6 to "Technical Analysis and the Random-Walk Theory." In brief, this chapter discusses the theory employed by stock market technical analysts to forecast the upward or downward movement of specific stocks or the market as a whole.

Technical analysts, or ***chartists*** as they are called, believe that "knowledge of a stock's past behavior can help predict its probable future behavior." According to Malkiel, chartists strongly favor stocks that have made an upward move, and they advocate selling stocks that have moved downward in price. Similarly, when the stock market averages have shown strength and moved upward, they forecast further upward movement in the market averages. Malkiel clearly has little faith in the forecasts of chartists: He states, "On close examination, technicians are often seen with holes in their shoes and frayed shirt collars. I, personally, have never known a successful technician, but I have seen wrecks of several unsuccessful ones. (This is, of course, in terms of following their own technical advice. Commissions from urging customers to act on their recommendations are very lucrative.)" In brief, Malkiel does not agree that knowledge of a stock's past performance can be used to predict its future behavior.

A theory that is the antithesis of that held by chartists is known as the ***random-walk theory.*** According to this theory, the price movement of a stock (or the stock market) today is completely independent of its movement in the past: the price will rise or fall today by a random amount. A sequence of these random increases or

*B. G. Malkiel, *A random walk down Wall Street.* New York: Norton, 1975.

FIGURE 6.15 Distribution of Changes in Standard & Poor's Index from January 2, 1975 through December 31, 1979

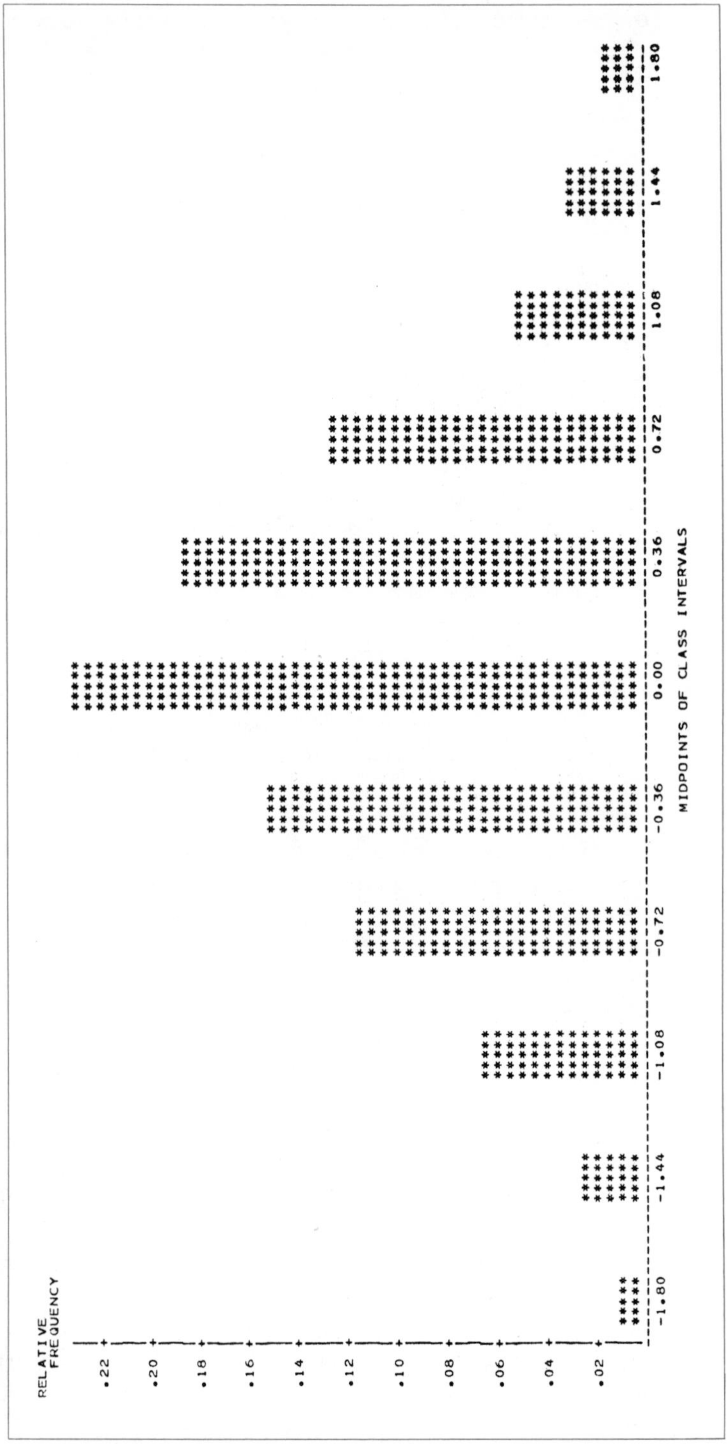

RELATIVE FREQUENCY

.22
.20
.18
.16
.14
.12
.10
.08
.06
.04
.02

-1.80 -1.44 -1.08 -0.72 -0.36 0.00 0.36 0.72 1.08 1.44 1.80

MIDPOINTS OF CLASS INTERVALS

decreases is known as a ***random walk.*** Note that this theory does not rule out the possibility of long-term trends, say a long-term upward trend in stock prices fueled by increased earnings, dividends, etc. It simply states that the ***daily*** change in price is independent of the changes that have occurred in the past.

To support his contention, Malkiel generated the stock price chart for several fictitious stocks by tossing a coin to decide whether a stock would move up (a head) or down (a tail) by a fixed amount, say ½ dollar, on a given day. This procedure was repeated for a large number of days. The movement of this fictitious stock was plotted, thus revealing its random walk. We performed an equivalent coin-tossing experiment in Chapter 4. Our fictitious stock charts are shown in Figure 4.2 (pages 106–107). You can see how the price seems to surge upward and downward in these "charts" (suggesting short-term trends), although the coin tossings were independent.

One way that we can refine the random-walk theory is to postulate the probability distribution of the daily price change in a stock (or the daily change in a market average). The most common assumption is that the change (even though stock prices change by discrete amounts) has a distribution that is approximately normal. To examine this theory, we recorded the daily change in the Standard & Poor's Stock Index for each market day during the period 1975 to 1979. The relative frequency distribution for these changes, shown in Figure 6.15, possesses a mean equal to \$.03 and a standard deviation equal to \$.72. Does this distribution differ markedly from a normal distribution with $\mu = .03$ and $\sigma = .72$?

Statistical tests are available to detect distributional departures from normality but they are beyond the scope of this text. Nevertheless, we can get an indication of the answer by fitting a normal curve to the data, using the method described in Case Study 6.1. Assign values to the daily change x, calculate the corresponding z-value, and then read the value of $f(x) = f(z)$ in the table, Case Study 6.1. Plot these points and sketch a smooth normal curve, $f(x)$, superimposed over the relative frequency distribution, Figure 6.15. Compare the two distributions. Does it appear that the daily change in the Standard & Poor's Stock Index possesses a distribution that is approximately normal?

CASE STUDY 6.3 Interpreting Those Wonderful EPA Mileage Estimates

One common ploy of advertisements for new automobiles is to list the EPA (Environmental Protection Agency) estimated miles per gallon (mpg) for the make of car being advertised. A recent advertisement in a national magazine boasts that the 1981 Dodge Aries-K wagon is "America's highest mileage 6-passenger wagon." The EPA estimated miles per gallon for this model is listed as 24 mpg. However, footnoted in fine print is the statement:

> Use EPA estimated mpg for comparison only. Your mileage may vary depending on speed, weather, and trip length. Actual highway mileage will probably be less.

How should the observant reader, in view of the footnote, interpret the 24-mpg figure? The EPA tests cars under conditions (weather, brand of gasoline, speed, terrain, etc.) ideally suited for maximum mileage performance. Nevertheless, even under identical conditions it is unreasonable (and impractical) to assume that all

Dodge Aries-K wagons tested will obtain the same gas mileage. If the EPA tested 50 Aries-K wagons, we would expect to observe 50 different mpg's. Conceivably, if all Dodge Aries-K wagons were tested under "ideal conditions," we would obtain a set of numbers which represents the population. Most likely then, the 24-mpg figure is the average miles per gallon obtained by the sample of wagons tested. The EPA uses this sample average mpg to estimate the population average mpg of all Dodge Aries-K wagons.

If you are an owner of (or are considering buying) a new Dodge Aries-K wagon, you may be interested in the likelihood that the wagon you own (or purchase) will perform "as advertised." The answer, of course, requires knowledge of the probability model for the continuous random variable Miles per gallon of Dodge Aries-K wagons.

Let us assume the EPA estimated mpg for this type of car is accurate, i.e., that the true mean mpg obtained under "ideal" conditions for all Dodge Aries-K wagons is, in fact, 24. Also, suppose that the distribution of mpg's is approximately normal with a standard deviation of 2.

a. What proportion of all Dodge Aries-K wagons tested under "ideal" conditions will obtain at least 29 mpg?
b. What is the probability that a Dodge Aries-K wagon, tested under "ideal" conditions, will obtain less than 20 mpg?
c. Fifteen percent of all Dodge Aries-K wagons tested will obtain an mpg rating above a particular value. Find this value.
d. Suppose that you test your new Dodge Aries-K wagon under "ideal" conditions and find that your car obtains 18 mpg. Does this result imply that you have bought a "lemon," or is it more likely that the EPA estimated mpg figure of 24 is too high? [*Hint:* Use your answer to part b.]

REFERENCES

Hogg, R. V., & Craig, A. T. *Introduction to mathematical statistics.* 4th ed. New York: Macmillan, 1978. Chapters 1 and 3.

Johnson, R. *Elementary statistics.* 3d ed. North Scituate, Mass.: Duxbury, 1980. Chapter 7.

Malkiel, B. G. *A random walk down Wall Street.* New York: Norton, 1975. Chapter 6.

Mendenhall, W. *Introduction to probability and statistics.* 5th ed. North Scituate, Mass.: Duxbury, 1979. Chapter 7.

Seven

Sampling and Sampling Distributions

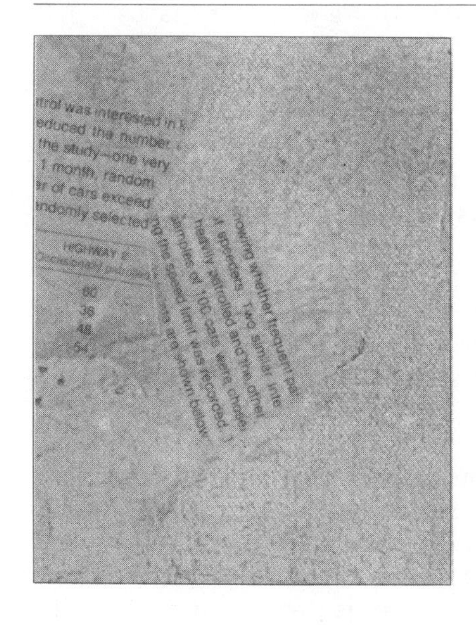

Do you think it advisable to buy that piece of timberland described in Case Study 3.3? Knowing the mean and standard deviation of the weights of the logs produced from a sample of 117 trees, how can we estimate the total weight of all logs that could be produced by the property? And, what is the reliability of this estimate? The behavior of sample statistics, described by their probability distributions, is the topic of Chapter 7.

Contents

7.1 Why the Method of Sampling Is Important

We now return to the objective of statistics, namely, the use of sample information to infer the nature of a population. Predicting your waiting time in a dentist's office, the annual amount of rainfall in your city, or an evening's winnings at blackjack based upon your previous (successful and unsuccessful) treks to the casino are all examples of sample inferences, and each involves an element of uncertainty. In this chapter we discuss a technique for measuring the uncertainty associated with sample inferences.

Suppose that we wish to make an inference about the mean, or average, starting salary of *all* June 1980–March 1981 University of Florida bachelor's degree graduates discussed in earlier chapters. In order to completely characterize this phenomenon, we would need the actual starting salary for each graduate. This complete listing of starting salaries constitutes the population of interest to us, and we are particularly interested in the parameter μ, the mean of the population. Unfortunately, in this and in most practical situations, the entire population of starting salaries is unavailable to us. However, we do have available (in Appendix A) a subset or ***sample*** of 948 pieces of data from the target population. But in order to use this sample to infer the characteristics of the population, the sample must be representative of the population about which inferences are to be made, i.e., the sample must possess characteristics similar to those which would be observed in the entire population, if it were available. In Example 1.2 (page 3), we explained why this sample of 948 starting salaries may not be characteristic of the much larger (and partly conceptual) population of all starting salaries of bachelor's degree graduates during this period.

EXAMPLE 7.1 Researchers have determined that the time patients spend waiting in physicians' offices plays an important role in an efficiently run practice (see Case Study 7.2). Suppose an orthodontist is interested in examining the waiting times of his patients over the past year as part of an annual evaluation of his practice. Now, unknown to the orthodontist, suppose the population relative frequency distribution for the waiting time for each of the orthodontist's 2,000 patients last year appears as in Figure 7.1. (We emphasize that this example is for illustration only. In actual practice, the entire population of 2,000 waiting times may not be easily accessible.) Now, assume that a member of his staff provides the orthodontist with the relative frequency distributions of waiting times for each of two samples of 50 patients (Figures 7.2(a) and 7.2(b)) selected from the 2,000 patients last year.

Compare the distributions of patient waiting times for the two samples. Which appears to better characterize the phenomenon of patient waiting time for the population?

Solution You can see that the two samples lead to quite different conclusions about the same population from which they were both selected. From Figure 7.2(b), 20% of the patients sampled waited in the orthodontist's office at least 50 minutes before receiving the orthodontist's services, whereas, from Figure 7.2(a), only 2% of the patients sampled had such a long wait. This may be compared to the relative

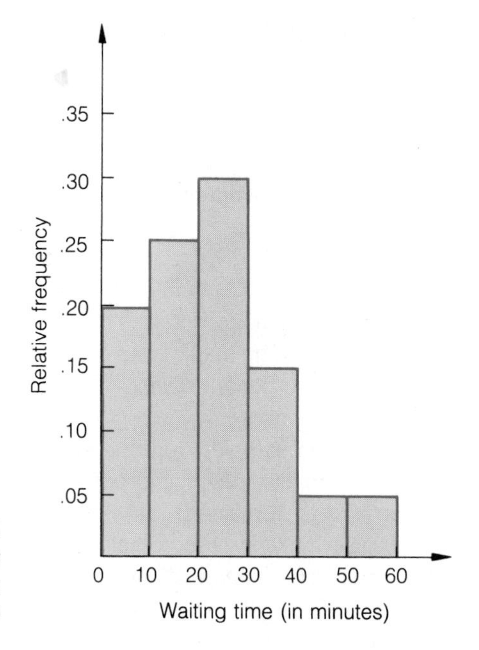

FIGURE 7.1 Relative Frequency Distribution of Waiting Times for 2,000 Patients

frequency distribution for the population, Figure 7.1, in which we observe that 5% of all the patients last year waited at least 50 minutes. In addition, note that none of the patients in the second sample, Figure 7.2(b), had waits of less than 10 minutes, whereas 18% of the patients sampled, Figure 7.2(a), had waits of less than 10

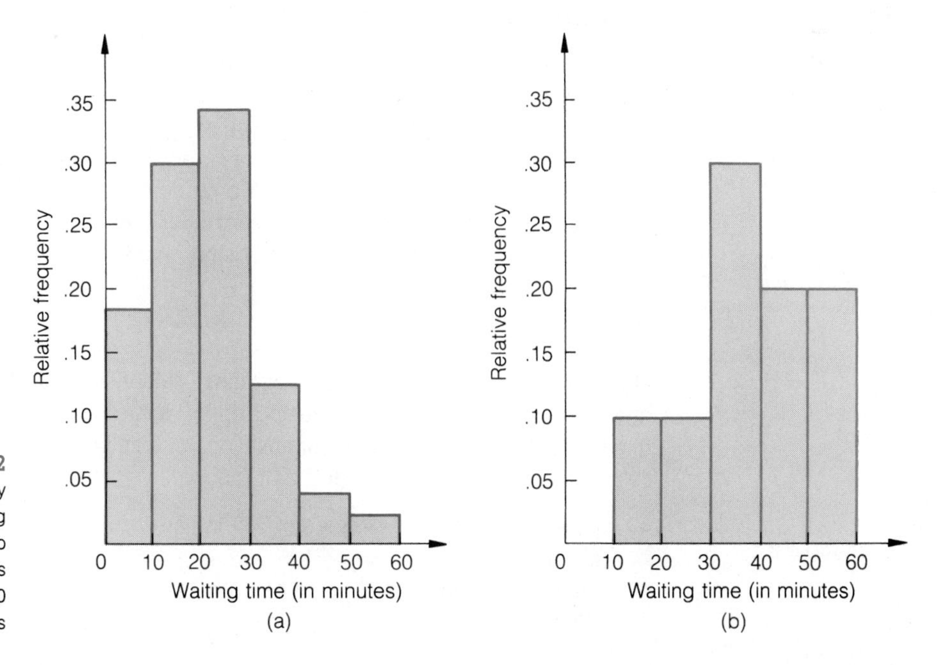

FIGURE 7.2 Relative Frequency Distributions of Waiting Times for Each of Two Samples of Fifty Patients Selected from 2,000 Patients

minutes. This value from the first sample, Figure 7.2(a), compares favorably with the 20% of the waiting times for the entire population, Figure 7.1, which were less than 10 minutes.

To rephrase the question posed in the example, we could ask: Which of the two samples is more representative of, or characteristic of, the phenomenon of patient waiting time for all 2,000 of the orthodontist's patients last year? Clearly, the information provided by the first sample, Figure 7.2(a), gives a better picture of the actual population phenomenon; its relative frequency distribution is more similar to that for the entire population, Figure 7.1, than is the one provided by the second sample, Figure 7.2(b). Thus, if the orthodontist were to rely upon information from the second sample (Figure 7.2(b)) only, he may have a distorted, or ***biased,*** impression of the true situation with respect to patient waiting time last year.

How is it possible that two samples from the same population can provide (apparently) contradictory information about the population? The key issue is the method by which the samples are obtained. The examples in this section demonstrate that great care must be taken in order to select a sample which will give an unbiased picture of the population about which inferences are to be made. One way to cope with this problem is to use random sampling. Random sampling will eliminate the possibility of bias in selecting a sample and, in addition, it will provide a probabilistic basis for evaluating the reliability of an inference. We will have more to say about random sampling in Section 7.2.

7.2 Obtaining a Random Sample

In the previous section, we demonstrated the importance of obtaining a sample which exhibits characteristics similar to those possessed by the population from which it came, the population about which we wish to make inferences. The selection of a random sample from the population is one way to satisfy this requirement. You will recall, Section 4.6, that we defined a random sample of n observations as one selected from a population in such a manner that each possible collection of n observations has the same chance of being chosen. In this section we will explain how to draw a random sample, and will then employ random sampling in the sections that follow.

EXAMPLE 7.2 The purchasing agent of a city can purchase stationery and office supplies from any of eight companies. If the purchasing agent decides to utilize three suppliers in a given year and wishes to avoid accusations of bias in their selection, the sample of three suppliers should be selected at random from among the eight.

a. How many different samples of 3 suppliers can be chosen from among the 8?
b. List them.
c. State the criterion which must be satisfied in order that the sample selected be random.

Solution **a.** In this example, the population of interest consists of 8 suppliers (call them *A*, *B*, *C*, *D*, *E*, *F*, *G*, H) from which we wish to select a sample of size $n=3$. You will recall (Chapter 4) that the number of different samples of $n=3$ elements that can be selected from a population of $N=8$ elements is

$$C_n^N = \frac{N!}{n!(N-n)!} = \frac{8!}{3!5!} = \frac{8\cdot7\cdot6\cdot5\cdot4\cdot3\cdot2\cdot1}{(3\cdot2\cdot1)(5\cdot4\cdot3\cdot2\cdot1)} = 56$$

b. The following is a listing of these 56 different samples:

A,B,C	A,C,F	A,E,G	B,C,G	B,E,H	C,E,F	D,E,H
A,B,D	A,C,G	A,E,H	B,C,H	B,F,G	C,E,G	D,F,G
A,B,E	A,C,H	A,F,G	B,D,E	B,F,H	C,E,H	D,F,H
A,B,F	A,D,E	A,F,H	B,D,F	B,G,H	C,F,G	D,G,H
A,B,G	A,D,F	A,G,H	B,D,G	C,D,E	C,F,H	E,F,G
A,B,H	A,D,G	B,C,D	B,D,H	C,D,F	C,G,H	E,F,H
A,C,D	A,D,H	B,C,E	B,E,F	C,D,G	D,E,F	E,G,H
A,C,E	A,E,F	B,C,F	B,E,G	C,D,H	D,E,G	F,G,H

c. Each sample must have the same chance of being selected in order to ensure that we have a random sample. Since there are 56 possible samples of size $n=3$, each must have a probability equal to $1/56$ of being selected by the sampling procedure.

What procedures may one use to generate a random sample? If the population is not too large, each observation may be recorded on a piece of paper and placed in a suitable container. After the collection of papers is thoroughly mixed, the researcher can remove *n* pieces of paper from the container; the elements named on these *n* pieces of paper would be the ones included in the sample. This simple method could be easily implemented by the purchasing agent of Example 7.2: Each

---

USING A TABLE OF RANDOM NUMBERS TO GENERATE A RANDOM SAMPLE OF SIZE *n* FROM A POPULATION OF *N* ELEMENTS

Step 1. Label the elements in the population from 1 to *N*.

Step 2. Begin at an arbitrary starting point in a table of random numbers (Table 9, Appendix E). The starting point may be obtained, for example, by closing your eyes and placing a pencil point haphazardly at an entry in the table.

Step 3. Record this number and then proceed in some direction (up or down, left or right, or diagonally), recording each number to identify the corresponding population element to be included in the sample. Continue in this manner until *n* elements have been selected. [*Note:* Use the necessary number of digits from the random numbers to accommodate the numbering of the elements in step 1.]

---

of the eight suppliers would be listed on a separate piece of paper and placed in a container for thorough mixing. The purchasing agent would then select three of the pieces of paper and contact the three suppliers named on the chosen papers. This method has its drawbacks, however: (1) It is not feasible when the population consists of a large number of observations; and (2) Since it is very difficult to achieve a thorough mixing, the procedure provides only an approximation to random sampling. A more practical method of generating a random sample, and one that may be used with large populations, is the use of a table of random numbers (Table 9, Appendix E). The details are summarized in the box on page 199.

EXAMPLE 7.3 Refer to Example 7.2. Use the portion of Table 9, Appendix E, reproduced in Table 7.1 to select a random sample of three suppliers for the purchasing agent.

Solution

Step 1. We first number the $N = 8$ suppliers from 1 to 8. For example, supplier *A* may be assigned the number 1, supplier *B*, the number 2, and so on. (This initial labelling is arbitrary; we are preserving the alphabetical ordering merely for convenience.)

Step 2. Let us begin in Column 4, Row 2 of the table. The random number entry given there is 85393. Since we need only a single digit to identify the elements in our population, we choose element number 8 (supplier *H*) for the sample.

Step 3. We will now proceed down Column 4, recording random numbers until we have $n = 3$ elements identified for the sample. The next entry in the column is 9; since we have no element numbered 9, we skip this entry and continue to move downward in the column. The next two entries are 6 and 1. Thus, the purchasing agent would select suppliers numbered 8, 6, and 1; i.e., suppliers *H*, *F*, and *A*.

TABLE 7.1 Reproduction of a Portion of Table 9, Appendix E

ROW \ COLUMN	1	2	3	4	5	6
1	10480	15011	01536	02011	81647	91646
2	22368	46573	25595	85393	30995	89198
3	24130	48360	22527	97265	76393	64809
4	42167	93093	06243	61680	07856	16376
5	37570	39975	81837	16656	06121	91782
6	77921	06907	11008	42751	27756	53498
7	99562	72905	56420	69994	98872	31016
8	96301	91977	05463	07972	18876	20922
9	89579	14342	63661	10281	17453	18103
10	85475	36857	53342	53988	53060	59533
11	28918	69578	88231	33276	70997	79936
12	63553	40961	48235	03427	49626	69445
13	09429	93969	52636	92737	88974	33488
14	10365	61129	87529	85689	48237	52267
15	07119	97336	71048	08178	77233	13916

A sample selected according to the procedure described above would eliminate the possibility of a biased selection. Every sample of three suppliers would have an equal probability of being selected.

EXAMPLE 7.4 Use the portion of Table 9, Appendix E, that is reproduced in Table 7.1 to select a random sample of size $n = 5$ from the 948 observations on starting salaries in Appendix A.

Solution **Step 1.** The observations have already been numbered from 001 to 948 in our listing of the data, Appendix A. This labelling implies that we will obtain random numbers of three digits from the table.

Step 2. Let us begin in Row 5, Column 1, and proceed horizontally to the right across the rows. The first element selected is numbered 375.

Step 3. The next four elements to be included in the sample are those numbered 703, 997 (skip), 581, 837, and 166.

The random numbers and the associated observations on starting salary are shown in Table 7.2.

TABLE 7.2
Random Sample of $n = 5$ Observations on Starting Salary Values from Appendix A

RANDOM NUMBER OBTAINED	STARTING SALARY FOR CORRESPONDING POPULATION ELEMENT
375	$18,800
703	11,300
581	11,100
837	14,700
166	9,400

Note in Example 7.4 that, for this sample of $n = 5$ observations on starting salary, the mean is

$$\bar{x} = \frac{\Sigma x}{n} = \frac{18{,}800 + 11{,}300 + 11{,}100 + 14{,}700 + 9{,}400}{5} = \$13{,}060$$

whereas the mean for all 948 observations is $15,909. In the next section, we will discuss how to judge the performance of a statistic computed from a random sample.

EXERCISES **7.1** Refer to Case Study 2.3. Use Table 9, Appendix E, to generate a random sample of $n = 15$ observations on carbon monoxide content in smoke of cigarettes based on Federal Trade Commission rankings, Table 2.6.

7.2 Suppose a lottery consists of 500 tickets. One ticket stub is to be chosen and the corresponding ticket holder will receive an all-expenses-paid trip for two to Acapulco, Mexico. How would you select this stub so that the prize will be awarded fairly?

7.3 Two soldiers are to be chosen from a platoon of ten men for a dangerous mission.

a. How many different samples of two soldiers may be chosen by the platoon sergeant?
b. List them.
c. State the criterion which must be satisfied in order that the sample of soldiers be random.

7.4 A clinical psychologist is asked to view tapes in which each of six experimental subjects is discussing his or her recent dreams. Three of the six subjects have been previously classified as "high anxiety" individuals, and the other three as "low anxiety." The psychologist is told only that there are three of each type and is asked to select the three high-anxiety subjects.
a. How many different samples of three subjects may be selected by the psychologist?
b. List them.
c. Do you think the sample chosen by the psychologist will be random? Explain.

7.5 A homeowner wishes to list his residential property for sale with the Multiple Listing Service (MLS) in the county. Since there are seven real estate agencies in the county which are affiliated with MLS, the homeowner has decided to randomly select three of them, and to personally interview a sales representative from each of the three selected agencies. Then he will decide with which of the three agencies the property will be listed.
a. How many different samples of three real estate agencies may be selected by the homeowner?
b. List them.
c. State the criterion which must be satisfied in order that the sample selected be random.

7.6 Many opinion surveys are conducted by mail. In this sampling procedure, a random sample of persons is selected from among a list of people who are supposed to constitute a target population (purchasers of a product, etc.). Each is sent a questionnaire and is requested to complete and return the questionnaire to the pollster. Why might this type of survey yield a sample that would produce biased inferences?

7.7 One of the most infamous examples of improper sampling was conducted in 1936 by the *Literary Digest* to determine the winner of the Landon—Roosevelt presidential election. The poll, which predicted Landon to be the winner, was conducted by sending ballots to a random sample of persons selected from among the names listed in the telephone directories of that year. In the actual election, Landon won in Maine and Vermont but lost in the remaining forty-six states. The *Literary Digest's* erroneous forecast is believed to be the major reason for its eventual failure.

What was the cause of the *Digest's* erroneous forecast; i.e., why might the sampling procedure described above yield a sample of people whose opinions might be biased in favor of Landon?

7.8 Use Table 9, Appendix E, to generate a random sample of $n = 20$ supermarket A customer service times from Appendix C. Construct a relative frequency distribution for the 20 customer service times. Does the relative frequency distribution you constructed possess a shape similar to the relative frequency distribution for the "population" of 500 supermarket A customer service times, Figure 3.8? If so, will the sample produce reliable inferences about the population?

7.9 A file clerk is assigned the task of selecting a random sample of 26 company accounts (from a total of 5,000) to be audited. The clerk is considering two sampling methods:

Method A. Organize the 5,000 company accounts in alphabetical order (according to the first letter of the client's last name), then randomly select one account card for each of the 26 letters of the alphabet.

Method B. Assign each company account a four-digit number from 0001 to 5000. From a table of random numbers, choose 26 four-digit numbers (in the range of 0001–5000) and match the numbers with the corresponding company account.

Which of the two methods would you recommend to the file clerk? Which sampling method could possibly yield a biased sample?

7.3 Sampling Distributions

In the previous section, we learned how to generate a random sample from a population of interest, the ultimate goal being to use information from the sample to make an inference about the nature of the population. In many situations, the objective will be to estimate a numerical characteristic of the population (called a ***parameter***), using information from the sample. To illustrate, in Example 7.4, we computed $\bar{x} = \$13,060$, the mean starting salary for a random sample of $n = 5$ observations from the data on the starting salaries recorded in Appendix A; that is, we used the sample information to compute a ***statistic,*** namely, the sample mean, $\bar{x}$.

DEFINITION 7.1

A numerical descriptive measure of a population is called a ***parameter.***

DEFINITION 7.2

A quantity computed from the observations in a sample is called a ***statistic.***

You may have observed that the value of a population parameter (for example, the mean μ) is constant (although it is usually unknown to us); its value does not vary from sample to sample. However, the value of a sample statistic (for example, the sample mean, $\bar{x}$) is highly dependent upon the particular sample which is selected. If, in Example 7.4, we had begun at a different point in the random number table, we would have obtained a different random sample of five observations, and thus a different value of $\bar{x}$.

Since statistics vary from sample to sample, any inferences based on them will necessarily be subject to some uncertainty. How, then, do we judge the reliability of a sample statistic as a tool in making an inference about the corresponding population parameter? Fortunately, the uncertainty of a statistic generally has characteristic properties which are known to us, and which are reflected in its ***sampling distribution.***

DEFINITION 7.3

The ***sampling distribution*** of a sample statistic (based on n observations) is the relative frequency distribution of the values of the statistic generated by taking repeated random samples of size n and computing the value of the statistic for each sample.

Knowledge of the sampling distribution of a particular statistic provides us with information about its performance over the long run.

We will illustrate the notion of a sampling distribution with an example, in which our interest focuses on the starting salaries of all June 1980–March 1981 University of Florida bachelor's degree graduates ***who indicated that they had secured a job*** on the CRC questionnaire described in Chapter 1. In particular, we wish to estimate the mean starting salary of all such graduates. Then, the target population consists of the 948 observations on starting salary contained in Appendix A. (We note again that the researcher is the one who defines the population. Depending on how the target population is defined, the data in Appendix A could represent only a sample, as described in Example 1.2. In this case, however, the 948 observations constitute the entire population, since we are interested only in those graduates who indicated on the questionnaire that they had secured employment, and starting salary data on *all* of these graduates are available in Appendix A. In addition, although the true value of μ, the mean of these 948 observations, is already known to us (Example 3.10), this example will serve to illustrate the concepts.)

EXAMPLE 7.5 Describe the generation of the sampling distribution of $\bar{x}$, the mean of a random sample of $n = 5$ observations from the population of 948 starting salaries in Appendix A.

Solution The sampling distribution for the statistic $\bar{x}$, based on a random sample of $n = 5$ measurements, would be generated in this manner: Select a random sample of 5 measurements from the population of 948 observations on starting salary in Appendix A; compute and record the value of $\bar{x}$ for this sample. Now return these 5 measurements to the population and repeat the procedure, i.e., draw another random sample of $n = 5$ measurements and record the value of $\bar{x}$ for this sample. Return these measurements and repeat the process. If this sampling procedure could be repeated an infinite number of times, the infinite number of values of $\bar{x}$ so obtained could be summarized in a relative frequency distribution, called the ***sampling distribution*** of $\bar{x}$.

This task, which may seem impractical if not impossible, is not performed in actual practice. Instead, the sampling distribution of a statistic is obtained by applying mathematical theory or computer simulation, as illustrated in the next example.

EXAMPLE 7.6 Use computer simulation to find the approximate sampling distribution of $\bar{x}$, the mean of a random sample of $n = 5$ observations from the population of 948 starting salaries in Appendix A.

Solution We obtained 100 computer-generated random samples of size $n = 5$ from the target population. The first ten of these samples are presented in Table 7.3.

TABLE 7.3 First Ten Samples of $n = 5$ Measurements from Population of Starting Salary Measurements

SAMPLE	MEASUREMENTS (IN DOLLARS)				
1	14,700	28,500	17,900	14,700	15,200
2	17,700	10,200	11,400	17,100	7,100
3	14,200	14,700	23,200	13,700	11,400
4	10,600	18,700	17,900	15,300	23,700
5	15,000	16,700	17,700	20,600	15,300
6	25,100	11,000	11,500	13,100	13,000
7	21,400	22,700	23,000	16,800	21,400
8	21,700	17,200	9,300	15,600	19,600
9	13,000	16,900	14,700	7,600	14,100
10	11,500	12,200	21,200	16,000	21,900

For example, the first computer-generated sample contained the following measurements:

14,700 28,500 17,900 14,700 15,200

The corresponding value of the sample mean is

$$\bar{x} = \frac{\Sigma x}{n} = \frac{14{,}700 + 28{,}500 + 17{,}900 + 14{,}700 + 15{,}200}{5} = \$18{,}200$$

For each sample of five observations, the sample mean $\bar{x}$ was computed. The 100 values of $\bar{x}$ are summarized in the relative frequency distribution, Figure 7.3.

Let us compare the relative frequency distribution for $\bar{x}$, Figure 7.3, with the relative frequency distribution for the population, Figure 3.7 (page 65). Note that the values of $\bar{x}$, Figure 7.3, tend to cluster around the population mean, $\mu = \$15{,}909$;

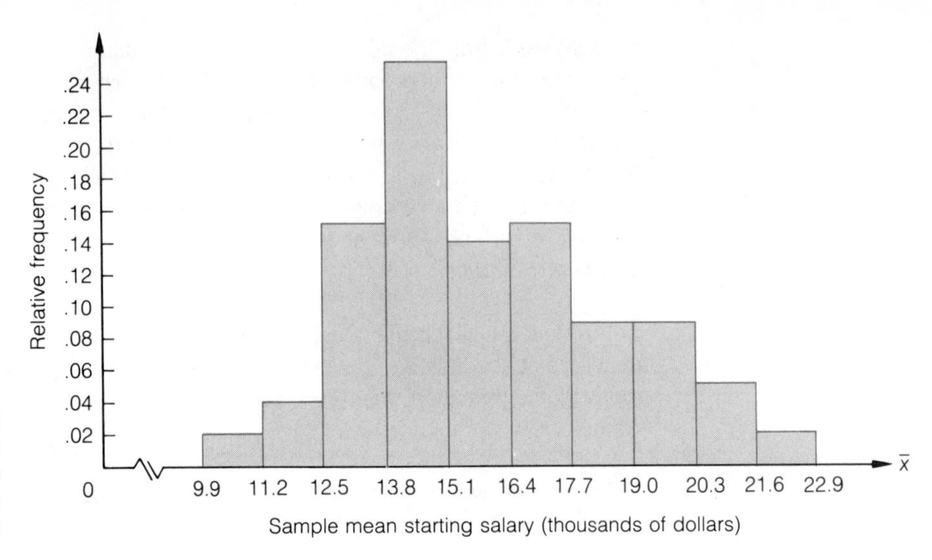

FIGURE 7.3 Sampling Distribution of $\bar{x}$: Relative Frequency Distribution of $\bar{x}$ Based on 100 Samples of Size $n = 5$

also, the values of the sample mean are less spread out (that is, they have less variation) than the population values shown in Figure 3.7. These two observations are borne out by comparing the means and standard deviations of the two sets of observations, as shown in Table 7.4.

TABLE 7.4 Comparison of the Population Distribution and the Approximate Sampling Distribution of $\bar{x}$, Based on 100 Samples of Size $n = 5$

	MEAN	STANDARD DEVIATION
Population of 948 starting salaries (Figure 3.7)	$\mu = \$15,909$	$\sigma = \$5,353$
100 values of $\bar{x}$ based on samples of size $n = 5$ (Figure 7.3)	\$16,014	\$2,617

EXAMPLE 7.7 Refer to Example 7.6. Simulate the sampling distribution of $\bar{x}$ for samples of size $n = 20$ from the population of 948 starting salary observations. Compare the result with the sampling distribution of $\bar{x}$, based on samples of size $n = 5$, obtained in Example 7.6.

Solution We obtained 100 computer-generated random samples of size $n = 20$ from the target population. A relative frequency distribution for the 100 corresponding values of $\bar{x}$ is shown in Figure 7.4.

It can be seen that, as with the sampling distribution based on samples of size $n = 5$, the values of $\bar{x}$ tend to center about the population mean. However, a visual inspection shows that the variation of the $\bar{x}$-values about their mean, Figure 7.4, is less than the variation in the values of $\bar{x}$ based on samples of size $n = 5$, Figure 7.3. The mean and standard deviation for these 100 values of $\bar{x}$ are shown in Table 7.5 for comparison with previous results.

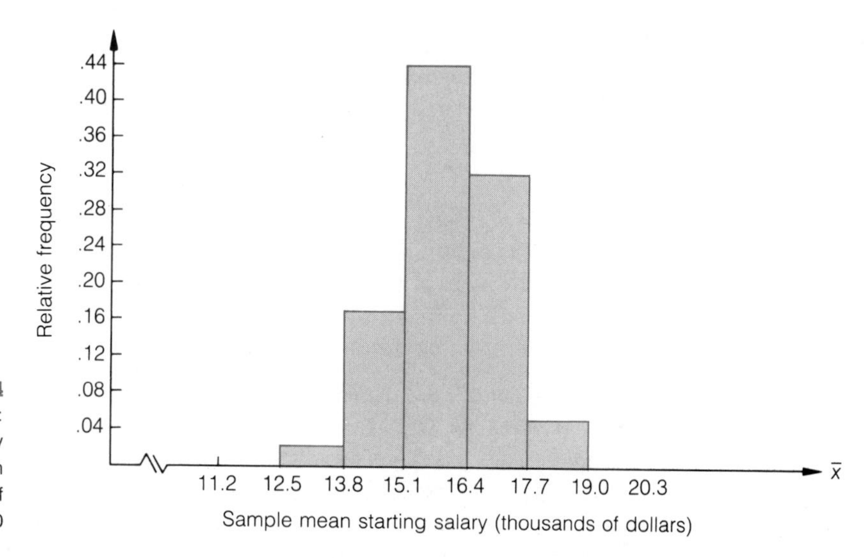

FIGURE 7.4 Sampling Distribution of $\bar{x}$: Relative Frequency Distribution of $\bar{x}$ Based on 100 Samples of Size $n = 20$

TABLE 7.5

	MEAN	STANDARD DEVIATION
Population of 948 starting salaries (Figure 3.7)	$\mu = \$15,909$	$\sigma = \$5,353$
100 values of $\bar{x}$ based on samples of size $n = 5$ (Figure 7.3)	\$16,014	\$2,617
100 values of $\bar{x}$ based on samples of size $n = 20$ (Figure 7.4)	\$15,990	\$1,107

From Table 7.5 we observe that, as the sample size increases, there is less variation in the sampling distribution of $\bar{x}$; that is, the values of $\bar{x}$ tend to cluster more closely about the population mean as n gets larger. This intuitively appealing result will be stated formally in the next section.

EXERCISES

7.10 Use computer simulation or Table 9, Appendix E, to obtain 30 random samples of size $n = 5$ from the "population" of 500 supermarket B customer service times in Appendix C. (Alternatively, each class member may generate several random samples and the results can be pooled.)

a. Calculate $\bar{x}$ for each of the 30 samples. Construct a relative frequency distribution for the 30 sample means. Compare with the population relative frequency distribution shown in Figure 3.9.

b. Compute the average of the 30 sample means.

c. Compute the standard deviation of the 30 sample means.

d. Locate the average of the 30 sample means, part b, on the relative frequency distribution. This value could be used as an estimate for μ, the mean of the entire population of 500 supermarket B customer service times.

7.11 Repeat parts a, b, c, and d of Exercise 7.10, using random samples of size $n = 10$. Compare the relative frequency distribution with that of Exercise 7.10(a). Do the values of $\bar{x}$ generated from samples of size $n = 10$ tend to cluster more closely about μ?

7.12 Generate the approximate sampling distribution of $\bar{x}$, the mean of a random sample of $n = 15$ observations from the population of DDT content in fish in Appendix D. [*Hint:* Obtain 50 random samples of size $n = 15$ from the data in Appendix D, compute $\bar{x}$ for each of the 50 samples, and then construct a relative frequency distribution for the 50 sample means. Again, each class member could generate several random samples and the results could be pooled.]

7.13 A research cardiologist wishes to examine the rate at which a person's heartbeat increases after 15 minutes of vigorous exercise. One method would be to consider the population of ratios of heart rate (in beats per minute) after exercising to heart rate before exercising for all human subjects. The cardiologist has reason to believe that the population relative frequency distribution for ratio of heart rate after exercise to heart rate before exercise would be markedly skewed to the right since no value can be less than zero, most values would presumably be greater than one, and, occasionally, very large values would be observed. Now, recalling from Section 3.4 that the mean is sensitive to very large observations, one could argue that the population median ratio would provide more information than would the mean about the increase in human heart rate after vigorous exercise.

Suppose a medical researcher has proposed two different statistics (call them *A* and *B*) for estimating the population median. In an attempt to judge which of the statistics is more suitable, you simulated the approximate sampling distributions for each of the statistics, based on random samples of size $n = 10$ human subjects. Your results were as shown in Figure 7.5. Comment on the two sampling distributions. Which of the statistics, *A* or *B*, would you recommend for use? (In the next section we will discuss desirable properties of a sampling distribution.)

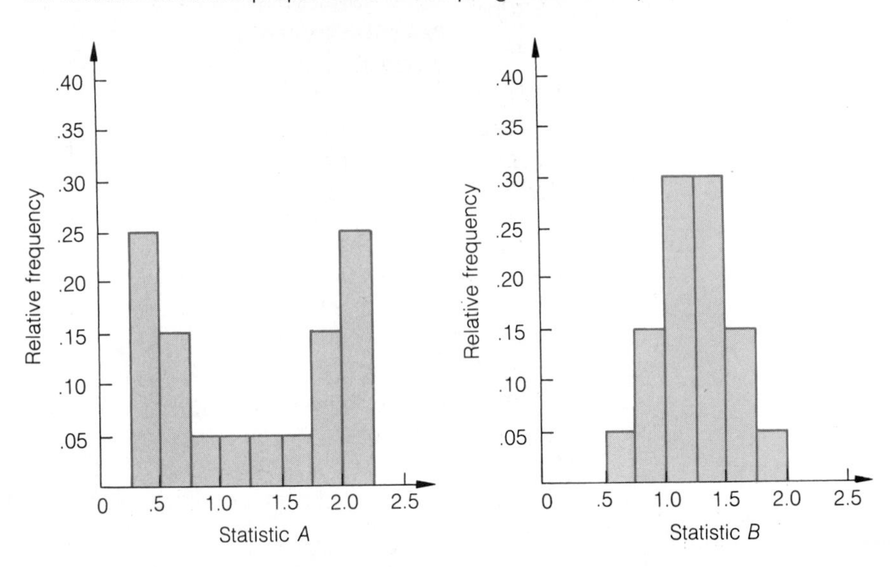

FIGURE 7.5
Approximate Sampling Distributions of Two Statistics

7.4 The Sampling Distribution of $\bar{x}$; the Central Limit Theorem

Estimating the mean starting salary for all college graduates of a certain university, or the average increase in heart rate for all human subjects after 15 minutes of vigorous exercise, or the mean yield per acre of farm land, are all examples of practical problems in which the goal is to make an inference about the mean, μ, of some target population. In previous sections, we have indicated that the sample mean $\bar{x}$ is often used as a tool for making an inference about the corresponding population parameter μ, and we have shown how to approximate its sampling distribution. The following theorem, of fundamental importance in statistics, provides information about the actual sampling distribution of $\bar{x}$:

THE CENTRAL LIMIT THEOREM

If the sample size is sufficiently large, then the mean $\bar{x}$ of a random sample from a population has a sampling distribution which is approximately normal, regardless of the shape of the relative frequency distribution of the target population. As the sample size increases, the better will be the normal approximation to the sampling distribution.

The sampling distribution of $\bar{x}$, in addition to being approximately normal, has other known characteristics which are summarized in the accompanying box.

PROPERTIES OF THE SAMPLING DISTRIBUTION OF $\bar{x}$

If $\bar{x}$ is the mean of a random sample of size n from a population with mean μ and standard deviation σ, then:

1. The sampling distribution of $\bar{x}$ has a mean which is equal to the mean of the population from which the sample was selected. That is, if we let $\mu_{\bar{x}}$ denote the mean of the sampling distribution of $\bar{x}$, then

   $$\mu_{\bar{x}} = \mu$$

2. The sampling distribution of $\bar{x}$ has a standard deviation which is equal to the standard deviation of the population from which the sample was selected, divided by the square root of the sample size. That is, if we let $\sigma_{\bar{x}}$ denote the standard deviation of the sampling distribution of $\bar{x}$, then

   $$\sigma_{\bar{x}} = \frac{\sigma}{\sqrt{n}}$$

EXAMPLE 7.8 Discuss the Central Limit Theorem and the two properties of the sampling distribution of $\bar{x}$, making reference to the empirical evidence obtained in Examples 7.6 and 7.7. Recall that, in Examples 7.6 and 7.7, we obtained repeated random samples of sizes $n = 5$ and $n = 20$ from the population of starting salaries recorded in Appendix A. For this target population, we know the values of the parameters μ and σ:

Population mean: $\mu = \$15{,}909$

Population standard deviation: $\sigma = \$5{,}353$

Solution In Figures 7.3 and 7.4, we noted that the values of $\bar{x}$ tended to cluster about the population mean, $\mu = \$15{,}909$. This is guaranteed by property 1, which implies that, in the long run, the average of ***all*** values of $\bar{x}$ which would be generated in infinite repeated sampling would be equal to μ.

We also observed, from Table 7.5, that the standard deviation of the sampling distribution of $\bar{x}$ decreased as the sample size increased from $n = 5$ to $n = 20$. Property 2 quantifies the decrease and relates it to the sample size. As an example, note that, for our approximate (simulated) sampling distribution based on samples of size $n = 5$, we obtained a standard deviation of \$2,617, whereas property 2 tells us that, for the actual sampling distribution of $\bar{x}$, the standard deviation is equal to

$$\sigma_{\bar{x}} = \frac{\sigma}{\sqrt{n}} = \frac{\$5{,}353}{\sqrt{5}} = \$2{,}394$$

Similarly, for samples of size $n = 20$, the sampling distribution of $\bar{x}$ actually has a standard deviation of

$$\sigma_{\bar{x}} = \frac{\sigma}{\sqrt{n}} = \frac{\$5{,}353}{\sqrt{20}} = \$1{,}197$$

The value we obtained by simulation was \$1,107.

Finally, the Central Limit Theorem guarantees an approximately normal distribution for $\bar{x}$, regardless of the shape of the original population. In our examples, the population from which the samples were selected is seen in Figure 3.7 (page 65) to be moderately skewed to the right. Note from Figures 7.3 and 7.4 that, although the sampling distribution of $\bar{x}$ tends to be mound-shaped in each case, the normal approximation improved when the sample size was increased from $n = 5$ (Figure 7.3) to $n = 20$ (Figure 7.4).

EXAMPLE 7.9 Three relative frequency distributions which provide reasonably accurate probability models for certain business phenomena are the ***normal distribution*** (which we discussed in Chapter 6), the ***uniform distribution,*** and the ***exponential distribution.*** Their vastly different shapes are shown in Figure 7.6(a). Simulate the sampling distributions of $\bar{x}$ by drawing 1,000 samples of $n = 5$ observations from populations that possess the relative frequency distributions shown in Figure 7.6(a). Repeat the procedure for $n = 10$, 15, 25, and 100. Does the Central Limit Theorem appear to provide adequate information about the shapes of the sampling distributions of $\bar{x}$?

Solution We obtained 1,000 computer-generated random samples of size $n = 5$ from each of the three populations. Relative frequency distributions for the 1,000 values of $\bar{x}$ obtained from each of the populations are shown in Figure 7.6(b) on page 212. Note their shapes for this small n.

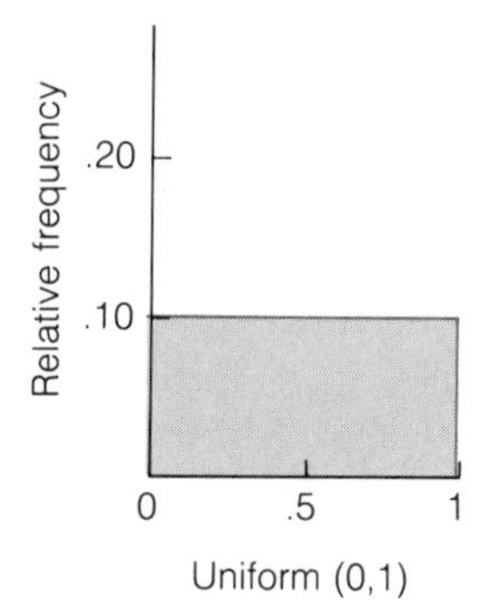

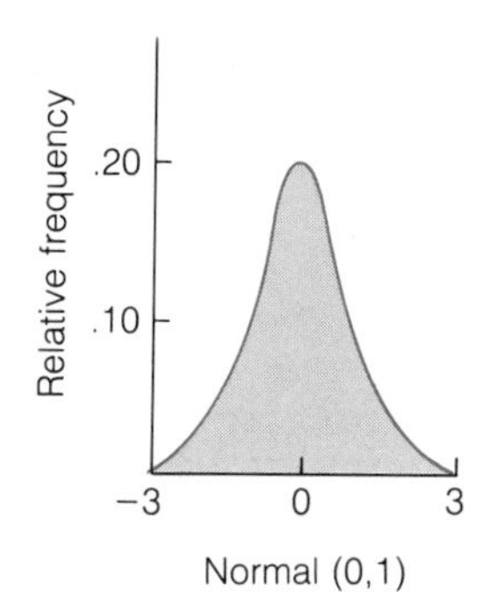

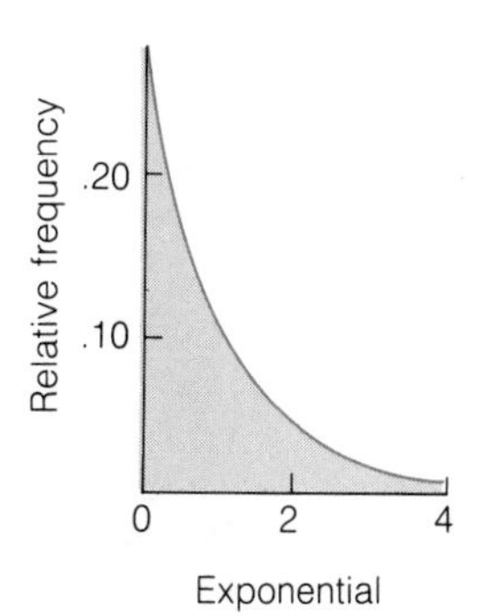

FIGURE 7.6(a)
Three Population Relative Frequency Distributions

The relative frequency distributions of $\bar{x}$ for $n = 10$, 15, 25, and 100, also simulated by computer, are shown in Figures 7.6(c), 7.6(d), 7.6(e), and 7.6(f) (pages 213–216), respectively. Note that the values of $\bar{x}$ tend to cluster about the mean of the probability distribution from which the sample was taken, and that as n increases, there is less variation in the sampling distribution. You can also see that as the sample size increases, the shape of the sampling distribution of $\bar{x}$ tends toward the shape of the normal distribution (symmetric and mound-shaped), regardless of the shape of the relative frequency distribution of the sampled population. The results of our computer simulations thus offer visual verification of the Central Limit Theorem and the other properties of the sampling distribution of $\bar{x}$ given in the box. (It is interesting to note that when sampling from a normal population, the sampling distribution of $\bar{x}$ is approximately normal for all values of n simulated in this example. In fact, it can be theoretically shown that when the relative frequency distribution of the target population is normal, the sample mean will have a normal sampling distribution, regardless of the sample size.)

EXAMPLE 7.10 A consumer magazine recently reported that for families residing in the northeast, the distribution of the weekly (per capita) expenditure for food consumed away from home has an average of \$3.28 and a standard deviation of \$1.12. In order to check this claim, an economist randomly samples 100 families residing in the northeast and monitors their expenditures for food consumed away from home for one week.

a. Assuming the consumer magazine's claim is true, describe the sampling distribution of the mean weekly (per capita) expenditure for food purchased away from home for a random sample of 100 families residing in the northeast.

b. Assuming the consumer magazine's claim is true, what is the probability that the sample mean weekly (per capita) expenditure for food purchased away from home will be at least \$3.60?

Solution **a.** Although we have no information about the shape of the relative frequency distribution of the weekly (per capita) expenditures for food purchased away from home for families residing in the area, we apply the Central Limit Theorem to conclude that the sampling distribution of the sample mean weekly (per

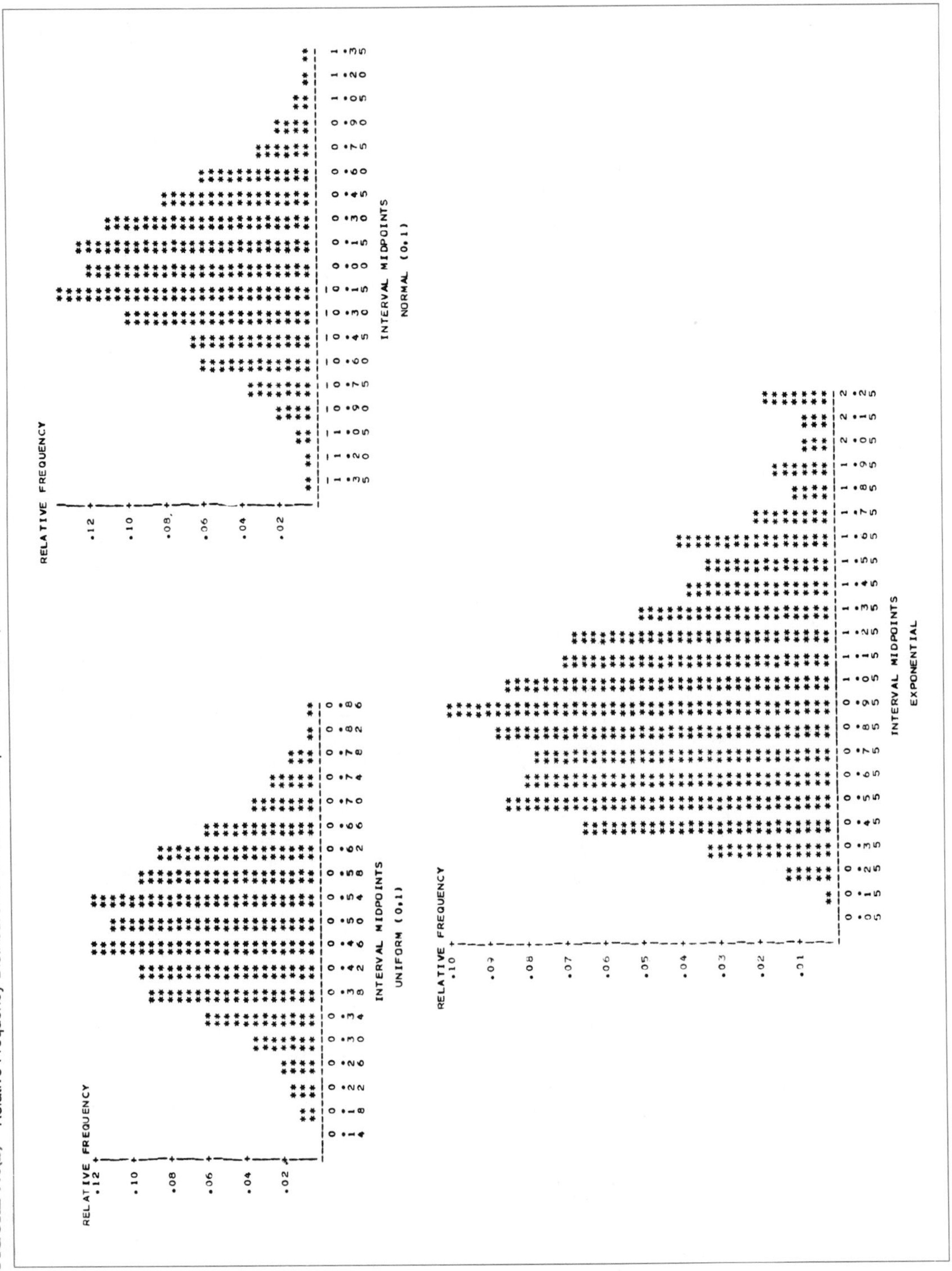

FIGURE 7.6(b) Relative Frequency Distribution of a Random Sample of 1,000 Sample Means ($n = 5$)

FIGURE 7.6(c) Relative Frequency Distribution of a Random Sample of 1,000 Sample Means ($n = 10$)

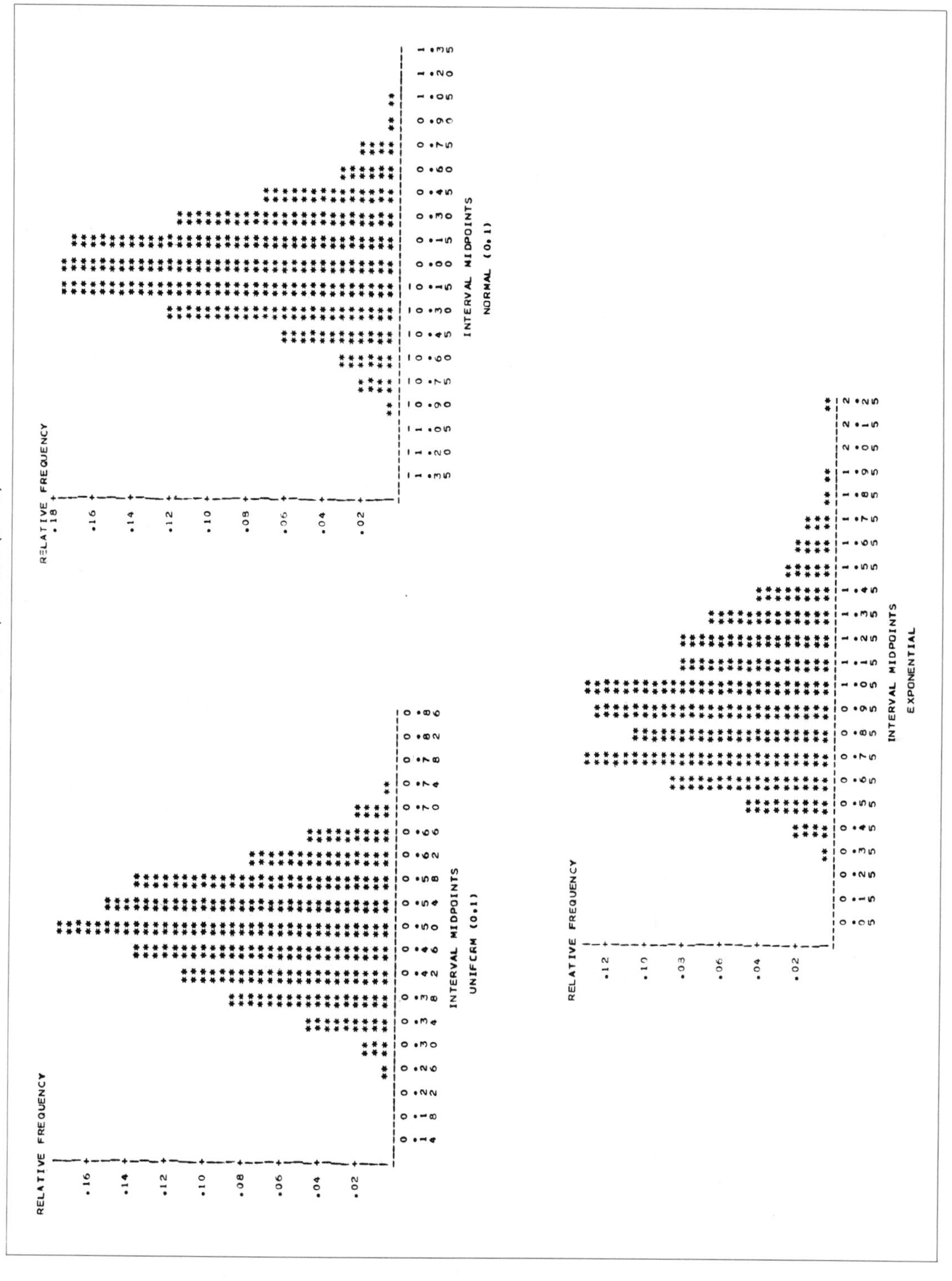

FIGURE 7.6(d) Relative Frequency Distribution of a Random Sample of 1,000 Sample Means ($n = 15$)

RELATIVE FREQUENCY

.20 .18 .16 .14 .12 .10 .08 .06 .04 .02

0.14 0.18 0.22 0.26 0.30 0.34 0.38 0.42 0.46 0.50 0.54 0.58 0.62 0.66 0.70 0.74 0.78 0.82 0.86

INTERVAL MIDPOINTS
UNIFORM (0,1)

RELATIVE FREQUENCY

.20 .18 .16 .14 .12 .10 .08 .06 .04 .02

-1.35 -1.20 -1.05 -0.90 -0.75 -0.60 -0.45 -0.30 -0.15 0.00 0.15 0.30 0.45 0.60 0.75 0.90 1.05 1.20 1.35

INTERVAL MIDPOINTS
NORMAL (0,1)

RELATIVE FREQUENCY

.15 .14 .12 .10 .08 .06 .04 .02

0.05 0.15 0.25 0.35 0.45 0.55 0.65 0.75 0.85 0.95 1.05 1.15 1.25 1.35 1.45 1.55 1.65 1.75 1.85 1.95 2.05 2.15 2.25

INTERVAL MIDPOINTS
EXPONENTIAL

FIGURE 7.6(e) Relative Frequency Distribution of a Random Sample of 1,000 Sample Means ($n = 25$)

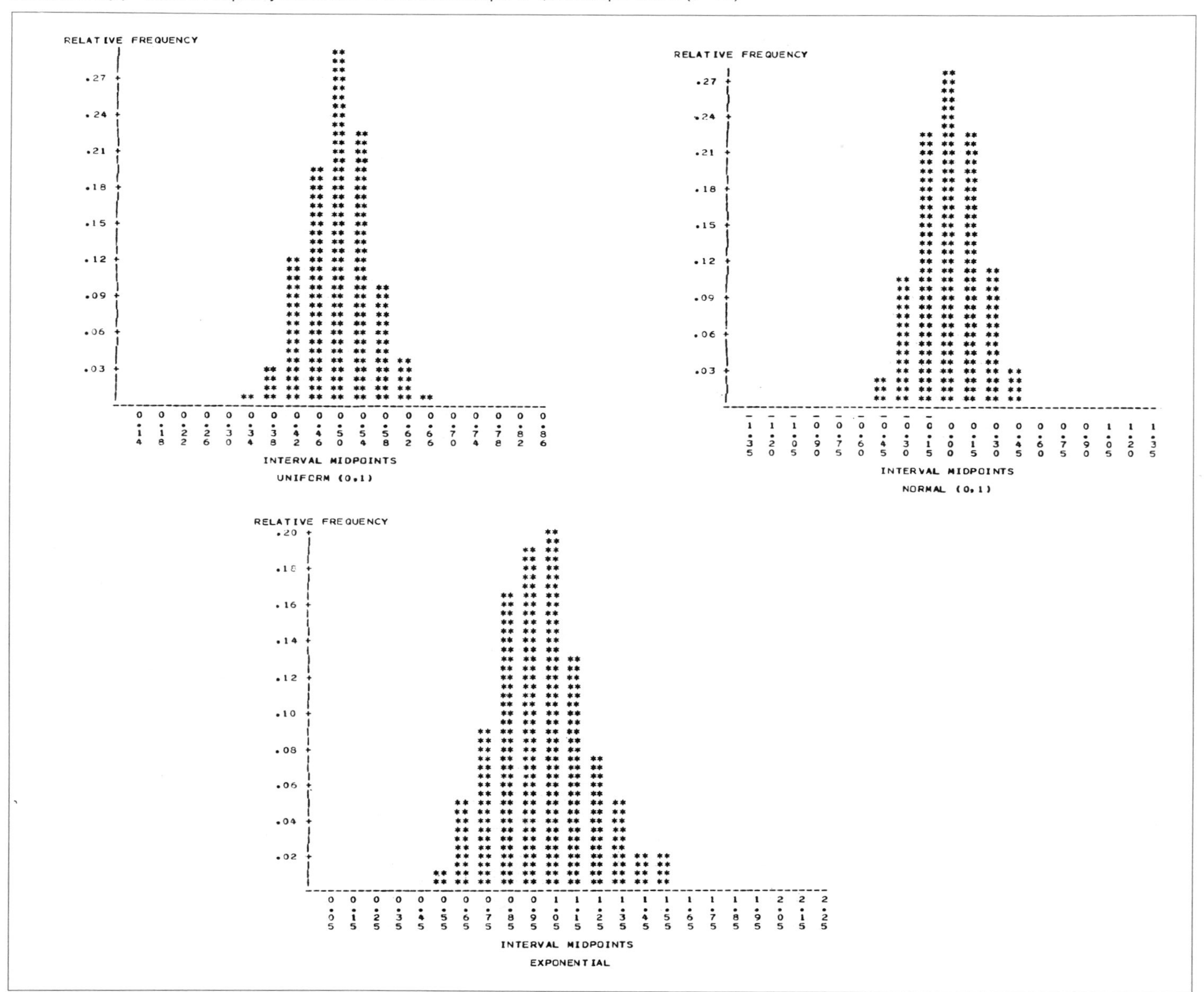

FIGURE 7.6(f) Relative Frequency Distribution of a Random Sample of 1,000 Sample Means ($n = 100$)

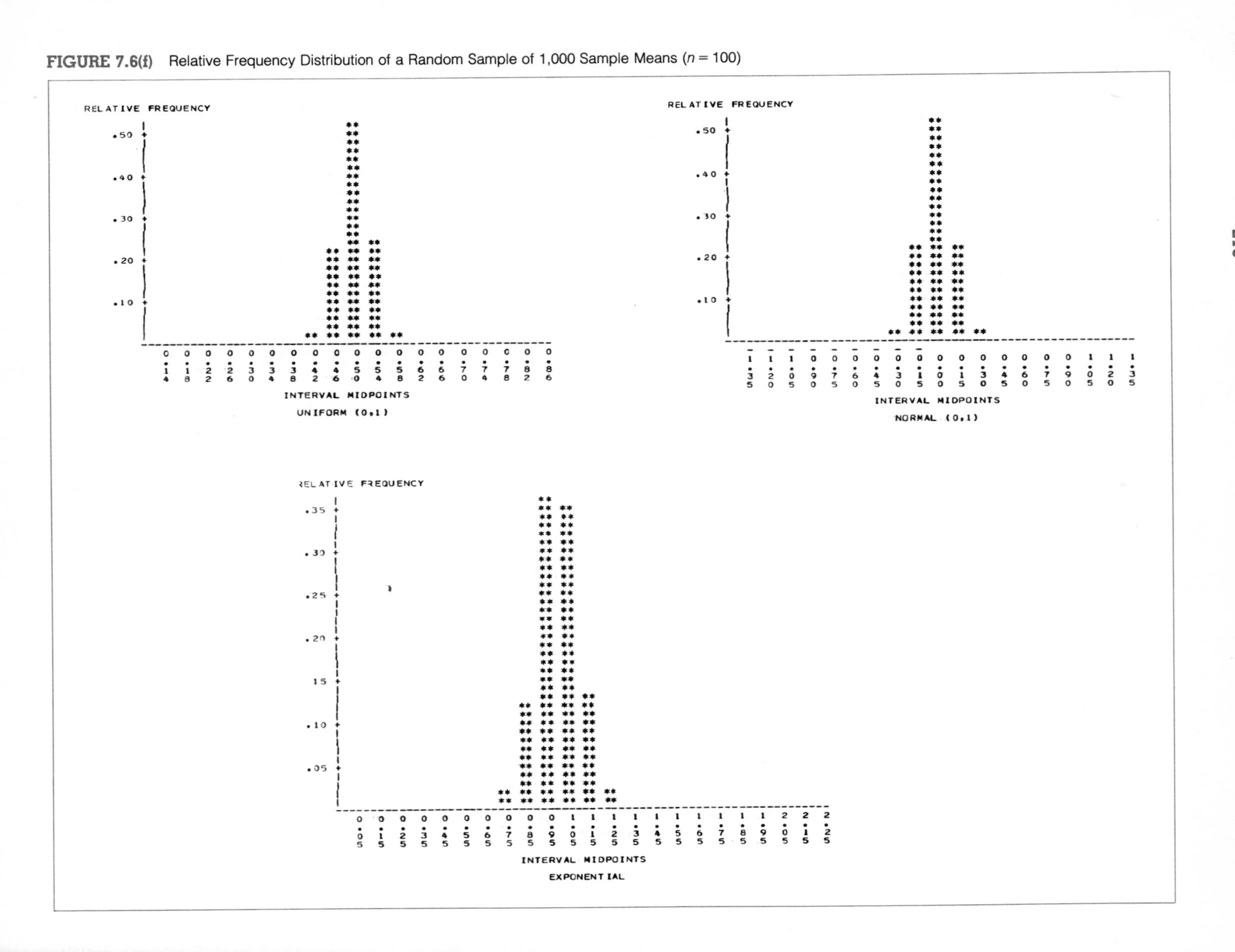

capita) expenditure, based on 100 observations, is approximately normally distributed. In addition, the mean, $\mu_{\bar{x}}$, and the standard deviation, $\sigma_{\bar{x}}$, of the sampling distribution are given by

$$\mu_{\bar{x}} = \mu = \$3.28$$

and

$$\sigma_{\bar{x}} = \frac{\sigma}{\sqrt{n}} = \frac{1.12}{\sqrt{100}} = .112$$

assuming that the consumer magazine's reported values of μ and σ are correct.

b. If the consumer magazine's claim is true, then $P(\bar{x} \geq 3.60)$, the probability of observing a mean weekly (per capita) expenditure of \$3.60 or more in the sample of 100 observations, is equal to the shaded area shown in Figure 7.7.

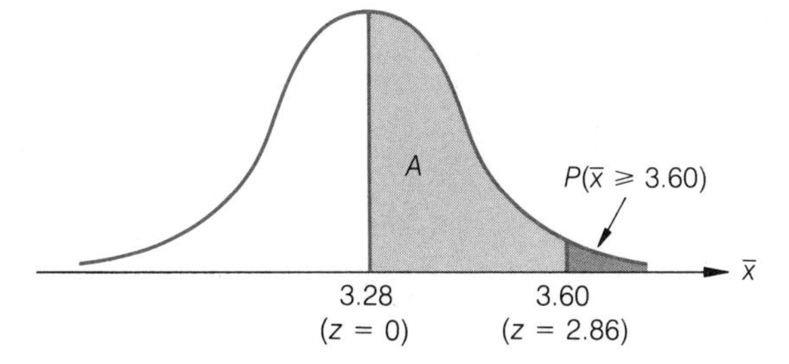

FIGURE 7.7
Sampling Distribution of $\bar{x}$ in Example 7.10

Since the sampling distribution is approximately normal, with mean and standard deviation as obtained in part a, we can compute the desired area by obtaining the z-score for $\bar{x} = 3.60$:

$$z = \frac{\bar{x} - \mu_{\bar{x}}}{\sigma_{\bar{x}}} = \frac{3.60 - 3.28}{.112} = 2.86$$

Thus, $P(\bar{x} \geq 3.60) = P(z \geq 2.86)$, and this probability (area) may be found using Table 3, Appendix E, and the methods of Chapter 6:

$$\begin{aligned} P(\bar{x} \geq 3.60) &= P(z \geq 2.86) \\ &= .5 - A \quad \text{(see Figure 7.7)} \\ &= .5 - .4979 \\ &= .0021 \end{aligned}$$

The probability that we would obtain a sample mean weekly (per capita) expenditure for food purchased away from home of \$3.60 or greater is only .0021, if the consumer magazine's claim is valid. If the 100 randomly selected families spend (per capita) an average of \$3.60 or more on food while away from home during the week, the economist would have strong evidence that the consumer magazine's claim is false, because such a large sample mean is very unlikely to occur if the claim is true.

In practical terms, the Central Limit Theorem and the two properties of the sampling distribution of $\bar{x}$ assure us that the sample mean $\bar{x}$ is a reasonable statistic to use in making inferences about the population mean μ, and they allow us to compute a measure of the reliability of inferences made about μ. (This topic will be treated more thoroughly in Chapter 8.)

As was noted earlier, we will not be required to obtain sampling distributions by simulation or by mathematical arguments. Rather, for all the statistics to be used in this course, the sampling distribution and its properties (which are a matter of record) will be presented as the need arises.

EXERCISES

7.14 Suppose we select a random sample of 40 wheat farmers in the midwest and record the amount, x, that it costs (per bushel) to produce their crop. Prior experience has shown that, in the midwest, the relative frequency distribution for the production cost per bushel of wheat has a mean of $\mu = \$3.25$ and a standard deviation of $\sigma = \$0.45$.

a. Describe the sampling distribution of $\bar{x}$, the mean production cost per bushel for a sample of 40 wheat farmers.
b. What is the probability that the mean production cost per bushel for the farmers sampled will be less than $3.00?
c. What is the probability that the mean production cost per bushel for the farmers sampled will be between $3.30 and $3.45?

7.15 The manufacturer of a new instant-picture camera claims that its product has "the world's fastest-developing color film by far." Extensive laboratory testing has shown that the relative frequency distribution for the time it takes the new instant camera to begin to reveal the image after shooting has a mean of 9.8 seconds and a standard deviation of 0.55 second. Suppose that 50 of these cameras are randomly selected from the production line and tested. The time until the image is first revealed, x, is recorded for each.

a. Describe the sampling distribution of $\bar{x}$, the mean time it takes the sample of 50 cameras to begin to reveal the image.
b. Find the probability that the mean time until the image is first revealed for the 50 sampled cameras is greater than 9.70 seconds.
c. If the mean and standard deviation of the population relative frequency distribution for the times until the cameras begin to reveal the image are correct, would you expect to observe a value of $\bar{x}$ below 9.55 seconds? Explain.

7.16 Refer to Exercise 7.15. Describe the changes in the sampling distribution of $\bar{x}$ if the sample size were:

a. Decreased from $n = 50$ to $n = 20$
b. Increased from $n = 50$ to $n = 100$

7.17 For weeks before the 1980 presidential election, public opinion pollsters were nearly unanimous in agreeing that the race between the then-incumbent President Carter and challenger Ronald Reagan was "too close to call." Reagan, of course, won comfortably. One possible explanation for the inaccurate projections is the following: Most of the private firms which conduct public opinion polls did not survey intensively right up until election time due to the high cost of interviewing voters.

Suppose that data from major national surveys (surveys which contact at least 1,500 people) for the past five years indicate that the relative frequency distribution of the cost of conducting a public opinion poll has a mean of $22,500 and a standard deviation of $6,000.

a. The financial records for thirty major national surveys conducted during the past five years are randomly selected and $\bar{x}$, the sample mean cost of the polls, is computed. Describe the sampling distribution of $\bar{x}$.

b. What is the probability that $\bar{x}$, the mean cost of the thirty surveys, will be between $20,000 and $25,000?

7.18 Refer to Exercise 7.17. Suppose that the financial records for 60 major national surveys conducted during the past five years are randomly selected, and $\bar{x}$ computed.

a. How will $P(20{,}000 < \bar{x} < 25{,}000)$ for $n = 60$ compare to the probability you computed in Exercise 7.17(b)? Do not compute the probability—base your answer on your mental images of the sampling distributions of $\bar{x}$ for $n = 30$ and $n = 60$.

b. Compute $P(20{,}000 < \bar{x} < 25{,}000)$ when $n = 60$. Was your answer to part a correct?

7.19 Research conducted by a tobacco company indicates that the relative frequency distribution of the tar content of their newly developed low-tar cigarette has a mean, μ, equal to 3.9 milligrams of tar per cigarette and a standard deviation, σ, equal to 1.0 milligram. Suppose that a sample of 100 low-tar cigarettes is randomly selected from a day's production and the tar content measured in each.

a. Find the probability that the mean tar content of the sample is greater than 4.15 milligrams, i.e., compute $P(\bar{x} > 4.15)$. Assume that the tobacco company's claim is true.

b. Suppose that the sample mean is computed to be $\bar{x} = 4.18$ milligrams. Based upon the probability of part a, do you think the tobacco company may have understated the true value of μ?

c. Refer to part b. If the tobacco company's figures are correct, give a plausible explanation for the observed value of $\bar{x} = 4.18$ milligrams.

7.20 The length of time between arrivals at a hospital clinic and the length of clinical service time are two random variables that play important roles in designing a clinic and deciding how many physicians and nurses are needed for its operation. Suppose the relative frequency distribution of the interarrival time (the time between the arrivals of two consecutive patients) at a clinic has a mean of 4.1 minutes and a standard deviation of 3.7 minutes.

a. A sample of 20 interarrival times is randomly selected and $\bar{x}$, the sample mean time between arrivals, is computed. Describe the sampling distribution of $\bar{x}$.

b. What is the probability that the mean interarrival time of the sample will be less than 2 minutes?

c. What is the probability that the mean interarrival time of the sample will exceed 6.5 minutes?

d. Would you expect $\bar{x}$ to exceed 6.5 minutes? Explain.

7.21 Water availability is of prime importance in the life cycle of most reptiles. To determine the rate of evaporative water loss of a certain species of lizard at a particular desert site, 34 such lizards were randomly collected, weighed, and placed under the appropriate experimental conditions. After 24 hours, each lizard was removed, reweighed, and its total water loss was calculated (Water loss = Initial body weight − Body weight after treatment). Previous studies have shown that the relative frequency distribution of water loss for the lizards has a mean of 3.1 grams and a standard deviation of .8 gram.

a. Compute the probability that the 34 lizards will have a mean water loss of less than 2.7 grams.

b. Suppose the sample mean water loss for the 34 lizards is computed to be 2.58 grams. Based upon the probability of part a, do you believe that the mean and standard deviation of the relative frequency distribution of water loss for this species of lizard may have changed since the previous studies? Explain.

7.5 Summary

The objective of most statistical investigations is to make an inference about a population parameter. Since we often base inferences upon information contained in a sample from the target population, it is essential that the sample be properly selected. A procedure for obtaining a random sample, using a table of random numbers, was described in this chapter.

After the sample has been selected, we compute a statistic which contains information about the target parameter. The sampling distribution of the statistic characterizes the relative frequency distribution of values of the statistic over an infinitely large number of samples.

The Central Limit Theorem provides information about the sampling distribution of the sample mean, $\bar{x}$. In particular, the sampling distribution will be approximately normal if the sample size is sufficiently large.

KEY WORDS

Population	Parameter
Sample	Statistic
Biased	Sampling distribution
Random sample	Computer simulation
Table of random numbers	Central Limit Theorem

SUPPLEMENTARY EXERCISES

7.22 The National Football League (NFL) is contemplating an expansion of the current 26-team league to 28 teams by 1985. Four cities are being seriously considered for expansion franchises: Phoenix, Arizona; Montreal, Quebec; Jacksonville, Florida; and Memphis, Tennessee.

a. How many different samples of two cities can be selected from among the four?
b. List the different samples.
c. Using Table 9, Appendix E, randomly select the two cities which will be awarded NFL expansion franchises.
d. Under what conditions should the NFL conduct a selection procedure similar to the one in part c?

7.23 The owners of a chain of coin-operated beverage machines will contract with three wholesalers to supply the 8-ounce paper cups used in the machines. A total of six wholesalers of paper cups are available for selection.

a. How many different samples of three wholesalers can be chosen from among the six?
b. List the different samples.
c. Using Table 9, Appendix E, randomly select the three wholesalers who will be contracted to supply the beverage machines with 8-ounce paper cups.

7.24 Before each year's Christmas shopping rush, a New York City department store sends Amount Due notices to each of its 10,000 customers who possess charge accounts. Last year's records revealed that the relative frequency distribution of the amount owed by the 10,000 customers had a mean of $170 and a standard deviation of $90. Suppose that a clerk at the store randomly selected 36 customers' charge accounts during the pre-Christmas period and recorded the amount owed on each account.

a. Describe the sampling distribution of $\bar{x}$, the mean amount owed for the 36 charge accounts sampled.
b. Find the probability that the mean amount owed for the 36 charge accounts falls between $190 and $200.
c. Find the probability that the mean amount owed for the 36 charge accounts is less than $160.

7.25 The administrator of a hospital for mentally ill patients reports that the mean score on a test designed to measure rehabilitation potential is 90, and the standard deviation is 4, for patients who have received treatment at the hospital. A clinical psychologist randomly samples 60 recent patient records and notes the score on the rehabilitation test for each patient. Of special interest is the sample mean of the 60 test scores.

a. Assuming the administrator's claim is true, describe the sampling distribution of the mean test score for the sample of 60 patients.
b. Assuming the administrator's claim is true, what is the probability that the sample mean test score will be 88 or less?

c. Suppose the clinical psychologist computes a sample mean test score of 85. Is this sufficient evidence to contradict the administrator's claim? Explain. [*Hint:* Use your answer to part b.]

7.26 Prior to the occurrence of a minor forest fire, the number of trees per square mile in a certain national forest had a relative frequency distribution with a mean of 7,500 trees and a standard deviation of 80 trees. In an effort to determine the extent of the fire's destruction, a forest ranger "cruised" (i.e., counted) the number of trees in 4 randomly selected square mile sectors and calculated the mean $\bar{x}$ of these 4 measurements.

a. If the relative frequency distribution of the number of trees per square mile in the forest after the fire is the same as before the fire, describe the sampling distribution of $\bar{x}$. Do you think that $\bar{x}$ will have a sampling distribution that is approximately normal? Explain.
b. Answer part a assuming the forest ranger "cruised" 100 randomly selected square mile sectors and calculated $\bar{x}$, the mean number of trees per square mile.

7.27 Every ten years the U.S. population census provides essential information about our nation and its people. The basic constitutional purpose of the census is to apportion the membership of the House of Representatives among the states. However, the census has many other important uses. For example, private business uses the census for plant location and marketing.

The 1980 census included questions on age, sex, race, marital status, family relationship, and income; this census was mailed to every household in the U.S. In some cities, however, a series of questions was added for a 5% sample of the city's households. That is, each of a random sample of the city's households was mailed a census form which included additional questions. Suppose that the city contained 100,000 households and, of these, 5,000 were selected and mailed the longer census form.

a. If you worked for the Bureau of the Census and were assigned the task of selecting a random sample of 5,000 of the city's households, describe how you would proceed. [*Hint:* Utilize Table 9, Appendix E.]
b. Suppose that one of the additional questions on the long form of the census concerned energy consumption. The city used this sample information to project the average energy consumption for the city's 100,000 households. Explain why it is important that the sample of 5,000 households be random.

7.28 Five hundred applicants are vying for three equivalent positions at a steel factory. The company has been able to narrow the field to seven equally qualified applicants, three of whom are minority candidates.

a. How many different samples of three applicants can be selected from the seven?
b. List the different samples.

c. If the three applicants who are chosen are selected at random from this final group of seven, would you expect them to be the three minority candidates?
d. Use Table 9, Appendix E, to randomly select the three applicants who will be awarded positions at the steel factory. How many minority candidates are included in your selection?

7.29 Suppose that the amount of heating oil used annually by households in a particular state has a relative frequency distribution with a mean of 200 gallons and a standard deviation of 40 gallons. The monthly utility bills for a random sample of 100 households in the state are examined and the amount of heating oil used annually by each is recorded.

a. Describe the sampling distribution of $\bar{x}$, the mean amount of heating oil used annually by the 100 randomly selected households.
b. Compute $P(\bar{x} < 196)$.
c. Compute $P(\bar{x} > 190)$.
d. Compute the probability that $\bar{x}$ falls within 1 gallon of the true mean of 200, i.e., compute $P(199 < \bar{x} < 201)$.

7.30 As part of a company's quality control program, it is common practice to monitor the quality characteristics of a product. For example, the amount of alkali in soap might be monitored by randomly selecting from the production process and analyzing $n = 30$ test quantities twice each day. If the mean, $\bar{x}$, of the sample falls within specified control limits, the process is deemed to be in control. If $\bar{x}$ is outside the limits, the monitor flashes a warning signal and suggests that something is wrong with the process. Suppose that the upper and lower control limits are located, respectively, $3\sigma_{\bar{x}}$ above and below μ, the true mean amount of alkali in the soap.

a. For the soap process, experience has shown that $\mu = 2\%$ and $\sigma = 1\%$. Specify the upper and lower control limits for the process. [*Hint:* Calculate $\mu - 3\sigma_{\bar{x}}$ and $\mu + 3\sigma_{\bar{x}}$.]
b. If the process is in control, what is the probability that $\bar{x}$ falls outside the control limits? Use the fact that the probability that $\bar{x}$ falls outside the control limits is given by

$$1 - P(\text{the process is in control}) = 1 - P(\mu - 3\sigma_{\bar{x}} < \bar{x} < \mu + 3\sigma_{\bar{x}})$$

7.31 A telephone company has determined that during non-holidays the number of phone calls which pass through the main branch office each hour has a relative frequency distribution with a mean, μ, of 80,000 and a standard deviation, σ, of 35,000.

a. Describe the shape of the sampling distribution of $\bar{x}$, the mean number of incoming phone calls per hour for a random sample of 60 non-holiday hours.
b. What is the mean of the sampling distribution, part a?
c. What is the standard deviation of the sampling distribution, part a?
d. Find the probability that $\bar{x}$, the mean number of incoming phone calls per hour for a random sample of 60 non-holiday hours, will be larger than 91,970.

e. Suppose that the telephone company wishes to determine whether the true mean number of incoming calls per hour during holidays is the same as for non-holidays. To accomplish this, the company randomly selects 60 hours during a holiday period, monitors the incoming phone calls each hour, and computes $\bar{x}$, the sample mean number of incoming phone calls. If the sample mean is computed to be $\bar{x} = 91{,}970$ calls per hour, do you believe that the true mean for holidays is $\mu = 80{,}000$ (the same as for non-holidays)? Assume that σ for holidays is 35,000 calls per hour. [*Hint:* If in fact $\mu = 80{,}000$ for holidays, the sampling distribution of $\bar{x}$ for holidays is identical to the sampling distribution, parts a, b, and c. Thus, the probability, part d, may be used to infer whether $\bar{x} = 91{,}970$ is a rare event.]

7.32 This past year, an elementary school began using a new method to teach arithmetic to first graders. A standardized test, administered at the end of the year, was used to measure the effectiveness of the new method. The relative frequency distribution of test scores in past years (before implementation of the new teaching method) had a mean of 75 and a standard deviation of 10. Consider the standardized test scores for a random sample of 36 first graders taught by the new method.

a. If the relative frequency distribution of test scores for first graders taught by the new method is no different from that of the old method, describe the sampling distribution of $\bar{x}$, the mean test score for the 36 sampled first graders.

b. If the sample mean test score was computed to be $\bar{x} = 79$, what would you conclude about the effectiveness of the new method of teaching arithmetic? Explain. [*Hint:* Calculate $P(\bar{x} > 79)$ using the sampling distribution described in part a].

7.33 Electric power plants that use water for cooling their condensers sometimes discharge heated water into rivers, lakes, or oceans. It is known that water heated above certain temperatures has a detrimental effect on plant and animal life in the water. Suppose it is known that the increased temperature of the heated water discharged by a certain power plant on any given day has a relative frequency distribution with a mean of 5.2°C and a standard deviation of 3.6°C.

a. An ecologist investigating the plant and animal life in the area recorded the increased temperature of the heated water discharged by the power plant for a total of 25 randomly selected days. What is the probability that $\bar{x}$, the mean increase in water temperature for the sample of 25 days, will fall between 5.0°C and 6.5°C?

b. What is the probability that $\bar{x}$, the mean increase in water temperature for the sample of 25 days, will exceed 6.0°C?

c. Within what limits should the ecologist expect $\bar{x}$ to fall?

7.34 The Chamber of Commerce of a certain city publishes a brochure which contains housing information for new-home buyers. The brochure states that the average cost of new homes built within the city's limits is $72,750. Believing this figure is too low, a local real estate appraiser will obtain the records of sale for 49 randomly selected new homes built in the city and compute $\bar{x}$, the mean sale price of the sample.

a. Suggest a method which will produce a random sample of 49 new homes built within the city limits.
b. Assuming that the relative frequency distribution of all new-home costs in the city has a standard deviation of $44,030, find the probability that the mean cost of a random sample of 49 new homes is larger than $91,650 (i.e., compute $P(\bar{x} > 91{,}650)$).
c. Suppose that the appraiser computes $\bar{x} = \$91{,}650$ for the sampled homes. What would you infer about the figure quoted in the Chamber of Commerce's brochure? Does it appear that the true mean cost of new homes built within the city's limits is greater than $72,750? [*Hint:* Use your answer to part b.]

7.35 Let $\bar{x}_{25}$ represent the mean of a random sample of size 25 obtained from a population with mean $\mu = 17$ and standard deviation $\sigma = 10$. Similarly, let $\bar{x}_{100}$ represent the mean of a random sample of size 100 selected from the same population.
a. Describe the sampling distribution of $\bar{x}_{25}$.
b. Describe the sampling distribution of $\bar{x}_{100}$.
c. Which of the probabilities, $P(15 < \bar{x}_{25} < 19)$ or $P(15 < \bar{x}_{100} < 19)$, would you expect to be the larger?
d. Calculate the two probabilities, part c. Was your answer to part c correct?

7.36 One of the monitoring methods the Environmental Protection Agency (EPA) uses to determine whether sewage treatment plants are conforming to standards is to take thirty-six 1-liter specimens from the plant's discharge during the period of investigation. Chemical methods are applied to determine the percentage of sewage in each specimen. If the sample data provide evidence to indicate that the true mean percentage of sewage exceeds a limit set by the EPA, the treatment plant must undergo mandatory repair and retooling. At one particular plant, the mean sewage discharge limit has been set at 15%. This plant is suspected of being in violation of the EPA standard.
a. Unknown to the EPA, the relative frequency distribution of sewage percentages in 1-liter specimens at the plant in question has a mean, μ, of 15.7% and a standard deviation, σ, of 2.0%. Thus, the plant is in violation of the EPA standard. What is the probability that the EPA will obtain a sample of thirty-six 1-liter specimens with a mean less than 15% even though the plant is violating the sewage discharge limit?
b. Suppose that the EPA computes $\bar{x} = 14.95\%$. Does this result lead you to believe that the sample of thirty-six 1-liter specimens obtained by the EPA was not random, but biased in favor of the sewage treatment plant? Explain. [*Hint:* Use your answer to part a.]

7.37 This year a large insurance firm began a program of compensating its salespeople for sick days not used. The firm decided to pay each salesperson a bonus for every unused sick day. In previous years, the number of sick days used per salesperson per year had a relative frequency distribution with a mean of 9.2 and a standard deviation of 1.8. To determine whether the compensation program has effectively reduced the mean number of sick days used, the firm randomly sampled

81 salespeople and recorded the number of sick days used by each at the year's end.

a. Assuming that the compensation program was not effective in reducing the average number of sick days used, find the probability that the 81 randomly selected salespeople produce a sample mean less than 8.76 days. [*Hint:* If the compensation program was not effective, then the mean and standard deviation of the relative frequency distribution of number of sick days used per salesperson this year is the same as in previous years, i.e., $\mu = 9.2$ and $\sigma = 1.8$.]

b. If the sample mean is computed to be $\bar{x} = 8.76$ days, is there sufficient evidence to conclude that the compensation program was effective, i.e., that the true mean number of sick days used per salesperson this year is less than 9.2, the mean for previous years?

7.38 A large freight elevator can transport a maximum of 9,800 pounds (4.9 tons). Suppose that a load of cargo containing 49 boxes needs to be transported via the elevator. Experience has shown that the weights of boxes for this type of cargo have a relative frequency distribution with $\mu = 205$ pounds and $\sigma = 14$ pounds. Given this information, what is the probability that all 49 boxes can be safely loaded onto the freight elevator and transported? [*Hint:* In order for all 49 boxes to be safely loaded onto the freight elevator, the total weight of the 49 boxes must not exceed the maximum of 9,800 pounds. This implies that $\bar{x}$, the average weight of the 49 boxes, must not exceed $9{,}800 \div 49 = 200$ pounds. Thus, the desired probability can be found by computing $P(\bar{x} < 200)$.]

7.39 Suppose you are in charge of student ticket sales for a major college football team. From past experience, you know that the number of tickets purchased by a student standing in line at the ticket window has a relative frequency distribution with a mean of 2.4 and a standard deviation of 2.0. For today's game, there are 100 eager students standing in line to purchase tickets. If only 250 tickets remain, what is the probability that all 100 students will be able to purchase the tickets they desire? [*Hint:* The solution is similar to that of Exercise 7.38. In order for all 100 students to get the tickets they desire, the total number of tickets purchased by the 100 students must not exceed 250. This implies that $\bar{x}$, the average number of tickets bought by students, must not exceed $250 \div 100 = 2.5$.]

CASE STUDY 7.1 Pollsters Blast ABC-TV's Survey

The ABC-TV poll of television viewers following the Carter–Reagan presidential debate produced cries of protest from professional pollsters. The cause of the pollsters' concern is described in the following article (Gainesville *Sun,* October 31, 1980):

> Some leading pollsters, recalling damage done to their industry by a Literary Digest survey that predicted Franklin D. Roosevelt would lose in 1936, are denouncing an ABC telephone survey taken after this week's presidential debate.
>
> "No credence at all should be given to the figures," said Dr. George Gallup Sr., whose surveys are among the nation's best known samples of public opinion.

In the ABC survey, nearly three-quarters of a million "votes" were cast by telephone Tuesday night during a 100 minute-period following the nationally televised debate between President Carter and Republican challenger Ronald Reagan.

Callers were given two numbers—one for Carter, the other for Reagan—to automatically register their opinion on who gained the most in the debate. Some 477,815 callers dialed the Reagan number; 243,554 dialed for Carter.

While he discounted the survey's likely impact on next Tuesday's election, White House spokesman Jody Powell called it "sort of a shame" that an unscientific survey with "no credibility" got a lot of attention.

Gallup, commenting Wednesday, derided the method as discredited and having "all the faults of the Literary Digest procedures of 1936, in which postcard ballots were sent to people who had car registrations and were listed in the telephone books, which biased the sample toward people with higher income levels."

That survey, a Waterloo for the soon-to-be defunct magazine and the darkest hour of the then-young polling industry, predicted Republican Alf Landon would whip Roosevelt.

Landon took Maine and Vermont and their eight electoral votes. The rest, 46 states and 523 electoral votes, went to Roosevelt in the most one-sided election of modern times.

ABC stressed that its survey was not scientific or statistically valid. Roone Arledge, the network's president for news, defended it, however, as providing "an early indication of what people generally thought about who came off better."

"I see it as comparable to reporting early election returns on election day. They don't prove anything, but there is an interest in them," said Arledge. He added that ABC was doing a separate scientific poll of reaction to the debate.

Albert Cantril, president of the National Council for Public Opinion Research, an association of 16 major pollers, said that despite ABC's disclaimers of the survey's reliability, "Getting 700,000 responses conveys a false impression of reliability, simply by the numbers."

He said that ABC's vigorous disclaimers "raises the question of why they did it at all."

ABC's scientific polling is done by Louis Harris Associates, which said it had nothing to do with the telephone survey.

Explain why the sample of opinions collected by the ABC-TV survey could have produced a biased forecast of the debate "winner."

CASE STUDY 7.2 The Role of Patient Waiting Time

F. A. Sloan and J. H. Lorant, in their article, "The Role of Patient Waiting Time: Evidence from Physicians' Practices" (*Journal of Business*, October 1977), report on the advantages and disadvantages of the time patients spend waiting in physicians' offices before they receive health care services. They write:

> Past studies by economists have considered waiting time to be fully unproductive in the provision of a particular service; their conceptual discussions have emphasized the dead-weight loss associated with a queue [i.e., a line waiting for services]. By contrast, operations researchers, especially in health care applications, have assessed productive aspects of waiting. The latter type of study is based on the premise that increasing patient waiting time is likely to reduce idle time of doctors and their staffs. The queue in the office serves at least three roles. First, with patients waiting, the pace of the

physicians' practice is less likely to be disturbed by late patient arrivals. If a patient arrives late for his appointment, the physician can draw from the queue of waiting patients. Second, given unanticipated variability in visit lengths (and visit complexity), the physician may use waiting patients to fill up unexpected idle moments in his schedule and that of staff when other patients are receiving X rays and the like. Finally, even if all patients were punctual and there were no variability in visit lengths, the patient may use waiting time to complete forms and/or undress prior to the medical examination. Were patient waiting to be reduced to an absolute minimum, the physician and/or his staff might have to wait for these tasks to be completed.

By maintaining queues, say Sloan and Lorant, patient demand for the physicians' services is reduced due to the higher patient (opportunity) time price. (For example, some patients will be reluctant to give up their opportunity to earn income during this possibly long waiting-time period.) It is up to physicians, then, to "determine the optimal mean wait in their practices by balancing the efficiency of their operations against patient demand considerations."

As part of a study to determine the relationship between the time patients wait in the physician's office and certain demand and cost factors, Sloan and Lorant obtained data on the typical patient waiting times for 4,500 physicians in the five largest specialties—general practice, general surgery, internal medicine, obstetrics/gynecology, and pediatrics. They reported a mean waiting time of 24.7 minutes and a standard deviation of 19.3 minutes. Sloan and Lorant note that the 4,500 observations in the data set represent estimates of waiting time based on recall by the physician. They warn that these "estimates may be biased downward because physicians may tend to underestimate the amount of time their patients wait."

For the purposes of this case study, let us assume that the 4,500 observations in the data set represent the *actual* waiting times of all patients who visited a particular pediatrician last year. Suppose also that the relative frequency distribution of the 4,500 patient waiting times has a mean $\mu = 24.7$ minutes and a standard deviation $\sigma = 19.3$ minutes, and that these figures are unavailable to the pediatrician.

In order to determine whether the "optimal" mean waiting time for his practice has been attained, the pediatrician has one of his staff monitor the waiting times for 100 randomly selected patients during the year. The sample average waiting time, $\bar{x}$, is computed. Applying the Central Limit Theorem, we know that the sampling distribution of $\bar{x}$ has a shape which is approximately normal. Also, the mean of the sampling distribution, $\mu_{\bar{x}}$, is

$$\mu_{\bar{x}} = \mu = 24.7 \text{ minutes}$$

and the standard deviation of the sampling distribution, $\sigma_{\bar{x}}$, is

$$\sigma_{\bar{x}} = \frac{\sigma}{\sqrt{n}} = \frac{19.3}{\sqrt{100}} = \frac{19.3}{10} = 1.93 \text{ minutes}$$

From experience, the pediatrician has learned that maximum operational efficiency for his practice is attained when the mean patient waiting time is $\mu = 22$ minutes. We, of course, having access to the records of all 4,500 patients, know that

$\mu = 24.7$ and that, if the pediatrician is correct, his practice is not operating at maximum efficiency. But since this fact is unknown to the pediatrician, he will utilize the following decision-making procedure: If the sample mean waiting time, $\bar{x}$, of the randomly selected patients falls between 19 minutes and 25 minutes, the pediatrician will assume that μ is approximately 22 minutes and that his practice is operating close to maximum efficiency. If $\bar{x}$ falls outside the interval (19, 25), the current patient appointment policy will be restructured.

a. Use the sampling distribution of $\bar{x}$ to find the probability that $\bar{x}$ falls between 19 minutes and 25 minutes.
b. The probability that you computed in part a is also the probability that the pediatrician will make the wrong decision and assume that the optimal mean waiting time has been nearly attained. If this probability is even as large as .15 to .20, the pediatrician is taking an unnecessary risk. Based on this probability, would you recommend that the pediatrician revise his strategy? (In Chapters 9 and 10 we will discuss a statistically valid decision-making process called a ***test of hypothesis.***)

CASE STUDY 7.3 Estimating the Total Weight of Short-Leaf Pine Trees

Throughout this chapter, we have demonstrated that the sample mean $\bar{x}$ provides a useful estimate of the population mean μ. In this case study, we will show how to use the quantity $\bar{x}$ to obtain an estimate of a population total for a finite population.

Refer to Case Study 1.2 (page 13). Suppose that a prospective investor desires an estimate of the total weight of short-leaf pine trees on the 40-acre tract of land located in western Arkansas. The population of interest to the investor, then, is the collection of weights of all trees on the property, and the target parameter is the sum of these weights, i.e., the total weight of the trees. How can we use the sample information provided by Table 1.1 to estimate the total weight? The first step is to determine the (approximate) number of trees on the entire 40-acre tract. Recall that the weights of the sample of 117 trees were obtained by "cruising" twenty ⅕-acre plots, a total of four acres. Since these four acres represent $\frac{1}{10}$ of the total area of the 40-acre tract of land, it is reasonable to assume that the total number of trees, N, on the entire 40-acre property is approximately 10 times the number of trees on the sampled four acres, i.e., $N \approx 10(117) = 1{,}170$ trees. For the purposes of this case study, we will assume that the total number of trees on the entire tract is in fact 1,170.

Now consider the following relationship between a population total and a population mean μ:

$$\text{Population total} = N\mu$$

where N is the total number of observations in the (finite) population. Since the value of μ is usually unknown, so too is the population total. To estimate a population total, we simply substitute our best estimate of μ, namely $\bar{x}$, into the equation. Thus, an estimate of a population total is given as

$$\text{Estimated population total} = N\bar{x}$$

where $\bar{x}$ is the mean of a sample of n elements selected at random from the population of N elements. Applying this to the case at hand, we obtain an estimate of the total weight of trees on the 40-acre tract of land:

$$\text{Estimated total weight} = N\bar{x} = (1{,}170)\bar{x}$$

where $\bar{x}$ is the mean weight of the sample of $n = 117$ trees selected from the entire population of $N = 1{,}170$ trees.

a. In Case Study 3.3 (page 96), you computed $\bar{x}$, the mean weight of the sample of 117 short-leaf pine trees. Use this value to estimate the total weight of trees on the entire 40-acre tract of land.
b. Since $N\bar{x}$ is a sample statistic, it possesses a sampling distribution. Explain how you could generate (by computer) the sampling distribution of $N\bar{x}$.
c. From the properties of the sampling distribution of $\bar{x}$, it can be shown (proof omitted) that the sampling distribution of $N\bar{x}$ is approximately normal for large n, with a mean of $N\mu$ and a standard deviation of $N\sigma/\sqrt{n}$. Compute an estimate of the standard deviation of the sampling distribution of the estimated total weight of short-leaf pine trees. [*Hint:* Use $Ns/\sqrt{n}$ as an estimate of $N\sigma/\sqrt{n}$, where s is the sample standard deviation that you calculated for Case Study 3.3.]
d. Find the probability that your estimate of the total weight of short-leaf pine trees on the 40-acre tract, $N\bar{x}$, falls within 100,000 pounds of the true total weight. [*Hint:* Compute $P(N\mu - 100{,}000 < N\bar{x} < N\mu + 100{,}000)$, using the properties of the sampling distribution of $N\bar{x}$ given in part c and Table 3, Appendix E.]

REFERENCES

Hogg, R. V., & Craig, A. T. *Introduction to mathematical statistics.* 4th ed. New York: Macmillan, 1978. Chapter 4.

McClave, J. T., & Dietrich, F. H. *Statistics.* 2d ed. San Francisco: Dellen, 1982. Chapter 6.

"Pollsters Blast ABC-TV's Survey." Gainesville *Sun,* October 31, 1980.

Sloan, F. A., & Lorant, J. H. "The role of patient waiting time: Evidence from physicians' practices." *Journal of Business,* October 1977, *50,* 486–507.

Yemane, T. *Statistics: An introductory analysis.* 2d ed. New York: Harper and Row, 1967. Chapter 7.

Eight Estimation of Means and Proportions

Are you a happy-go-lucky person or do you tend to worry a lot? The June 10, 1981, Gallup Youth Survey asked a similar question to a sample of teenagers 13–18 years old. The purpose of the survey was to estimate the true proportion of all American teenagers 13–18 years old who "take life pretty much as it comes," based on the sample data. How many of the millions of American teenagers should be sampled? Do you need a large sample because the population, consisting of the responses of all American teenagers, is so large? How does the sample size affect the accuracy of the estimate? These and other questions will be answered in this chapter. Particularly, you will learn how sample size and other factors affect the behavior of sample statistics and, in Exercise 8.27, you will obtain an estimate of the proportion of happy-go-lucky American teenagers.

Contents

8.1 Introduction

In preceding chapters we learned that populations are characterized by numerical descriptive measures (parameters), and that inferences about parameter values are based on statistics computed from the information in a sample selected from the population of interest. In this chapter, we will demonstrate how to estimate population means or proportions, and how to estimate the difference between two population means or proportions. We will also be able to assess the reliability of our estimates, based on knowledge of the sampling distributions of the statistics being used.

EXAMPLE 8.1 Suppose we are interested in estimating the average starting salary of all bachelor's degree graduates of the University of Florida (during June 1980–March 1981) who indicated they had secured employment on the CRC questionnaire described in Chapter 1. (Recall that the target population consists of the 948 observations on starting salary in Appendix A. Although we already know the value of the population mean, this example will be continued to illustrate the concepts involved in estimation.) How could one estimate the parameter of interest in this situation?

Solution An intuitively appealing estimate of a population mean, μ, is the sample mean, $\bar{x}$, computed from a random sample of n observations from the target population. Assume, for example, that we obtain a random sample of size $n = 30$ from the starting salary measurements in Appendix A, and then compute the sample mean and find its value to be $\bar{x} = \$16,200$. This value of $\bar{x}$ provides a ***point estimate*** of the population mean.

DEFINITION 8.1

A ***point estimate*** of a parameter is a statistic, a single value computed from observations in a sample, that is used to estimate the value of the target parameter.

How reliable is a point estimate for a parameter? In order to be truly practical and meaningful, an inference concerning a parameter (in this case, estimation of the value of μ) must consist not only of a point estimate, but also must be accompanied by a measure of the reliability of the estimate; that is, we need to be able to state how close our estimate is likely to be to the true value of the population parameter. This can be done by using the characteristics of the sampling distribution of the statistic which was used to obtain the point estimate; the procedure will be illustrated in the next section.

8.2 Estimation of a Population Mean: Large-Sample Case

Recall from Section 7.4 that, for sufficiently large sample sizes, the sampling distribution of the sample mean, $\bar{x}$, is approximately normal, as indicated in Figure 8.1.

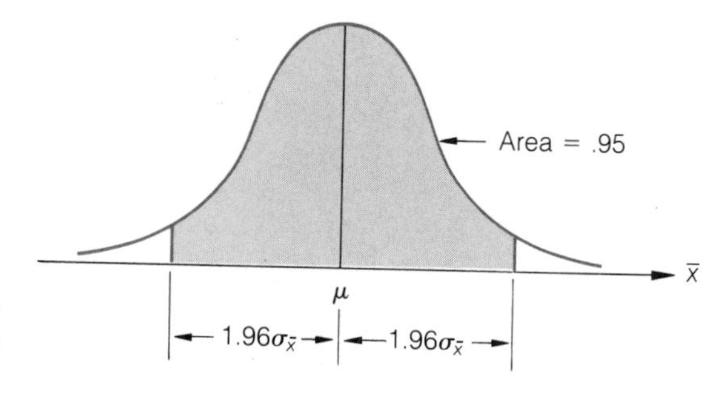

FIGURE 8.1
Sampling Distribution of $\bar{x}$

EXAMPLE 8.2 Suppose we plan to take a sample of $n = 30$ measurements from the population of starting salaries in Appendix A and construct the interval

$$\bar{x} \pm 1.96\sigma_{\bar{x}} = \bar{x} \pm 1.96\left(\frac{\sigma}{\sqrt{n}}\right)$$

where σ is the population standard deviation of the 948 starting salary values. In other words, we will construct an interval 1.96 standard deviations around the sample mean, $\bar{x}$. What can we say about how likely it is that this interval will contain the true value of the population mean, μ?

Solution We arrive at a solution by the following three-step process:

Step 1. First note that the area beneath the sampling distribution of $\bar{x}$ between $\mu - 1.96\sigma_{\bar{x}}$ and $\mu + 1.96\sigma_{\bar{x}}$ is approximately .95. (This area, shaded in Figure 8.1, is obtained from Table 3, Appendix E.) This implies that before the sample of 30 measurements is drawn, the probability that $\bar{x}$ will fall in the interval $\mu \pm 1.96\sigma_{\bar{x}}$ is .95.

Step 2. If in fact the sample yields a value of $\bar{x}$ that falls within the interval $\mu \pm 1.96\sigma_{\bar{x}}$, then it is also true that the interval $\bar{x} \pm 1.96\sigma_{\bar{x}}$ will contain μ. This point is demonstrated in Figure 8.2. For a particular value of $\bar{x}$ (shown with an arrow) which falls within the interval $\mu \pm 1.96\sigma_{\bar{x}}$, a distance of $1.96\sigma_{\bar{x}}$ is drawn both to the left and to the right of $\bar{x}$. You can see that the value of μ must fall between $\bar{x} \pm 1.96\sigma_{\bar{x}}$.

Step 3. Steps 1 and 2 combined imply that, before the sample is drawn, the probability that the interval $\bar{x} \pm 1.96\sigma_{\bar{x}}$ will enclose μ is approximately .95.

The interval $\bar{x} \pm 1.96\sigma_{\bar{x}} = \bar{x} \pm 1.96(\sigma/\sqrt{n})$ is called a large-sample 95% ***confidence interval*** for the population mean μ. The term *large-sample* refers to the

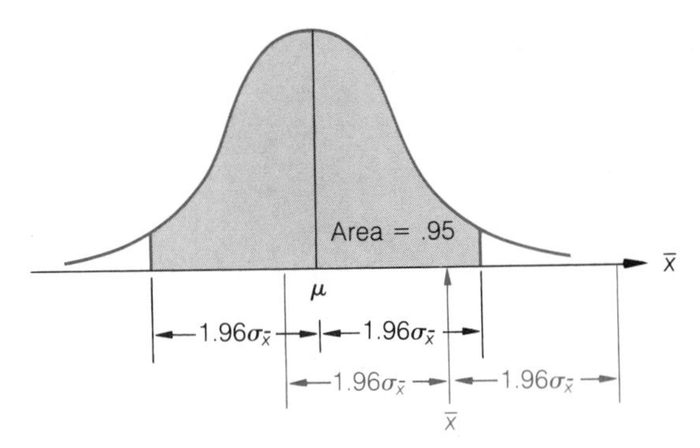

FIGURE 8.2
Sampling Distribution of $\bar{x}$, Example 8.2

sample being of such a size that we can apply the Central Limit Theorem to determine the form of the sampling distribution of $\bar{x}$. Although it is arbitrary, the conventional rule of thumb is that a sample size of $n \geq 30$ is required to employ large-sample confidence interval procedures.

DEFINITION 8.2

A ***confidence interval*** for a parameter is an interval of numbers within which we expect the true value of the population parameter to be contained. The endpoints of the interval are computed based on sample information.

EXAMPLE 8.3 Suppose that a random sample of 30 observations from the population of starting salaries yielded the following sample statistics:

$$\bar{x} = \$16{,}200$$
$$s = \$2{,}850$$

Construct a 95% confidence interval for μ, the population mean starting salary, based on this sample information.

Solution A 95% confidence interval for μ, based on a sample of size $n = 30$, is given by

$$\bar{x} \pm 1.96\sigma_{\bar{x}} = \bar{x} \pm 1.96\left(\frac{\sigma}{\sqrt{n}}\right)$$
$$= 16{,}200 \pm 1.96\left(\frac{\sigma}{\sqrt{30}}\right)$$

In most practical applications, the value of the population standard deviation, σ, will be unknown. However, for large samples ($n \geq 30$), the sample standard deviation, s, provides a good approximation to σ, and may be used in the formula. Then the confidence interval becomes

$$16{,}200 \pm 1.96\frac{2{,}850}{\sqrt{30}} = 16{,}200 \pm 1{,}020$$

or (15,180, 17,220). Hence, we estimate that the population mean starting salary falls within the interval between \$15,180 and \$17,220.

How much confidence do we have that μ, the true population mean starting salary, lies within the interval (\$15,180, \$17,220)? Although we cannot be certain whether the sample interval contains μ (unless we calculate the true value of μ for all 948 observations), we can be reasonably sure that it does. This confidence is based on the interpretation of the confidence interval procedure: If we were to select repeated random samples of size $n = 30$ starting salaries, and form a 1.96 standard deviation interval around $\bar{x}$ for each sample, then approximately 95% of the intervals so constructed would contain μ. Thus, we are 95% confident that the particular interval (\$15,180, \$17,220) contains μ, and this is our measure of the reliability of the point estimate $\bar{x}$.

EXAMPLE 8.4 To illustrate the classical interpretation of a confidence interval, we generated 40 random samples, each of size $n = 30$, from the population of starting salaries in Appendix A. For each sample, the sample mean and standard deviation are presented in Table 8.1. We then constructed the 95% confidence interval for μ, using the information from each sample. Interpret the results, which are shown in Table 8.2.

TABLE 8.1

SAMPLE	MEAN	STANDARD DEVIATION	SAMPLE	MEAN	STANDARD DEVIATION
1	16866.7	6262.50	21	14303.3	5846.57
2	15716.7	4713.30	22	16333.3	5058.85
3	16530.0	4623.27	23	15136.7	4756.88
4	15580.0	4276.19	24	17043.3	5645.24
5	16510.0	4751.43	25	16246.7	4922.69
6	15033.3	6012.02	26	16820.0	4467.69
7	15083.3	5174.35	27	14546.7	5190.76
8	15453.3	5313.79	28	17443.3	6229.28
9	15233.3	5404.43	29	18240.0	4499.24
10	16156.7	4741.36	30	16386.7	5246.39
11	15733.3	6962.03	31	15416.7	5162.07
12	15860.0	3713.87	32	15140.0	3987.36
13	15393.3	4311.89	33	15193.3	5448.44
14	17580.0	4774.53	34	15813.3	5230.73
15	16333.3	6305.02	35	16070.0	4963.95
16	14550.0	4488.05	36	16333.3	4788.37
17	14690.0	4265.20	37	15480.0	5000.79
18	14390.0	4730.19	38	16160.0	4190.35
19	16683.3	5331.37	39	14983.3	5470.11
20	13870.0	4574.68	40	16600.0	6006.09

Solution For the target population of 948 starting salaries, we have previously obtained the population mean value, $\mu = \$15,909$. In the 40 repetitions of the confidence interval procedure described above, note that only 2 of the intervals (those based on

TABLE 8.2

SAMPLE	CORRESPONDING 95% CONFIDENCE INTERVAL FOR μ	SAMPLE	CORRESPONDING 95% CONFIDENCE INTERVAL FOR μ
1	(14625.7, 19107.7)	21	(12211.2, 16395.5)
2	(14030.0, 17403.3)	22	(14523.0, 18143.6)
3	(14875.6, 18184.4)	23	(13434.4, 16838.9)
4	(14049.8, 17110.2)	24	(15023.2, 19063.5)
5	(14809.7, 18210.3)	25	(14485.1, 18008.2)
6	(12882.0, 17184.7)	26	(15221.3, 18418.7)
7	(13231.7, 16934.9)	27	(12689.2, 16404.2)
8	(13551.8, 17354.8)	28	(15214.2, 19672.5)
9	(13299.4, 17167.3)	29	(16630.0, 19850.0)
10	(14460.0, 17853.3)	30	(14509.3, 18264.1)
11	(13242.0, 18224.7)	31	(13569.4, 17263.9)
12	(14531.0, 17189.0)	32	(13713.1, 16566.9)
13	(13850.3, 16936.3)	33	(13243.6, 17143.0)
14	(15871.5, 19288.5)	34	(13941.5, 17685.1)
15	(14077.1, 18589.6)	35	(14293.7, 17846.3)
16	(12944.0, 16156.0)	36	(14619.8, 18046.8)
17	(13163.7, 16216.3)	37	(13690.5, 17269.5)
18	(12697.3, 16082.7)	38	(14660.5, 17659.5)
19	(14775.5, 18591.1)	39	(13025.9, 16940.8)
20	(12233.0, 15507.0)	40	(14450.7, 18749.3)

samples 20 and 29) do not contain the value of μ, whereas the remaining 38 of the 40 intervals (or 95% of the intervals) do contain the true value of μ.

Keep in mind that, in actual practice, you would not know the true value of μ and you would not perform this repeated sampling; rather you would select a single random sample and construct the associated 95% confidence interval. The one confidence interval you form may or may not contain μ, but you can be fairly sure it does because of your confidence in the statistical procedure, the basis for which was illustrated in this example.

Suppose you wish to construct an interval which you believe will contain μ with some degree of confidence other than 95%; that is, you want to choose a confidence coefficient other than .95.

DEFINITION 8.3

The ***confidence coefficient*** is the proportion of times that a confidence interval encloses the true value of the population parameter if the confidence interval procedure is used repeatedly a very large number of times.

The first step in learning how to construct a confidence interval with any desired confidence coefficient is to notice from Figure 8.1 that, for a 95% confidence interval, the confidence coefficient of .95 is equal to the total area under the sampling distribution, less .05 of the area which is divided equally between the two tails of the distribution. Thus, each of the tails has an area of .025. Secondly, consider that the tabulated value of z from Table 3, Appendix E, which cuts off an area of .025 in the right tail of the standard normal distribution is 1.96 (see Figure 8.3). The value $z = 1.96$ is also the distance, in terms of standard deviations, that $\bar{x}$ is from the upper endpoint of the 95% confidence interval. Thus, this z-value provides the key to constructing a confidence interval with any desired confidence coefficient.

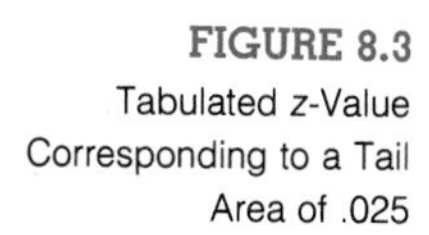
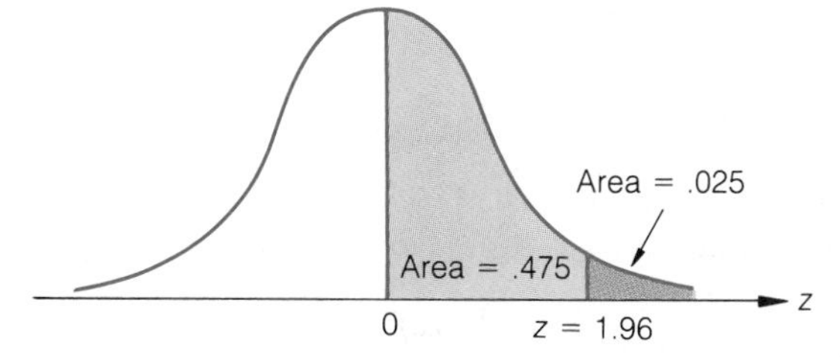

FIGURE 8.3
Tabulated z-Value Corresponding to a Tail Area of .025

By assigning a confidence coefficient other than .95 to the confidence interval, we change the area under the sampling distribution between the endpoints of the interval, which in turn changes the tail area associated with z. In general, we have:

DEFINITION 8.4

$z_{\alpha/2}$ is defined to be the z-value such that an area of $\alpha/2$ lies to its right (see Figure 8.4).

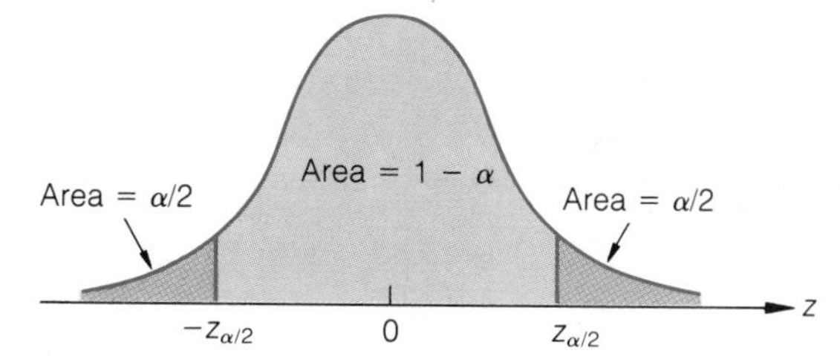

FIGURE 8.4
Locating $z_{\alpha/2}$ on the Standard Normal Curve

Now, if an area of $\alpha/2$ lies beyond $z_{\alpha/2}$ in the right tail of the distribution, then an area of $\alpha/2$ lies to the left of $-z_{\alpha/2}$ in the left tail (Figure 8.4) because of the symmetry of the distribution. The remaining area, $(1 - \alpha)$, is equal to the confidence coefficient;

i.e., the probability that $\bar{x}$ falls within $z_{\alpha/2}$ standard deviations of μ is $(1 - \alpha)$. Thus, a large-sample confidence interval for μ, with confidence coefficient equal to $(1 - \alpha)$, is given by

$$\bar{x} \pm z_{\alpha/2}\sigma_{\bar{x}}$$

EXAMPLE 8.5 In published works which employ confidence interval techniques, a very common confidence coefficient is .90. Determine the value of $z_{\alpha/2}$ which would be used in constructing a 90% confidence interval for a population mean based on large samples.

Solution For a confidence coefficient of .90, we have

$$1 - \alpha = .90$$
$$\alpha = .10$$
$$\alpha/2 = .05$$

and we need to obtain the value $z_{\alpha/2} = z_{.05}$ which locates an area of .05 in the upper tail of the standard normal distribution. Since the total area to the right of 0 is .50, $z_{.05}$ will be the value such that the area between 0 and $z_{.05}$ will be $.50 - .05 = .45$. From the body of Table 3, Appendix E, we find $z_{.05} = 1.645$ (see Figure 8.5). We conclude that a large-sample 90% confidence interval for a population mean is given by

$$\bar{x} \pm 1.645\sigma_{\bar{x}}$$

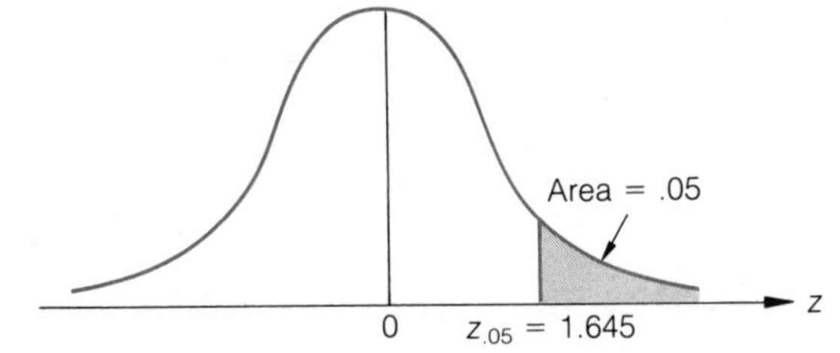

FIGURE 8.5 Location of $z_{\alpha/2}$ for Example 8.5

In Table 8.3, we present the values of $z_{\alpha/2}$ for the most commonly used confidence coefficients.

TABLE 8.3 Commonly Used Confidence Coefficients

CONFIDENCE COEFFICIENT $1 - \alpha$	$\alpha/2$	$z_{\alpha/2}$
.90	.05	1.645
.95	.025	1.96
.98	.01	2.33
.99	.005	2.58

A summary of the large-sample confidence interval procedure for estimating a population mean appears in the accompanying box.

LARGE-SAMPLE $(1 - \alpha)$ 100% CONFIDENCE INTERVAL FOR A POPULATION MEAN, μ

$$\bar{x} \pm z_{\alpha/2}\sigma_{\bar{x}} = \bar{x} \pm z_{\alpha/2}\frac{\sigma}{\sqrt{n}}$$

where $z_{\alpha/2}$ is the z-value which locates an area of $\alpha/2$ to its right, σ is the standard deviation of the population from which the sample was selected, n is the sample size, and $\bar{x}$ is the value of the sample mean.
[*Note:* When the value of σ is unknown (as will usually be the case), the sample standard deviation s may be used to approximate σ in the formula for the confidence interval.]

EXAMPLE 8.6 A fact long known but little understood is that twins, in their early years, tend to have lower IQ's and pick up language more slowly than nontwins. Recently, psychologists have found that the slower intellectual growth of most twins may be caused by benign parental neglect. Suppose we desire to investigate this phenomenon. A random sample of $n = 50$ sets of 2½-year-old twin boys is selected, and the total parental attention time given to each pair during one week is recorded. This experiment produced the following statistics:

$$\bar{x} = 22.4 \text{ hours}$$
$$s = 15.1 \text{ hours}$$

Estimate μ, the mean weekly attention time given to all 2½-year-old twin boys by their parents, using a 99% confidence interval. Interpret the interval in terms of the problem.

Solution The general form of a large-sample 99% confidence interval for μ is

$$\bar{x} \pm 2.58\frac{\sigma}{\sqrt{n}}$$

We will substitute the value of s for σ to obtain

$$22.4 \pm 2.58\frac{15.1}{\sqrt{50}} = 22.4 \pm 5.51$$

or (16.89, 27.91).

We can be 99% confident that the interval (16.89, 27.91) encloses the true mean weekly attention time given to all 2½-year-old twin boys by their parents. Since all of the values fall below 28 hours, we conclude that there is a general tendency for 2½-year-old twin boys to receive less than 4 hours of parental attention time per day, on the average. Further investigation would be required to relate this phenomenon to the intellectual growth of the twins.

EXAMPLE 8.7 Refer to Example 8.6.

a. Using the sample information provided in Example 8.6, construct a 95% confidence interval for the mean weekly attention time given to all 2½-year-old twin boys by their parents.

b. For a fixed sample size, how does the width of the confidence interval relate to the confidence coefficient?

Solution a. The form of a large-sample 95% confidence interval for a population mean μ is

$$\bar{x} \pm 1.96\frac{\sigma}{\sqrt{n}} \approx \bar{x} \pm 1.96\frac{s}{\sqrt{n}}$$

$$= 22.4 \pm 1.96\frac{15.1}{\sqrt{50}}$$

$$= 22.4 \pm 4.19$$

or (18.21, 26.59).

b. The 99% confidence interval for μ was determined in Example 8.6 to be (16.89, 27.91). The 95% confidence interval, obtained in part a and based on the same sample information, is narrower than the 99% confidence interval. This will hold true in general, as noted in the accompanying box.

RELATION OF WIDTH OF CONFIDENCE INTERVAL TO THE CONFIDENCE COEFFICIENT

For a given sample size, the width of the confidence interval for a parameter increases as the confidence coefficient increases. Intuitively, the interval must become wider for us to have greater assurance (confidence) that it contains the true parameter value.

EXAMPLE 8.8 Refer to Example 8.6.

a. Assume that the given values of the statistics $\bar{x}$ and s were based on a sample of size $n = 100$ instead of a sample of size $n = 50$. Construct a 99% confidence interval for μ, the population mean weekly attention time given to 2½-year-old twin boys by their parents.

b. For a fixed confidence coefficient, how does the width of the confidence interval relate to the sample size?

Solution a. Substitution of the values of the sample statistics into the general formula for a 99% confidence interval for μ yields

$$\bar{x} \pm 2.58\frac{\sigma}{\sqrt{n}} \approx 22.4 \pm 2.58\frac{15.1}{\sqrt{100}}$$

$$= 22.4 \pm 3.90$$

or (18.50, 26.30).

b. The 99% confidence interval based on a sample of size $n = 100$, part a, is narrower than the 99% confidence interval constructed from a sample of size $n = 50$, Example 8.6. This will also hold true in general, as noted in the box.

RELATION OF WIDTH OF CONFIDENCE INTERVAL TO THE SAMPLE SIZE

For a fixed confidence coefficient, the width of the confidence interval decreases as the sample size increases. In other words, larger samples generally provide more information about the target population than do smaller samples.

In this section, we have introduced the concepts of point and interval estimation for the population mean, μ, based on large samples. The general theory appropriate for the estimation of μ also carries over to the estimation of other population parameters. Hence, in subsequent sections, we will present only the point estimate, its sampling distribution, the general form of a confidence interval for the parameter of interest, and any assumptions required for the validity of the procedure.

EXERCISES

8.1 In a large-sample confidence interval for a population mean, what does the confidence coefficient represent?

8.2 Use Table 3, Appendix E, to determine the value of $z_{\alpha/2}$ which would be used to construct a large-sample confidence interval for μ, for each of the following confidence coefficients:

a. .85 **b.** .95 **c.** .975

8.3 Give a precise interpretation of the statement, "We are 95% confident that the interval estimate contains μ."

8.4 Insurance companies often require that an applicant for life insurance submit the results of a complete physical examination. One simple method the insurance companies use to assess the cardiovascular fitness of an individual is to request that the examining physician monitor the applicant's heart rate before and after moderate exercise. Of particular interest is the number of premature contractions per minute after hopping on one foot for 30 seconds. Information on 36 randomly selected male applicants aged 40–55 yielded the following summary statistics on premature contractions per minute after 30 seconds of exercise: $\bar{x} = 2.5$, $s = 1.4$. Estimate the mean number of premature contractions per minute for all male life insurance applicants aged 40–55, using a 98% confidence interval.

8.5 Refer to Exercise 8.4. What steps could you take to reduce the width of the confidence interval for the population mean?

8.6 A private elementary school is considering relocating its facility to an outlying residential area. Since the school does not provide bus transportation for its students, school board officials wish to estimate the average distance between the proposed site and the students' homes. A random sample of 64 current students yielded an average distance of 20 blocks and a standard deviation of 5 blocks. Construct a 90% confidence interval for the average distance between the proposed new site and students' homes for all current students. Interpret the interval.

8.7 You are interested in purchasing a new automobile which is designed to run most efficiently when fueled by gasohol. Before you decide whether or not to buy this model, you desire an estimate of the mean highway mileage obtained by the car. Information pertaining to highway mileages for 35 randomly selected gasohol-fueled models tested by the EPA is summarized as follows:

$\bar{x} = 37.1$ miles per gallon

$s = 1.4$ miles per gallon

Construct a 95% confidence interval for the mean highway gasohol mileage obtained by all cars of this model. Interpret the interval.

8.8 By law, a manufacturer of a food product is required to list Food and Drug Administration (FDA) estimates of the contents of the packaged product. Suppose that the FDA wished to estimate the mean sugar content (by weight) in a 16-ounce box of corn flakes. The FDA randomly selected 100 boxes of corn flakes and measured the sugar content in each, with the following results: $\bar{x} = 3.2$ ounces, $s = .5$ ounce.

a. Estimate the true mean sugar content in the 16-ounce boxes of corn flakes cereal with a 90% confidence interval.

b. Interpret the interval, part a.

c. How could the FDA reduce the width of the confidence interval in part a? Are there any drawbacks to reducing the interval width? Explain.

8.9 Refer to the data given in Appendix C. The manager of supermarket B, which uses automated checkers, desires an estimate of the average customer service time at the supermarket. Select a random sample of 30 checkout service times and use this information to construct a 99% confidence interval for the true average customer service time at supermarket B. Interpret your interval.

8.10 As part of a Department of Energy survey, 240 American families were randomly selected and interviewed concerning the amount of money they spent last year on home heating oil or gas. The survey results indicated that the amount spent by the 240 families had an average of \$425 and a standard deviation of \$130.

a. Give a point estimate for the true mean amount spent, per family, last year on home heating oil or gas.

b. Use the sample information to construct a 97% confidence interval for the true mean amount of money, per family, spent last year on home heating oil or gas.

c. What is the confidence coefficient for the interval, part b? Interpret this value.

d. Based upon your interval, part b, would you expect the true mean annual expenditure per family on home heating oil or gas to fall below \$400?

8.3 Estimation of a Population Mean: Small-Sample Case

In the previous section, we discussed estimation of a population mean based on large samples (samples of size 30 or greater). However, time or cost limitations may often restrict the number of sample observations which may be obtained, so that the estimation procedures of Section 8.2 would not be applicable.

With small samples, the following two problems arise:

1. Since the Central Limit Theorem applies only to large samples, we are not able to assume that the sampling distribution of $\bar{x}$ is approximately normal. For small samples, the sampling distribution of $\bar{x}$ depends on the particular form of the relative frequency distribution of the population being sampled.
2. The sample standard deviation s may not be a satisfactory approximation to the population standard deviation σ if the sample size is small.

Fortunately, we may proceed with estimation techniques based on small samples if we can make the following assumption:

ASSUMPTION REQUIRED FOR ESTIMATION OF μ BASED ON SMALL SAMPLES ($n < 30$)

The population from which the sample is selected has an approximate normal distribution.

If this assumption is valid, then we may again use $\bar{x}$ as a point estimate for μ, and the general form of a small-sample confidence interval for μ is as shown in the next box.

SMALL-SAMPLE CONFIDENCE INTERVAL FOR μ

$$\bar{x} \pm t_{\alpha/2}\frac{s}{\sqrt{n}}$$

where the distribution of t is based on $(n-1)$ degrees of freedom.

Upon comparing this to the large-sample confidence interval for μ, you will observe that the sample standard deviation s replaces the population standard deviation σ. Also, the sampling distribution upon which the confidence interval is based is no longer normal, but is known as a ***Student's t distribution.*** Consequently, we must replace the value of $z_{\alpha/2}$ used in a large-sample confidence interval by a value obtained from the t distribution.

The t distribution is very much like the z distribution. In particular, both are symmetric, mound-shaped, and have a mean of zero. However, the distribution of t

depends on a quantity called its ***degrees of freedom (df),*** which is equal to $(n-1)$ when estimating a population mean based on a small sample of size n. Intuitively, we can think of the number of degrees of freedom as the amount of information available for estimating, in addition to μ, the unknown quantity σ^2. Table 4, Appendix E, a portion of which is reproduced in Figure 8.6, gives the value of t_α which locates an area of α in the upper tail of the t distribution for various values of α and for degrees of freedom ranging from 1 to 29.

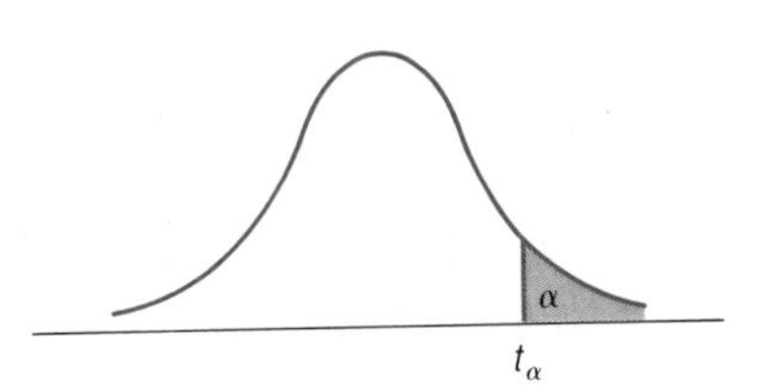

DEGREES OF FREEDOM	$t_{.100}$	$t_{.050}$	$t_{.025}$	$t_{.010}$	$t_{.005}$
1	3.078	6.314	12.706	31.821	63.657
2	1.886	2.920	4.303	6.965	9.925
3	1.638	2.353	3.182	4.541	5.841
4	1.533	2.132	2.776	3.747	4.604
5	1.476	2.015	2.571	3.365	4.032
6	1.440	1.943	2.447	3.143	3.707
7	1.415	1.895	2.365	2.998	3.499
8	1.397	1.860	2.306	2.896	3.355
9	1.383	1.833	2.262	2.821	3.250
10	1.372	1.812	2.228	2.764	3.169
11	1.363	1.796	2.201	2.718	3.106
12	1.356	1.782	2.179	2.681	3.055
13	1.350	1.771	2.160	2.650	3.012
14	1.345	1.761	2.145	2.624	2.977
15	1.341	1.753	2.131	2.602	2.947

FIGURE 8.6
Reproduction of a Portion of Table 4, Appendix E

EXAMPLE 8.9 Use Table 4, Appendix E, to determine the t-value which would be used in constructing a 95% confidence interval for μ based on a sample of size $n = 14$.

Solution For a confidence coefficient of .95, we have

$$1-\alpha = .95$$
$$\alpha = .05$$
$$\alpha/2 = .025$$

We thus require the value of $t_{.025}$ for a t distribution based on $n-1 = 14-1 = 13$ degrees of freedom. Now, in Table 4, at the intersection of the column labelled $t_{.025}$ and the row corresponding to df = 13, we find the entry 2.160 (see Figure 8.7).

Hence, a 95% confidence interval for μ, based on a sample of 14 observations, would be given by

$$\bar{x} \pm 2.160 \frac{s}{\sqrt{14}}$$

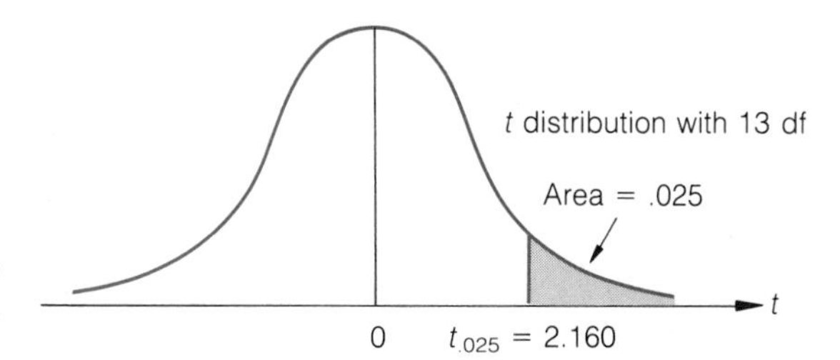

FIGURE 8.7
Location of $t_{.025}$ for Example 8.9

At this point, the reasoning for the arbitrary cutoff point of $n = 30$ for distinguishing between large and small samples may be better understood. Observe that the values in the last row of Table 4 (corresponding to df = infinity) are the values from the standard normal z distribution. This phenomenon occurs because, as the sample size increases, the t distribution becomes more and more like the z distribution. By the time n reaches 30, i.e., df = 29, there is very little difference between tabulated values of t and z.

EXAMPLE 8.10 The Chamber of Commerce of a large city is interested in monthly rental rates of 3-bedroom, 2-bath apartments in the area. The managers of six apartment complexes submitted the following monthly rental rates for their 3-bedroom, 2-bath units:

$340, $405, $260, $500, $380, $390

a. Construct a 90% confidence interval for μ, the average monthly rental rate of a 3-bedroom, 2-bath apartment in this area.

b. State the assumption(s) required for the validity of the procedure you used in part a.

Solution **a.** It is first required to compute the values of the sample mean and standard deviation.

$$\bar{x} = \frac{\Sigma x}{n} = \frac{340 + 405 + 260 + 500 + 380 + 390}{6}$$

$$= 379.17$$

$$s = \sqrt{\frac{\Sigma x^2 - (\Sigma x)^2/n}{n - 1}}$$

$$= \sqrt{\frac{(340)^2 + (405)^2 + (260)^2 + (500)^2 + (380)^2 + (390)^2 - (2{,}275)^2/6}{5}}$$

$$= \sqrt{\frac{893{,}725 - 862{,}604.17}{5}} = \sqrt{6{,}224.1667} = 78.89$$

A small-sample 90% confidence interval for μ will be based on $t_{\alpha/2} = t_{.05}$, where the distribution of t has $n - 1 = 6 - 1 = 5$ df. From Table 4, we obtain the value

$$t_{.05} = 2.015$$

Substitution of these values into the general form for a 90% confidence interval yields

$$\bar{x} \pm t_{.05}\frac{s}{\sqrt{n}} = 379.17 \pm 2.015\frac{78.89}{\sqrt{6}}$$
$$= 379.17 \pm 64.90$$

or (314.27, 444.07).

We are 90% confident that the interval \$314.27 to \$444.07 encloses the true average rental rate for a 3-bedroom, 2-bath apartment in this area. If we were to employ this estimation procedure for repeated random samples, then 90% of the intervals constructed in this manner would contain the true value of μ.

b. The procedure requires the assumption that the relative frequency distribution of the prices for all 3-bedroom, 2-bath apartments in the area is approximately normal.

EXAMPLE 8.11 The National Aeronautics and Space Administration (NASA) is continually testing the components of its spacecraft. Suppose NASA desires to estimate the mean lifetime of a particular mechanical component used in the space shuttle Columbia. Due to the prohibitive cost, only ten components can be tested under simulated space conditions. The lifetimes (in hours) of the components were recorded with the following results:

$$\bar{x} = 1{,}173.6$$
$$s = 36.3$$

Estimate μ, the mean lifetime of the mechanical components, with a 95% confidence interval.

Solution Since we must base our estimation procedure on a small sample ($n = 10$), it is necessary to assume that the population of lifetimes of the particular mechanical component has a relative frequency distribution which is approximately normal.

The desired confidence interval is based on a t distribution with $n - 1 = 10 - 1 = 9$ degrees of freedom; we obtain the value of $t_{\alpha/2} = t_{.025} = 2.262$ from Table 4, Appendix E. Then we have

$$\bar{x} \pm t_{.025}\frac{s}{\sqrt{n}} = 1{,}173.6 \pm 2.262\frac{36.3}{\sqrt{10}}$$
$$= 1{,}173.6 \pm 25.97$$

or (1,147.63, 1,199.57).

Thus, we are reasonably confident that the interval from 1,147.63 hours to 1,199.57 hours contains the true mean lifetime of the mechanical components.

EXERCISES

8.11 Use Table 4, Appendix E, to determine the values of $t_{\alpha/2}$ which would be used in the construction of a confidence interval for a population mean for each of the following combinations of confidence coefficient and sample size:

a. Confidence coefficient .99, $n = 18$
b. Confidence coefficient .95, $n = 10$
c. Confidence coefficient .90, $n = 15$

8.12 Give two reasons why the interval estimation procedure of Section 8.2 may not be applicable when the sample size is small, i.e., when $n < 30$.

8.13 A major utilities firm is currently working with the U.S. Department of Energy in developing an electric car. A prototype, which needs absolutely no gasoline, has already been developed. Tests on a random sample of $n = 8$ of these prototype electric cars produced the following statistics on maximum speed attained:

$\bar{x} = 54.8$ miles per hour

$s = 10.3$ miles per hour

a. Estimate the average maximum speed of all prototype electric cars manufactured by this utilities firm, using a 99% confidence interval.
b. What assumption is required for the confidence interval procedure of part a to be valid?

8.14 How are the t distribution and z distribution similar? How are they different?

8.15 One of the problems encountered by fresh fruit growers is spoilage of the fruit during transport to market. Each day during the harvest, a Florida citrus grower transports 100 bushels of fruit by truck to the nearest market to be sold. In order to estimate the mean number of bushels of fruit per truckload lost to spoilage, the citrus grower randomly selects 6 departing truckloads (each carrying 100 bushels of fresh fruit) and counts the number of bushels of spoiled fruit in each at the end of the trip. Suppose that these 6 sample measurements produce a mean of 10.1 bushels and a standard deviation of 2.8 bushels.

a. State the assumption, in the words of the problem, which is required for a small-sample confidence interval technique to be valid.
b. Construct a 98% confidence interval for the true mean number of bushels of spoiled fruit per truckload.
c. Interpret the interval, part b.

8.16 Designer jeans (Jordache, Calvin Klein, Brittania, Sassoon, etc.) became best-selling items despite their high retail price. The jeans were very marketable because they were designed to fit both males and females and came to represent a symbol of status among the younger generation. In order to estimate the mean retail price of designer jeans, a buyer for a major clothing chain sampled 19 retailers in the New York City area and recorded the selling price of the particular brand of designer jeans sold by each. The following information was obtained:

$\bar{x} = \$41.75$

$s = \$\ 5.50$

a. Form a 95% confidence interval for the true mean retail price of designer jeans in the New York City area.
b. What assumption is required for the interval estimation procedure, part a, to be valid?
c. Should the buyer for the major clothing chain use the interval to infer the value of the mean price of designer jeans at retail outlets across the entire U.S.? Explain.

8.17 A zoologist is interested in determining the average lifespan of African bull elephants. After five years of field experiments in Africa, the zoologist has collected data on 15 bull elephants. The average lifespan of these bull elephants was found to be 27.8 years and the standard deviation was found to be 5.5 years. Estimate the true average lifespan of African bull elephants with a 90% confidence interval. State any assumptions which are required for the interval estimation procedure to be valid.

8.18 More than 12 million tin-coated steel cans are removed from the municipal waste streams of our cities and recycled each day, according to a study released by the American Iron and Steel Institute (*American City and County,* April 1981). Suppose it is desired to estimate the mean number of tin cans recovered from mixed refuse per year in American cities. A random sample of 8 American cities yielded the following summary statistics on number of tin cans (in millions) recovered per city last year:

$$\bar{x} = 105.7$$

$$s = 9.3$$

a. Construct a 95% confidence interval for the true mean number of tin cans removed annually from mixed refuse for recycling in American cities.
b. Interpret the interval, part a.
c. What assumption is required for the interval estimate, part a, to be valid?

8.19 Rising interest rates on loans to new automobile purchasers are pushing potential customers out of the market. Suppose that it is desired to estimate the average interest rate charged for a 48-month loan to new car buyers in Michigan last year. The 48-month loan statements for a random sample of $n = 20$ new car buyers in Michigan were selected, and the interest rate charged was recorded for each. The following summary statistics were computed:

$$\bar{x} = 16.05\%$$

$$s = 1.25\%$$

Estimate μ, the average interest rate charged for a 48-month loan last year to new car buyers in Michigan, with a 95% confidence interval.

8.20 Refer to the data on DDT concentrations in fish inhabiting the Tennessee River and its tributary creeks, Appendix D. Suppose you wish to estimate the mean DDT content of all fish in the river and its tributary creeks.
a. Using Table 9, Appendix E, select a random sample of size $n = 5$ from the DDT measurements, Appendix D.

b. Compute $\bar{x}$ and s for this sample.

c. Using the summary statistics you computed in part a, construct a 99% confidence interval for the true mean DDT content of all fish in the river.

d. Now select a random sample of size $n = 30$ from the DDT measurements, Appendix D. Repeat parts b and c.

e. Compare the widths of the two intervals. Which interval gives a more reliable estimate of the true mean DDT content? Explain.

8.4 Estimation of a Population Proportion: Large-Sample Case

We will now consider the method for estimating the proportion of elements in a population which have a certain characteristic. For example, a sociologist may be interested in the proportion of urban New York City residents who are black; a pollster may be interested in the proportion of Americans who favor the President's policy of limited government spending; or a supplier of heating oil may be interested in the proportion of homes in its service area which are heated by natural gas. How would you estimate a population proportion p (e.g., the proportion of urban New York City residents who are black), based on information contained in a sample from the population?

EXAMPLE 8.12 The United States Commission on Crime is interested in estimating the proportion of crimes related to firearms in an area with one of the highest crime rates in the country. The commission selects a random sample of 300 files of recently committed crimes in the area and finds 180 in which a firearm was reportedly used. Estimate the true proportion p of all crimes committed in the area in which some type of firearm was reportedly used.

Solution A logical candidate for a point estimate of the population proportion p is the proportion of observations in the sample which have the characteristic of interest; we will call this sample proportion $\hat{p}$ (read "p hat"). In this example, the sample proportion of crimes with firearms is given by

$$\hat{p} = \frac{\text{Number of crimes in sample in which a firearm was reportedly used}}{\text{Total number of crimes in the sample}}$$

$$= \frac{180}{300} = .60$$

That is, 60% of the crimes in the sample were related to firearms; the value $\hat{p} = .60$ serves as our point estimate of the population proportion, p.

To assess the reliability of the point estimate $\hat{p}$, we need to know its sampling distribution. This information may be derived by an application of the Central Limit Theorem (details are omitted here). Properties of the sampling distribution of $\hat{p}$ (Figure 8.8) are given in the box at the top of page 250.

SAMPLING DISTRIBUTION OF $\hat{p}$

For sufficiently large samples, the sampling distribution of $\hat{p}$ is approximately normal, with

Mean $\mu_{\hat{p}} = p$

and

Standard deviation $\sigma_{\hat{p}} = \sqrt{\dfrac{pq}{n}}$

where $q = 1 - p$.

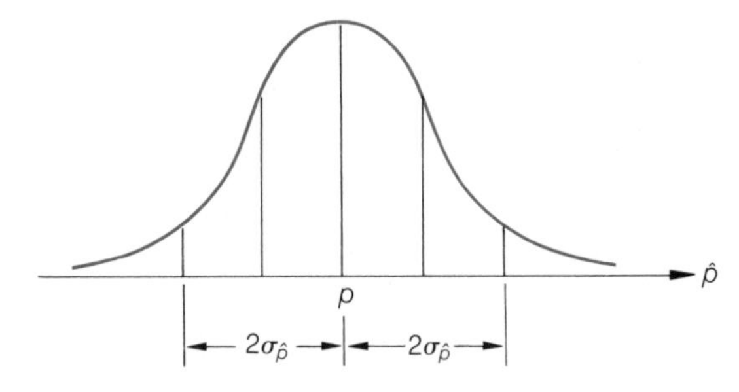

FIGURE 8.8
Sampling Distribution of $\hat{p}$

A large-sample confidence interval for p may be constructed by using a procedure which is analogous to that used for estimating a population mean.

LARGE-SAMPLE $(1 - \alpha)$ 100% CONFIDENCE INTERVAL FOR A POPULATION PROPORTION, p

$$\hat{p} \pm z_{\alpha/2}\sigma_{\hat{p}} \approx \hat{p} \pm z_{\alpha/2}\sqrt{\frac{\hat{p}\hat{q}}{n}}$$

where $\hat{p}$ is the sample proportion of observations with the characteristic of interest, and $\hat{q} = 1 - \hat{p}$.
[*Note:* We have substituted the sample values $\hat{p}$ and $\hat{q}$ for the corresponding population values required for $\sigma_{\hat{p}}$. This approximation is valid for large sample sizes.]

As a rule of thumb, the condition of a "sufficiently large" sample size will be satisfied if the interval $\hat{p} \pm 2\sigma_{\hat{p}}$ does not contain 0 or 1.

EXAMPLE 8.13 Refer to Example 8.12. Construct a 95% confidence interval for p, the population proportion of the crimes committed in the area in which some type of firearm was reportedly used.

Solution For a confidence coefficient of .95, we have $1 - \alpha = .95$; $\alpha = .05$; $\alpha/2 = .025$; and the required z-value is $z_{.025} = 1.96$. In Example 8.12, we obtained $\hat{p} = 180/300 = .60$. Thus, $\hat{q} = 1 - .60 = .40$. Substitution of these values into the general formula for a confidence interval yields

$$\hat{p} \pm z_{\alpha/2}\sqrt{\frac{\hat{p}\hat{q}}{n}} = .60 \pm 1.96\sqrt{\frac{(.60)(.40)}{300}}$$
$$= .60 \pm .06$$

or (.54, .66).

We are 95% confident that the interval from .54 to .66 contains the true proportion of the crimes committed in the area which were related to firearms. That is, in repeated construction of 95% confidence intervals, 95% of all samples would produce confidence intervals that enclose p.

EXAMPLE 8.14 Each winter the struggle between the natural gas and oil industries to supply heating fuel to American homes increases. Recently, the American Gas Association (AGA) has undertaken a multimillion-dollar advertising campaign in the hopes of persuading the owners of homes heated with oil that gas is "the most efficient way to heat" (*Time,* December 1, 1980). Suppose that the AGA is interested in estimating the proportion of homes in the Northern states (where oil usage is heavy) which are heated by gas. A random sample of 80 Northern homes indicated that 42 were heated by gas and 38 were heated by oil or other fuels. Estimate p, the proportion of all Northern homes heated by gas, using a 90% confidence interval.

Solution The sample proportion of homes which are heated by gas is

$$\hat{p} = \frac{\text{Number of homes in sample which are heated by gas}}{\text{Number of homes in sample}}$$
$$= \frac{42}{80} = .525$$

Thus, $\hat{q} = 1 - .525 = .475$.

The 90% confidence interval is then

$$\hat{p} \pm z_{.05}\sqrt{\frac{\hat{p}\hat{q}}{n}} = .525 \pm 1.645\sqrt{\frac{(.525)(.475)}{80}} = .525 \pm .092$$

or (.433, .617).

We are 90% confident that the interval from .433 to .617 encloses the true proportion of Northern homes which are heated by gas. If we repeatedly selected random samples of $n = 80$ Northern homes and constructed a 90% confidence interval based on each sample, then we would expect 90% of the confidence intervals constructed to contain p.

It should be noted that small-sample procedures are available for the estimation of a population proportion, p. The details are not included in our discussion, however, because most surveys in actual practice use samples that are large enough to employ the procedure of this section.

EXERCISES

8.21 A recent development in do-it-yourself home improvement is a special type of foam insulation which may be installed to make a home more energy efficient. However, it has been suggested that, in some instances, the foam (which usually contains a derivative of formaldehyde) produces a vapor which may be carcinogenic. The National Cancer Institute wishes to estimate the proportion of recently insulated homes in a particular state which have this type of foam insulation. In order to obtain this estimate, a random selection of 100 homes in which insulation had been recently added was taken and the type of insulation was noted. The results show that 24 of the 100 homes had foam insulation.

a. Construct a 95% confidence interval for the true proportion of recently insulated homes in this state which have the foam insulation.

b. Interpret the interval in terms of the problem.

c. How would the width of the confidence interval in part a change if the confidence coefficient were increased from .95 to .99?

8.22 It has been observed that persons who suffer a myocardial infarction (heart attack) often suffer a second one within a year because of damage to the heart muscle during the first episode. A cardiologist desires an estimate of the proportion of heart patients who suffer a second myocardial infarction within one year after the first. In a random sample of 138 patient records, 26 patients had suffered a second heart attack within one year after the first. Construct a 99% confidence interval for the true proportion of heart patients who suffer a second myocardial infarction within one year after the first. Interpret the interval.

8.23 Potential advertisers value television's well-known Nielsen ratings as a barometer of a TV show's popularity among viewers. The Nielsen rating of a certain TV program is an estimate of the proportion of viewers, expressed as a percentage, who tune their sets to the program on a given night. In a random selection of 165 families who regularly watch television, a Nielsen survey indicated that 101 of the families were tuned to a certain TV program on the night of its premiere. Estimate the true proportion of all TV-viewing families who watched the premiere. Use a 90% confidence interval.

8.24 As part of its normal audit procedure, a bank randomly selects 30 customer checking accounts and mails a statement of the balance showing in the bank's records to each customer. The customer is asked to compare his or her own statement of the balance with the bank's statement. If the statements disagree, the customer is requested to return to the bank a form describing the differences. If the statements agree, no reply is necessary. In last month's audit, 9 of the 30 customers mailed in the forms, i.e., 9 of the 30 received balance statements which disagreed with their own records. Estimate the true proportion of the bank's checking account customers who had balance statements which disagreed with the bank's records last month. Use a 98% confidence interval.

8.25 One piece of information of great interest to an individual who wishes to contract for the services of a home builder is the proportion of the builder's construction projects which are completed on or before the target date. The records for a random sample of 60 recent construction projects for a particular contractor indicated that 27 of the projects were completed on time and 33 extended beyond the estimated completion date.

a. Estimate p, the proportion of this builder's projects which are completed by the target date, using a 95% confidence interval.

b. Explain how the width of the interval constructed in part a could be decreased.

8.26 The jobless figure compiled by the U.S. Bureau of the Census showed that in January 1981, 7.4% of the labor force was unemployed. However, some critics believe that this figure underestimates actual unemployment since so-called "discouraged workers," those who have given up hope of finding a job, are sometimes not counted as part of the labor force by the Bureau. Suppose that in a random sample of 1,000 members of the labor force, some of whom are discouraged workers, 86 are found to be unemployed.

a. Give a point estimate for the unemployment rate when discouraged workers are included.

b. Construct a 92% confidence interval for the true unemployment rate when discouraged workers are included.

c. Based upon your interval, part b, does it appear that the figure reported by the Census Bureau is actually an underestimate of the true unemployment rate when discouraged workers are included?

8.27 Do American teenagers tend to be a "happy-go-lucky" group, or are they generally worriers? A recent Gallup Youth Survey (reported in the Gainesville *Sun,* June 10, 1981) confronted a random sample of 1,030 teenagers 13–18 years old with the following question: "Would you say you worry a lot, or that you take life pretty much as it comes?" The survey found that 711 of the teenagers questioned feel that they take life as it comes. Estimate the true proportion of all teenagers 13–18 years old who believe that they "take life pretty much as it comes" with a 98% confidence interval. Interpret the interval.

8.5 Estimation of the Difference between Two Population Means: Large-Sample Case

In Section 8.2, we learned how to estimate the parameter μ based on a large sample from a single population. We now proceed to a technique for using the information in two samples to estimate the difference between two population means. For example, we may wish to compare the mean starting salaries for bachelor's degree graduates of the University of Florida in the colleges of Engineering and Education, or the mean gasoline consumptions that may be expected this year for drivers in two areas of the country, or the mean reaction times of men and women to a visual stimulus. The technique to be presented is a straightforward extension of that used for large-sample estimation of a single population mean.

EXAMPLE 8.15 It is desired to estimate the difference between the mean starting salaries for all bachelor's degree graduates of the University of Florida in the colleges of Engineering and Education during June 1980–March 1981. The following information is available:

1. A random sample of 40 starting salaries for College of Engineering graduates produced a sample mean of \$19,653 and a standard deviation of \$4,172.
2. A random sample of 30 starting salaries for College of Education graduates produced a sample mean of \$11,291 and a standard deviation of \$2,864.

Calculate a point estimate for the difference between mean starting salaries for graduates of the two colleges.

Solution We will let the subscript 1 refer to the College of Engineering and the subscript 2 to the College of Education. We will also define the following notation:

μ_1 = The population mean starting salary of all bachelor's degree graduates of the College of Engineering during June 1980–March 1981

μ_2 = The population mean starting salary of all bachelor's degree graduates of the College of Education during June 1980–March 1981

Similarly, let $\bar{x}_1$ and $\bar{x}_2$ denote the respective sample means; s_1 and s_2, the respective sample standard deviations; and n_1 and n_2, the respective sample sizes. The given information may be summarized as in Table 8.4.

TABLE 8.4
Summary of Information for Example 8.15

	COLLEGE OF ENGINEERING	COLLEGE OF EDUCATION
Sample size	$n_1 = 40$	$n_2 = 30$
Sample mean	$\bar{x}_1 = \$19,653$	$\bar{x}_2 = \$11,291$
Sample standard deviation	$s_1 = \$4,172$	$s_2 = \$2,864$

Now, to estimate $(\mu_1 - \mu_2)$, it seems logical to use the difference between the sample means

$$(\bar{x}_1 - \bar{x}_2) = (\$19,653 - \$11,291) = \$8,362$$

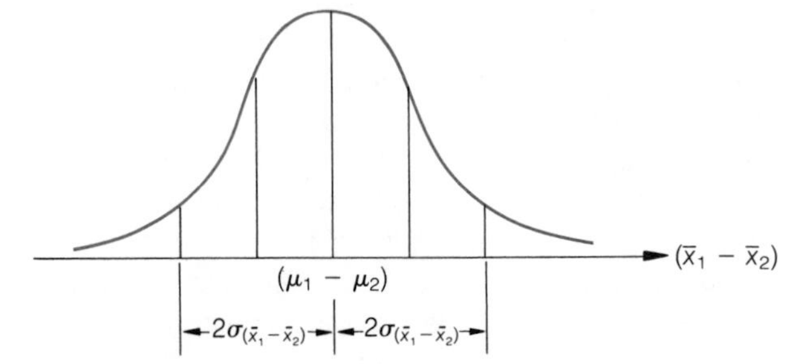

FIGURE 8.9
Sampling Distribution of $(\bar{x}_1 - \bar{x}_2)$

as our point estimate of the difference between the population means. The properties of the point estimate $(\bar{x}_1 - \bar{x}_2)$ are summarized by its sampling distribution, shown in the accompanying box (see also Figure 8.9).

SAMPLING DISTRIBUTION OF $(\bar{x}_1 - \bar{x}_2)$

For sufficiently large sample sizes (say, $n_1 \geq 30$ and $n_2 \geq 30$), the sampling distribution of $(\bar{x}_1 - \bar{x}_2)$, based on independent random samples from two populations, is approximately normal, with

Mean $\mu_{(\bar{x}_1 - \bar{x}_2)} = (\mu_1 - \mu_2)$

Standard deviation $\sigma_{(\bar{x}_1 - \bar{x}_2)} = \sqrt{\frac{\sigma_1^2}{n_1} + \frac{\sigma_2^2}{n_2}}$

where σ_1^2 and σ_2^2 are the variances of the two populations from which the samples were selected.

As was the case with large-sample estimation of a single population mean, the requirement of "large" sample sizes enables us to apply the Central Limit Theorem to obtain the sampling distribution of $(\bar{x}_1 - \bar{x}_2)$; it also justifies the use of s_1^2 and s_2^2 as approximations to the respective population variances, σ_1^2 and σ_2^2.

The procedure for forming a confidence interval for $(\mu_1 - \mu_2)$ appears in the accompanying box.

LARGE-SAMPLE $(1 - \alpha)$ 100% CONFIDENCE INTERVAL FOR $(\mu_1 - \mu_2)$

$$(\bar{x}_1 - \bar{x}_2) \pm z_{\alpha/2}\sigma_{(\bar{x}_1 - \bar{x}_2)} = (\bar{x}_1 - \bar{x}_2) \pm z_{\alpha/2}\sqrt{\frac{\sigma_1^2}{n_1} + \frac{\sigma_2^2}{n_2}} \approx (\bar{x}_1 - \bar{x}_2) \pm z_{\alpha/2}\sqrt{\frac{s_1^2}{n_1} + \frac{s_2^2}{n_2}}$$

[*Note:* We have employed the sample variances s_1^2 and s_2^2 as approximations to the corresponding population parameters.]

The assumptions upon which the above procedure is based are the following:

ASSUMPTIONS REQUIRED FOR LARGE-SAMPLE ESTIMATION OF $(\mu_1 - \mu_2)$

1. The two random samples are selected in an ***independent*** manner from the target populations. That is, the choice of elements in one sample does not affect, and is not affected by, the choice of elements in the other sample.
2. The sample sizes n_1 and n_2 are sufficiently large. (We recommend $n_1 \geq 30$ and $n_2 \geq 30$.)

EXAMPLE 8.16 Refer to Example 8.15. Construct a 95% confidence interval for $(\mu_1 - \mu_2)$, the difference between the mean starting salaries of all bachelor's degree graduates of the colleges of Engineering and Education during June 1980–March 1981. Interpret the interval.

Solution The general form of a 95% confidence interval for $(\mu_1 - \mu_2)$, based on large samples from the target populations, is given by

$$(\bar{x}_1 - \bar{x}_2) \pm z_{.025}\sqrt{\frac{\sigma_1^2}{n_1} + \frac{\sigma_2^2}{n_2}}$$

Recall that $z_{.025} = 1.96$ and use the information in Table 8.4 to make the following substitutions to obtain the desired confidence interval:

$$(19{,}653 - 11{,}291) \pm 1.96\sqrt{\frac{\sigma_1^2}{40} + \frac{\sigma_2^2}{30}}$$

$$\approx (19{,}653 - 11{,}291) \pm 1.96\sqrt{\frac{(4{,}172)^2}{40} + \frac{(2{,}864)^2}{30}}$$

$$= 8{,}362 \pm 1{,}650$$

or ($6,712, $10,012).

The use of this method of estimation produces confidence intervals which will enclose $(\mu_1 - \mu_2)$, the difference between population means, 95% of the time. Hence, we can be reasonably confident that the mean starting salary of College of Engineering bachelor's degree graduates in June 1980–March 1981 was between $6,712 and $10,012 higher than the mean starting salary of College of Education bachelor's degree graduates during this period.

EXAMPLE 8.17 The personnel manager for a large steel company suspects a difference between the mean amounts of work time lost due to sickness for blue-collar and white-collar workers at the plant. She randomly samples the records of 45 blue-collar workers and 38 white-collar workers and records the number of days lost due to sickness within the past year. Summary statistics were computed, with the results shown in Table 8.5. Estimate $(\mu_1 - \mu_2)$, the difference between the population mean days lost to sickness for blue-collar and white-collar workers at the steel company last year, using a 90% confidence interval.

TABLE 8.5 Comparison of Lost Sick Days, Example 8.17

	BLUE COLLAR	WHITE COLLAR
Sample size	$n_1 = 45$	$n_2 = 38$
Sample mean	$\bar{x}_1 = 10.4$	$\bar{x}_2 = 7.8$
Sample standard deviation	$s_1 = 12.8$	$s_2 = 5.5$

Solution The general form of a large-sample 90% confidence interval for $(\mu_1 - \mu_2)$ is

$$(\bar{x}_1 - \bar{x}_2) \pm z_{.05}\sqrt{\frac{\sigma_1^2}{n_1} + \frac{\sigma_2^2}{n_2}}$$

Substitution of the sample variances, s_1^2 and s_2^2, for the corresponding population values, σ_1^2 and σ_2^2, together with $z_{.05} = 1.645$ and the statistics provided in Table 8.5, yields the approximate 90% confidence interval

$$(10.4 - 7.8) \pm 1.645\sqrt{\frac{(12.8)^2}{45} + \frac{(5.5)^2}{38}} = 2.6 \pm 3.47$$

or $(-.87, 6.07)$.

Thus, we estimate that the difference between the mean days lost to sickness for the two groups of workers falls in the interval $-.87$ to 6.07. In other words, we estimate that μ_2, the mean days lost to sickness for white-collar workers, could be larger than μ_1, the mean days lost to sickness for blue-collar workers, by as much as .87, or it could be less than μ_1 by as much as 6.07. Since the interval contains the value zero, we are unable to conclude that there is a real difference between the mean numbers of sick days lost by the two groups. If, in fact, such a difference exists, we would have to increase the sample sizes to be able to detect it. This would reduce the width of the confidence interval and provide more information about the phenomenon under investigation.

EXERCISES

8.28 Chemical scientists are currently working to develop a substance to serve as a substitute for sucrose (sugar) in the diet. In addition to producing a sweet sensation, the compound must also meet other requirements, such as stability and nontoxicity. In one series of experiments, two compounds were tested for their ability to survive the high temperatures of cooking. Samples of each compound were heated and the temperature was recorded at which the substance melted. The results are summarized in Table 8.6. Estimate the difference between the mean melting temperatures for compound 1 and compound 2, using a 95% confidence interval. Interpret the interval.

TABLE 8.6 Comparison of Melting Temperatures, Exercise 8.28

COMPOUND 1	COMPOUND 2
$n_1 = 50$	$n_2 = 40$
$\bar{x}_1 = 287°F$	$\bar{x}_2 = 348°F$
$s_1 = 23°F$	$s_2 = 31°F$

8.29 Lack of motivation is a problem of many students in inner-city schools. To cope with this problem, an experiment was conducted to determine whether motivation could be improved by allowing students greater choice in the structure of their curricula. Two schools with similar student populations were chosen and thirty students were randomly selected from each to participate in the experiment. School A permitted its thirty students to choose only the courses they wanted to take. School B permitted its students to choose their courses and also to choose when and from which instructors to take the courses. The measure of student motivation was the number of times each student was absent from or late for a class during a 20-day period. A summary of the results is shown in Table 8.7.

TABLE 8.7
Student Absentees or Tardiness, Exercise 8.29

SCHOOL A	SCHOOL B
$n_1 = 30$	$n_2 = 30$
$\bar{x}_1 = 21.5$	$\bar{x}_2 = 18.6$
$s_1 = 5.2$	$s_2 = 4.9$

a. Estimate the difference between the mean number of times students are absent or late for a class during a 20-day period at the two schools, using a 90% confidence interval.

b. Does it appear that allowing students greater choice in the structure of their curricula improves student motivation? Explain. [*Hint:* Does the interval of part a contain only positive numbers?]

8.30 Every six months, Management Centre Europe (MCE), a Brussels-based consulting firm, measures living costs in 16 European cities for comparison with the cost of living in New York City. As a yardstick, MCE uses the dollar value of a group of 101 common items randomly selected from among foods, clothing, taxi rides, etc. In one survey, MCE wished to compare the costs of items in Stockholm, Sweden, to the costs in New York. MCE randomly selected 101 items and recorded their costs (in dollars) in Stockholm and in New York. The data are summarized in Table 8.8.

TABLE 8.8
Dollar Value of Common Items, Exercise 8.30

STOCKHOLM	NEW YORK
$n_1 = 101$	$n_2 = 101$
$\bar{x}_1 = \$526$	$\bar{x}_2 = \$312$
$s_1 = \$214$	$s_2 = \$170$

a. Construct a 99% confidence interval for the difference between the mean costs of the common items in Stockholm and New York.

b. Interpret the interval.

c. Explain how MCE could decrease the width of the interval obtained in part a.

8.31 A restaurant specializing in all-natural foods has experienced a reduction in daily sales during the past year. The manager of the restaurant hopes to offset these losses by instituting a new advertising campaign. On an experimental basis, the campaign is begun, and the daily sales of the restaurant are recorded for 40 operating days. This information is to be compared with data on daily sales compiled before the campaign was initiated. The results are shown in Table 8.9.

a. Give a point estimate for the difference between the mean daily sales before and after the advertising campaign.

TABLE 8.9
Comparison of Daily Sales, Exercise 8.31

BEFORE CAMPAIGN	DURING CAMPAIGN
$n_1 = 50$	$n_2 = 40$
$\bar{x}_1 = \$487$	$\bar{x}_2 = \$548$
$s_1 = \$23$	$s_2 = \$31$

b. Construct a 93% confidence interval for the difference between the means, part a.

c. What assumptions are necessary for the validity of the interval estimation procedure?

d. Suppose that the new advertising campaign will cost the restaurant $50 per day to run. Thus, the manager must decide whether the new campaign will increase mean daily sales over and above the cost of the campaign. Based on the interval, part b, should the manager adopt the new advertising campaign on a full-time basis? [*Hint:* The manager will adopt the campaign only if there is evidence that μ_2, the mean daily sales during the campaign, is greater than μ_1, the mean daily sales before the campaign, by an amount greater than $50, i.e., if $(\mu_1 - \mu_2) < -\$50$.]

8.32 A psychologist is studying the effect of different aspects of the working environment on productivity. Recently, he conducted an experiment to determine whether productivity is affected when workers are allowed to operate without formal supervision. Two groups of 60 workers each were randomly and independently selected from a large factory to take part in the study. For two weeks, one group worked under formal supervision, while the second group was allowed to operate without formal supervision. The productivity per worker (i.e., the number of items produced per worker) for this time period was recorded for both groups of employees. A summary of the data is given in Table 8.10.

TABLE 8.10
Productivity per Worker (Number of Items), Exercise 8.32

FORMAL SUPERVISION	NO FORMAL SUPERVISION
$n_1 = 60$	$n_2 = 60$
$\bar{x}_1 = 85.3$	$\bar{x}_2 = 89.9$
$s_1 = 6.2$	$s_2 = 18.1$

a. Use a 90% confidence interval to estimate the difference between mean productivities per worker for the two groups.

b. Interpret the interval, part a. Is there evidence of a difference between the mean productivities of workers under the two types of supervision?

8.6 Estimation of the Difference between Two Population Means: Small-Sample Case

This section presents a method for estimating the difference between two population means, based on small samples from each population. As was the case with estimating a single population mean from information in a small sample, specific assumptions about the relative frequency distribution of the two populations must be made, as indicated in the box.

ASSUMPTIONS REQUIRED FOR SMALL-SAMPLE ESTIMATION OF $(\mu_1 - \mu_2)$

1. The populations from which the samples are selected both have relative frequency distributions which are approximately normal.
2. The variances σ_1^2 and σ_2^2 of the two populations are equal.
3. The random samples are selected in an independent manner from the two populations.

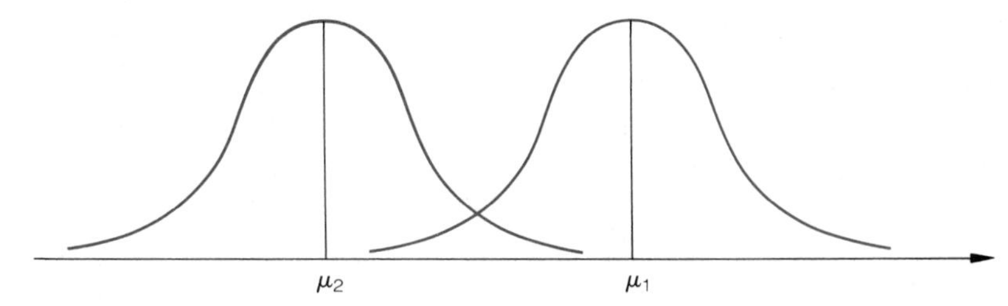

FIGURE 8.10 Assumptions Required for Small-Sample Estimation of $(\mu_1 - \mu_2)$: Normal Distributions with Equal Variances

Figure 8.10 illustrates the form of the population distributions implied by assumptions 1 and 2. Observe that both populations have relative frequency distributions which are approximately normal. Although the means of the two populations may differ, we require the variances σ_1^2 and σ_2^2, which measure the spreads of the two distributions, to be equal.

When the above assumptions are satisfied, we may use the following procedure to construct a confidence interval for $(\mu_1 - \mu_2)$, based on small samples (say, $n_1 < 30$ and $n_2 < 30$) from the respective populations.

SMALL-SAMPLE $(1 - \alpha)$ 100% CONFIDENCE INTERVAL FOR $(\mu_1 - \mu_2)$

$$(\bar{x}_1 - \bar{x}_2) \pm t_{\alpha/2} \sqrt{s_p^2\left(\frac{1}{n_1} + \frac{1}{n_2}\right)}$$

where

$$s_p^2 = \frac{(n_1 - 1)s_1^2 + (n_2 - 1)s_2^2}{n_1 + n_2 - 2}$$

and the value of $t_{\alpha/2}$ is based on $(n_1 + n_2 - 2)$ degrees of freedom.

Since we assume that the two populations have equal variances (i.e., $\sigma_1^2 = \sigma_2^2 = \sigma^2$), we construct an estimate of σ^2 which is based on the information contained in both samples. This ***pooled estimate*** is denoted by s_p^2 and is computed as shown in the box.

EXAMPLE 8.18 Table 8.11 shows summary statistics for random samples selected from two normal populations which are assumed to have the same variance. Estimate this common variance.

TABLE 8.11 Summary Information, Example 8.18

DATA FROM POPULATION 1	DATA FROM POPULATION 2
$n_1 = 12$	$n_2 = 17$
$\bar{x}_1 = 10.6$	$\bar{x}_2 = 9.5$
$s_1 = 2.4$	$s_2 = 4.7$

Solution We have made the following assumptions:

1a. Population 1 has a relative frequency distribution which is normal, with mean μ_1 and variance σ^2.

1b. Population 2 has a relative frequency distribution which is normal, with mean μ_2 and variance σ^2.

In order to estimate the variance, σ^2, common to both populations, we pool the information available from both samples and compute

$$s_p^2 = \frac{(n_1 - 1)s_1^2 + (n_2 - 1)s_2^2}{n_1 + n_2 - 2}$$

$$= \frac{(12 - 1)(2.4)^2 + (17 - 1)(4.7)^2}{12 + 17 - 2}$$

$$= \frac{11(5.76) + 16(22.09)}{27} = \frac{416.8}{27} = 15.44$$

Our pooled estimate of the common population variance is 15.44.

EXAMPLE 8.19 Refer to Example 8.18. Use the information provided in Table 8.11 to construct a 95% confidence interval for $(\mu_1 - \mu_2)$.

Solution In order to properly apply the small-sample confidence interval procedure for $(\mu_1 - \mu_2)$, it is necessary to make the assumptions 1a and 1b in Example 8.18. In addition, we assume that the samples were randomly and independently selected from the two populations.

The 95% confidence interval for $(\mu_1 - \mu_2)$ will be based on the value of $t_{.025}$, where t has $n_1 + n_2 - 2 = 12 + 17 - 2 = 27$ degrees of freedom. From Table 4, Appendix E, we obtain $t_{.025} = 2.052$. We now substitute the appropriate quantities into the general formula:

$$(\bar{x}_1 - \bar{x}_2) \pm t_{.025}\sqrt{s_p^2\left(\frac{1}{n_1}+\frac{1}{n_2}\right)}$$

$$= (10.6 - 9.5) \pm 2.052\sqrt{15.44\left(\frac{1}{12}+\frac{1}{17}\right)}$$

$$= 1.1 \pm 3.0$$

or $(-1.9, 4.1)$.

We estimate the difference $(\mu_1 - \mu_2)$ to fall in the interval from -1.9 to 4.1. In other words, we estimate the mean for population 1 to be anywhere from 1.9 less than to 4.1 greater than the mean for population 2.

EXAMPLE 8.20 In order to study the effectiveness of a new type of dental anesthetic, a dentist conducted an experiment with 10 randomly selected patients. Five patients received the standard anesthetic (Novocain) while the remaining five patients received the proposed new anesthetic. While being treated, each patient was asked to give a measure of his or her discomfort, on a scale of 0 to 100. (Higher scores indicate greater discomfort.) A summary of the results is shown in Table 8.12.

TABLE 8.12 Discomfort Scores, Example 8.20

NOVOCAIN	NEW ANESTHETIC
$n_1 = 5$	$n_2 = 5$
$\bar{x}_1 = 60.33$	$\bar{x}_2 = 32.21$
$s_1 = 15.82$	$s_2 = 12.77$

a. Construct a 98% confidence interval for the true difference between the mean discomfort levels for dental patients who receive Novocain and those who receive the new anesthetic.

b. What assumptions are required for the validity of the procedure you used in part a?

Solution a. The first step is to compute the pooled estimate of variance:

$$s_p^2 = \frac{(n_1 - 1)s_1^2 + (n_2 - 1)s_2^2}{n_1 + n_2 - 2}$$

$$= \frac{(5-1)(15.82)^2 + (5-1)(12.77)^2}{5+5-2}$$

$$= \frac{1653.38}{8} = 206.67$$

Then, the 98% confidence interval for $(\mu_1 - \mu_2)$, the difference between the mean discomfort levels for the two groups of patients, is

$$(\bar{x}_1 - \bar{x}_2) \pm t_{.01}\sqrt{s_p^2\left(\frac{1}{n_1}+\frac{1}{n_2}\right)}$$

where the value of t is based on $(n_1 + n_2 - 2) = 8$ degrees of freedom. Thus, we have

$$(60.33 - 32.21) \pm 2.896\sqrt{206.67\left(\frac{1}{5} + \frac{1}{5}\right)}$$
$$= 28.12 \pm 26.33$$

or (1.79, 54.45). We can be 98% confident that the mean discomfort level for patients receiving Novocain exceeds the mean discomfort level for patients treated with the new anesthetic by an amount which lies in the interval between 1.79 and 54.45.

b. The following assumptions must be satisfied:

1. The relative frequency distribution for the discomfort level of dental patients is approximately normal for both groups of patients (those receiving Novocain and those treated with the new anesthetic).
2. The variance in the discomfort levels is the same for both groups of patients.
3. The samples are randomly and independently selected from the two target populations.

EXERCISES

8.33 To use the t statistic in a confidence interval for the difference between the means of two populations, what assumptions must be made about the two populations? About the two samples?

8.34 Refer to Exercise 8.28. Suppose that time restrictions would have prevented the selection of such large samples, and that melting temperatures could be obtained for only 10 samples of each compound. Assume that the sample means and standard deviations based on these samples of size $n_1 = n_2 = 10$ remain as shown in Table 8.6.

a. Construct a 95% confidence interval for the difference between the mean melting temperatures for compound 1 and compound 2. Compare the result with the 95% confidence interval you obtained in Exercise 8.28.

b. What assumptions do you need to make for the validity of the procedure used in part a? Do they seem reasonable?

8.35

a. Use a table of random numbers and the data in Appendix C to generate random samples of size $n_1 = n_2 = 6$ from the customer checkout service times for supermarkets A and B. Compute the mean and standard deviation for each sample.

b. Use the information from part a to construct a 90% confidence interval for the difference between the mean customer checkout service times at supermarkets A and B. Interpret the interval.

c. State the assumptions necessary for the estimation procedure you used in part b to be valid.

d. State how you could construct a confidence interval with a smaller width than the interval you constructed in part b.

8.36 A federal traffic safety researcher was hired to ascertain the effect of wearing safety devices (shoulder harnesses, seat belts) on reaction times to peripheral stimuli. To investigate this question, he randomly selected thirty subjects from the students enrolled in a driver education program. Fifteen of these students were randomly assigned to the Restrained treatment group (i.e., those wearing safety devices) and the other fifteen to the Unrestrained treatment group (i.e., those wearing no safety devices). Subjects from both groups performed a simulated driving task which allowed reaction times to be recorded. Each subject received a total reaction time score. The summary data (in hundredths of a second) are shown in Table 8.13.

TABLE 8.13 Reaction Time Scores, Exercise 8.36

RESTRAINED	UNRESTRAINED
$n_1 = 15$	$n_2 = 15$
$\bar{x}_1 = 38.0$	$\bar{x}_2 = 36.4$
$s_1 = 3.1$	$s_2 = 2.2$

a. Construct a 98% confidence interval for the difference between mean reaction time scores for the Restrained and Unrestrained drivers.

b. What assumptions are necessary for the validity of the interval estimation procedure, part a?

8.37 Whom would you trust to charge the fairer price for automobile service and labor, gas station mechanics or mechanics employed by car dealers? Data collected on the hourly labor rates of mechanics working at gas stations and at car dealerships in the city of Detroit are summarized in Table 8.14. Estimate the difference between the mean hourly labor rates of Detroit auto mechanics employed at gas stations and at car dealerships, using a 95% confidence interval. Interpret the interval. State any assumptions that are needed to make the interval estimate valid. [*Hint:* Recall that as df increases, *t*-values and *z*-values are nearly identical.]

TABLE 8.14 Hourly Labor Rates, Exercise 8.37

	GAS STATIONS	CAR DEALERS
Sample size	20	15
Mean ($)	26.75	30.50
Standard deviation ($)	5.17	2.76

8.38 Two alloys, A and B, are used in the manufacture of steel bars. Suppose that a steel producer wishes to compare the two alloys on the basis of average load capacity, where the load capacity of a steel bar is defined as the maximum load (weight) it can support without breaking. Steel bars containing alloy A and steel bars containing alloy B were randomly selected and tested for load capacity. The results are summarized in Table 8.15.

a. Find a 99% confidence interval for the difference between the true average loading capacities for the two alloys.

b. For the interval, part a, to be valid, what assumptions must be satisfied?

TABLE 8.15
Load Capacities of Steel Bars (in Tons), Exercise 8.38

ALLOY A	ALLOY B
$n_1 = 11$	$n_2 = 17$
$\bar{x}_1 = 43.7$	$\bar{x}_2 = 48.5$
$s_1^2 = 24.4$	$s_2^2 = 19.9$

c. Interpret the interval, part a. Can you conclude with reasonable confidence that the average load capacities for the two alloys are different?

8.39

a. Use a table of random numbers to generate independent random samples of size $n_1 = 10$ male graduates and $n_2 = 10$ female graduates from the data in Appendix A. Record the starting salaries of the graduates and then compute the mean and standard deviation for each sample.

b. Use the information from part a to construct a 95% confidence interval for the difference between the mean starting salaries of male and female graduates. Interpret the interval.

c. If you have access to a computer, compute the mean starting salary of *all* male graduates and the mean starting salary of *all* female graduates in the listing of 948 graduates, Appendix A. Does the difference in mean starting salaries of the two populations fall within the interval you constructed in part b?

8.7 Estimation of the Difference between Two Population Proportions: Large-Sample Case

This section extends the method of Section 8.4 to the case in which we wish to estimate the difference between two population proportions. For example, we may be interested in comparing the proportions of married and unmarried persons who are overweight, or the proportions of homes in two states which are heated by natural gas, or the proportions of adults today and in 1974 who smoke, etc.

EXAMPLE 8.21 Is the United States becoming increasingly Republican? Democrats still outnumber Republicans in this country, but according to a recent New York Times–CBS News Poll, more Americans are calling themselves Republicans since the inauguration of President Reagan.* The Times–CBS News Poll was conducted in January 1980 and again in April 1981. In both surveys, a random sample of 1,400 American adults was asked whether they considered themselves Republicans, Democrats, or Independents. The results of the survey are reported in Table 8.16. Construct a point estimate for the difference between the proportions of American adults in January 1980 and in April 1981 who considered themselves Republicans or Republican-leaning Independents.

*Gainesville *Sun*, May 3, 1981.

TABLE 8.16 Political Party Preferences, Example 8.21

	JANUARY 1980	APRIL 1981
Number surveyed	$n_1 = 1{,}400$	$n_2 = 1{,}400$
Number in sample who said they were Republicans or Republican-leaning Independents	462	574

Solution Let us define

p_1 = The population proportion of American adults who considered themselves Republicans or Republican-leaning Independents in January 1980

and

p_2 = The population proportion of American adults who considered themselves Republicans or Republican-leaning Independents in April 1981

As a point estimate of $(p_1 - p_2)$, we will use the difference between the corresponding sample proportions, $(\hat{p}_1 - \hat{p}_2)$, where

$$\hat{p}_1 = \frac{\text{Number of surveyed Americans in January 1980 who considered themselves Republicans}}{\text{Number of Americans surveyed in January 1980}}$$

$$= \frac{462}{1{,}400} = .33$$

and

$$\hat{p}_2 = \frac{\text{Number of surveyed Americans in April 1981 who considered themselves Republicans}}{\text{Number of Americans surveyed in April 1981}}$$

$$= \frac{574}{1{,}400} = .41$$

Thus, the point estimate of $(p_1 - p_2)$ is

$$(\hat{p}_1 - \hat{p}_2) = .33 - .41 = -.08$$

To judge the reliability of our point estimate $(\hat{p}_1 - \hat{p}_2)$, we need to know the characteristics of its performance in repeated independent sampling from two populations. This information is provided by the sampling distribution of $(\hat{p}_1 - \hat{p}_2)$, illustrated in Figure 8.11 and shown in the box.

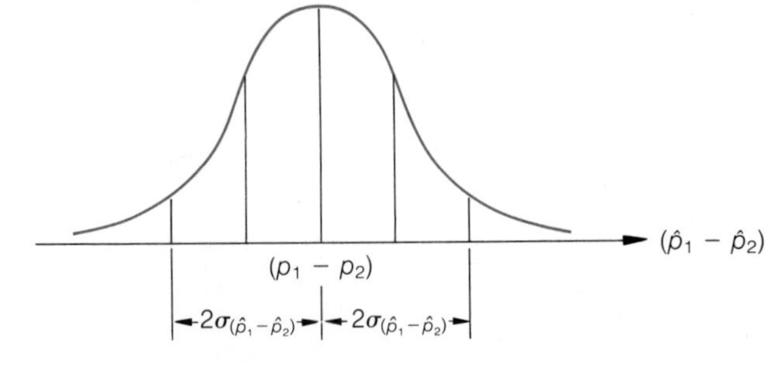

FIGURE 8.11 Sampling Distribution of $(\hat{p}_1 - \hat{p}_2)$

SAMPLING DISTRIBUTION OF $(\hat{p}_1 - \hat{p}_2)$

For sufficiently large sample sizes, n_1 and n_2, the sampling distribution of $(\hat{p}_1 - \hat{p}_2)$, based on independent random samples from two populations, is approximately normal with

Mean $\mu_{(\hat{p}_1-\hat{p}_2)} = (p_1 - p_2)$

and

Standard deviation $\sigma_{(\hat{p}_1-\hat{p}_2)} = \sqrt{\frac{p_1q_1}{n_1} + \frac{p_2q_2}{n_2}}$

where $q_1 = 1 - p_1$ and $q_2 = 1 - p_2$.

It follows that a large-sample confidence interval for $(p_1 - p_2)$ may be obtained as shown in the box.

LARGE-SAMPLE $(1 - \alpha)$ 100% CONFIDENCE INTERVAL FOR $(p_1 - p_2)$

$$(\hat{p}_1 - \hat{p}_2) \pm z_{\alpha/2}\sigma_{(\hat{p}_1-\hat{p}_2)} \approx (\hat{p}_1 - \hat{p}_2) \pm z_{\alpha/2}\sqrt{\frac{\hat{p}_1\hat{q}_1}{n_1} + \frac{\hat{p}_2\hat{q}_2}{n_2}}$$

where $\hat{p}_1$ and $\hat{p}_2$ are the sample proportions of observations with the characteristic of interest.

[*Note:* We have followed the usual procedure of substituting the sample values $\hat{p}_1$, $\hat{q}_1$, $\hat{p}_2$, and $\hat{q}_2$ for the corresponding population values required for $\sigma_{(\hat{p}_1-\hat{p}_2)}$. The approximation is quite satisfactory for large sample sizes.]

EXAMPLE 8.22 Refer to Example 8.21. Estimate the difference between the proportions of Americans in January 1980 and in April 1981 who considered themselves Republicans or Republican-leaning Independents, using a 95% confidence interval.

Solution For a confidence coefficient of .95, we will use $z_{.025} = 1.96$ in constructing the confidence interval. From Example 8.21, we have $n_1 = 1{,}400$, $n_2 = 1{,}400$, $\hat{p}_1 = .33$, and $\hat{p}_2 = .41$. Thus, $\hat{q}_1 = 1 - .33 = .67$, $\hat{q}_2 = 1 - .41 = .59$, and the approximate 95% confidence interval for $(p_1 - p_2)$ is given by

$$(\hat{p}_1 - \hat{p}_2) \pm z_{.025}\sqrt{\frac{\hat{p}_1\hat{q}_1}{n_1} + \frac{\hat{p}_2\hat{q}_2}{n_2}} = (.33 - .41) \pm 1.96\sqrt{\frac{(.33)(.67)}{1{,}400} + \frac{(.41)(.59)}{1{,}400}}$$
$$= .08 \pm .04$$

or $(-.12, -.04)$. Thus, we estimate that the interval $(-.12, -.04)$ encloses the difference $(p_1 - p_2)$ with 95% confidence. It appears that there are between 4% and 12% more Americans in April 1981 than in January 1980 who considered themselves Republicans or Republican-leaning Independents.

EXAMPLE 8.23 Refer to Example 8.14. Suppose the AGA would like to compare the proportion of homes heated by gas in the North with the corresponding proportion in the South. AGA selected a random sample of 60 homes located in the South and found that 34 of the homes use gas as a heating fuel. This information, together with that provided in Example 8.14, is summarized in Table 8.17. Construct a 90% confidence interval for the difference between the proportions of Northern homes and Southern homes which are heated by gas.

TABLE 8.17
Homes Heated by Gas, Example 8.23

	NORTH	SOUTH
Number of homes sampled	$n_1 = 80$	$n_2 = 60$
Number of sampled homes which are heated by gas	42	34

Solution In Example 8.14 we obtained

$$\hat{p}_1 = .525, \qquad \hat{q}_1 = .475, \qquad n_1 = 80$$

Now, for Southern homes, $n_2 = 60$ and the sample proportion of homes heated by gas is

$$\hat{p}_2 = \frac{34}{60} = .567$$

Hence, $\hat{q}_2 = 1 - .567 = .433$.

The 90% confidence interval is then

$$(\hat{p}_1 - \hat{p}_2) \pm z_{.05}\sqrt{\frac{\hat{p}_1\hat{q}_1}{n_1} + \frac{\hat{p}_2\hat{q}_2}{n_2}}$$

$$= (.525 - .567) \pm 1.645\sqrt{\frac{(.525)(.475)}{80} + \frac{(.567)(.433)}{60}}$$

$$= -.042 \pm .140$$

or $(-.182, .098)$.

The 90% confidence interval for $(p_1 - p_2)$ contains the value 0; thus, our samples have not indicated a real difference between the proportions of Northern homes and Southern homes which are heated by gas. The proportion of Northern homes heated by gas could be smaller than the corresponding proportion for Southern homes by as much as .182, or the proportion of Northern homes heated by gas could exceed the proportion for Southern homes by as much as .098.

Small-sample estimation procedures for $(p_1 - p_2)$ will not be discussed here for the reasons outlined at the end of Section 8.4.

EXERCISES **8.40** In a recent clinical study, it was learned that the use of aspirin to alleviate the symptoms of viral infections in children may lead to serious complications (Reyes' Syndrome). A random sample of 500 children with viral infections received no aspirin to alleviate symptoms, and 12 developed Reyes' Syndrome. In a random sample of

450 children with viral infections who were given aspirin, 23 developed Reyes' Syndrome. Construct a 95% confidence interval for the difference in the proportions of children who develop Reyes' Syndrome between those who receive no aspirin and those who receive aspirin during a viral infection. Interpret the interval.

8.41 A Pentagon statistician is evaluating a prototype bomber to see if it can strike on target more often than the existing bomber can. Two independent samples of size 50 each are obtained with the results shown in Table 8.18.

TABLE 8.18
Bomber Accuracy Data, Exercise 8.41

	PROTOTYPE BOMBER	EXISTING BOMBER
Number of bomber runs	$n_1 = 50$	$n_2 = 50$
Number of times the target was hit	42	31

a. Estimate the difference between the proportions of runs which result in target hits for the prototype bomber and the existing bomber, using a 90% confidence interval.

b. Does it appear that the new prototype bomber can strike on target more often than the existing bomber can? Explain.

8.42 One of American industry's most fundamental problems—the stagnation in productivity—has economic experts seeking methods of "reindustrializing" the U.S. One possible answer to the productivity stagnation may be industrial robots (*Time*, December 8, 1980). An industrial robot has a control and memory system, often in the form of a minicomputer, which enables it to be programmed to carry out a number of work routines faster and more efficiently than a human. Since the Japanese now operate most of the robots in the world, it is decided to estimate the difference between the proportions of U.S. and Japanese firms which currently employ at least one industrial robot. Random samples of U.S. and Japanese firms were selected, and the number of firms employing at least one industrial robot recorded. The sample sizes and results are summarized in Table 8.19.

TABLE 8.19
Industrial Robot Study, Exercise 8.42

	U.S.	JAPAN
Number of firms sampled	$n_1 = 75$	$n_2 = 50$
Number of sampled firms which employ at least one industrial robot	16	22

a. Form a 99% confidence interval for the difference between the proportions of U.S. and Japanese firms which currently employ at least one industrial robot.

b. Based on the interval in part a, which country has the higher proportion of firms which employ industrial robots?

c. If the confidence coefficient were decreased to .95, would you expect the width of the interval to increase, decrease, or stay the same?

8.43 A new type of chicken feed, called Ration A, is now on the market. Ration A contains an unusually large amount of a certain feed ingredient which enables farmers to raise heavier chickens. However, farmers are warned that the new feed may be too strong and that the mortality rate for chickens on this special feed may be higher than that with the usual feed. One farmer wished to compare the mortality rate of chickens fed the new Ration A with the mortality rate of chickens fed the current best-selling feed, Ration B. The farmer fed the different feeds to two groups of 50 chickens each, for a one-week period. (Assume that all factors other than feed are the same for both groups.) Of those fed Ration A, 17 died within a week. Of those fed Ration B, 8 died within a week.

a. Assuming that all chicken deaths in the experiment are due to type of feed, find a 92% confidence interval for the difference between the true mortality rates for the two feeds.

b. Based on your interval, part a, can you conclude that there is a difference between the mortality rates for the two feeds? If so, which ration appears to kill the larger proportion of chickens?

8.44 The Environmental Protection Agency (EPA) wishes to compare the proportions of industrial plants which are in violation of government air pollution standards in two different industries: steel and utility. Independent random samples of 130 industrial plants are selected and monitored for each industry, and the number deemed by the EPA to be in violation of air pollution standards is recorded. This information is given in Table 8.20. Construct a 90% confidence interval for the difference between the true proportions of industrial plants violating air pollution standards in the two industries. Interpret the interval.

TABLE 8.20
Industrial Plants Violating Air Pollution Standards, Exercise 8.44

	STEEL	UTILITY
Number of plants	130	130
Number of violations	8	12

8.45 Recently, the Florida Board of Regents proposed that courses offered at all state-supported universities or colleges be indexed under a common four-digit numbering system. In an attempt to determine faculty opinions regarding the common course-numbering system, the Board decided to survey opinions at the state's two largest universities: Florida State University (FSU) and the University of Florida (UF). At FSU, a random sample of 68 faculty yielded 16 in favor of the system and 52 opposed. At UF, a random sample of 103 faculty yielded 29 for the system and 74 against.

a. Establish a 95% confidence interval for the difference between the proportions of FSU and UF faculty who favor the common course-numbering system.

b. One member of the Florida Board of Regents has made the claim that there is no difference between the proportions of FSU and UF faculty who favor the system. Is this a reasonable claim? Explain.

c. How could the Board reduce the width of the confidence interval, part a?

8.8 Summary

This chapter presented the technique of estimation: using sample information to make an inference about the value of a population parameter, or the difference between two population parameters. In each instance, we presented the point estimate of the parameter of interest, its sampling distribution, the general form of a confidence interval, and any assumptions required for the validity of the procedure. These results are collected in Tables 8.21(a) and 8.21(b) on the next two pages.

KEY WORDS

Point estimate
Confidence interval
Confidence coefficient
t distribution
Degrees of freedom
Proportion
Independent samples
Pooled estimate of variance

SUPPLEMENTARY EXERCISES

[*Note:* List the assumptions necessary to ensure the validity of the interval estimation procedures you use to work these exercises.]

8.46 List the assumptions necessary for each of the following inferential techniques:

a. Large-sample confidence interval for a population mean μ.
b. Small-sample confidence interval for a population mean μ.
c. Large-sample confidence interval for the difference $(\mu_1 - \mu_2)$ between population means.
d. Small-sample confidence interval for the difference $(\mu_1 - \mu_2)$ between population means.
e. Large-sample confidence interval for a population proportion p.
f. Large-sample confidence interval for the difference $(p_1 - p_2)$ between population proportions.

8.47 Suppose that a 95% confidence interval for μ is calculated to be (12.31, 19.55). Give a precise interpretation of the phrase, "We are 95% confident that the interval (12.31, 19.55) encloses the true value of μ."

8.48 The U.S. Social Security system, which now collects taxes from approximately 115 million workers and pays benefits to 36 million people, faces serious financial problems because of high inflation, high unemployment, and low worker productivity. Consequently, many Americans are worried that the system will go broke before they can receive their retirement benefits. An Associated Press–NBC News

TABLE 8.21(a)

Summary of Estimation Procedures: One-Population Cases

PARAMETER	POINT ESTIMATE	SAMPLING DISTRIBUTION OF POINT ESTIMATE	$(1-\alpha)$ 100% CONFIDENCE INTERVAL	ASSUMPTIONS
μ Population mean	$\bar{x}$ Sample mean	Approximately normal Mean μ Standard deviation $\sigma/\sqrt{n}$ where σ = Standard deviation of sampled population	$\bar{x} \pm z_{\alpha/2}(\sigma/\sqrt{n})$ $\approx \bar{x} \pm z_{\alpha/2}(s/\sqrt{n})$	$n \geq 30$ (large sample)
μ Population mean	$\bar{x}$ Sample mean	t distribution with $(n-1)$ degrees of freedom	$\bar{x} \pm t_{\alpha/2}(s/\sqrt{n})$ where $t_{\alpha/2}$ is based on $(n-1)$ degrees of freedom	$n < 30$ (small sample) Relative frequency distribution of population is approximately normal
p Proportion of population with specified characteristic(s)	Sample proportion with specified characteristic(s): $\hat{p} = \frac{\text{Number in sample with characteristic}}{n}$ where n = Number of observations sampled	Approximately normal Mean p Standard deviation $\sqrt{\frac{pq}{n}}$ where $q = 1 - p$	$\hat{p} \pm z_{\alpha/2}\sqrt{\frac{pq}{n}}$ $\approx \hat{p} \pm z_{\alpha/2}\sqrt{\frac{\hat{p}\hat{q}}{n}}$ where $\hat{q} = 1 - \hat{p}$	The interval $\hat{p} \pm 2\sqrt{\frac{\hat{p}\hat{q}}{n}}$ does not contain 0 or 1 (large sample)

TABLE 8.21(b)

Summary of Estimation Procedures: Two-Population Cases

PARAMETER	POINT ESTIMATE	SAMPLING DISTRIBUTION OF POINT ESTIMATE	$(1-\alpha)$ 100% CONFIDENCE INTERVAL	ASSUMPTIONS
$(\mu_1 - \mu_2)$ Difference between population means	$(\bar{x}_1 - \bar{x}_2)$ Difference between sample means	Approximately normal Mean $(\mu_1 - \mu_2)$ Standard deviation $\sqrt{\frac{\sigma_1^2}{n_1} + \frac{\sigma_2^2}{n_2}}$ where σ_1^2 and σ_2^2 are the variances of the sampled populations	$(\bar{x}_1 - \bar{x}_2) \pm z_{\alpha/2}\sqrt{\frac{\sigma_1^2}{n_1} + \frac{\sigma_2^2}{n_2}}$ $\approx (\bar{x}_1 - \bar{x}_2) \pm z_{\alpha/2}\sqrt{\frac{s_1^2}{n_1} + \frac{s_2^2}{n_2}}$	$n_1 \geq 30$ (large samples) $n_2 \geq 30$ Samples are randomly and independently selected from the two populations
$(\mu_1 - \mu_2)$ Difference between population means	$(\bar{x}_1 - \bar{x}_2)$ Difference between sample means	t distribution with $(n_1 + n_2 - 2)$ degrees of freedom	$(\bar{x}_1 - \bar{x}_2) \pm t_{\alpha/2}\sqrt{s_p^2\left(\frac{1}{n_1} + \frac{1}{n_2}\right)}$ where $s_p^2 = \frac{(n_1 - 1)s_1^2 + (n_2 - 1)s_2^2}{n_1 + n_2 - 2}$ and $t_{\alpha/2}$ is based on $(n_1 + n_2 - 2)$ degrees of freedom	$n_1 < 30$ (small samples) $n_2 < 30$ **1.** Relative frequency distributions of both populations are approximately normal **2.** Variances of both populations are equal **3.** Samples are randomly and independently selected from the two populations
$(p_1 - p_2)$ Difference between population proportions	$(\hat{p}_1 - \hat{p}_2)$ Difference between sample proportions	Approximately normal Mean $(p_1 - p_2)$ Standard deviation $\sqrt{\frac{p_1q_1}{n_1} + \frac{p_2q_2}{n_2}}$ where $q_1 = 1 - p_1$ and $q_2 = 1 - p_2$	$(\hat{p}_1 - \hat{p}_2) \pm z_{\alpha/2}\sqrt{\frac{p_1q_1}{n_1} + \frac{p_2q_2}{n_2}}$ $\approx (\hat{p}_1 - \hat{p}_2) \pm z_{\alpha/2}\sqrt{\frac{\hat{p}_1\hat{q}_1}{n_1} + \frac{\hat{p}_2\hat{q}_2}{n_2}}$ where $\hat{q}_1 = 1 - \hat{p}_1$ and $\hat{q}_2 = 1 - \hat{p}_2$	The intervals $\hat{p}_1 \pm 2\sqrt{\frac{\hat{p}_1\hat{q}_1}{n_1}}$ and $\hat{p}_2 \pm 2\sqrt{\frac{\hat{p}_2\hat{q}_2}{n_2}}$ do not contain 0 or 1

poll (reported in the Gainesville *Sun,* May 22, 1981) found that 1,184 of the 1,600 American adults interviewed have little or no confidence that the Social Security system will have the funds available to pay them retirement benefits. Use this information to construct a 97% confidence interval for the true proportion of all American adults who have little or no confidence that the Social Security system will have the funds available to pay them retirement benefits. Interpret this interval.

8.49 A power corporation is considering building a floating nuclear power plant a few miles offshore of the Gulf of Mexico. Because there is concern about the possibility of a ship colliding with the floating (but anchored) plant, an estimate of the density of ship traffic in the area is needed. For each of 30 randomly selected days during the summer months of June, July, and August, the number of ships passing within 10 miles of the proposed power plant location was recorded. The sample had a mean of 7.5 ships and a standard deviation of 3.1 ships.

a. Using a 95% confidence interval, estimate the mean number of ships per day passing within 10 miles of the proposed power plant location during the summer months.

b. Ships were also monitored for a random sample of 30 days during the winter months of December, January, and February. This sample produced a mean of 3.8 and a standard deviation of 3.2. Use this additional information to construct a 95% confidence interval for the difference between the mean densities of ship traffic near the proposed power plant location in the summer and winter months.

8.50 A sociologist is interested in estimating the average number of children per licensed family day-care center in the United States. A random sample of 200 licensed family day-care centers is selected, and the number of children cared for in each is recorded. The results are summarized as follows:

$$\bar{x} = 3.8$$

$$s = 1.5$$

Estimate the population mean number of children per licensed family day-care center in the United States, using a 99% confidence interval. Interpret the interval.

8.51 Manufacturers of golf balls are continually improving the durability of their product. Suppose that a manufacturer wishes to compare the durability of its golf balls with that of a competitor's. Ten balls of each brand were randomly selected for the experiment. Each ball is put into a machine which consistently hits the ball with the same force that a typical golfer uses on the course. The number of hits required until the outer covering is cracked is recorded for each ball. The results of the durability test are presented in Table 8.22.

TABLE 8.22
Golf Ball Durability Data, Exercise 8.51

	MANUFACTURER	COMPETITOR
Number of balls	10	10
Average number of hits until cracked	270.3	240.7
Standard deviation	44.5	31.3

a. Construct a 98% confidence interval for the difference between the mean numbers of hits until cracking for the manufacturer's and competitor's golf balls.

b. Does the interval, part a, indicate that a difference in average durability exists between the two brands of golf balls? Explain.

c. If the respective sample sizes were increased to 100 golf balls each, do you think the width of the interval, part a, would increase or decrease? Explain.

8.52 What proportion of "repaired" trucks need to return to the maintenance shop with the same mechanical problem? In order to investigate this phenomenon, an automotive engineer randomly sampled 150 trucks that had been repaired in a maintenance shop and released for use. He found that 16 of the trucks were back in the shop with the same problem within 30 days of being released. Estimate the true proportion of repaired trucks which return to the maintenance shop with the same problem within 30 days. Use a 90% confidence interval.

8.53 One method of evaluating treatments for cancer in clinical experiments is to compare the remission rates under each treatment. (Remission occurs if the growth of cancerous cells is stopped completely.) At a large clinic, 200 patients suffering from a particular type of cancer have agreed to participate in such a clinical experiment. (All patients are at approximately the same stage in the development of the disease, all begin the experiment at the same time, and it can be reasonably assumed that the 200 patients represent a random sample of all people afflicted with this particular form of cancer.) Each patient was administered one of two treatments: the standard treatment currently in use or a recently developed treatment not previously tried on human subjects. The conduct of the study was *blind,* i.e., the patients did not know which of the drugs they were receiving. In actuality, half of the patients were receiving the standard drug and half the new drug. At the end of six weeks of treatment, the frequency of remission for each treatment was recorded, as shown in Table 8.23.

TABLE 8.23
Frequency of Remission, Exercise 8.53

	STANDARD	NEW
Number of patients	$n_1 = 100$	$n_2 = 100$
Number of remissions	47	53

a. Construct a 99% confidence interval for the difference between the proportions of cancer patients receiving the standard drug and those receiving the new drug who experience remission.

b. Interpret the interval, part a. Is there evidence of a difference between the remission rates of the two treatments?

8.54 Chemical plants must be regulated to prevent poisoning of fish in nearby rivers or streams. One of the measurements made on fish to evaluate potential toxicity of chemicals is the total length reached by adults. If a river or stream is inhabited by an abundance of adult fish with total lengths less than the average adult length of their species, we have strong evidence that the river is being chemically contaminated. A chemical plant, under investigation for chlorine poisoning of a stream, has hired a biologist to estimate the mean length of fathead minnows (the main inhabitants of the stream) exposed to 20 micrograms of chlorine per liter of water. The biologist captures 11 newborn fish of this species from the stream and rears them in aquaria with this chlorine concentration. The length of each is measured after ten weeks maturation. The results are $\bar{x} = 27.5$ mm, $s = 2.6$ mm. Construct a 90% confidence interval for the true mean length of fathead minnows reared in chlorine-contaminated water. Interpret the interval.

8.55 Two preservatives, tested and determined safe for use in red meats, are to be compared for their effects on retarding spoilage. Fifteen cuts of fresh red meat are treated with preservative A and fifteen cuts of fresh red meat are treated with preservative B. The cuts are placed in a chilled container and the number of hours until spoilage begins recorded for each. The results are summarized in Table 8.24.

TABLE 8.24 Meat Spoilage Data, Exercise 8.55

PRESERVATIVE A	PRESERVATIVE B
$n_1 = 15$	$n_2 = 15$
$\bar{x}_1 = 257.3$ hours	$\bar{x}_2 = 249.8$ hours
$s_1 = 12.8$ hours	$s_2 = 9.7$ hours

a. Estimate the difference between the true mean times until spoilage for the two preservatives using a 98% confidence interval.
b. Can you detect a real difference between mean times until spoilage for the two preservatives from the 98% confidence interval, part a? Explain.
c. Give two ways in which you could decrease the width of the interval, part a. Which of these do you recommend?

8.56 When food prices began their rapid increase in the early 1970's, some of the media began periodically to purchase a grocery basket full of food at supermarkets around the country. The same items were bought at each store so that the food prices could be compared. Suppose you, the consumer, wish to estimate the average price for a grocery basket of food in your home town. You purchase the specified items at twelve supermarkets selected at random from among those supermarkets in your home town. The mean and standard deviation of the costs at the twelve supermarkets are: $\bar{x} = \$47.17$, $s = \$5.88$.
a. Find a 95% confidence interval for the true average cost of a grocery basket of food in your home town.
b. Prior to your survey, a consumer report claimed that the average cost of a basket of food (specifying the same items you bought) purchased in your home town is \$43. Based on your interval, part a, is the consumer report claim reasonable? Explain.

8.57 An economist is interested in comparing the proportions of eligible workers who are currently receiving unemployment compensation in the states of Pennsylvania and Ohio. Independent random samples of 50 eligible workers in each state are selected and interviewed. The economist found that 3 of the Pennsylvania workers interviewed and 7 of the Ohio workers interviewed are currently receiving unemployment compensation. Construct a 90% confidence interval for the difference between the proportions of eligible workers currently receiving unemployment compensation in the two states. Interpret the interval.

8.58 The *Wall Street Journal* (February 9, 1981) reported that Tyco Industries, Inc., the largest producer of electrical trains in the U.S., will begin to supplement its standard line of toy trains and racing cars with electrical trucks. Market research has convinced the corporation that "modern youngsters prefer the cab of an 18-wheeler to a coal car or caboose." This trend is reflected in Tyco's national advertising campaign slogan: "Some day your child may want to be a doctor. Right now, he wants to drive a truck." Suppose that one goal of Tyco's market research was to obtain an accurate estimate of the true proportion of children in the 6–10 year age group who would prefer playing with a toy truck rather than a toy train. Each of 1,000 youngsters was presented with a Tyco toy truck and a Tyco toy train, and the number who selected the toy truck was recorded. If 615 youngsters selected the toy truck, estimate the true proportion of children 6–10 years old who prefer toy trucks to toy trains with a 97% confidence interval. Interpret the interval.

8.59 A new type of band has been developed by a dental laboratory for children who wear braces. The bands are designed to be more comfortable and better looking, but some dental researchers fear that they are slower in realigning teeth than the old braces. The parents of children who require braces may then choose not to purchase the new bands in order to avoid paying for the dentist's costly services over a longer period of time. With this in mind, the dental researchers conducted an experiment to compare the mean wearing times necessary for the new bands and old braces to correct a specific type of misalignment. Two hundred children were randomly assigned, one hundred to each group. A summary of the data is shown in Table 8.25.

TABLE 8.25 Wearing Time of Braces until Correction, Exercise 8.59

OLD BRACES	NEW BANDS
$n_1 = 100$	$n_2 = 100$
$\bar{x}_1 = 412$ days	$\bar{x}_2 = 437$ days
$s_1 = 87$ days	$s_2 = 55$ days

a. Find a 99% confidence interval for the difference between the mean wearing times for the two types of braces.
b. Based on the interval, part a, can the dental researchers conclude that either type of band has a lower mean wearing time than the other?

8.60 Despite gloomy news about drought, sinkholes, forest fires, and the economy, a recent Florida opinion poll (as reported in the Gainesville *Sun*, May 21, 1981) says that one in four Floridians believes crime is now the worst problem in the state. This result is based on a random sample of 568 Florida residents, 145 of whom indicated that, in their opinion, crime is the state's worst problem.

a. Construct a 95% confidence interval for the true proportion of Floridians who believe that crime is the worst problem in the state.

b. Consider the following statement extracted from the Gainesville *Sun* article. "For this survey, it is 95 percent certain that [the] percentage listed for all 568 respondents [will not] be more than 4.1 percentage points—plus or minus—off the actual mark for the [entire] state." Interpret this statement.

8.61 There have now been seven space probes made by rockets bearing scientific equipment with the intent of studying the planets closer to the sun than Earth. Suppose four of the probes have studied Venus, while three have studied Mercury. One major point of interest to scientists is the comparison of surface temperatures for the two planets. Suppose the seven probes produced the summary information in Table 8.26.

TABLE 8.26 Surface Temperatures, Exercise 8.61

MERCURY	VENUS
$n_1 = 4$	$n_2 = 3$
$\bar{x}_1 = 250°F$	$\bar{x}_2 = 400°F$
$s_1 = 11°F$	$s_2 = 14°F$

a. Construct a 90% confidence interval for the difference between the mean surface temperatures of the two planets.
b. Interpret the interval. Is there evidence of a difference between the mean surface temperatures of the two planets?

8.62 A public utilities company is considering increasing the price of electricity during peak-load periods of the day (9:00 AM—4:00 PM) and reducing the price during off-peak periods. The company hopes that this revised pricing structure will force customers to conserve energy during the period when electrical consumption is the highest, and that it will eventually lead to an overall reduction in monthly consumption. To determine the effectiveness of the plan, the company randomly selected and notified 45 customers of the change in pricing policy, effective during the month of August. A random sample of 60 customers, independent of the first, was also selected, but these customers were billed under the regular pricing schedule during August. The total electric consumption (in kilowatt-hours) during the month was recorded for each. The data are summarized in Table 8.27. Estimate the difference between the true mean August electrical consumptions for the two groups of customers, using a 90% confidence interval. Interpret the interval. Is the revised pricing policy effective in reducing mean monthly electrical consumption?

TABLE 8.27 August Electrical Consumption, (Kilowatt-hours) Exercise 8.62

	REGULAR PRICING POLICY	REVISED POLICY
Sample size	60	45
Mean consumption	2,115	2,003
Standard deviation	450	388

8.63 Unoccupied seats on flights cause airlines to lose revenue. Suppose a large airline wants to estimate its average number of unoccupied seats per flight over the past year. To accomplish this, the records of 225 flights are randomly selected from the files, and the number of unoccupied seats is noted for each of the sampled flights. The sample mean and standard deviation are

$$\bar{x} = 11.6 \text{ seats}$$
$$s = 4.1 \text{ seats}$$

Choose a confidence coefficient and then construct an interval estimate of the mean number of unoccupied seats per flight over the past year.

8.64 A traffic engineer conducted a study of vehicular speeds on a segment of street that had the posted speed limit changed several times. When the posted speed limit on the street was 35 miles per hour (mph), the engineer monitored the speeds of 50 randomly selected vehicles traversing the street, and counted the number of violations of the speed limit. After the speed limit had been lowered to 30 miles per hour, the engineer again monitored the speeds of 50 randomly selected vehicles. The results of the study are shown in Table 8.28. Construct a 99% confidence interval for the difference between the proportions of vehicles exceeding the two posted speed limits. Interpret the interval.

TABLE 8.28
Speed Limit Violations, Exercise 8.64

	35 MPH	30 MPH
Number of vehicles sampled	$n_1 = 50$	$n_2 = 50$
Number exceeding the posted speed limit	13	40

8.65 In this age of escalating professional sports salaries, the salaries of players' agents have also increased tremendously. One agent, who represents both professional football and basketball players, desires to estimate the difference between the average annual salaries of National Football League (NFL) players' agents and National Basketball Association (NBA) players' agents. Independent random samples of $n_1 = 17$ NFL players' agents and $n_2 = 21$ NBA players' agents are selected and the annual salaries of each recorded. The data are summarized in Table 8.29.

TABLE 8.29
Player Agent Salaries, Exercise 8.65

NFL	NBA
$n_1 = 17$	$n_2 = 21$
$\bar{x}_1 = \$187{,}330$	$\bar{x}_2 = \$100{,}440$
$s_1 = \$53{,}610$	$s_2 = \$32{,}720$

a. Estimate the difference between the true average annual salaries of NFL players' agents and NBA players' agents, using a 95% confidence interval.

b. A comment by an NFL team owner in a national publication reads as follows: "There is no significant difference between the salaries of NFL players' agents and NBA players' agents." Comment on this claim.

CASE STUDY 8.1 The Crazy Daisy Shave

"Many adolescent girls can't wait until they are old enough to shave their legs: They consider it a sign of growing up. But by the time they are adults—and have been shaving a while—the thrill is gone," writes Judy Hill in her weekly newspaper column on consumer affairs, "Watch This Space" (Gainesville *Sun,* April 26, 1981). Hill continues: "Shaving is a pain. There's the soap or the cream all over the place, there's the stubble, the nicks, the cuts—and the cost. But a magazine advertisement for [a new disposable shaver] promises to end at least some of those problems."

According to the advertisement, the new twin-blade Daisy shaver from Gillette shaves legs smoother, closer, and safer than any single-blade shaver. And, says the ad, "The Crazy Daisy Shave costs less than 25 cents. That's really crazy!"

"How reliable is Gillette's claim?" Hill asked Greg Niblet, assistant manager of public relations for the Gillette Company. Niblet explained that the claim was substantiated after lengthy but uncontrolled consumer testing. "Since so many millions (of women) shave (their legs), and they all shave differently, the consumers used the product in their own homes and shaved according to their own routine."

In order to substantiate the advertisement's claim, Hill conducted her own survey. Using a test similar to Gillette's, she asked 13 women ranging in age from 19 to 50 to shave one leg with a Daisy and the other leg with a Lady Bic (a single-blade disposable razor for women). After shaving according to their own routine, they each gave their opinion as to which of the two shavers gave them a "smoother, closer, and safer" shave. Hill reports the results as follows: "Nine of the women chose the Daisy as being superior in all three categories; two said they could tell no difference; one chose the Lady Bic as being superior in all three categories, and one said the Bic gave a close shave while the Daisy gave a smoother shave."

a. Give a point estimate of the true proportion of women (tested under conditions similar to Hill's survey) who prefer the Daisy shaver to the Lady Bic shaver in all three categories.

b. Using your answer to part a, construct a 98% confidence interval for the true proportion of women who prefer the Daisy shaver to the Lady Bic shaver.

c. Why might the interval estimate, part b, lead to unreliable inferences about the true population proportion? Explain.

d. Give two ways in which you could reduce the width of the interval in part b. Which of the two do you recommend?

CASE STUDY 8.2 Consumer Attitudes toward Automated Supermarket Checkers

In Chapter 1, Exercise 1.1, we introduced you to the data of Appendix C, the checkout times for 500 grocery shoppers at each of two supermarkets, supermarket A and supermarket B. Supermarket B, you will recall, employs automated checkers in contrast to the familiar manual checkers of supermarket A. After five years of experimentation and evaluation, these automated checkout systems, more formally known as Universal Product Code (UPC) symbol-scanning systems, are on the threshold of widespread use by retail food marketers. The system was originally developed (in 1972) to benefit both retailers and customers. Retailers who install scanning equipment can expect higher labor productivity in the form of faster checkouts and more efficient labor scheduling, labor savings from not price-marking individual items, and the accumulation of valuable marketing information by the scanner's microcomputer. The consumer who shops at a store with an automated checker is expected to benefit in four ways: (1) decreased checkout time; (2) increased accuracy; (3) a detailed receipt tape; and (4) lower prices as cost savings are passed on by retailers.

Pommer, Berkowitz, and Walton* conducted a study designed to elicit the true nature of consumer feelings on the benefits of the UPC scanning system. Questionnaires were distributed to a sample of shoppers at three different stores that employ automated checkers. (Two stores were located in Illinois, and the third was in Minnesota.) A total of 161 questionnaires were returned in readable form. The following is a brief description of some of the more important results of the survey:

1. In response to questions concerning the removal of prices from individual food items at scanning-equipped stores, 67% agreed that price removal would make it difficult to shop for the best buy, 51% agreed that price removal would allow store owners to raise prices without the consumer's knowledge, and 45% believed price removal would allow stores to take advantage of consumers.
2. The checkout service advantages of the scanning systems were recognized by the consumers sampled in that 78% believed that less time was spent waiting in line, 71% agreed that checkout service was better, and 62% wished to see automated checkers installed at other stores.
3. Only 17% of the sample indicated that the presence of scanning systems would influence their store-selection decisions, and even fewer, 15%, reported that this technological change caused them to shop more at a particular store.
4. Although 60% agreed with the statement that the scanner does not make mistakes, a full 50% of the sampled consumers still pay attention to automated checker accuracy.
5. The detailed receipt tape was deemed both easy to understand (98% agreed) and helpful in verifying purchases (89% agreed).

The survey results of (2) indicate that a large majority of consumers believe that automated checking systems speed up the supermarket checkout process (at least for the three stores used in the survey). Let us use the data, Appendix C, to compare the mean checkout service times at the two supermarkets, A and B.

a. In Exercise 8.35, you constructed a 90% confidence interval for the difference between the mean checkout service times at the two supermarkets, based on samples of size $n_1 = n_2 = 6$. Does the interval provide evidence that the mean checkout service time at supermarket B (automated checkers) is less than the mean checkout service time at supermarket A (manual checkers)?
b. Now select independent random samples of size $n_1 = n_2 = 50$ from the checkout service time values for supermarkets A and B, Appendix C. Compute $\overline{x}$ and s for the two samples.

---

*M. D. Pommer, E. N. Berkowitz, and J. R. Walton. "UPC scanning: An assessment of shopper response to technological change." *Journal of Retailing*, 1980, *56*, 25–44.

c. Use the sample information, part b, to construct a 90% confidence interval for the difference between the mean checkout service times at the two supermarkets. Compare this interval to the interval you found in Exercise 8.35. How have the increased sample sizes affected the width of the 90% confidence interval?

d. Does the interval, part c, provide evidence that the mean checkout service time at supermarket B is less than the mean checkout service time at supermarket A? Which interval, the interval of Exercise 8.35 or the interval of part c of this case study, would you recommend for making inferences concerning the difference between the mean checkout service times at the two supermarkets? Why?

CASE STUDY 8.3 Sample Surveys: The Foot in the Door and the Door in the Face Approach

Sample surveys often suffer from a lack of a suitable number of respondents; any inferences derived from surveys with low response rates could very well be biased. Many strategies have been devised for the purpose of increasing survey response rates. Although these compliance-gaining tactics originated in the nonbusiness behavioral sciences (social psychology, personality, etc.), much attention has recently been given to them in business and marketing literature. Marketing researchers are just beginning to investigate and understand how these behavioral influence techniques can be used successfully in a business setting.

Mowen and Cialdini* give brief descriptions of various manipulative strategies. The most popular of these among business and marketing researchers is the "foot-in-the-door" or, more simply, the "foot" principle. Mowen and Cialdini write: "In using this compliance-gaining tactic, a requester first makes a request so small that nearly anyone would comply, in effect getting a 'foot in the door.' After compliance with the first request occurs, a second, larger request is made—actually the one desired from the outset." For example, Hansen and Robinson** conducted an experiment in which a random group of subjects were contacted by phone and initially asked whether they had purchased a new car within the last three years. If they had, they were asked some basic questions on general perceptions toward automobile dealers, such as, "All car dealers overcharge on their repair work; do you agree or disagree?" After the brief (no longer than five minutes) "foot-in-the-door" interview, the subject was asked if he/she would be willing to participate in the mail portion of the survey (the desired, larger request). This "foot" technique has been shown to increase response rates in a number of business settings, typical of the one described above. The key to the success of the "foot" principle, say Hansen and Robinson, is that it allows the respondent to become involved in the subject area, which eventually leads to a greater degree of participation in the subsequent larger request.

A second strategy discussed by Mowen and Cialdini is labelled the "door-in-the-face" principle. In the "face" approach, "the [person administering the survey] begins with an initial request so large that nearly everyone refuses it (i.e., the door is

---

*J. C. Mowen and R. B. Cialdini. *Journal of Marketing Research,* May 1980, *17,* 253–258.

**R. A. Hansen and L. M. Robinson. *Journal of Marketing Research,* August 1980, *17,* 359–363.

slammed in his face). [After the first refusal,] the requester then retreats to a smaller favor—actually the one desired from the outset." The "face" principle is based upon the social rule of reciprocation that states, "One should make concessions to those who make concessions to oneself." Mowen and Cialdini explain: "The requester's movement from the initial, extreme favor to the second, more moderate one is seen by the [potential respondent] as a concession. To reciprocate this concession, the [respondent] must move from his or her initial position of noncompliance with the large request to a position of compliance with the smaller request." The key to the successful "face" approach is that the respondent perceive the original request as being legitimate, and that a concession was clearly made in the movement from the large to the small request.

An example of the "door-in-the-face" technique is given by Mowen and Cialdini. Subjects were approached by experimenters representing a fictitious corporation, the California Mutual Insurance Company. The experimenters' initial request went as follows:

> Hello, I'm doing a survey for the California Mutual Insurance Company. For each of the last twelve years, we have been on campus to gather survey information on safety in the home or dorm. The survey takes about one hour to administer. Would you be willing to take an hour, right now, to answer the questions?

After the subject declined to participate, the experimenter would make the second, smaller request:

> Oh, . . . well, look, one part of the survey is particularly important and is fairly short. It will take only fifteen minutes to administer. If you take fifteen minutes right now to complete this short survey, it would really help us out.

The "foot-in-the-door" and "door-in-the-face" strategies present an interesting contrast in sample survey designs. The "foot" approach uses an initial, small request to enhance the likelihood of compliance with a second, larger (desired) request; the "face" approach uses an initial, large request to increase the response rate on a second, smaller (desired) request. Suppose that we wish to compare the response rates of the "foot" and "face" techniques for a sample survey on insurance coverage in the home (similar to the experiment devised by Mowen and Cialdini). Two sample surveys are designed for our experiment, both intended to gather identical information on home insurance. However, one utilizes the "foot-in-the-door" principle, the other the "door-in-the-face" principle. The critical second request (i.e., the desired request) is the same in each. Suppose that we randomly and independently select two groups of subjects, 210 in the first group and 180 in the second. Each of the 210 subjects in the first group is interviewed using the "face" approach, while each of the 180 subjects in the second group is interviewed using the "foot" approach. We are interested in comparing the response rates (i.e., the proportions of subjects who agree to the critical second request) in each group.

a. What is the parameter of interest, in the words of the problem?

b. Suppose that 84 "face" subjects and 78 "foot" subjects responded affirmatively to the critical second request. Give a point estimate for the difference between the true response rates for the two groups.

c. Construct a 95% confidence interval for the difference between the response rates.

d. Interpret the interval, part c. Is there evidence of a difference between the response rates for the two groups of subjects?

REFERENCES

Book, S. *Statistics: Basic techniques for solving applied problems.* New York: McGraw-Hill, 1977. Chapter 4.

Crawford, C. "Floridians favor a war on crime." Gainesville *Sun,* May 21, 1981, 1A and 10A.

Hansen, R. A., & Robinson, L. M. "Testing the effectiveness of alternative foot-in-the-door manipulations." *Journal of Marketing Research,* August 1980, *17,* 359–363.

Hill, J. "Watch this space—Gillette Daisy shaver does a job on the legs." Gainesville *Sun,* April 26, 1981.

Johnson, R. *Elementary statistics.* 3d ed. North Scituate, Mass.: Duxbury, 1980. Chapters 9, 10, and 11.

McClave, J. T., & Dietrich, F. H. *Statistics.* 2d ed. San Francisco: Dellen, 1982. Chapters 7 and 8.

Mendenhall, W., Scheaffer, R., & Wackerly, D. *Mathematical statistics with applications.* 2d ed. Boston: Duxbury, 1981. Chapter 8.

Mowen, J. C., & Cialdini, R. B., "On implementing the door-in-the-face compliance technique in a business context." *Journal of Marketing Research,* May 1980, *17,* 253–258.

Pommer, M. D., Berkowitz, E. N., & Walton, J. R. "UPC scanning: An assessment of shopper response to technological change." *Journal of Retailing,* 1980, *56,* 25–44.

Snedecor, G. W., & Cochran, W. G. *Statistical methods.* 6th ed. Ames, Iowa: Iowa State University Press, 1967. Chapters 2, 3, 10, and 11.

Nine

Collecting Evidence to Support a Theory: General Concepts of Hypothesis Testing

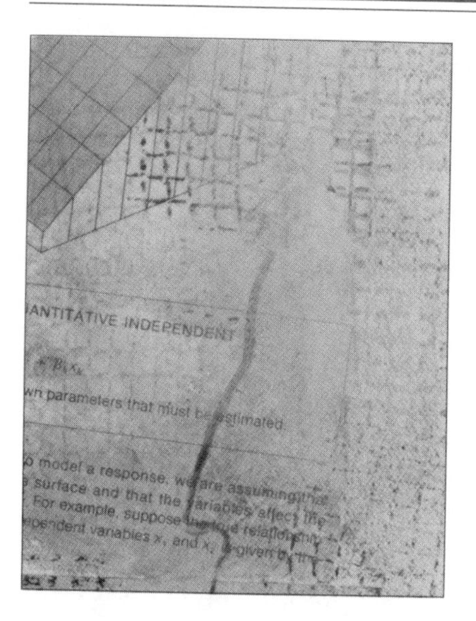

The famous *Schlitz versus Budweiser* confrontation (Mug-to-Mug) was viewed by sports enthusiasts across the country during the half-time break of the December 28, 1980, NFL–AFC wildcard football game between the Houston Oilers and the Oakland Raiders. According to Schlitz, 100 "loyal" Budweiser drinkers were selected to taste each of two unmarked mugs of beer, one Budweiser and the other Schlitz. Live, on television, 46 of the Budweiser drinkers chose the mug containing Schlitz. Ignoring the favorable publicity obtained from this live taste test, do the test results, 46 out of 100, suggest that Schlitz might carve a larger share of the market for itself? If all confirmed Budweiser drinkers were to conduct a similar nonlabelled taste test, would as many as 40% prefer Schlitz? In this chapter we will learn how sample data can be used to make decisions about population parameters, and we will examine the Mug-to-Mug confrontation in greater detail in Case Study 9.1.

Contents

9.1 Introduction

In the next two chapters we turn our attention to another method of inference-making, called ***hypothesis testing.*** The procedures to be discussed are useful in situations where we are interested in making a decision about a parameter value, rather than obtaining an estimate of its value. For example, we may be interested in deciding whether the mean tar content, μ, of a particular brand of cigarette exceeds a certain value, say, 4 milligrams; whether the proportion of Americans who believe the President is doing a good job exceeds .5; whether the mean life of a product manufactured by industry A is less than the mean life of a similar product manufactured by industry B; or whether the proportion of blackjack games won by a professional card counter is larger than the proportion of games won by an experienced, but typical player; etc.

This chapter will treat the general concepts involved in hypothesis testing; specific applications will be demonstrated in Chapter 10.

9.2 Formulation of Hypotheses

When a researcher in any field sets out to test a new theory, he or she first formulates an ***hypothesis,*** or claim, which he or she believes to be true. For example, a college recruiter may claim that the mean starting salary of graduates of the College of Liberal Arts is less than the mean starting salary of graduates of the College of Engineering. In statistical terms, the hypothesis that the researcher tries to establish is called the ***alternative hypothesis,*** or ***research hypothesis.*** To be paired with this alternative hypothesis, which the researcher believes is true, is the ***null hypothesis,*** which is the opposite of the alternative hypothesis. In this way, the null and alternative hypotheses, both stated in terms of the appropriate population parameters, describe two possible states of nature which cannot simultaneously be true. When the researcher begins to collect information about the phenomenon of interest, he or she generally tries to present evidence which lends support to the alternative hypothesis. As you will subsequently learn, we take an indirect approach to obtaining support for the alternative hypothesis. Instead of trying to show that the alternative hypothesis is true, we attempt to produce evidence to show that the null hypothesis (which may often be interpreted as "no change from the status quo") is false.

DEFINITION 9.1

A statistical ***hypothesis*** is a statement about the value of a population parameter.

DEFINITION 9.2

The hypothesis which we hope to disprove or reject is called the ***null hypothesis,*** denoted by H_0.

DEFINITION 9.3

The hypothesis for which we wish to gather supporting evidence is called the ***alternative hypothesis,*** denoted by H_a.

EXAMPLE 9.1 Formulate appropriate null and alternative hypotheses for testing the college recruiter's theory that the mean starting salary of graduates of the College of Liberal Arts is less than the mean starting salary of graduates of the College of Engineering.

Solution The hypotheses must be stated in terms of a population parameter or parameters. We will thus define

μ_1 = The mean starting salary of graduates of the College of Liberal Arts

and

μ_2 = The mean starting salary of graduates of the College of Engineering

The recruiter wishes to support the claim that μ_1 is less than μ_2; therefore, the null and alternative hypotheses, in terms of these parameters, are

H_0: $(\mu_1 - \mu_2) = 0$ (i.e., $\mu_1 = \mu_2$; there is no difference between the mean starting salaries of graduates of the Colleges of Liberal Arts and Engineering)

H_a: $(\mu_1 - \mu_2) < 0$ (i.e., $\mu_1 < \mu_2$; the mean starting salary of graduates of the College of Liberal Arts is less than that for the College of Engineering)

EXAMPLE 9.2 Since 1970, cigarette advertisements are required by law to carry the following statement: "Warning: The Surgeon General has determined that cigarette smoking is dangerous to your health." However, this warning is often put in inconspicuous corners of the advertisements and in small type. Consequently, a spokesperson for the Federal Trade Commission (FTC) believes that over 80% of those who read cigarette advertisements fail to see the warning. Specify the null and alternative hypotheses which would be used in testing the spokesperson's theory.

Solution The FTC spokesperson wishes to make an inference about p, where p is the proportion of all readers of cigarette advertisements who fail to see the Surgeon

General's warning. In particular, the FTC spokesperson wishes to collect evidence to support his or her claim that p is greater than .80; thus, the null and alternative hypotheses are

H_0: $p = .80$

H_a: $p > .80$

Observe that the statement of H_0, in these examples and in general, is written with an equality (=) sign. In Example 9.2, you may have been tempted to write the null hypothesis as H_0: $p \leq .80$. However, since the alternative of interest is that $p > .80$, then any evidence which would cause you to reject the null hypothesis H_0: $p = .80$ in favor of H_a: $p > .80$ would also cause you to reject H_0: $p = p'$, for any value of p' which is ***less*** than .80. In other words, H_0: $p = .80$ represents the worst possible case, from the researcher's point of view, if in fact the alternative hypothesis is ***not*** correct. Thus, for mathematical ease, we combine all possible situations for describing the opposite of H_a into one statement involving an equality.

EXAMPLE 9.3 Periodically, a metal lathe is checked to determine if it is producing machine bearings with a mean diameter of ½ inch. If the mean diameter of the bearings is larger or smaller than ½ inch, then the process is out of control and will need to be adjusted. Formulate the null and alternative hypotheses which could be used in testing whether the bearing production process is out of control.

Solution We define the following parameter:

μ = The true mean diameter (in inches) of all bearings produced by the metal lathe

If either $\mu > ½$ or $\mu < ½$, then the metal lathe's production process is out of control. Since we wish to be able to detect either possibility, the null and alternative hypotheses would be

H_0: $\mu = ½$ (i.e., the process is in control)

H_a: $\mu \neq ½$ (i.e., the process is out of control)

An alternative hypothesis may hypothesize a change from H_0 in a particular direction, or it may merely hypothesize a change without specifying a direction. In Examples 9.1 and 9.2, the researcher is interested in detecting departure from H_0 in one particular direction: interest focuses on whether the mean starting salary for graduates of the College of Liberal Arts is ***less than*** the mean starting salary for graduates of the College of Education in Example 9.1, and on whether the proportion of cigarette advertisement readers who fail to see the Surgeon General's warning is ***greater than*** .80 in Example 9.2. These two tests are called ***one-tailed tests.*** In contrast, Example 9.3 illustrates a ***two-tailed test*** in which we are interested in whether the mean diameter of the machine bearings differs in either direction from ½ inch, i.e., whether the process is out of control.

DEFINITION 9.4

A ***one-tailed test*** of an hypothesis is one in which the alternative is directional, and includes either the symbol "$<$" or "$>$."

DEFINITION 9.5

A ***two-tailed test*** of an hypothesis is one in which the alternative does not specify departure from H_0 in a particular direction; such an alternative will be written with the symbol "$\neq$."

EXAMPLE 9.4 A large mail-order company has placed an order for 10,000 electric blenders with a supplier on condition that no more than 1% of the blenders will be defective. In order to check whether the shipment contains too many defectives, the company will test the null hypothesis

$$H_0: \quad p = .01$$

where p is the true proportion of defective electric blenders in the shipment of 10,000. Formulate the appropriate alternative hypothesis for the company.

Solution The mail-order company is interested in detecting whether the true proportion of defectives in the shipment of 10,000 blenders is larger than 1%, for if it is the case that $p > .01$, then the supplier has violated the contractual agreement. Thus, the alternative hypothesis of interest to the company is

$$H_a: \quad p > .01$$

Note that the null hypothesis

$$H_0: \quad p = .01$$

actually represents all possible situations for which the supplier has met the contractual obligation that no more than 1% of the blenders are defective, i.e., $p \leq .01$. Since the alternative is directional, i.e., since the company is interested in detecting a departure from H_0 in the direction of p-values larger than .01, a one-tailed test is to be performed.

EXAMPLE 9.5 The economy of the state of Nevada depends heavily upon tourists, especially those who visit the city of Las Vegas. A state representative, wishing to determine whether there is a difference between the mean amounts of money spent by tourists visiting the state during the years 1980 and 1981, decides to test the null hypothesis

$$H_0: \quad (\mu_1 - \mu_2) = 0$$

where μ_1 and μ_2 represent the mean amounts of money tourists spent while in Nevada during 1980 and 1981, respectively. Specify the appropriate alternative hypothesis for this test.

Solution The state representative is interested only in detecting whether there is a difference between the mean amounts spent by tourists in Nevada during 1980 and during 1981. If there is a difference, then $\mu_1 \neq \mu_2$ or, equivalently, the difference between means $(\mu_1 - \mu_2)$ differs from 0. Thus, the alternative hypothesis of interest to the representative is the two-tailed alternative

$$H_a: \quad (\mu_1 - \mu_2) \neq 0$$

EXERCISES

9.1 Explain the difference between an alternative hypothesis and a null hypothesis.

In Exercises 9.2–9.6, formulate the appropriate null and alternative hypotheses. Define all notation used.

9.2 It is desired to test whether the mean price of straight-leg jeans at all retail outlets in New York City is greater than $35.00 per pair.

9.3 Cannibalism among chickens is common when the birds are confined in small areas. A breeder and seller of live chickens wants to test whether the mortality rate due to cannibalism of a certain breed of chickens is less than .04.

9.4 In order to determine whether car ownership is detrimental to academic achievement, a university investigator wishes to test whether there is a difference between the mean grade-point averages of car owners and non-car owners.

9.5 A medical researcher would like to determine whether the proportion of males admitted to a hospital because of heart disease is greater than the corresponding proportion of females.

9.6 Federal scientists tracking storms in Florida must forecast whether the storms will become hurricanes. Suppose we wish to test whether the accuracy rate of the forecasts (i.e., the proportion of times the scientists correctly forecast the outcome of the storm) is greater than .90.

9.7 State whether the tests in Exercises 9.2–9.6 are one-tailed or two-tailed.

9.3 Conclusions and Consequences for an Hypothesis Test

The goal of any hypothesis-testing situation is to make a decision; in particular, we will decide whether to reject the null hypothesis, H_0, in favor of the alternative hypothesis, H_a. Although we would like to be able to make a correct decision always, we must remember that the decision will be based on sample information, and thus we are subject to make one of two types of error.

DEFINITION 9.6

A ***Type I error*** occurs if we reject a null hypothesis which is in fact true. The probability of committing a Type I error is usually denoted by α.

DEFINITION 9.7

A ***Type II error*** occurs if we fail to reject a null hypothesis which is in fact false. The probability of making a Type II error is usually denoted by β.

The alternative hypothesis can be either true or false; further, we will make a decision either to reject or not reject the null hypothesis. Thus, there are four possible situations which may arise in testing an hypothesis; these are summarized in Table 9.1.

TABLE 9.1 Conclusions and Consequences for Testing an Hypothesis

		TRUE STATE OF NATURE	
		H_a false	H_a true
DECISION	Do not reject H_0	Correct decision	Type II error
	Reject H_0	Type I error	Correct decision

Note that we risk a Type I error only if the null hypothesis is rejected, and we risk a Type II error only if the null hypothesis is not rejected. Thus, we may make no error, or we may make either a Type I error (with probability α), or a Type II error (with probability β), but not both. There is an intuitively appealing relationship between the probabilities for the two types of error: As α increases, β decreases; similarly, as β increases, α decreases. The only way to reduce α and β simultaneously is to increase the amount of information available in the sample, i.e., to increase the sample size.

EXAMPLE 9.6 Refer to Example 9.3. Specify what Type I and Type II errors would represent, in terms of the problem.

Solution A Type I error is that of incorrectly rejecting the null hypothesis. In our example, this would occur if we concluded that the process is out of control if in fact the process is in control, i.e., that the mean bearing diameter is different from ½ inch, if in fact the mean is equal to ½ inch. The consequence of making such an error would be that unnecessary time and effort would be expended to repair the metal lathe.

A Type II error, that of incorrectly failing to reject the null hypothesis, would occur if we concluded that the mean bearing diameter is equal to ½ inch, if in fact the mean differs from ½ inch. The practical significance of making a Type II error is that the metal lathe would not be repaired, when in fact the process is out of control.

Since the probability of making a Type I error is controlled by the researcher (how to do this will be explained in Section 9.4), it is often used as a measure of the reliability of the conclusion and thus has a special name.

DEFINITION 9.8

The probability, α, of making a Type I error is called the ***level of significance*** for an hypothesis test.

You may note that we have carefully avoided stating a decision in terms of "accept the null hypothesis H_0." Instead, if the sample does not provide enough evidence to support the alternative hypothesis H_a, we prefer a decision "not to reject H_0." This is because, if we were to "accept H_0," the reliability of the conclusion would be measured by β, the probability of a Type II error. However, the value of β is not constant, but depends on the specific alternative value of the parameter and is difficult to compute in most testing situations.

In summary, we recommend the following procedure for formulating hypotheses and stating conclusions: State the hypothesis you wish to support as the alternative hypothesis, H_a. The null hypothesis, H_0, will be the opposite of H_a and will contain an equality sign. Then, if the sample evidence supports the alternative hypothesis, you will reject the null hypothesis and will know that the probability of having made an incorrect decision (if in fact H_0 is true) is α, a quantity which you can manipulate to be as small as you wish. If the sample does not provide sufficient evidence to support the alternative, then conclude that the null hypothesis cannot be rejected on the basis of your sample. In this situation, you may wish to obtain a larger sample in order to collect more information about the phenomenon under study.

EXAMPLE 9.7 The logic used in hypothesis testing has often been likened to that used in the courtroom in which a defendant is on trial for committing a crime.

a. Formulate appropriate null and alternative hypotheses for judging the guilt or innocence of the defendant.

b. Interpret the Type I and Type II errors in this context.

c. If you were the defendant, would you want α to be small or large? Explain.

Solution **a.** Under our judicial system, a defendant is "innocent until proven guilty." That is, the burden of proof is *not* on the defendant to prove his or her innocence; rather, the court must collect sufficient evidence to support the claim that the defendant is guilty. Thus, the null and alternative hypotheses would be

H_0: The defendant is innocent.

H_a: The defendant is guilty.

b. The four possible outcomes are shown in Table 9.2. A Type I error would be to conclude that the defendant is guilty, if in fact he or she is innocent; a Type II error would be to conclude that the defendant is innocent, if in fact he or she is guilty.

TABLE 9.2
Conclusions and Consequences, Example 9.7

		TRUE STATE OF NATURE	
		Defendant is innocent	**Defendant is guilty**
DECISION OF COURT	**Defendant is innocent**	Correct decision	Type II error
	Defendant is guilty	Type I error	Correct decision

c. Most would agree that, in this example, the Type I error is by far the more serious. Thus, we would want α, the probability of committing a Type I error, to be very small indeed.

A convention that is generally observed when formulating the null and alternative hypotheses of any statistical test is to state H_0 so that the possible error of incorrectly rejecting H_0 (Type I error) is considered more serious than the possible error of incorrectly failing to reject H_0 (Type II error). In many cases, the decision as to which error, Type I or Type II, is more serious is admittedly not as clear-cut as that of Example 9.7; a little experience will help to minimize this potential difficulty.

EXERCISES

9.8 Refer to Exercise 9.4. Interpret the Type I and Type II errors in the context of the exercise. What would be the practical consequences of each for the university investigator?

9.9 Explain why each of the following statements is incorrect:

a. The probability that the null hypothesis is correct is equal to α.

b. If the null hypothesis is rejected, then the test proves that the alternative hypothesis is correct.

c. $\alpha + \beta = 1$ in all statistical tests of hypothesis.

9.10 Refer to Exercise 9.6. Specify what Type I and Type II errors would represent, in terms of the problem.

9.11 Why do we avoid stating a decision in terms of "accept the null hypothesis H_0"?

9.12 Over the last month, a large supermarket chain received many consumer complaints about the quantity of chips in 16-ounce bags of a particular brand of potato chips. Suspecting that the complaints were merely the result of the potato chips settling to the bottom of the bags during shipping, but wanting to be able to assure its customers they were getting their money's worth, the chain decided to test the following hypotheses concerning μ, the mean weight (in ounces) of a bag of potato chips in the next shipment of chips received from their largest supplier.

$$H_0: \mu = 16$$
$$H_a: \mu < 16$$

If there is evidence that $\mu < 16$, then the shipment would be refused and a complaint registered with the supplier.

a. What is a Type I error, in terms of the problem?
b. What is a Type II error, in terms of the problem?
c. Which type of error would the chain's customers view as more serious? Which type of error would the chain's supplier view as more serious?

9.4 Test Statistics and Rejection Regions

In this section we will describe how to arrive at a decision in an hypothesis-testing situation. Recall that when making any type of statistical inference (of which hypothesis testing is a special case), we collect information by obtaining a random sample from the population(s) of interest. In all our applications, we will assume that the appropriate sampling process has already been carried out.

EXAMPLE 9.8 Suppose we wish to test the hypotheses

$$H_0: \mu = 72$$
$$H_a: \mu < 72$$

What is the general format for carrying out a statistical test of hypothesis?

Solution The first step is to obtain a random sample from the population of interest. The information provided by this sample, in the form of a sample statistic, will help us decide whether to reject the null hypothesis. The sample statistic upon which we base our decision is termed the ***test statistic.***

The second step, then, is to determine a test statistic which is reasonable in the context of a given hypothesis test. For this example, we are hypothesizing about the value of the population mean μ. Since our best guess at the value of μ is the sample mean $\bar{x}$ (see Section 8.2), it seems reasonable to use $\bar{x}$ as a test statistic. We will learn how to choose the test statistic for other hypothesis-testing situations in the examples that follow.

The third step is to specify the range of computed values of the test statistic for which the null hypothesis will be rejected. That is, what specific values of the test statistic will lead you to reject the null hypothesis in favor of the alternative hypothesis? These specific values are collectively known as the ***rejection region*** for the test. For this example, we would need to specify the values of $\bar{x}$ which would lead us to believe that H_a is true, i.e., that μ is less than 72. Again, we will learn how to find the appropriate rejection region in later examples.

DEFINITION 9.9

The ***test statistic*** is a sample statistic, computed from the information provided by the sample, upon which the decision concerning the null and alternative hypotheses is based.

DEFINITION 9.10

The ***rejection region*** is the set of computed values of the test statistic for which the null hypothesis will be rejected.

Once the rejection region has been specified, the fourth step is to use the data in the sample to compute the value of the test statistic. Finally, we make our decision by observing whether the computed value of the test statistic lies within the rejection region. If in fact the computed value falls within the rejection region, we will reject the null hypothesis; otherwise, we fail to reject the null hypothesis.

An outline of this hypothesis-testing procedure is given in the box. Each step in this approach will be explained in greater detail as we proceed.

OUTLINE FOR TESTING AN HYPOTHESIS

1. Obtain a random sample from the population(s) of interest. (In all our applications, we will assume that the appropriate sampling process has already been carried out.)
2. Determine a ***test statistic*** which is reasonable in the context of the given hypothesis test.
3. Specify the ***rejection region,*** the range of computed values of the test statistic for which the null hypothesis will be rejected.
4. Use the data in the sample to compute the value of the test statistic.
5. Observe whether the computed value of the test statistic lies within the rejection region. If so, reject the null hypothesis; otherwise, fail to reject the null hypothesis.

Recall that the null and alternative hypotheses will be stated in terms of specific population parameters. Thus, in step 2, we decide on a test statistic, to be computed from the sample, which will provide information about the target parameter.

EXAMPLE 9.9 Refer to Example 9.2, in which the spokesperson for the FTC wishes to test

H_0: $p = .80$

H_a: $p > .80$

where p is the proportion of all readers of cigarette advertisements who fail to notice the Surgeon General's warning. Suggest a test statistic which may be useful in deciding whether to reject H_0.

Solution Since the target parameter is a population proportion, p, it would be logical to use the sample proportion, $\hat{p}$, as a tool in the decision-making process. Recall from Section 8.4 that $\hat{p}$ is the point estimate of p used in the interval estimation procedure.

EXAMPLE 9.10 Refer to Example 9.1, in which we wish to test

H_0: $(\mu_1 - \mu_2) = 0$

H_a: $(\mu_1 - \mu_2) < 0$

where μ_1 and μ_2 are the population mean starting salaries of all graduates of the Colleges of Liberal Arts and Engineering, respectively. Suggest an appropriate test statistic in the context of this problem.

Solution The parameter of interest is $(\mu_1 - \mu_2)$, the difference between two population means. Therefore, we will use $(\bar{x}_1 - \bar{x}_2)$, the difference between corresponding sample means, as a basis for deciding whether to reject H_0. If the difference between the sample means, $(\bar{x}_1 - \bar{x}_2)$, falls greatly below the hypothesized value of $(\mu_1 - \mu_2) = 0$, then we have evidence that disagrees with our null hypothesis. In fact, it would support the alternative hypothesis that $(\mu_1 - \mu_2) < 0$. Again, we are using the point estimate of the target parameter as the test statistic in the hypothesis-testing approach.

GUIDELINE FOR STEP 2 OF HYPOTHESIS TESTING

In general, when the hypothesis test involves a specific population parameter, the test statistic to be used is the conventional ***point estimate*** of that parameter.

In step 3, we divide all possible values of the test statistic (or a standardized version of it) into two sets: the ***rejection region*** and its complement. If the computed value of the test statistic falls within the rejection region, we reject the null hypothesis. If the computed value of the test statistic does not fall within the rejection region, we fail to reject the null hypothesis.

EXAMPLE 9.11 Refer to Example 9.9. For the hypothesis test

H_0: $p = .80$

H_a: $p > .80$

indicate which decision you may make for each of the following values of the test statistic:

a. $\hat{p} = .99$ b. $\hat{p} = .65$ c. $\hat{p} = .84$

Solution a. If 99% of the cigarette ad readers in the sample failed to notice the Surgeon General's warning, then much doubt is cast upon the null hypothesis. In other words, ***if the null hypothesis were true*** (i.e., if p is in fact equal to .80), then we would be very unlikely to observe a sample proportion $\hat{p}$ as large as .99. We would thus tend to reject the null hypothesis on the basis of information contained in this sample.

b. Since the alternative of interest is $p > .80$, this value of the sample proportion, $\hat{p} = .65$, provides no support for H_a. Thus we would ***not*** reject H_0 in favor of H_a: $p > .80$, based on this sample.

c. Does a sample value of $\hat{p} = .84$ cast sufficient doubt on the null hypothesis to warrant its rejection? Although the sample proportion $\hat{p} = .84$ is larger than the null hypothesized value of $p = .80$, is this due to chance variation, or does it provide strong enough evidence to conclude in favor of H_a? We think you will agree that the decision is not as clear-cut as in parts a and b, and that we need a more formal mechanism for deciding what to do in this situation.

We now illustrate how to determine a rejection region which takes into account such factors as the sample size and the maximum probability of a Type I error that you are willing to tolerate.

EXAMPLE 9.12 Refer to Example 9.9. Specify completely the form of the rejection region for a test of

H_0: $p = .80$

H_a: $p > .80$

at a significance level of $\alpha = .05$.

Solution We are interested in detecting a directional departure from H_0; in particular, we are interested in the alternative that p is ***greater than*** .80. Now, what values of the sample proportion $\hat{p}$ would cause us to reject H_0 in favor of H_a? Clearly, values of $\hat{p}$ which are "sufficiently greater" than .80 would cast doubt on the null hypothesis. But how do we decide whether a value, say $\hat{p} = .84$, is "sufficiently greater" than .80 to reject H_0? A convenient measure of the distance between $\hat{p}$ and .80 is the z-score, which "standardizes" the value of the test statistic $\hat{p}$:

$$z = \frac{\hat{p} - \mu_{\hat{p}}}{\sigma_{\hat{p}}} = \frac{\hat{p} - .80}{\sqrt{\frac{(.80)(.20)}{n}}}$$

(The z-score is obtained by using the values of $\mu_{\hat{p}}$ and $\sigma_{\hat{p}}$ which would be valid if the null hypothesis were true, i.e., if $p = .80$.) The z-score then gives us a measure of how many standard deviations the observed $\hat{p}$ is from what we would expect to observe ***if H_0 were true.***

Now examine Figure 9.1 and observe that the chance of obtaining a value of $\hat{p}$ more than 1.645 standard deviations above .80 is only .05, ***if in fact the true value of p is .80.*** (We are assuming that the sample size is large enough to ensure that the sampling distribution of $\hat{p}$ is approximately normal.) Thus, if we observe a sample proportion which is more than 1.645 standard deviations above .80, then either H_0 is true and a relatively rare (with probability .05) event has occurred, ***or*** H_a is true and the population proportion exceeds .80. We would tend to favor the latter explanation for obtaining such a large value of $\hat{p}$, and would then reject the null hypothesis.

In summary, our rejection region for this example consists of all values of z which are greater than 1.645 (i.e., all values of $\hat{p}$ which are more than 1.645 standard deviations above .80). The ***critical value*** 1.645 is shown in Figure 9.1. In this situation, the probability of a Type I error, that is, deciding in favor of H_a if in fact H_0 is true, is equal to $\alpha = .05$.

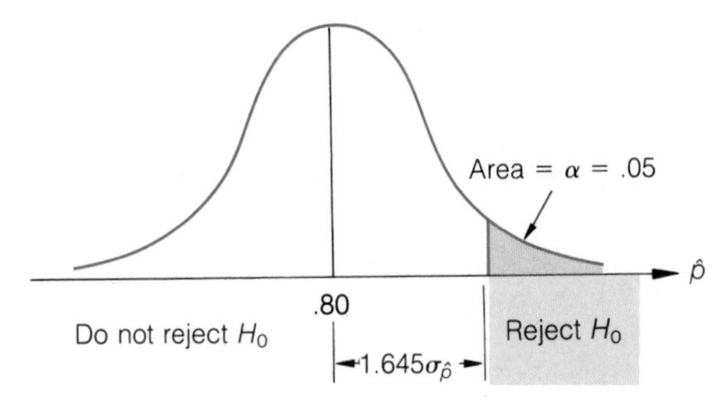

FIGURE 9.1
Location of Rejection Region, Example 9.12

DEFINITION 9.11

In the specification of the rejection region for a particular test of hypothesis, the value at the boundary of the rejection region is called the ***critical value.***

EXAMPLE 9.13 Refer to Example 9.8. Specify the form of the rejection region for a test of

H_0: $\mu = 72$

H_a: $\mu < 72$

at significance level $\alpha = .01$.

Solution Here, we wish to be able to detect the directional alternative that μ is ***less than*** 72; in this case it is "sufficiently small" values of the test statistic $\bar{x}$ which would cast doubt on the null hypothesis. As in Example 9.12, we will standardize the value of the test statistic to obtain a measure of the distance between $\bar{x}$ and the null hypothesized value of 72:

$$z = \frac{(\bar{x} - \mu_{\bar{x}})}{\sigma_{\bar{x}}} = \frac{\bar{x} - 72}{\sigma/\sqrt{n}} \approx \frac{\bar{x} - 72}{s/\sqrt{n}}$$

This z-value tells us how many standard deviations the observed $\bar{x}$ is from what would be expected ***if H_0 were true.*** (Again, we have assumed that $n \geq 30$ so that the sampling distribution of $\bar{x}$ will be approximately normal. The appropriate modifications for small samples will be indicated in Chapter 10.)

Figure 9.2 shows us that, ***if in fact the true value of μ is 72,*** then the chance of observing a value of $\bar{x}$ more than 2.33 standard deviations below 72 is only .01. Thus, at significance level (probability of Type I error) equal to .01, we would reject the null hypothesis for all values of z which are less than -2.33, i.e., for all values of $\bar{x}$ which lie more than 2.33 standard deviations below 72.

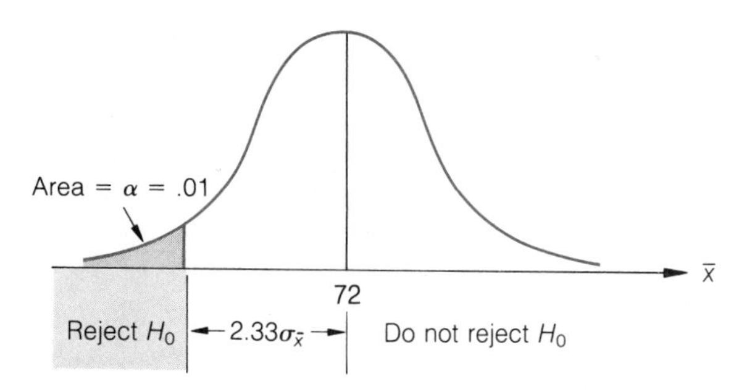

FIGURE 9.2 Location of Rejection Region, Example 9.13

EXAMPLE 9.14 Specify the form of the rejection region for a test of

$$H_0\text{:} \quad (\mu_1 - \mu_2) = 0$$
$$H_a\text{:} \quad (\mu_1 - \mu_2) \neq 0$$

where we are willing to tolerate a .05 chance of making a Type I error.

Solution For this two-sided (nondirectional) alternative, we would reject the null hypothesis for "sufficiently small" ***or*** "sufficiently large" values of the test statistic, $(\bar{x}_1 - \bar{x}_2)$. We will standardize the value of $(\bar{x}_1 - \bar{x}_2)$, assuming $n_1 \geq 30$ and $n_2 \geq 30$, to obtain a measure of how far the observed difference $(\bar{x}_1 - \bar{x}_2)$ lies from zero, the value which would be expected ***if H_0 were true:***

$$z = \frac{(\bar{x}_1 - \bar{x}_2) - \mu_{(\bar{x}_1 - \bar{x}_2)}}{\sigma_{(\bar{x}_1 - \bar{x}_2)}} = \frac{(\bar{x}_1 - \bar{x}_2) - 0}{\sqrt{\dfrac{\sigma_1^2}{n_1} + \dfrac{\sigma_2^2}{n_2}}} \approx \frac{(\bar{x}_1 - \bar{x}_2)}{\sqrt{\dfrac{s_1^2}{n_1} + \dfrac{s_2^2}{n_2}}}$$

Now, from Figure 9.3, we note that the chance of observing a difference between the sample means, $(\bar{x}_1 - \bar{x}_2)$, more than 1.96 standard deviations below 0 ***or*** more than 1.96 standard deviations above 0, ***if in fact H_0 is true,*** is only $\alpha = .05$. Thus, the rejection region consists of two sets of values: We will reject H_0 if z is either less than -1.96 or greater than 1.96. For this rejection rule, the probability of a Type I error is .05.

The three previous examples all exhibit certain common characteristics regarding the rejection region, as indicated in the box on page 300.

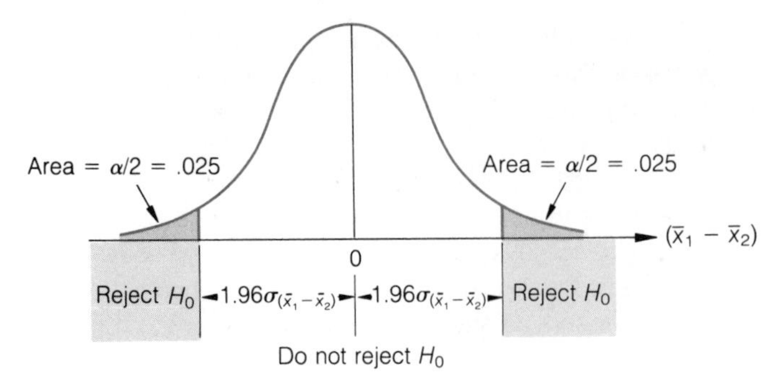

FIGURE 9.3 Location of Rejection Region, Example 9.14

GUIDELINES FOR STEP 3 OF HYPOTHESIS TESTING

1. The value of α, the probability of a Type I error, is specified in advance by the researcher. It can be made as small or as large as desired; typical values are $\alpha = .01$, .02, .05, and .10. For a fixed sample size, the size of the rejection region decreases as the value of α decreases (see Figure 9.4). That is, for smaller values of α, more extreme departures of the test statistic from the null hypothesized parameter value are required to permit rejection of H_0.
2. The test statistic (i.e., the point estimate of the target parameter) is standardized to provide a measure of how great is its departure from the null hypothesized value of the parameter. The standardization is based on the sampling distribution of the point estimate, ***assuming H_0 is true.*** (It is through the standardization that the rejection rule takes into account the sample sizes.)

$$\text{Standardized test statistic} = \frac{\text{Point estimate} - \text{Hypothesized value}}{\text{Standard deviation of point estimate}}$$

3. The location of the rejection region depends upon whether the test is one-tailed or two-tailed, and upon the prespecified significance level, α.
   a. For a one-tailed test in which the symbol ">" occurs in H_a, the rejection region will consist of values in the upper tail of the sampling distribution of the standardized test statistic. The critical value is selected so that the area to its right is equal to α.
   b. For a one-tailed test in which the symbol "<" appears in H_a, the rejection region will consist of values in the lower tail of the sampling distribution of the standardized test statistic. The critical value is selected so that the area to its left is equal to α.
   c. For a two-tailed test, in which the symbol "≠" occurs in H_a, the rejection region will consist of two sets of values. The critical values are selected so that the area in each tail of the sampling distribution of the standardized test statistic is equal to $\alpha/2$.

Steps 4 and 5 of the hypothesis-testing approach require the computation of a test statistic from the sample information. Then we determine if its standardized value lies within the rejection region in order to make a decision about whether to reject the null hypothesis.

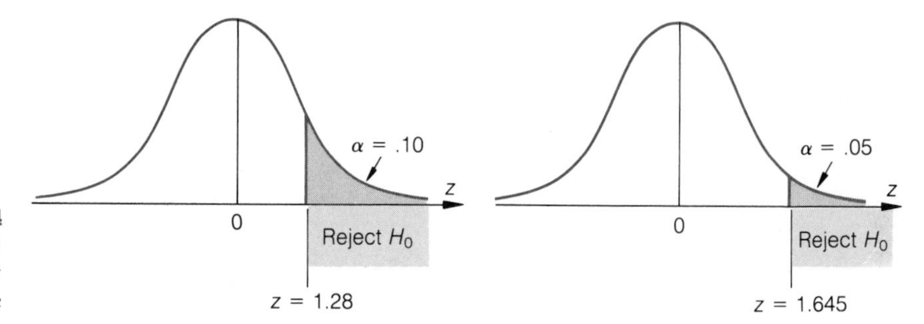

FIGURE 9.4 Size of the Upper-Tail Rejection Region for Different Values of α

EXAMPLE 9.15 Refer to Examples 9.9 and 9.12. Suppose that, in a random sample of $n = 30$ people who were asked to read a cigarette advertisement, 25 indicated that they failed to notice the Surgeon General's warning. Perform a test of

$$H_0: \quad p = .80$$
$$H_a: \quad p > .80$$

at a significance level of $\alpha = .05$.

Solution In Example 9.12, we determined the following rejection rule for the given value of α and the alternative hypothesis of interest:

Reject H_0 if $z > 1.645$.

Now, the test statistic is $\hat{p}$, the sample proportion of cigarette ad readers who failed to notice the Surgeon General's warning, and

$$\hat{p} = \frac{25}{30} = .83$$

The test statistic is standardized, assuming H_0 is true:

$$z = \frac{\hat{p} - \mu_{\hat{p}}}{\sigma_{\hat{p}}} = \frac{\hat{p} - .80}{\sqrt{\frac{(.80)(.20)}{30}}} = \frac{.83 - .80}{\sqrt{\frac{(.80)(.20)}{30}}} = .41$$

This value does not lie within the rejection region (see Figure 9.5 on page 302). We thus fail to reject H_0 and conclude there is insufficient evidence to support the FTC spokesperson's claim that over 80% of all cigarette ad readers fail to notice the Surgeon General's warning. (Note that we do ***not*** conclude that H_0 is true; rather, we state that we have insufficient evidence to reject H_0.)

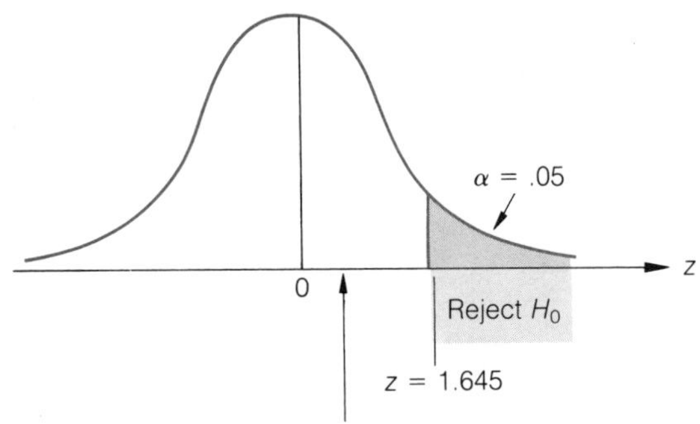

FIGURE 9.5 Location of Rejection Region and Test Statistic, Example 9.15

EXAMPLE 9.16 Refer to Example 9.14. Suppose that random samples of sizes $n_1 = 50$ and $n_2 = 80$ from the target populations yielded the information shown in Table 9.3. (Data may represent scores for males and females on a statistics examination.) Perform a test, at significance level .05, of

$$H_0: \ (\mu_1 - \mu_2) = 0$$
$$H_a: \ (\mu_1 - \mu_2) \neq 0$$

TABLE 9.3 Data for Example 9.16

DATA FROM POPULATION 1	DATA FROM POPULATION 2
$n_1 = 50$	$n_2 = 80$
$\bar{x}_1 = 79$	$\bar{x}_2 = 82$
$s_1 = 5$	$s_2 = 3$

Solution In Example 9.14, we determined the form of the rejection region for this two-tailed test at significance level $\alpha = .05$:

Reject H_0 if $z < -1.96$ or if $z > 1.96$.

For a large-sample test about the difference between two means, the point estimate is $(\bar{x}_1 - \bar{x}_2)$, which is standardized as follows:

$$z = \frac{(\bar{x}_1 - \bar{x}_2) - \mu_{(\bar{x}_1 - \bar{x}_2)}}{\sigma_{(\bar{x}_1 - \bar{x}_2)}} \approx \frac{(\bar{x}_1 - \bar{x}_2) - 0}{\sqrt{\dfrac{s_1^2}{n_1} + \dfrac{s_2^2}{n_2}}} = \frac{79 - 82}{\sqrt{\dfrac{(5)^2}{50} + \dfrac{(3)^2}{80}}} = -3.8$$

This value lies within the rejection region shown in Figure 9.6; we therefore conclude that there is a significant difference between the means of the two populations (i.e., the mean scores for males and females on this statistics examination are significantly different). We acknowledge that we may be making a Type I error, with probability $\alpha = .05$.

In the sequel, we will not differentiate between the test statistic (point estimate) and its standardized value. We will employ the common usage, in which "test statistic" refers to the standardized value of the point estimate for the target parameter. Thus, in Example 9.16, the value of the ***test statistic*** was computed to be $z = -3.8$.

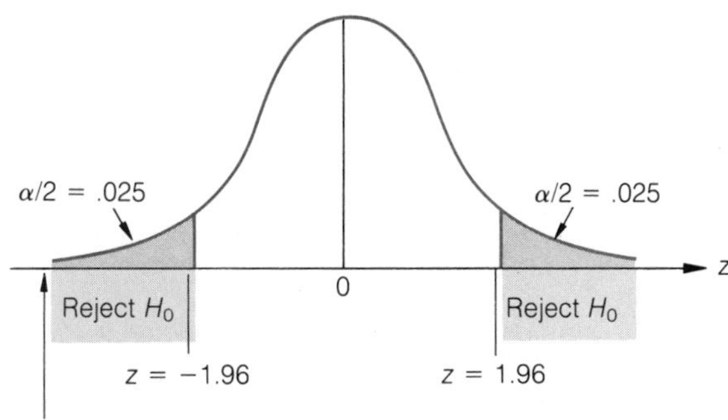

FIGURE 9.6 Location of Rejection Region and Test Statistic, Example 9.16

EXERCISES

9.13 Suppose it is desired to test

$$H_0: \mu = 65$$
$$H_a: \mu \neq 65$$

at significance level $\alpha = .02$. Specify the form of the rejection region. (You may assume that the sample size will be sufficient to guarantee the approximate normality of the sampling distribution of $\bar{x}$.)

9.14 Indicate the form of the rejection region for a test of

$$H_0: (p_1 - p_2) = 0$$
$$H_a: (p_1 - p_2) > 0$$

Assume that the sample sizes will be appropriate to apply the normal approximation to the sampling distribution of $(\hat{p}_1 - \hat{p}_2)$, and that the maximum tolerable probability of committing a Type I error is .05.

9.15 For each of the following rejection regions, determine the value of α, the probability of a Type I error:

a. $z < -1.96$ **b.** $z > 1.645$ **c.** $z < -2.58$ or $z > 2.58$

9.16 Assuming that the sample sizes will be appropriate to apply the normal approximation to the sampling distribution of $(\bar{x}_1 - \bar{x}_2)$, specify the form of the rejection region for a test of

$$H_0: (\mu_1 - \mu_2) = 0$$
$$H_a: (\mu_1 - \mu_2) < 0$$

at significance level:

a. $\alpha = .01$ **b.** $\alpha = .02$ **c.** $\alpha = .05$ **d.** $\alpha = .10$

Locate the rejection region, α, and the critical value on a sketch of the standard normal curve for each part of the exercise.

9.17 Refer to Exercise 9.12. The supermarket chain randomly samples $n = 50$ bags of potato chips from the shipment and measures the weight of the chips in each. The mean weight of the sample was determined to be $\bar{x} = 15.7$ ounces and the sample standard deviation was $s = .8$ ounce.

a. Calculate the appropriate (standardized) test statistic for this test.

b. Specify the form of the rejection region if the level of significance is $\alpha = .01$. Locate the rejection region, α, and the critical value on a sketch of the standard normal curve.

c. Use the results of parts a and b to make the proper conclusion in terms of the problem.

9.18 A certain species of beetle produces offspring with either blue eyes or black eyes. Suppose a biologist wishes to determine which, if either, of the two eye colors is dominant for this species of beetle. Let p represent the true proportion of offspring possessing blue eyes. If in fact the beetles produce blue-eyed and black-eyed offspring at an equal rate, then $p = .5$. Thus, the biologist desires to test

$$H_0\colon\ p = .5$$
$$H_a\colon\ p \neq .5$$

a. Give the form of the rejection region if the biologist is willing to tolerate a Type I error probability of $\alpha = .05$. Locate the rejection region, α, and the critical value(s) on a sketch of the standard normal distribution. (Assume that the sample size will be sufficient to guarantee approximate normality of the sampling distribution of $\hat{p}$.)

b. In a random sample of 75 beetle offspring, 41 were found to have blue eyes. Calculate the value of the appropriate test statistic.

c. In terms of the problem, what is the proper conclusion for the biologist to make?

9.5 Summary

In this chapter, we have introduced the logic and general concepts involved in the statistical procedure of hypothesis testing. The techniques will be illustrated more fully with practical applications in Chapter 10.

KEY WORDS

Hypothesis testing
Null hypothesis
Alternative hypothesis
One-tailed test
Two-tailed test
Type I error
Type II error
Significance level
Rejection region
Critical value
Test statistic

SUPPLEMENTARY EXERCISES

9.19 Explain the difference between the null hypothesis and the alternative hypothesis in a statistical test.

9.20 Define each of the following:

a. Type I error
b. Type II error
c. α
d. β
e. Critical value
f. Level of significance
g. One-tailed test
h. Two-tailed test

9.21 What are the two possible conclusions in a statistical test of hypothesis?

9.22 In a test of hypothesis, is the size of the rejection region increased or decreased when α, the level of significance, is reduced?

9.23 When do you risk making a Type I error? A Type II error?

9.24 If the calculated value of the test statistic falls in the rejection region, we reject H_0 in favor of H_a. Does this prove that H_a is correct? Explain.

9.25 Specify the form of the rejection region for a two-tailed test of hypothesis conducted at significance level:

a. $\alpha = .01$ **b.** $\alpha = .02$ **c.** $\alpha = .04$

Locate the rejection region, α, and the critical values on a sketch of the standard normal curve for each part of the exercise. (Assume the sampling distribution of the test statistic is approximately normal.)

9.26 For each of the following rejection regions, determine the value of α, the probability of a Type I error:

a. $z > 2.576$ **b.** $z < -1.29$ **c.** $z < -1.645$ or $z > 1.645$

Locate the rejection region, α, and the critical value(s) on a sketch of the standard normal curve.

In Exercises 9.27–9.30, formulate the appropriate null and alternative hypotheses. Define all notation used.

9.27 A manufacturer of fishing line wishes to show that the mean breaking strength of a competitor's 22-pound line is really less than 22 pounds.

9.28 As part of a study of the relationship between birth order and college success, an investigator wants to determine whether there is a difference in the proportions of college graduates and nongraduates who were first-born or only children.

9.29 An environmentalist will conduct an investigation to determine if the mean acidity of rainfall in the city of Los Angeles is greater than the mean acidity of rainfall in the city of Denver.

9.30 A craps player who has experienced a long run of bad luck at the craps table wants to test whether the casino dice are "loaded," i.e., whether the proportion of "sevens" occurring in many tosses of the two dice is different from $\frac{1}{6}$ (if the dice are fair, the probability of tossing a "seven" is $\frac{1}{6}$).

9.31 Recently, Fiat Motors of North America, Inc., has been advertising its new 2-year, 24,000-mile warranty. The warranty covers the engine, transmission, and drive train of all new Fiat-made cars for up to two years or 24,000 miles, whichever comes first. However, one Fiat dealer believes the 2-year part of the warranty is unnecessary since μ, the true mean number of miles driven by Fiat owners in two years, is greater than 24,000 miles. Suppose that the dealer wishes to test

$$H_0: \mu = 24{,}000$$

$$H_a: \mu > 24{,}000$$

at a significance level of $\alpha = .01$.

a. Give the form of the rejection region for this test. Locate the rejection region, α, and the critical value on a sketch of the standard normal curve. (Assume the sample will be sufficient to guarantee normality of the test statistic.)

b. A random sample of 32 new Fiat owners produced the following statistics on number of miles driven after two years: $\bar{x} = 24{,}517$ and $s = 1{,}866$. Calculate the appropriate test statistic.

c. Make the appropriate conclusion in terms of the problem.

d. Describe a Type I error in terms of the problem.

e. Describe a Type II error in terms of the problem.

9.32 Refer to Exercise 9.30. In the next 100 tosses of the two dice at the craps table, 5 resulted in the outcome of "seven."

a. Compute the test statistic appropriate for testing the hypothesis of Exercise 9.30.

b. Set up the rejection region for the test if the craps player is willing to tolerate a Type I error probability of $\alpha = .10$. Locate the pertinent quantities on a sketch of the standard normal curve.

c. Give a full conclusion in terms of the problem.

d. What are the consequences of a Type I error for the craps player?

9.33 A physiologist is studying the effect of birth control pills on exercise capacity. One way to determine a person's exercise capacity is to measure their maximal oxygen uptake (in milliliters per kilogram of body weight) during a treadmill session. To determine whether a difference exists between the average maximal oxygen uptake of female subjects who have never taken the pill and female subjects who have been on the pill for one year, the physiologist will test

$$H_0: (\mu_1 - \mu_2) = 0$$

$$H_a: (\mu_1 - \mu_2) \neq 0$$

where μ_1 is the average maximal oxygen uptake of females who have never taken the pill and μ_2 is the average maximal oxygen uptake of females who have been on the pill for one year.

a. If the test is performed at significance level $\alpha = .02$, specify the form of the rejection region. Locate the rejection region, α, and the critical values on a sketch of the standard normal curve. (Assume that the samples are sufficiently large to guarantee that the sampling distribution of $(\bar{x}_1 - \bar{x}_2)$ is approximately normal.)

b. The physiologist randomly sampled $n_1 = 37$ females who have never taken the pill and $n_2 = 33$ females who have been on the pill for a year, and measured the maximal oxygen uptake of each during a treadmill session. The results are given in the table. Use the sample data to calculate the appropriate test statistic.

NO BIRTH-CONTROL PILLS	ON THE PILL FOR A YEAR
$n_1 = 37$	$n_2 = 33$
$\bar{x}_1 = 36.8$	$\bar{x}_2 = 33.3$
$s_1 = 1.52$	$s_2 = 2.71$

c. Can the physiologist conclude that a difference exists between the average maximal oxygen uptakes of the two groups of females?

9.34 Refer to the data of Appendix C. Suppose that we wish to test the null hypothesis that μ_1, the average checkout service time of customers at supermarket A (manual checkers), is identical to μ_2, the average checkout service time of customers at supermarket B (automated checkers), i.e.,

$$H_0: \ (\mu_1 - \mu_2) = 0$$

against the alternative that the average at supermarket A is greater than the average at supermarket B, i.e.,

$$H_a: \ (\mu_1 - \mu_2) > 0$$

a. Interpret Type I and Type II errors in the context of the problem.

b. Which error has the more serious consequences for the manager of supermarket A? The manager of supermarket B?

CASE STUDY 9.1 Schlitz versus Budweiser—Mug to Mug

In a "bold gamble to revive depressed sales," the Joseph Schlitz Brewing Co. announced that it would broadcast on live television a taste test featuring 100 beer drinkers during half time of the December 28, 1980, National Football League AFC wildcard playoff game between the Houston Oilers and the Oakland Raiders.* During the live broadcast, Schlitz claimed that the 100 beer drinkers selected for the taste test were "loyal" drinkers of Budweiser, the industry's best-selling beer. Each of the participants was served two beers, one Schlitz and one Budweiser, in unlabelled ceramic mugs. Tasters were then told to make a choice by pulling an electronic switch left or right in the direction of the beer they preferred. (Prior to the test, the tasters were informed that one of the mugs contained their regular beer, Budweiser, and the other contained Schlitz, but the ordering was not revealed.) The percentage

*Orlando *Sentinel Star*, December 11, 1980.

of the 100 "loyal" Budweiser drinkers who preferred Schlitz was then tabulated live, in front of millions of football fans. The newspaper report went on to say:

> One beer industry observer was quoted as calling the test "a giant roll of the dice" in Schlitz' effort to gain a bigger slice of the $8.5 billion beer industry, where consumption increased 25% from 1972 to 1979. Schlitz, a one-time brewery giant, has seen its sales tumble from 16 million barrels in 1974 to between 7 and 9 million barrels [in 1980]. However, Frank Sellinger, the newly appointed Chief Executive at Schlitz, disagrees that the move was a gamble: "Some people thought it was risky to do live TV taste tests. But it didn't take nerve, it just took confidence."

The results of the live TV taste test showed that 46 of the 100 "loyal" Budweiser beer drinkers preferred Schlitz. Schlitz, of course, labelled the outcome "an impressive showing" in a magazine advertisement following the test. For the purposes of this case study, let us suppose that market experts hired by Schlitz informed the company that the taste test would be successful in boosting sales if more than 40 of the 100 Budweiser drinkers selected Schlitz as their favorite. Since 46 tasters pulled the switch in the direction of Schlitz, the brewer called the outcome "impressive," and anxiously awaited sales of Schlitz beer to increase. However, do these sample results indicate that the true proportion of "loyal" Budweiser drinkers who prefer Schlitz is larger than 40%? We can obtain an answer to this question by applying the statistical methods outlined in this chapter.

a. Set up the null and alternative hypotheses of a test to determine whether the true proportion of "loyal" Budweiser drinkers who prefer Schlitz over Budweiser in a similar taste test is larger than .40.

b. For a significance level of $\alpha = .05$, specify the form of the rejection region. Locate the rejection region, α, and the critical value on a sketch of the standard normal curve.

c. Use the results of the live taste test to determine the value of the appropriate test statistic.

d. What is the proper conclusion, in terms of the problem?

e. A valid test of hypothesis, of course, requires that the 100 tasters actually represent a random sample from the segment of the beer-drinking population who are truly "loyal" Budweiser drinkers. Discuss the problems with obtaining a truly random sample from the target population of "loyal" Budweiser drinkers. Do you think that it is possible to select such a sample? In what way(s) could Schlitz have selected the sample (either intentionally or unintentionally) in order to bias the results in their favor?

CASE STUDY 9.2 Drug Screening: A Statistical Decision Problem

Pharmaceutical companies are continually searching for new drugs. Charles W. Dunnett, in his essay,* "Drug Screening: The Never-Ending Search for New and Better Drugs," writes that "research chemists often know what types of chemical structures to look for to treat a particular disease, and the chemists can set about synthesizing compounds of the desired type. Sometimes, however, their knowledge may be vague, resulting in such a wide range of possibilities that many, many

*From J. M. Tanur et al., eds. *Statistics: A guide to the unknown.* San Francisco: Holden-Day, 1978.

compounds have to be made and tested. In such a case, the search is very lengthy and requires years of effort by many people to develop a useful new drug." Testing these thousands of compounds for the few that might be effective is known in the pharmaceutical industry as *drug screening*. Because of the obvious impact on human health, drug screening requires highly organized, efficient testing methods, and "anything that improves the efficiency of the testing procedure," writes Dunnett, "increases the chance of discovering a new cure."

Drug-screening techniques have improved tremendously over the years, and one of the major contributors to this continual improvement is the discipline of statistics. In fact, Dunnett views the drug-screening procedure in its preliminary stage in terms of a statistical decision problem: "In drug screening, two actions are possible: (1) to 'reject' the drug, meaning to conclude that the tested drug has little or no effect, in which case it will be set aside and a new drug selected for screening; and (2) to 'accept' the drug provisionally, in which case it will be subjected to further, more refined experimentation." Since it is the goal of the researcher to find a drug which effects a cure, the null and alternative hypotheses in a statistical test would take the following form:

H_0: The drug is ineffective in treating a particular disease.

H_a: The drug is effective in treating a particular disease.

Dunnett comments on the possible errors associated with the drug-screening procedure: "To abandon a drug when in fact it is a useful one (a *false negative*) is clearly undesirable, yet there is always some risk in that. On the other hand, to go ahead with further, more expensive testing of a drug that is in fact useless (a *false positive*) wastes time and money that could have been spent on testing other compounds." Thus, to a statistician, a false positive result corresponds to a Type I error (i.e., to reject H_0 if in fact H_0 is true), and a false negative result corresponds to a Type II error (i.e., to fail to reject H_0 if in fact H_0 is false).

For this case study, we will consider the following hypothetical drug-screening experiment: A drug developed by a pharmaceutical company for possible treatment of cancerous tumors is to be screened. An investigator implants cancer cells in 100 laboratory mice. From this group, 50 mice are randomly selected and treated with the drug. The remaining 50 are left untreated, and comprise what is known as the *control group*. After a fixed length of time, the actual tumor weights of all the mice in the experiment are measured. If μ_1, the mean tumor weight of the treated mice, is significantly less than μ_2, the mean tumor weight of the control group, then the drug will be provisionally accepted and subjected to further testing; otherwise, the drug will be rejected.

a. Give the appropriate null and alternative hypotheses for the drug-screening test.

b. What are the Type I and Type II errors for this test? (Explain in terms of false positive and false negative results.)

c. Using a significance level of $\alpha = .05$, set up the rejection region for the test.

d. From the experimental results given in the table on page 310, calculate the required test statistic.

TREATED GROUP	CONTROL GROUP
$\bar{x}_1 = 1.23$ grams	$\bar{x}_2 = 1.77$ grams
$s_1 = .55$ gram	$s_2 = .21$ gram

e. Should the pharmaceutical company provisionally accept the drug and subject it to further testing?

CASE STUDY 9.3 The Marriage Tax: Double Trouble for Working Couples

Do married individuals face a heavier income tax burden than singles? For married couples where one spouse reports considerably less income than the other or no income at all, the answer is a definite "no." For example, a one-income couple filing jointly, with standard deductions on a $22,000 income, will pay $3,219 in taxes to the Internal Revenue Service (IRS).* In contrast, a single person earning the same amount has to pay $4,517, a difference of $1,298. And as a married couple's taxable income increases, so do the savings. For an earned income of $30,000, the married couple pays $1,724 less than a single person; for an earned income of $50,000, the married couple pays $3,289 less.

However, where both spouses report similar taxable incomes, marriage becomes a costly proposition, at least from a tax angle. For example, if both spouses earn $22,000 in taxable income, the IRS requires that they pay $11,086 in taxes. Compare this figure to the $9,034 that is due if the couple is unmarried but residing in the same home. And as the dual-income married couple's incomes rise, so too do the additional tax payments (over and above what single income earners would pay).

This "marriage penalty" has many couples thinking of ways to "beat" the tax system, and the following tax rule has enabled them to do so, at least until now: The IRS considers a person married for the entire year, even if that person marries on December 31, the last day of the year. Likewise, the IRS will consider a person unmarried for the entire year, even if that person divorces or obtains a legal separation on December 31. This latter interpretation has many dual-income couples divorcing each other in December and then remarrying in January to avoid paying extra taxes. However, the IRS recently issued a ruling that says it will disregard a divorce obtained solely to save taxes and require the couple to recalculate their taxes as if they had stayed married for the entire year.

Is there any relief in sight for the working married couple? Yes, if a bill suggested by Representative Barber B. Conable, Jr. (R-N.Y.) is passed. The bill aims to create a marriage-neutral system that would impose equal tax burdens on singles, married couples, and heads of households. However, the earliest the bill could go into effect is for the tax year of 1981.

In view of the existing tax laws, and the proposed new bill, let us consider the following statistical decision problem. We wish to determine if there is a difference between the proportions of one-income and dual-income married couples who favor passage of the new tax "equalizer" bill.

*The values reported in this case study are based on 1980 tax laws.

a. What is the target parameter for this problem?

b. Specify the appropriate null and alternative hypotheses for a statistical test to determine if a difference between the proportions exists.

c. Independent random samples of $n_1 = 100$ single-income married couples and $n_2 = 100$ two-income married couples are surveyed. If 42 single-income earners and 74 dual-income earners favor passage of the new "equalizer" bill, calculate the appropriate test statistic. [*Hint:* Utilize the mean and standard deviation of the sampling distribution of $(\hat{p}_1 - \hat{p}_2)$, discussed in Section 8.7.]

d. Set up the rejection region if we are willing to tolerate a Type I error probability of $\alpha = .05$.

e. Give the proper conclusion in terms of the problem.

REFERENCES

Johnson, R. *Elementary statistics.* 3d ed. North Scituate, Mass.: Duxbury, 1980. Chapter 9.

McClave, J. T., & Dietrich, F. H. *Statistics.* 2d ed. San Francisco: Dellen, 1982. Chapters 7 and 8.

Mendenhall, W. *Introduction to probability and statistics.* 5th ed. North Scituate, Mass.: Duxbury, 1979. Chapters 8 and 9.

Tanur, J. M., Mosteller, F., Kruskal, W. H., Link, R. F., Pieters, R. S., & Rising, G. R. *Statistics: A guide to the unknown.* San Francisco: Holden-Day, 1978.

Zuwaylif, F. H. *General applied statistics.* 3d ed. Reading, Mass.: Addison-Wesley, 1979. Chapters 7, 8, and 10.

Ten

Hypothesis Testing: Applications

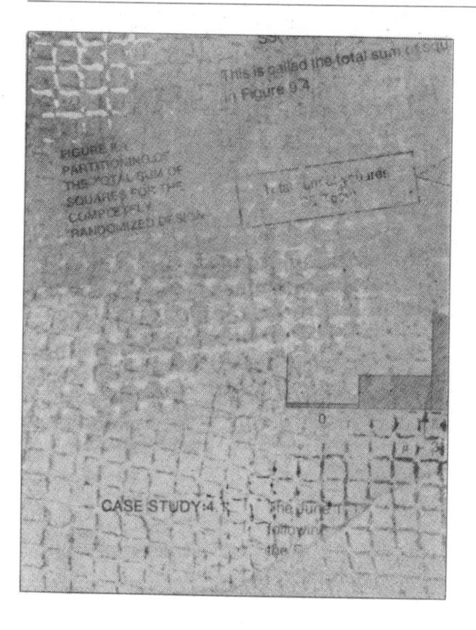

Is a motorcycle helmet law really effective in reducing fatal or serious head injuries? A sampling of hospital records in Michigan (which has a helmet law) revealed that 8 of 31 motorcyclists who received medical treatment suffered fatal or serious head injuries. In contrast, a similar sample of 55 cases in Illinois (which has no helmet law) indicated that 41 of the 55 received fatal or serious head injuries. Do the data imply a difference between the proportions of motorcycle accidents in the two states that result in serious head injuries? We will learn how to answer this and similar questions in this chapter, and will examine the motorcycle accident data in Exercises 10.58 and 10.59.

Contents

10.1 Introduction

In this chapter we will present applications of the hypothesis-testing logic developed in Chapter 9. The cases to be considered are those for which we developed estimation procedures in Chapter 8: large-sample test about μ; small-sample test about μ; large-sample test about p; large (independent) samples test about $(\mu_1 - \mu_2)$; small (independent) samples test about $(\mu_1 - \mu_2)$; and large (independent) samples test about $(p_1 - p_2)$.

Since the theory and reasoning involved are derived from the developments of Chapters 8 and 9, we will present only a summary of the hypothesis-testing procedure for one-tailed and two-tailed tests in each situation.

10.2 Hypothesis Test about a Population Mean: Large-Sample Case

Suppose that we wish to determine whether the mean level of billing per private customer per month for long-distance telephone calls is in excess of $30.00. That is, we will test

H_0: $\mu = 30$

H_a: $\mu > 30$

where

μ = Mean expenditure per private customer per month for long-distance telephone calls

We are conducting this study in an attempt to gather support for H_a; we hope that the sample data will lead to the rejection of H_0. Now, the point estimate of the population mean μ is the sample mean $\bar{x}$. Will the value of $\bar{x}$ which we obtain from our sample be large enough for us to safely conclude that μ is greater than 30? In order to answer this question, we need to perform each step of the hypothesis-testing procedure developed in Chapter 9. The box on page 315 contains the elements of a large-sample hypothesis test about a population mean, μ.

In this large-sample case, only one assumption is required for the validity of the procedure:

ASSUMPTION REQUIRED

The sample size must be sufficiently large (say, $n \geq 30$) so that the sampling distribution of $\bar{x}$ is approximately normal and that s provides a good approximation to σ.

LARGE-SAMPLE TEST OF HYPOTHESIS ABOUT A POPULATION MEAN

a. One-tailed test

H_0: $\mu = \mu_0$

H_a: $\mu > \mu_0$

(or H_a: $\mu < \mu_0$)

Test statistic:

$$z = \frac{\bar{x} - \mu_0}{\sigma_{\bar{x}}} \approx \frac{\bar{x} - \mu_0}{s/\sqrt{n}}$$

Rejection region:

$z > z_\alpha$ (or $z < -z_\alpha$)

b. Two-tailed test

H_0: $\mu = \mu_0$

H_a: $\mu \neq \mu_0$

Test statistic:

$$z = \frac{\bar{x} - \mu_0}{\sigma_{\bar{x}}} \approx \frac{\bar{x} - \mu_0}{s/\sqrt{n}}$$

Rejection region:

$z < -z_{\alpha/2}$ or $z > z_{\alpha/2}$

where z_α is the z-value such that $P(z > z_\alpha) = \alpha$; and $z_{\alpha/2}$ is the z-value such that $P(z > z_{\alpha/2}) = \alpha/2$. [*Note:* μ_0 is our symbol for the particular numerical value specified for μ in the null hypothesis.]

EXAMPLE 10.1 The long-distance telephone charges during a given month for a random sample of $n = 37$ private customers were obtained from the billing files of a telephone company. The results are summarized below:

$\bar{x} = \$33.15$

$s = \$21.21$

Test the hypothesis that μ, the population mean monthly billing level for long-distance telephone calls, is equal to \$30.00 against the alternative that μ is larger than \$30.00, using a significance level of $\alpha = .05$.

Solution We have previously formulated the hypotheses as

H_0: $\mu = 30$

H_a: $\mu > 30$

Note that the sample size $n = 37$ is sufficiently large so that the sampling distribution of $\bar{x}$ is approximately normal and that s provides a good approximation to σ. Having satisfied the required assumption, we may proceed with a large-sample test about μ.

Using a significance level of $\alpha = .05$, we will reject the null hypothesis for this one-tailed test if

$$z > z_\alpha = z_{.05}$$

i.e., if $z > 1.645$. (This rejection region is shown in Figure 10.1.)

Computing the value of the test statistic, we obtain

$$z = \frac{\bar{x} - \mu_0}{s/\sqrt{n}} = \frac{33.15 - 30}{21.21/\sqrt{37}} = .903$$

Since this value does not fall within the rejection region (see Figure 10.1), we fail to reject H_0. We say that there is insufficient evidence (at $\alpha = .05$) to conclude that the mean billing level for long-distance calls per private customer during the given month is greater than \$30.00. We would need to take a larger sample before we could detect whether $\mu > 30$, if in fact this were the case.

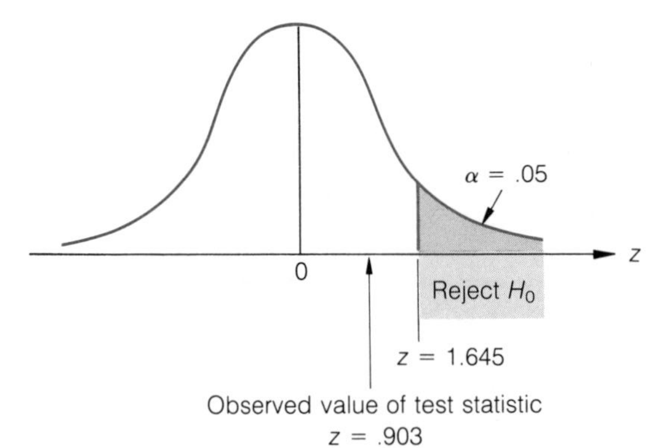

FIGURE 10.1 Rejection Region, Example 10.1

EXAMPLE 10.2 Humerus bones from the same species of animal tend to have approximately the same length-to-width ratios. When fossils of humerus bones are discovered, archeologists can often determine the species of animal by examining the length-to-width ratios of the bones. It is known that species A has a mean ratio of 8.5. Suppose that 41 fossils of humerus bones were unearthed at an archeological site in East Africa, where species A is believed to have inhabited. (Assume that the 41 unearthed bones are all from the same unknown species.) The length-to-width ratios of the bones were measured and are summarized as follows:

$$\bar{x} = 9.25$$

$$s = 1.16$$

We wish to test the hypothesis that μ, the population mean ratio of all bones of this particular species, is equal to 8.5 against the alternative that it is different from 8.5, i.e., we wish to test whether the unearthed bones are from species A. Suppose that we also want a very small chance of rejecting H_0, if in fact μ is equal to 8.5. That is, it is important that we avoid making a Type I error. The hypothesis-testing procedure that we have developed gives us the advantage of choosing any significance level that we desire. Since the significance level, α, is also the probability of a Type I error, we will choose α very small. In general, researchers who consider a Type I error to have very serious practical consequences should perform the test at a very low α-value, say $\alpha = .01$. Other researchers may be willing to tolerate an α-value as high as .10 if a Type I error is not deemed a serious error to make in practice.

Test whether μ, the population mean ratio, is different from 8.5, using a significance level of $\alpha = .01$.

Solution We formulate the following hypotheses:

$$H_0: \quad \mu = 8.5$$

$$H_a: \quad \mu \neq 8.5$$

Since we wish to avoid making a Type I error, we choose a very small significance level of $\alpha = .01$. The sample size exceeds 30, thus we may proceed with the large-sample test about μ.

At significance level $\alpha = .01$, we will reject the null hypothesis for this two-tailed test if

$$z < -z_{\alpha/2} = -z_{.005} \quad \text{or if} \quad z > z_{\alpha/2} = z_{.005}$$

i.e., if $z < -2.58$ or if $z > 2.58$. (This rejection region is shown in Figure 10.2.)

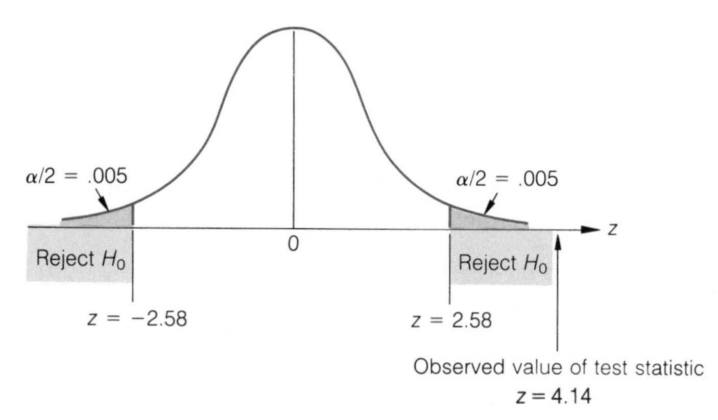

FIGURE 10.2
Rejection Region, Example 10.2

The value of the test statistic is computed as follows:

$$z \approx \frac{\bar{x} - \mu_0}{s/\sqrt{n}} = \frac{9.25 - 8.5}{1.16/\sqrt{41}} = 4.14$$

Since this value lies within the rejection region (see Figure 10.2), we reject H_0 and conclude that the mean length-to-width ratio of all humerus bones of this particular species is significantly different from 8.5. If the null hypothesis is in fact true (i.e., if $\mu = 8.5$), then the probability that we have incorrectly rejected it is equal to $\alpha = .01$.

The practical implications of this result remain to be studied further. Perhaps the animal discovered at the archeological site is of some species other than A. Alternatively, the humerus bones unearthed may have larger than normal length-to-width ratios due to unusual feeding habits of species A. It is not always the case that a ***statistically*** significant result implies a ***practically*** significant result. The researcher must retain his or her objectivity and judge the practical significance using, among other criteria, his or her knowledge of the subject matter and the phenomenon under investigation.

EXAMPLE 10.3 Prior to the institution of a new safety program, the average number of on-the-job accidents per day at a factory was 4.5. To determine if the safety program has been effective in reducing the average number of accidents per day, a random sample of 30 days is taken after the institution of the new safety program and the number of accidents per day is recorded. The sample mean and standard deviation were computed as follows:

$$\bar{x} = 3.7$$

$$s = 1.3$$

a. Is there sufficient evidence to conclude (at significance level .01) that the average number of on-the-job accidents per day at the factory has decreased since the institution of the safety program?

b. What is the practical interpretation of the test statistic computed in part a?

Solution **a.** In order to determine whether the safety program was effective, we will conduct a large-sample test of

H_0: $\mu = 4.5$ (i.e., no change in average number of on-the-job accidents per day)

H_a: $\mu < 4.5$ (i.e., average number of on-the-job accidents per day has decreased)

where μ represents the average number of on-the-job accidents per day at the factory after institution of the new safety program. For a significance level of $\alpha = .01$, we will reject the null hypothesis if

$$z < -z_{.01} = -2.33$$

(See Figure 10.3.) The computed value of the test statistic is

$$z \approx \frac{\bar{x} - \mu_0}{s/\sqrt{n}} = \frac{3.7 - 4.5}{1.3/\sqrt{30}} = -3.37$$

Since this value does fall within the rejection region (see Figure 10.3), there is sufficient evidence (at $\alpha = .01$) to conclude that the average number of on-the-job accidents per day at the factory has decreased since the institution of the safety program. It appears that the safety program was effective in reducing the average number of accidents per day.

b. If the null hypothesis is true, $\mu = 4.5$. Recall that for large samples, the sampling distribution of $\bar{x}$ is approximately normal, with mean $\mu_{\bar{x}} = \mu$ and standard deviation $\sigma_{\bar{x}} = \sigma/\sqrt{n}$. Then the z-score for $\bar{x}$, under the assumption that H_0 is true, is given by

$$z = \frac{\bar{x} - 4.5}{\sigma/\sqrt{n}}$$

You can see that the test statistic computed in part a is simply the z-score for the sample mean $\bar{x}$, if in fact $\mu = 4.5$. A calculated z-score of -3.37 indicates that the value of $\bar{x}$ computed from the sample falls a distance of 3.37 standard deviations below the hypothesized mean of $\mu = 4.5$. Of course, we would not expect to observe a z-score this extreme if in fact $\mu = 4.5$. Without benefit of a formal test of hypothesis, this fact alone indicates strongly that the true mean is less than 4.5.

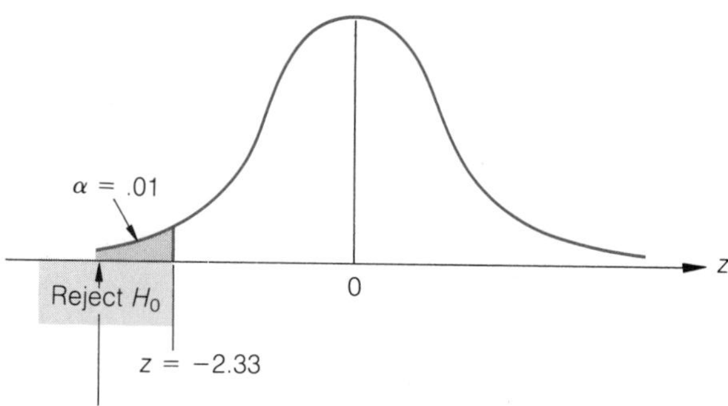

FIGURE 10.3 Rejection Region, Example 10.3

EXERCISES

[*Note:* In all the exercises for this chapter, you should carefully define any notation used, perform all steps of the relevant hypothesis test, state a conclusion in terms of the problem, and specify any assumptions required for the validity of the procedure.]

10.1 The daily newspaper at a particular university endeavors to print editorial letters by students regarding current issues. Last year, the newspaper published an average of 4 student editorial letters per issue. However, a random sample of 35 newspapers published this year was selected, and the number of student editorial letters appearing in each was recorded. The mean and standard deviation were computed as follows:

$$\bar{x} = 2.1$$

$$s = 1.3$$

Is there sufficient evidence to conclude (at significance level $\alpha = .05$) that the mean number of student editorial letters published per issue has decreased since last year?

10.2

a. Use a random number table to generate a random sample of $n = 40$ observations on DDT concentration in fish from the data of Appendix D. Compute $\bar{x}$ and s.

b. Recall that the Food and Drug Administration (FDA) sets the limit for DDT content in individual fish at 5 parts per million (ppm) (see Case Study 1.1). Does the sample, part a, provide sufficient evidence to conclude that the average

DDT content of individual fish inhabiting the Tennessee River and its creek tributaries exceeds 5 ppm? Test using a significance level of $\alpha = .01$.

10.3 To protect U.S. steel producers against competition from low-priced foreign imports, the government has adopted a trigger price mechanism. If steel imports enter this country at less than the minimum or trigger price (based on the cost of production in Japan, which has the world's most efficient steel factories), an investigation into the illegal dumping of cheap steel in the U.S. by foreign countries is launched. Last year, the average trigger price was set at $358.31 per ton. This year, a random sample of $n = 40$ days was selected, and the trigger price on each of the days was recorded. The data are summarized as follows:

$\bar{x} = \$390.88$

$s = \$106.19$

Does the sample evidence indicate that the mean trigger price during this year differs significantly from $358.31? Use a significance level of $\alpha = .10$.

10.4 A recent "Fuel Gauge" report by the Automobile Association of America showed that the average price (per gallon) for all grades of gasoline selling at full-service pumps in Florida was $1.429 (Gainesville *Sun,* February 27, 1981). Suppose that a Florida service station manager wants to compare the average price of gasoline in his state to the national average. A random sample of $n = 250$ full-service stations from various locations across the U.S. revealed the following summary statistics on price per gallon: $\bar{x} = \$1.475$, $s = \$.508$. Can the Florida service station manager conclude that the average price for all grades of gasoline selling at full-service pumps across the U.S. is higher than his statewide average of $1.429? Test at $\alpha = .02$.

10.5 The application of adrenalin is the prevailing treatment to reduce eye pressure in glaucoma patients. Theoretically, a new synthetic drug will cause the same mean drop in blood pressure (5.5 units) without the side effects caused by adrenalin. The new drug is given to $n = 50$ glaucoma patients and the reduction in pressure for each patient is measured. The results are summarized as follows:

$\bar{x} = 4.68$

$s = .82$

Is this sufficient evidence (at significance level $\alpha = .05$) to conclude that the mean reduction in pressure due to the new drug is different from that produced by adrenalin?

10.3 Hypothesis Test about a Population Mean: Small-Sample Case

An item of interest to the United States Defense Department is the maximum speed that can be reached by the Soviet Union's fastest and most powerful tank, the T-72. In order to obtain this information, the Defense Department would need data on the top speeds reached by several Soviet T-72's. A sample of maximum speeds could be used to make an inference about the mean maximum speed, μ, of all Soviet T-72 tanks. However, time, cost, and security considerations would probably limit the sample of maximum speeds to a small number. Consequently, the assumption required for a large-sample test of hypothesis about μ will be violated. We need, then, an hypothesis-testing procedure which is appropriate for use with small samples.

An hypothesis test about a population mean, μ, based on a small sample ($n < 30$), consists of the following elements:

SMALL-SAMPLE TEST OF HYPOTHESIS ABOUT A POPULATION MEAN

a. One-tailed test

H_0: $\mu = \mu_0$

H_a: $\mu > \mu_0$

(or H_a: $\mu < \mu_0$)

Test statistic:

$$t = \frac{\bar{x} - \mu_0}{s/\sqrt{n}}$$

Rejection region:

$t > t_\alpha$ (or $t < -t_\alpha$)

b. Two-tailed test

H_0: $\mu = \mu_0$

H_a: $\mu \neq \mu_0$

Test statistic:

$$t = \frac{\bar{x} - \mu_0}{s/\sqrt{n}}$$

Rejection region:

$t < -t_{\alpha/2}$ or $t > t_{\alpha/2}$

where the distribution of t is based on $(n-1)$ degrees of freedom; t_α is the t-value such that $P(t > t_\alpha) = \alpha$; and $t_{\alpha/2}$ is the t-value such that $P(t > t_{\alpha/2}) = \alpha/2$.

As we noticed in the development of estimation procedures, when we are making inferences based on small samples, more restrictive assumptions are required than when making inferences from large samples. In particular, this hypothesis test requires the following assumption:

ASSUMPTION REQUIRED

The relative frequency distribution of the population from which the sample was selected is approximately normal.

EXAMPLE 10.4 What is the practical significance of the test statistic for a small-sample test of hypothesis about μ?

Solution Notice that the test statistic given in the box is a t statistic and is calculated exactly as our approximation to the large-sample test statistic, z, given in Section 10.2. Therefore, just like z, the computed value of t will indicate the direction and approximate distance (in units of standard deviations) that the sample mean, $\bar{x}$, is from the hypothesized population mean, μ_0.

EXAMPLE 10.5 The building specifications in a certain city require that the sewer pipe used in residential areas has a mean breaking strength of more than 2,500 pounds per lineal foot. A manufacturer who would like to supply the city with sewer pipe has submitted a bid and provided the following additional information: An independent contractor randomly selected 7 sections of the manufacturer's pipe and tested each for breaking strength. The results (pounds per lineal foot) are shown below:

2,610 2,750 2,420 2,510 2,540 2,490 2,680

Do we have sufficient evidence to conclude that the manufacturer's sewer pipe meets the required specifications? Use a significance level of $\alpha = .10$.

Solution The relevant hypothesis test has the following elements:

H_0: $\mu = 2{,}500$ (i.e., the manufacturer's pipe does not meet the city's specifications)

H_a: $\mu > 2{,}500$ (i.e., the pipe meets the specifications)

where μ represents the true mean breaking strength (in pounds per lineal foot) for all sewer pipe produced by this manufacturer.

This small-sample ($n = 7$) test requires the assumption that the relative frequency distribution of the population values of breaking strength for the manufacturer's pipe is approximately normal. Then the test will be based upon a t distribution with $(n - 1) = 6$ degrees of freedom. We will thus reject H_0 if

$$t > t_{.10} = 1.440$$

(See Figure 10.4.)

In order to compute the value of the test statistic, we need to obtain the values of $\bar{x}$ and s:

$$\bar{x} = \frac{\Sigma x}{n} = \frac{2{,}610 + 2{,}750 + 2{,}420 + 2{,}510 + 2{,}540 + 2{,}490 + 2{,}680}{7}$$

$$= 2{,}571.43$$

$$s = \sqrt{\frac{\Sigma x^2 - (\Sigma x)^2/n}{n-1}}$$

$$= \sqrt{\frac{(2{,}610)^2 + (2{,}750)^2 + (2{,}420)^2 + (2{,}510)^2 + (2{,}540)^2 + (2{,}490)^2 + (2{,}680)^2 - (18{,}000)^2/7}{6}}$$

$$= \sqrt{\frac{46{,}365{,}200 - 46{,}285{,}714}{6}} = 115.10$$

Substitution of these values yields the test statistic:

$$t = \frac{\bar{x} - \mu_0}{s/\sqrt{n}} = \frac{2{,}571.43 - 2{,}500}{115.10/\sqrt{7}} = 1.64$$

Since this value of t is larger than the critical value of 1.440 (see Figure 10.4), we reject H_0. There is sufficient evidence (at significance level .10) that the manufacturer's pipe meets the city's building specifications.

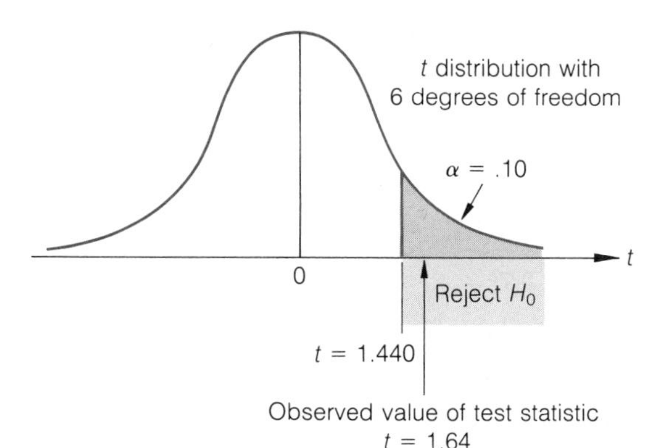

FIGURE 10.4 Rejection Region, Example 10.5

EXAMPLE 10.6 In an effort to offset the Soviet Union's growing armored force, the U.S. Defense Department has selected a new Army tank designed by Chrysler Corp. Extensive testing has shown that the tank, called the XM-1, can reach an average top speed of 45 miles per hour. In order to compare the XM-1 to the Soviet T-72 tank, the Defense Department gained access to data on $n = 16$ T-72's and computed the following summary statistics on the maximum speed reached by each:

$\bar{x} = 43.5$ miles per hour

$s = 3.0$ miles per hour

At the .05 significance level, does the sample evidence indicate that the average top speed of the T-72 is less than 45 miles per hour?

Solution Define the parameter of interest:

μ = The true average maximum speed reached by all Soviet T-72 tanks

We are interested in a test of

H_0: $\mu = 45$

H_a: $\mu < 45$

(The alternative which we wish to detect is that the average top speed of the Soviet T-72 is less than that of the new American tank, the XM-1.)

Since we are restricted to a small sample, we must make the assumption that the maximum speeds reached by all Soviet T-72 tanks have a relative frequency distribution which is approximately normal. Under this assumption, the test statistic will have a t distribution with $(n - 1) = (16 - 1) = 15$ degrees of freedom. The rejection rule is then to reject the null hypothesis for values of t such that

$$t < -t_{.05} = -1.753$$

(See Figure 10.5 on page 324.)

The value of the test statistic is

$$t = \frac{\bar{x} - \mu_0}{s/\sqrt{n}} = \frac{43.5 - 45}{3.0/\sqrt{16}} = -2.0$$

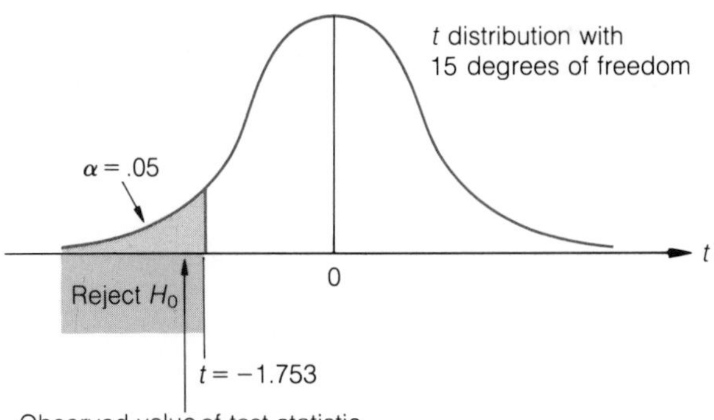

FIGURE 10.5
Rejection Region, Example 10.6

The computed value of the test statistic, $t = -2.0$, falls below the critical value of -1.753 (see Figure 10.5). We thus reject H_0 and conclude that the average top speed reached by the Soviet T-72 tank is significantly less than 45 miles per hour, the average top speed of the American XM-1. If in fact the null hypothesis is true (i.e., if the average top speeds of the two tanks are identical), the probability of our incorrectly rejecting it using this procedure is equal to $\alpha = .05$.

EXERCISES

10.6 Refer to Example 10.5. If you were the city building inspector, why might it be desirable from your viewpoint to use a smaller value of α, say, $\alpha = .01$, in conducting the test?

10.7 Recently, a new miniaturized component for use in desk-top calculators has been developed. These components greatly reduce the size of the calculators, but may not last long enough to be practical. It is known that the mean life length of the larger component used in a desk-top calculator manufactured by a successful firm is 500 hours. However, a random sample of 18 newly built miniaturized components yielded the following summary statistics on life length:

$\bar{x} = 482$ hours

$s = 59$ hours

Is there evidence, at significance level $\alpha = .01$, that the mean life length of the new miniaturized component is significantly less than the average of 500 hours for the larger components? What assumption is required for the hypothesis test to be valid?

10.8 One of the most feared predators in the ocean is the great white shark. Although it is known that the white shark grows to a mean length of 21 feet, a marine biologist believes that great white sharks off the Bermuda coast grow much longer due to unusual feeding habits. To test this claim, a number of full-grown great white sharks are captured off the Bermuda coast, measured, then set free. However, because the capture of sharks is difficult, costly, and very dangerous, only three are sampled. Their lengths are 24, 20, and 22 feet.

a. Do the data provide sufficient evidence to support the marine biologist's claim? Test at a significance level of $\alpha = .05$.

b. What assumptions are required for the hypothesis test, part a, to be valid? Do you think these assumptions are likely to be satisfied in this particular sampling situation?

10.9 A psychologist is studying the effects of lack of sleep on the performance of various perceptual-motor tasks. After a given period of sleep deprivation, a measurement of reaction time to an auditory stimulus was taken for each of 6 adult male subjects. The reaction times (in seconds) are summarized as follows:

$$\bar{x} = 1.82$$

$$s = .22$$

Previous psychological studies show that the true mean reaction time for non–sleep-deprived male subjects is 1.70 seconds. Does the sample evidence indicate (at significance level $\alpha = .10$) that the mean reaction time for sleep-deprived male subjects is longer than 1.70 seconds? (Make any assumptions that are necessary for the hypothesis test to be valid.)

10.10 In any canning process, a manufacturer will lose money if the cans contain either more or less than is claimed on the label. Accordingly, canners pay close attention to the amount of their product being dispensed by the can-filling machines. Consider a company that produces a fast-drying rubber cement in 32-ounce aluminum cans. A quality control inspector is interested in testing whether the average number of ounces of rubber cement dispensed into the cans is really 32 ounces. Since inspection of the canning process requires that the dispensing machines be shut down, and shutdowns for any lengthy period of time cost the company thousands of dollars in lost revenue, the inspector is able to obtain a random sample of only 10 cans for testing. After measuring the weights of their contents, the inspector computes the following summary statistics:

$$\bar{x} = 31.55 \text{ ounces}$$

$$s = .48 \text{ ounce}$$

a. Does the sample evidence indicate that the dispensing machines are in need of adjustment? Test at significance level $\alpha = .05$.

b. What assumption is necessary for the hypothesis test, part a, to be valid?

10.11 Refer to Exercise 10.2.

a. Suppose the test of hypothesis concerning μ, the true mean DDT content of individual fish inhabiting the Tennessee River and its creek tributaries, was based on a random sample of only $n = 8$ fish. What are the disadvantages of conducting this small-sample test?

b. Repeat part b of Exercise 10.2 using only the information on the DDT content of a sample of 8 fish (randomly selected from the 40 observations of Exercise 10.2(a)). Compare the results of the large- and small-sample tests.

10.4 Hypothesis Test about a Population Proportion: Large-Sample Case

In this country, advertisements for automobiles have traditionally been directed towards males. However, according to researchers at the Ford Motor Co. (*Time*, October 27, 1980), women bought 39% of all new cars sold in the U.S. in 1979. Suppose that we wish to test the null hypothesis that the true proportion of new car buyers in 1980 who were female is equal to .39 (i.e., H_0: $p = .39$) against the alternative H_a: $p > .39$.

The following procedure is used to test an hypothesis about a population proportion, p, based on a large sample from the target population:

LARGE-SAMPLE TEST OF HYPOTHESIS ABOUT A POPULATION PROPORTION

a. One-tailed test

H_0: $p = p_0$

H_a: $p > p_0$

(or H_a: $p < p_0$)

Test statistic:

$$z = \frac{\hat{p} - p_0}{\sqrt{p_0 q_0 / n}}$$

Rejection region:

$z > z_\alpha$ (or $z < -z_\alpha$)

b. Two-tailed test

H_0: $p = p_0$

H_a: $p \neq p_0$

Test statistic:

$$z = \frac{\hat{p} - p_0}{\sqrt{p_0 q_0 / n}}$$

Rejection region:

$z < -z_{\alpha/2}$ or $z > z_{\alpha/2}$

where $q_0 = 1 - p_0$

In order to validly apply the procedure, the sample size must be sufficiently large to guarantee approximate normality of the sampling distribution of the sample proportion, $\hat{p}$:

ASSUMPTION REQUIRED

The interval $\hat{p} \pm 2\sqrt{\hat{p}\hat{q}/n}$ does not contain 0 or 1.

EXAMPLE 10.7 Suppose that in a random sample of $n = 120$ new car buyers in 1980, 57 were women. Does this evidence indicate that the true proportion of new car buyers in 1980 who were women is significantly larger than .39, the 1979 proportion? Test at significance level $\alpha = .05$.

Solution We wish to perform a large-sample test about a population proportion, p. We define

H_0: $p = .39$ (i.e., no change from 1979 to 1980)

H_a: $p > .39$ (i.e., proportion of new car buyers who were women increased in 1980)

where p represents the true proportion of all new car buyers in 1980 who were women.

At significance level $\alpha = .05$, the rejection region for this one-tailed test consists of all values of z for which

$$z > z_{.05} = 1.645$$

(See Figure 10.6.)

The test statistic requires the calculation of the sample proportion, $\hat{p}$, of new car buyers who were women:

$$\hat{p} = \frac{\text{Number of sampled new car buyers who were women}}{\text{Number of new car buyers sampled}}$$

$$= 57/120 = .475$$

Noting that $q_0 = 1 - p_0 = 1 - .39 = .61$, we obtain the following value of the test statistic:

$$z = \frac{\hat{p} - p_0}{\sqrt{p_0 q_0 / n}} = \frac{.475 - .39}{\sqrt{(.39)(.61)/120}} = 1.91$$

This value of z lies within the rejection region (see Figure 10.6); we thus conclude that the proportion of new car buyers in 1980 who were women increased significantly from .39. The probability of our having made a Type I error (rejecting H_0 if, in fact, it is true) is $\alpha = .05$.

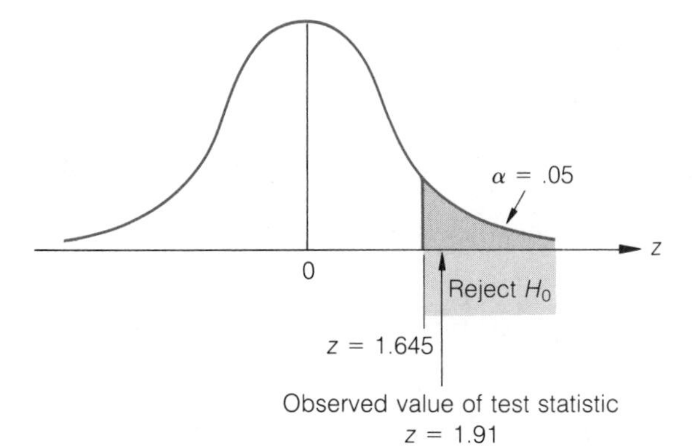

FIGURE 10.6 Rejection Region, Example 10.7

EXAMPLE 10.8 Refer to Example 8.14. Does the information given there provide evidence (at $\alpha = .01$) that the proportion of Northern homes which are heated by gas differs significantly from .5?

Solution The parameter of interest is a population proportion, p. We define

$$H_0: \quad p = .5$$

$$H_a: \quad p \neq .5$$

where p is the true proportion of all Northern homes which are heated by gas. Note that we wish to be able to detect a departure in either direction from the null hypothesized value of $p = .5$; hence, the test is two-tailed.

At significance level $\alpha = .01$, the null hypothesis will be rejected if

$$z < -z_{.005} \quad \text{or} \quad z > z_{.005}$$

that is, H_0 will be rejected if

$$z < -2.58 \quad \text{or} \quad z > 2.58$$

This rejection region is shown in Figure 10.7. The sample proportion of Northern homes which are heated by gas is

$$\hat{p} = 42/80 = .525$$

Thus, the test statistic has the value

$$z = \frac{\hat{p} - p_0}{\sqrt{p_0 q_0 / n}} = \frac{.525 - .5}{\sqrt{(.5)(.5)/80}} = .45$$

The null hypothesis cannot be rejected (at $\alpha = .01$), since the computed value of z does not lie within the rejection region (see Figure 10.7). There is insufficient evidence to support the hypothesis that the proportion of Northern homes heated by gas differs significantly from .5.

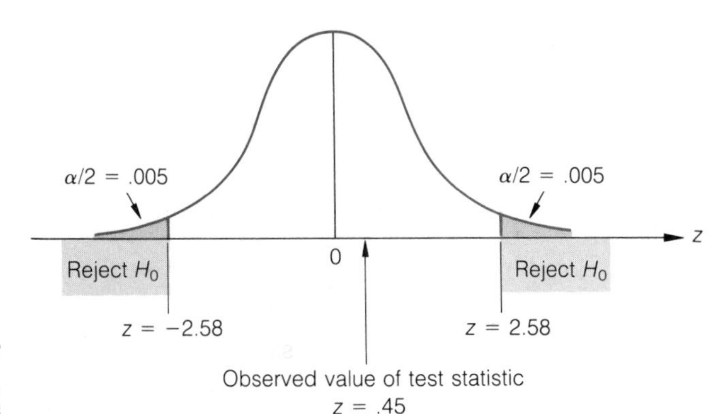

FIGURE 10.7
Rejection Region, Example 10.8

Although small-sample procedures are available for testing hypotheses about a population proportion, the details are omitted from our discussion. It is our experience that they are of limited utility, since most surveys of binomial populations (e.g., opinion polls) performed in the real world use samples which are large enough to employ the techniques of this section.

EXERCISES

10.12 A recent report compiled by the American Dental Association claimed that only 40% of all school-age children in a particular area receive regular dental care. The head of the local Health Department, believing that this figure is too low, randomly sampled 400 school-age children in the area and discovered that 180 receive regular dental care. Test the hypothesis of interest to the Health Department head, using a significance level of $\alpha = .02$.

10.13 During the summer of 1981, a strike by Major League baseball players left baseball fans without the national pastime for a period of seven weeks. The strike was brought about by a failure of the team owners and the Major League Players' Association to agree on a form of compensation for teams that lose a player to another team through the free agent market. In order to gauge the fans' opinions on the issue, NBC television conducted a telephone poll in which viewers were asked to respond to the question: "Which side do you favor in the baseball strike, the owners or the players?" (Gainesville *Sun,* June 15, 1981). The sample of respondents, however, was not selected at random. Any viewer who wished to participate in the poll could do so by dialing long distance at a charge of 50 cents, plus tax. NBC spokesman Mike Cohen admitted that "even though we felt it (the poll) was unscientific, and Bryant Gumbel (an NBC sports commentator) mentioned that on the air, there might be some degree of accuracy in doing it because ABC tried it prior to the last (presidential) election and had predicted a landslide victory for Ronald Reagan."

Of the 162,802 viewers who responded, 85,666 said they favored the team owners' position.

a. Conduct a test of hypothesis (at significance level $\alpha = .01$) to determine whether the true proportion of baseball fans favoring the team owners exceeds .5.
b. What are some of the problems associated with conducting an "unscientific" telephone opinion poll such as the one discussed above? Explain why any inferences derived from the poll may be biased.

10.14 The Food Marketing Institute (FMI) was formed in 1977 to guard against forces outside the supermarket industry (including government) which impede the efficiency and productivity of the industry. With some 1,100 national members, FMI is the voice of 55% of the retail food business, including one-store operators, large food chains, and grocery wholesalers. However, in a random sample of $n = 40$ food retailers located in rural areas of the country, only 18 are registered members of FMI. Is there evidence, at $\alpha = .05$, that the true proportion of food retailers located in rural areas who are members of FMI is significantly less than the national proportion of .55?

10.15 Usually, when trees grown in greenhouses are replanted in their natural habitat, there is only a 50% survival rate. However, a recent General Telephone and Electronics (GTE) advertisement claimed that trees grown in a particular environment ideal for plant growth have a 95% survival rate when replanted. These trees are grown inside a mountain in Idaho where the air temperature, carbon dioxide content, and humidity are all constant, and there are no major disease or insect problems. A key growth ingredient—light—is supplied by specially made GTE Sylvania Super-Metalarc lamps. These lights help the young trees develop a more fibrous root system which aids in the transplantation. Suppose that we wish to challenge GTE's claim, i.e., we wish to test whether the true proportion of all trees grown inside the Idaho mountain which survive when replanted in their natural habitat is less than .95. We randomly sample 50 of the trees grown in the controlled environment, replant the trees in their natural habitat, and observe that 46 of the trees survive. Perform the test at a level of significance of $\alpha = .01$.

10.16 The strength of a pesticide dosage is often measured by the proportion of pests the dosage will kill. A particular dosage of rat poison is fed to 250 rats. Of these rats, 215 died due to the poison. Test the hypothesis that the true proportion of rats that will succumb to the dosage is larger than .85. Use a significance level of $\alpha = .10$.

10.5 Hypothesis Test about the Difference between Two Population Means: Large-Sample Case

General Foods and Procter & Gamble, the nation's two largest roasters of coffee beans, have recently slashed their wholesale prices for ground coffee. Suppose that a consumer group wishes to determine whether the mean price per pound, μ_1, of Procter & Gamble's ground coffee exceeds the mean price per pound, μ_2, of General Foods' ground coffee. That is, the consumer group will test the null hypothesis H_0: $(\mu_1 - \mu_2) = 0$ against the alternative H_a: $(\mu_1 - \mu_2) > 0$. The following large-sample procedure is applicable for testing an hypothesis about $(\mu_1 - \mu_2)$, the difference between two population means:

LARGE-SAMPLE TEST OF HYPOTHESIS ABOUT $(\mu_1 - \mu_2)$

a. One-tailed test

H_0: $(\mu_1 - \mu_2) = D_0$

H_a: $(\mu_1 - \mu_2) > D_0$

(or H_a: $(\mu_1 - \mu_2) < D_0$)

Test statistic:

$$z = \frac{(\bar{x}_1 - \bar{x}_2) - D_0}{\sigma_{(\bar{x}_1 - \bar{x}_2)}} \approx \frac{(\bar{x}_1 - \bar{x}_2) - D_0}{\sqrt{\frac{s_1^2}{n_1} + \frac{s_2^2}{n_2}}}$$

Rejection region:

$z > z_\alpha$ (or $z < -z_\alpha$)

b. Two-tailed test

H_0: $(\mu_1 - \mu_2) = D_0$

H_a: $(\mu_1 - \mu_2) \neq D_0$

Test statistic:

$$z = \frac{(\bar{x}_1 - \bar{x}_2) - D_0}{\sigma_{(\bar{x}_1 - \bar{x}_2)}} \approx \frac{(\bar{x}_1 - \bar{x}_2) - D_0}{\sqrt{\frac{s_1^2}{n_1} + \frac{s_2^2}{n_2}}}$$

Rejection region:

$z < -z_{\alpha/2}$ or $z > z_{\alpha/2}$

[*Note:* D_0 is our symbol for the particular numerical value specified for $(\mu_1 - \mu_2)$ in the null hypothesis. In many practical applications, we wish to hypothesize that there is no difference between the population means; in such cases, $D_0 = 0$.]

The following assumptions are required about the sample sizes and the sampling procedure:

ASSUMPTIONS REQUIRED

1. The sample sizes n_1 and n_2 are sufficiently large, say, $n_1 \geq 30$ and $n_2 \geq 30$.
2. The two samples are selected randomly and independently from the target populations.

EXAMPLE 10.9 A consumer group selected independent random samples of supermarkets located throughout the country for the purpose of comparing the retail prices per pound of General Foods and Procter & Gamble brands of ground coffee. The results of the investigation are summarized in Table 10.1. Does this evidence indicate that the mean retail price per pound of Procter & Gamble's ground coffee is significantly higher than the mean retail price per pound of General Foods' ground coffee? Use a significance level of $\alpha = .01$.

TABLE 10.1 Ground Coffee Prices, Example 10.9

PROCTER & GAMBLE	GENERAL FOODS
$n_1 = 63$	$n_2 = 58$
$\bar{x}_1 = \$2.98$	$\bar{x}_2 = \$2.93$
$s_1 = \$.11$	$s_2 = \$.07$

Solution The consumer group wishes to test the hypotheses

H_0: $(\mu_1 - \mu_2) = 0$ (i.e., no difference in mean retail price)

H_a: $(\mu_1 - \mu_2) > 0$ (i.e., mean retail price per pound of Procter & Gamble brand is higher than that of the General Foods brand)

where

μ_1 = Mean retail price per pound of Procter & Gamble's ground coffee at all supermarkets

μ_2 = Mean retail price per pound of General Foods' ground coffee at all supermarkets

This one-tailed, large-sample test is based on a z statistic. Thus, we will reject H_0 if $z > z_\alpha = z_{.01}$. Since $z_{.01} = 2.33$, the rejection region, as shown in Figure 10.8 on page 332, is given by

$$z > 2.33$$

We compute the test statistic as follows:

$$z \approx \frac{(\bar{x}_1 - \bar{x}_2) - D_0}{\sqrt{\frac{s_1^2}{n_1} + \frac{s_2^2}{n_2}}} = \frac{(2.98 - 2.93) - 0}{\sqrt{\frac{(.11)^2}{63} + \frac{(.07)^2}{58}}} = 3.01$$

Since this computed value of $z = 3.01$ lies in the rejection region (see Figure 10.8), there is sufficient evidence (at $\alpha = .01$) to conclude that the mean retail price per pound of Procter & Gamble's ground coffee is significantly higher than the mean retail price per pound of General Foods' ground coffee. The probability of our having committed a Type I error is $\alpha = .01$

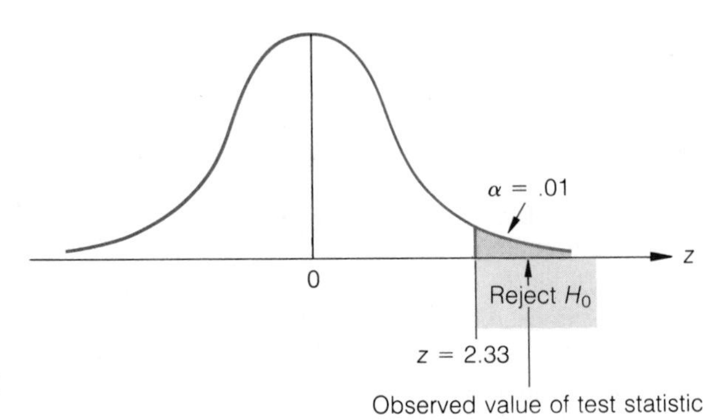

FIGURE 10.8 Rejection Region, Example 10.9

EXAMPLE 10.10 The Federal Trade Commission (FTC) conducts periodic tests on the tar and nicotine content and carbon monoxide emission of each brand of cigarette sold in the United States (see Case Study 2.3). Suppose an investigator for the FTC suspects there is a significant difference between the mean amounts of carbon monoxide in smoke emitted from Marlboro and Kool brand cigarettes. To test his theory, he selects independent random samples of cigarettes of each brand, determines the amount (in milligrams) of carbon monoxide in the smoke emitted from each, and summarizes his results as shown in Table 10.2. Test the investigator's belief, using significance level $\alpha = .01$.

TABLE 10.2 Carbon Monoxide in Smoke, Example 10.10

MARLBORO	KOOL
$n_1 = 30$	$n_2 = 40$
$\bar{x}_1 = 16.4$	$\bar{x}_2 = 15.5$
$s_1 = 1.2$	$s_2 = 1.1$

Solution The relevant hypothesis test consists of the elements

$$H_0: \quad (\mu_1 - \mu_2) = 0$$

$$H_a: \quad (\mu_1 - \mu_2) \neq 0$$

where μ_1 and μ_2 are the mean amounts of carbon monoxide in smoke from Marlboro and Kool brand cigarettes, respectively.

For this large-sample test, we will reject H_0 if $z < -z_{\alpha/2} = -z_{.005}$ or if $z > z_{\alpha/2} = z_{.005}$; in other words, the rejection region (see Figure 10.9) consists of the following sets of z-values:

$$z < -2.58 \quad \text{or} \quad z > 2.58$$

The test statistic is computed as follows:

$$z \approx \frac{(\bar{x}_1 - \bar{x}_2) - D_0}{\sqrt{\dfrac{s_1^2}{n_1} + \dfrac{s_2^2}{n_2}}} = \frac{(16.4 - 15.5) - 0}{\sqrt{\dfrac{(1.2)^2}{30} + \dfrac{(1.1)^2}{40}}} = 3.22$$

This z-value lies in the upper-tail rejection region (see Figure 10.9); there is sufficient evidence (at $\alpha = .01$) to support the investigator's claim of a significant difference between the mean amounts of carbon monoxide in smoke from Marlboro and Kool brand cigarettes.

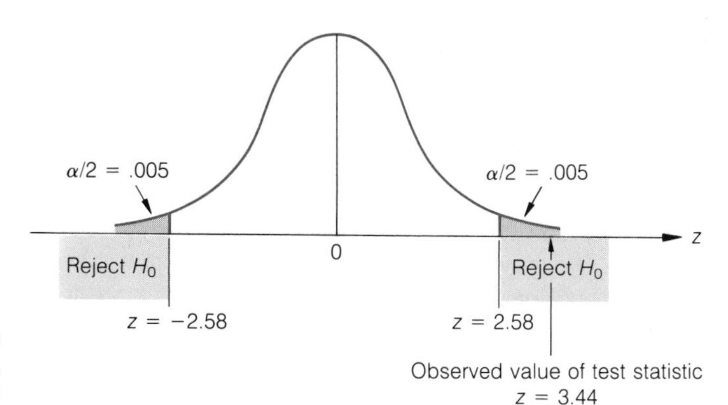

FIGURE 10.9 Rejection Region, Example 10.10

EXERCISES

10.17 Refer to Exercise 8.28, in which the data (Table 10.3) relating to the melting temperatures of two sucrose substitute compounds were obtained. Does the sample information provide sufficient evidence (at $\alpha = .05$) that the mean melting temperature of compound 1 is less than the mean melting temperature of compound 2?

TABLE 10.3 Comparison of Melting Temperatures, Exercise 10.17

COMPOUND 1	COMPOUND 2
$n_1 = 50$	$n_2 = 40$
$\bar{x}_1 = 287°F$	$\bar{x}_2 = 348°F$
$s_1 = 23°F$	$s_2 = 31°F$

10.18 The Metro Atlanta Rapid Transit Authority (MARTA) has recently implemented a six-week bus driver training program designed to reduce tardiness of buses on regularly scheduled routes. In order to gauge the effectiveness of the training program, a study was conducted before and after the program was instituted.

MARTA authorities were stationed at 30 randomly selected bus stops (involving 30 different buses) in the metro-Atlanta area before the program was implemented and observed tardiness (in minutes) of the scheduled bus arrivals. Similar data were collected at 35 randomly selected bus stops (involving 35 different buses) after the six-week training program. The results are summarized in Table 10.4. MARTA authorities are interested in determining if the mean tardiness of bus arrivals at metro-Atlanta bus stops has decreased significantly since the implementation of the training program. Perform a test of hypothesis for MARTA. Use a significance level of $\alpha = .02$.

TABLE 10.4 Bus Tardiness Times, Exercise 10.18

BEFORE TRAINING PROGRAM	AFTER TRAINING PROGRAM
$n_1 = 30$	$n_2 = 35$
$\bar{x}_1 = 5.25$ minutes	$\bar{x}_2 = 2.37$ minutes
$s_1 = 1.88$ minutes	$s_2 = 1.45$ minutes

[*Note:* The tardiness of a bus which arrived early or on time would be recorded as 0 minutes.]

10.19 A major oil company has developed a new gasoline additive that is designed to increase average gas mileage in subcompact cars. Before marketing the new additive, the company conducts the following experiment: One hundred subcompact cars are randomly selected and divided into two groups of 50 cars each. The gasoline additive is dispensed into the tanks of the cars in one group but not the other. The miles per gallon obtained by each car in the study is then recorded. The data are summarized in Table 10.5. Is there sufficient evidence for the oil company to claim that the average gas mileage obtained by subcompact cars with the additive is greater than the average gas mileage obtained by subcompact cars without the additive? Test at $\alpha = .10$.

TABLE 10.5 Gas Mileages, Exercise 10.19

WITHOUT ADDITIVE	WITH ADDITIVE
$n_1 = 50$	$n_2 = 50$
$\bar{x}_1 = 28.4$ miles per gallon	$\bar{x}_2 = 32.1$ miles per gallon
$s_1 = 9.5$ miles per gallon	$s_2 = 10.7$ miles per gallon

10.20 Pediatricians have recently theorized that infants who receive a particular form of reflex stimulation exercise during the first months of life will begin to walk more than 1.5 months earlier, on the average, than those children who receive no such exercise. To test this theory, a random sample of 50 newborns was selected to receive a particular reflex stimulation exercise on a regular basis; an independent random sample of 60 newborns received no such exercise. The age (in months) at which each child began to walk was recorded; the data are summarized in Table 10.6. Test the pediatricians' theory at a significance level of $\alpha = .05$.

TABLE 10.6 Walking Ages, Exercise 10.20

EXERCISE GROUP	NO EXERCISE GROUP
$n_1 = 50$	$n_2 = 60$
$\bar{x}_1 = 10.75$	$\bar{x}_2 = 12.50$
$s_1 = 3.05$	$s_2 = 2.50$

10.6 Hypothesis Test about the Difference between Two Population Means: Small-Sample Case

The Environmental Protection Agency (EPA) often conducts studies designed to estimate highway and city gas mileages for automobiles (see Case Study 6.3). Suppose that the EPA is interested in comparing the mean highway mileages for cars using leaded and unleaded gasoline. That is, the EPA will test the hypothesis that μ_1, the mean highway mileage for cars using leaded gasoline, differs from μ_2, the mean highway mileage for cars using unleaded gasoline, i.e., $(\mu_1 - \mu_2) \neq 0$. However, the EPA is able to obtain independent random samples of only $n_1 = 11$ cars which use leaded gasoline and $n_2 = 10$ cars which require unleaded gasoline. When the sample sizes n_1 and n_2 are inadequate to permit use of the large-sample procedure of Section 10.5, modifications may be made to perform a small-sample test of an hypothesis about the difference between two population means (see box below). The test procedure is based on assumptions that are, again, more restrictive than in the large-sample case.

SMALL-SAMPLE TEST OF HYPOTHESIS ABOUT $(\mu_1 - \mu_2)$

a. One-tailed test

H_0: $(\mu_1 - \mu_2) = D_0$

H_a: $(\mu_1 - \mu_2) > D_0$

(or H_a: $(\mu_1 - \mu_2) < D_0$)

Test statistic:

$$t = \frac{(\bar{x}_1 - \bar{x}_2) - D_0}{\sqrt{s_p^2\left(\frac{1}{n_1} + \frac{1}{n_2}\right)}}$$

Rejection region:

$t > t_\alpha$ (or $t < -t_\alpha$)

b. Two-tailed test

H_0: $(\mu_1 - \mu_2) = D_0$

H_a: $(\mu_1 - \mu_2) \neq D_0$

Test statistic:

$$t = \frac{(\bar{x}_1 - \bar{x}_2) - D_0}{\sqrt{s_p^2\left(\frac{1}{n_1} + \frac{1}{n_2}\right)}}$$

Rejection region:

$t < -t_{\alpha/2}$ or $t > t_{\alpha/2}$

where

$$s_p^2 = \frac{(n_1 - 1)s_1^2 + (n_2 - 1)s_2^2}{n_1 + n_2 - 2}$$

and the distribution of t is based on $(n_1 + n_2 - 2)$ degrees of freedom

ASSUMPTIONS REQUIRED

1. The populations from which the samples are selected both have relative frequency distributions which are approximately normal.
2. The variances of the two populations are equal.
3. The random samples are selected in an independent manner from the two populations.

EXAMPLE 10.11 Each of the 21 cars (11 using leaded gas and 10 using unleaded gas) selected for the EPA study was tested and the number of miles per gallon obtained by each recorded. The results are summarized in Table 10.7. Is there evidence (at $\alpha = .02$) that the mean number of miles per gallon obtained by all cars using leaded gasoline differs significantly from the mean number of miles per gallon obtained by all cars using unleaded gasoline?

TABLE 10.7 Gas Mileages, Example 10.11

LEADED	UNLEADED
$n_1 = 11$	$n_2 = 10$
$\bar{x}_1 = 17.2$ miles per gallon	$\bar{x}_2 = 19.9$ miles per gallon
$s_1 = 2.1$ miles per gallon	$s_2 = 2.0$ miles per gallon

Solution The EPA desires to test the following hypotheses:

H_0: $(\mu_1 - \mu_2) = 0$ (i.e., no difference in mean miles per gallon)

H_a: $(\mu_1 - \mu_2) \neq 0$ (i.e., mean miles per gallon values for leaded and unleaded gas differ)

where

μ_1 = True mean miles per gallon for cars using leaded gasoline

μ_2 = True mean miles per gallon for cars using unleaded gasoline

Since the samples selected for the study are small ($n_1 = 11$, $n_2 = 10$), the following assumptions are required:

1. The populations of number of miles per gallon obtained by cars using leaded gasoline and cars using unleaded gasoline both have approximately normal distributions.
2. The variances of the populations of miles per gallon values for the two types of gasoline are equal.
3. The samples were independently and randomly selected.

If these assumptions are valid, the test statistic will have a t distribution with $(n_1 + n_2 - 2) = (11 + 10 - 2) = 19$ degrees of freedom. Using a significance level of $\alpha = .02$, the rejection region is given by

$$t < -t_{.01} = -2.539 \quad \text{or} \quad t > t_{.01} = 2.539$$

(See Figure 10.10.)

Since we have assumed that the two populations have equal variances (i.e., that $\sigma_1^2 = \sigma_2^2 = \sigma^2$), we need to compute an estimate of this common variance. Our pooled estimate is given by

$$s_p^2 = \frac{(n_1 - 1)s_1^2 + (n_2 - 1)s_2^2}{n_1 + n_2 - 2} = \frac{(11 - 1)(2.1)^2 + (10 - 1)(2.0)^2}{11 + 10 - 2} = 4.216$$

Using this pooled sample variance in the computation of the test statistic, we obtain

$$t = \frac{(\bar{x}_1 - \bar{x}_2) - D_0}{\sqrt{s_p^2\left(\frac{1}{n_1} + \frac{1}{n_2}\right)}} = \frac{(17.2 - 19.9) - 0}{\sqrt{4.216\left(\frac{1}{11} + \frac{1}{10}\right)}} = -3.01$$

Now the computed value of t falls within the rejection region (see Figure 10.10), thus we reject the null hypothesis (at $\alpha = .02$) and conclude that there is a significant difference between the mean number of miles per gallon obtained by cars using leaded gasoline and the mean number of miles per gallon obtained by cars using unleaded gasoline. The probability that we will incorrectly reject the null hypothesis (i.e., conclude that there is a difference in mean miles per gallon if in fact there is no difference) is only .02.

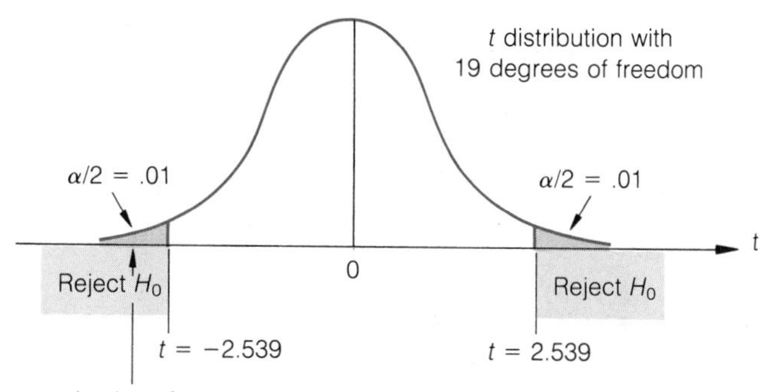

FIGURE 10.10 Rejection Region, Example 10.11

EXAMPLE 10.12 Two relatively new energy-saving concepts in home building are solar-powered homes and earth-sheltered homes. An individual is drawing up plans for a new home and wishes to compare his expected annual heating costs for the two types of innovation. Independent random samples of solar-powered homes (which receive 50% of their energy from the sun) and earth-sheltered homes yielded the summary data, Table 10.8, on annual heating costs. (You may assume the homes were comparable with respect to size, climatic conditions, etc.) Is there evidence (at $\alpha = .05$) that the mean annual cost of heating an earth-sheltered home is significantly less than the corresponding cost of heating a 50% solar-powered home?

TABLE 10.8 Heating Costs, Example 10.12

SOLAR-POWERED	EARTH-SHELTERED
$n_1 = 12$	$n_2 = 6$
$\bar{x}_1 = \$285$	$\bar{x}_2 = \$234$
$s_1 = \$55$	$s_2 = \$26$

Solution Of interest to the potential home builder is a test of the hypotheses

$$H_0: \ (\mu_1 - \mu_2) = 0$$
$$H_a: \ (\mu_1 - \mu_2) > 0$$

where

μ_1 = True mean annual cost of heating a solar-powered home

μ_2 = True mean annual cost of heating an earth-sheltered home

Since we are restricted to small samples ($n_1 = 12$, $n_2 = 6$) from the target populations, we must make the following assumptions:

1. The populations of annual heating costs for both solar-powered and earth-sheltered homes have approximately normal distributions.
2. The variances of the populations of annual heating costs for the two types of homes are equal.
3. The samples were randomly and independently selected.

Under these assumptions, the test statistic will have a t distribution with $(n_1 + n_2 - 2) = (12 + 6 - 2) = 16$ degrees of freedom. Thus, at significance level $\alpha = .05$, the null hypothesis will be rejected if

$$t > t_{.05} = 1.746$$

This rejection region is shown in Figure 10.11.

We have assumed that the two populations have equal variances (i.e., that $\sigma_1^2 = \sigma_2^2 = \sigma^2$); our pooled estimate of this common variance is given by

$$s_p^2 = \frac{(n_1 - 1)s_1^2 + (n_2 - 1)s_2^2}{n_1 + n_2 - 2} = \frac{(12 - 1)(55)^2 + (6 - 1)(26)^2}{12 + 6 - 2} = 2{,}290.9$$

Now, the value of the test statistic may be computed as follows:

$$t = \frac{(\bar{x}_1 - \bar{x}_2) - D_0}{\sqrt{s_p^2\left(\frac{1}{n_1} + \frac{1}{n_2}\right)}} = \frac{(285 - 234) - 0}{\sqrt{2{,}290.9\left(\frac{1}{12} + \frac{1}{6}\right)}} = 2.13$$

The computed value of t exceeds the critical value of $t = 1.746$. We thus reject the null hypothesis (at significance level .05) and conclude that the mean annual cost of heating a solar-powered home exceeds the mean cost of heating an earth-sheltered home; i.e., the mean annual cost of heating an earth-sheltered home is significantly less than the cost of heating a solar-powered home.

We wish to note that the two-sample statistic used here (and in Section 8.6 for interval estimation) relies heavily upon the assumptions that the two target populations are (at least approximately) normally distributed and that their respective variances are equal. However, studies have shown that, when the sample sizes n_1 and n_2 are equal, the requirement of equal population variances can be relaxed somewhat. In other words, when $n_1 = n_2$, σ_1^2 and σ_2^2 can assume different values and the test statistic t will still be approximated by a t distribution with appropriate degrees of freedom.

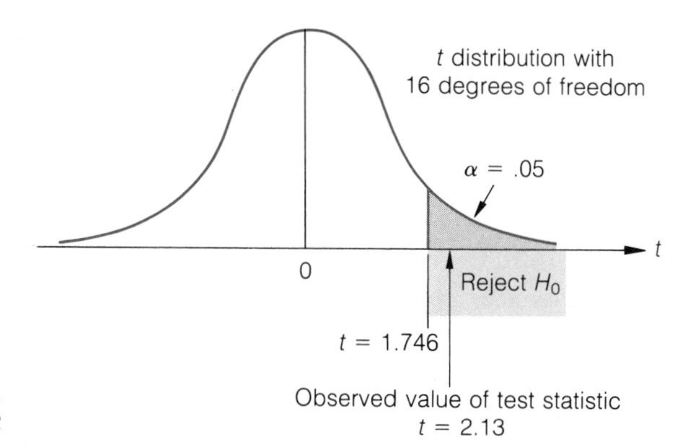

FIGURE 10.11 Rejection Region, Example 10.12

EXERCISES

10.21 The main ingredient in a reputable car repair service is a good mechanic. However, many repair service owners in the city of Chicago are finding it difficult to hire good mechanics without charging exorbitant labor prices to the customer. Suppose that we wish to compare the hourly labor rates of mechanics working at gas stations and at car dealerships in Chicago. Independent random samples of 15 gas stations and 10 car dealerships are selected and the hourly labor rate of mechanics at each is recorded. The data are summarized in Table 10.9. Test to see whether there is a difference between the true mean hourly labor rates of mechanics working at gas stations and mechanics working at car dealerships in the city of Chicago. Use a significance level of $\alpha = .05$.

TABLE 10.9 Hourly Labor Rates, Exercise 10.21

GAS STATIONS	CAR DEALERS
$n_1 = 15$	$n_2 = 10$
$\bar{x}_1 = \$25.77$	$\bar{x}_2 = \$32.50$
$s_1 = \$4.10$	$s_2 = \$5.60$

10.22 Many college and university professors have been accused of "grade promotion" over the past several years. This means they assign higher grades now than in the past, even though students' work is of the same caliber. If grade promotion has occurred, the mean grade-point average (GPA) of today's students should exceed the mean of 10 years ago. To test the grade promotion theory at one university, a psychology professor randomly selects 30 psychology majors who are graduating with the present class and records the GPA of each. However, the professor is able to obtain the GPA's for a sample of only 8 psychology majors who graduated 10 years ago. A summary of the results is shown in Table 10.10. Test whether the data support the hypothesis of grade promotion in the psychology department of this university. Use a significance level of $\alpha = .10$.

TABLE 10.10 Grade-Point Averages, Exercise 10.22

10 YEARS AGO	PRESENT
$n_1 = 8$	$n_2 = 30$
$\bar{x}_1 = 2.82$	$\bar{x}_2 = 3.04$
$s_1 = .43$	$s_2 = .38$

10.23 An industrial plant wants to determine which of two types of fuel—gas or electric—will produce more useful energy at the lower cost. One measure of economical energy production, called the "plant investment per delivered quad," is calculated by taking the amount of money (in dollars) invested in the particular utility by the plant and dividing it by the amount of delivered energy (in quadrillion British thermal units). The smaller this ratio, the less an industrial plant pays for its delivered energy.

Random samples of eleven plants using electrical utilities and sixteen plants using gas utilities were taken, and the plant investment per quad was calculated for each. The resulting data are summarized in Table 10.11.

TABLE 10.11
Plant Investment/Quad ($ Billions), Exercise 10.23

ELECTRIC	GAS
$n_1 = 11$	$n_2 = 16$
$\bar{x}_1 = \$22.5$	$\bar{x}_2 = \$17.5$
$s_1 = \$4.18$	$s_2 = \$3.87$

a. Do these data provide sufficient evidence at the $\alpha = .01$ level of significance to indicate a difference between the average investments per quad for plants using gas and those using electrical utilities?
b. What assumptions are required for the hypothesis test, part a, to be valid?

10.24 Since certain preservatives may have an effect on the nutritional quality of food into which they are introduced, food packaging plants must continually test their products. A preliminary experiment to determine the nutritional effect of a certain preservative used 24 guinea pigs. The guinea pigs were randomly split into two equal groups so that 12 guinea pigs received a diet without the preservative and 12 were fed a diet with the preservative. The data yielded the summary information shown in Table 10.12. Test whether guinea pigs on a diet with this preservative gain less weight, on the average, than those on a diet without this preservative. Use a significance level of $\alpha = .02$. (List any assumptions necessary for the validity of the test.)

TABLE 10.12
Weight Gains, Exercise 10.24

WITHOUT PRESERVATIVE	WITH PRESERVATIVE
$n_1 = 12$	$n_2 = 12$
$\bar{x}_1 = 6.8$ grams	$\bar{x}_2 = 5.3$ grams
$s_1 = 1.5$ grams	$s_2 = 0.9$ grams

10.7 Hypothesis Test about the Difference between Two Proportions: Large-Sample Case

Suppose we are interested in comparing the proportion, p_1, of working mothers who feel they have enough "free" time for themselves, with the proportion p_2, of working fathers who feel they have enough "free" time for themselves. Then the target parameter about which we will test an hypothesis is $(p_1 - p_2)$. The method for performing a large-sample test of hypothesis about the difference, $(p_1 - p_2)$, between two population proportions is outlined in the box.

LARGE-SAMPLE TEST OF HYPOTHESIS ABOUT $(p_1 - p_2)$

a. One-tailed test

H_0: $(p_1 - p_2) = D_0$

H_a: $(p_1 - p_2) > D_0$

(or H_a: $(p_1 - p_2) < D_0$)

Test statistic:

$$z = \frac{(\hat{p}_1 - \hat{p}_2) - D_0}{\sigma_{(\hat{p}_1 - \hat{p}_2)}} \approx \frac{(\hat{p}_1 - \hat{p}_2) - D_0}{\sqrt{\frac{\hat{p}_1\hat{q}_1}{n_1} + \frac{\hat{p}_2\hat{q}_2}{n_2}}}$$

Rejection region:

$z > z_\alpha$ (or $z < -z_\alpha$)

b. Two-tailed test

H_0: $(p_1 - p_2) = D_0$

H_a: $(p_1 - p_2) \neq D_0$

Test statistic:

$$z = \frac{(\hat{p}_1 - \hat{p}_2) - D_0}{\sigma_{(\hat{p}_1 - \hat{p}_2)}} \approx \frac{(\hat{p}_1 - \hat{p}_2) - D_0}{\sqrt{\frac{\hat{p}_1\hat{q}_1}{n_1} + \frac{\hat{p}_2\hat{q}_2}{n_2}}}$$

Rejection region:

$z < -z_{\alpha/2}$ or $z > z_{\alpha/2}$

where $\hat{q}_1 = 1 - \hat{p}_1$ and $\hat{q}_2 = 1 - \hat{p}_2$

The sample sizes n_1 and n_2 must be sufficiently large to ensure that the sampling distributions of $\hat{p}_1$ and $\hat{p}_2$, and hence of the difference $(\hat{p}_1 - \hat{p}_2)$, are approximately normal. [*Note:* If the sample sizes are not sufficiently large, p_1 and p_2 can be compared using the technique discussed in Chapter 14.]

ASSUMPTIONS REQUIRED

The intervals

$$\hat{p}_1 \pm 2\sqrt{\frac{\hat{p}_1\hat{q}_1}{n_1}} \quad \text{and} \quad \hat{p}_2 \pm 2\sqrt{\frac{\hat{p}_2\hat{q}_2}{n_2}}$$

do not contain 0 or 1.

EXAMPLE 10.13 In recent years there has been a trend toward both parents working outside the home. Do working mothers experience the same burdens and family pressures as their spouses? A popular belief is that the proportion of working mothers who feel they have enough spare time for themselves is significantly less than the corresponding proportion of working fathers. In order to test this claim, independent random samples of 100 working mothers and 100 working fathers were selected and their views on spare time for themselves were recorded. A summary of the data is given in Table 10.13. (Assume that the spouses of all individuals sampled were also working outside the home.) Is the belief that the proportion of working mothers who feel they have enough spare time for themselves is less than the corresponding proportion of working fathers supported (at significance level $\alpha = .01$) by the sample information?

TABLE 10.13
Data on Working Parents, Example 10.13

	WORKING MOTHERS	WORKING FATHERS
Number sampled	100	100
Number in sample who feel they have enough spare time for themselves	37	56

Solution We wish to perform a test of

$$H_0: \ (p_1 - p_2) = 0$$

$$H_a: \ (p_1 - p_2) < 0$$

where

p_1 = The proportion of all working mothers who feel they have enough spare time for themselves

p_2 = The proportion of all working fathers who feel they have enough spare time for themselves

For this large-sample, one-tailed test, the null hypothesis will be rejected if

$$z < -z_{.01} = -2.33$$

(See Figure 10.12.)

The sample proportions $\hat{p}_1$ and $\hat{p}_2$ are computed for substitution into the formula for the test statistic:

$\hat{p}_1$ = Sample proportion of working mothers who feel they have enough spare time for themselves

$= 37/100 = .37$

$\hat{p}_2$ = Sample proportion of working fathers who feel they have enough spare time for themselves

$= 56/100 = .56$

Hence,

$$\hat{q}_1 = 1 - \hat{p}_1 = 1 - .37 = .63$$

$$\hat{q}_2 = 1 - \hat{p}_2 = 1 - .56 = .44$$

Then the value of the test statistic is

$$z = \frac{(\hat{p}_1 - \hat{p}_2) - D_0}{\sqrt{\frac{\hat{p}_1\hat{q}_1}{n_1} + \frac{\hat{p}_2\hat{q}_2}{n_2}}} = \frac{(.37 - .56) - 0}{\sqrt{\frac{(.37)(.63)}{100} + \frac{(.56)(.44)}{100}}} = -2.74$$

This value falls below the critical value of −2.33 (see Figure 10.12). Thus, at $\alpha = .01$, we reject the null hypothesis; there is sufficient evidence to conclude that the proportion of working mothers who feel they have enough spare time for themselves is significantly less than the corresponding proportion of working fathers.

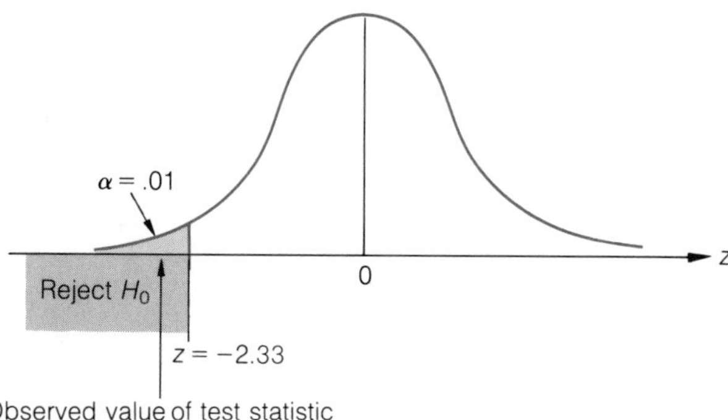

FIGURE 10.12 Rejection Region, Example 10.13

EXERCISES

10.25 Refer to Example 8.21, in which the results of a survey of the political party preferences of independent random samples of American adults in January 1980 and April 1981 are reported. (The information is reproduced in Table 10.14.) Does the sample information indicate a significant difference between the proportions of American adults in January 1980 and in April 1981 who call themselves Republicans or Republican-leaning Independents? Perform an appropriate hypothesis test at significance level $\alpha = .05$.

TABLE 10.14 Party Preferences, Exercise 10.25

	JANUARY 1980	APRIL 1981
Number surveyed	1,400	1,400
Number in sample who said they were Republicans or Republican-leaning Independents	462	574

10.26 To market a new white wine, a winemaker decides to use two different advertising agencies, one operating in the East, one in the West. After the white wine has been on the market for eight months, independent random samples of wine drinkers are taken from each of the two regions and questioned concerning their white wine preference. The numbers favoring the new brand are shown in Table 10.15. Is there evidence that the proportion of wine drinkers in the West who prefer the new white wine is larger than the corresponding proportion of wine drinkers in the East? Test at significance level $\alpha = .02$.

TABLE 10.15 Data on White Wine Preference, Exercise 10.26

	EAST	WEST
Number of wine drinkers sampled	516	438
Number of sampled wine drinkers who prefer the new white wine	18	23

10.27 A large shipment of produce contains McIntosh and Red Delicious apples. To determine whether there is a difference between the percentages of nonmarketable fruit for the two varieties, random samples of 800 McIntosh apples and 800 Red Delicious apples were independently selected and the number of nonmarketable apples of each type were counted. It was found that 33 McIntosh and 58 Red Delicious apples from these samples were nonmarketable. Do these data provide sufficient evidence to indicate a difference between the percentages of nonmarketable McIntosh and Red Delicious apples in the entire shipment? Test at the $\alpha = .10$ level of significance.

10.28 Scientists have been studying the use of heroin to relieve the pain of cancer patients. Heroin has not been available for legitimate medical use in the United States since 1924. Currently, morphine is used as a pain reliever for cancer patients. However, if heroin is shown to be effective, it could be approved for use in terminal cases. In one of a series of trials (approved by the Food and Drug Administration), 60 terminal cancer patients with moderate to severe pain were divided into two groups of 30 each. Each patient in the first group received a single intramuscular injection of morphine, while those in the second group received heroin. One hour after the injection the patients rated the intensity of their pain. (In this "double blind trial," neither the patients nor the doctors knew which drug had been administered until after the trial so that the results would not be biased.) Of the morphine group, 18 patients rated the intensity of their pain as low or none; of the heroin group, 25 patients rated the intensity of their pain as low or none. Is there evidence of a difference between the proportions of terminal patients who rate their pain intensity as low or none for the two drugs? Test using a significance level of $\alpha = .05$.

10.8 Reporting Test Results: *p*-Values

The statistical hypothesis-testing technique which we have developed in Chapter 9 and in this chapter requires that we choose the significance level α (i.e., the maximum probability of a Type I error that we are willing to tolerate) prior to obtaining the data and computing the test statistic. By choosing α a priori, we in effect fix the rejection region of the test. Thus, no matter how large or how small the observed value of the test statistic, our decision regarding H_0 is clear-cut: reject H_0 (i.e., conclude that the test results are statistically significant) if the observed value of the test statistic falls into the rejection region; fail to reject H_0 otherwise (i.e., conclude that the test results are insignificant). This "fixed" significance level, α, then serves as a measure of the reliability of our inference. However, there is one drawback to a test conducted in this manner: a measure of the ***degree*** of significance of the test results is not readily available. That is, if in fact the value of the test statistic falls into the rejection region, we have no measure of the extent to which the data disagree with the null hypothesis, H_0.

EXAMPLE 10.14 A large-sample test of H_0: $\mu = 80$ versus H_a: $\mu > 80$ is to be conducted at a fixed significance level of $\alpha = .05$. Consider the following possible values of the computed test statistic:

$$z = 1.82$$
$$z = 5.66$$

a. Which of the above values of the test statistic gives the stronger evidence for the rejection of H_0?

b. How can we measure the extent of disagreement between the sample data and H_0 for each of the computed values?

Solution **a.** The appropriate rejection region for this test, at $\alpha = .05$, is given by

$$z > z_{.05} = 1.645$$

Clearly, for either of the test statistic values given above, $z = 1.82$ or $z = 5.66$, we will reject H_0; hence, the result in each case is statistically significant. Recall, however, that the appropriate test statistic for a large-sample test concerning μ is simply the z-score for the observed sample mean, $\bar{x}$, calculated by using the hypothesized value of μ in H_0 (in this case, $\mu = 80$). The larger the z-score, the greater the distance (in units of standard deviations) that $\bar{x}$ is from the hypothesized value of $\mu = 80$. Thus, a z-score of 5.66 would present stronger evidence that the true mean is larger than 80 than would a z-score of 1.82. Our reasoning stems from our knowledge of the sampling distribution of $\bar{x}$; if in fact $\mu = 80$, we would certainly not expect to observe an $\bar{x}$ with a z-score as large as 5.66.

b. One way of measuring the amount of disagreement between the observed data and the value of μ in the null hypothesis is to calculate the probability of observing a value of the test statistic equal to or greater than the actual computed value, if in fact H_0 were true. That is, if z_c is the computed value of the test statistic, calculate

$$P(z \geq z_c)$$

assuming the null hypothesis is true. This "disagreement" probability, or ***p*-value,** is calculated below for each of the computed test statistics, $z = 1.82$ and $z = 5.66$, using Table 3, Appendix E.

$$P(z \geq 1.82) = .5 - .4656 = .0344$$
$$P(z \geq 5.66) \approx .5 - .5 = 0$$

From our discussion in part a, you can see that the smaller the p-value, the greater the extent of disagreement between the data and the null hypothesis, i.e., the more significant the result.

In general, p-values are computed as shown in the box at the top of page 346.

MEASURING THE DISAGREEMENT BETWEEN THE DATA AND H_0: p-VALUES

Large-sample one-tailed test: $p\text{-value} = P(z \geq |z_c|)$

Large-sample two-tailed test: $p\text{-value} = 2 \cdot P(z \geq |z_c|)$

where z_c is the computed value of the test statistic.

Small-sample one-tailed test: $p\text{-value} = P(t \geq |t_c|)$

Small-sample two-tailed test: $p\text{-value} = 2 \cdot P(t \geq |t_c|)$

where t_c is the computed value of the test statistic.

[*Note:* $|z_c|$ and $|t_c|$ denote the *absolute values* of z_c and t_c and will always be positive.]

Notice that the p-value for a two-tailed test is twice the probability for the one-tailed test. This is because the disagreement between the data and H_0 can be in two directions.

When publishing the results of a statistical test of hypothesis in journals, case studies, reports, etc., many researchers make use of p-values. Instead of selecting α a priori and then conducting a test as outlined in this chapter, the researcher will compute and report the value of the appropriate test statistic and its associated p-value. It is left to the reader of the report to judge the significance of the result, i.e., the reader must determine whether to reject the null hypothesis in favor of the alternative, based upon the reported p-value. This p-value is often referred to as the ***attained significance level*** of the test. Usually, the null hypothesis will be rejected if the attained significance level is ***less*** than the fixed significance level, α, chosen by the reader. The inherent advantages of reporting test results in this manner are twofold: (1) each reader is permitted to select the maximum value of α that they would be willing to tolerate if they actually carried out a standard test of hypothesis in the manner outlined in this chapter; and (2) a measure of the degree of significance of the result (i.e., the p-value) is provided.

REPORTING TEST RESULTS AS p-VALUES: HOW TO DECIDE WHETHER TO REJECT H_0

1. Choose the maximum value of α that you are willing to tolerate.
2. If the attained significance level (p-value) of the test is less than the maximum value of α, then reject the null hypothesis.

EXAMPLE 10.15 Refer to Example 10.1. Compute the attained significance level of the test. Interpret this value.

Solution In this large-sample test concerning a population mean μ, the computed value of the test statistic was $z_c = .903$. Since the test is one-tailed, the associated p-value is given by

$$P(z \geq |z_c|) = P(z \geq .903)$$
$$\approx .5 - .3159 = .1841$$

Thus, the attained significance level of the test is approximately .184. In order to reject the null hypothesis H_0: $\mu > 30$, we would have to be willing to risk a Type I error probability, α, of at least .184. Most researchers would not be willing to take this risk and would deem the result insignificant (i.e., conclude that there is insufficient evidence to reject H_0).

EXAMPLE 10.16 Refer to Example 10.10.

a. Compute the attained significance level of the test.
b. Make the appropriate conclusion if you are willing to tolerate a Type I error probability of $\alpha = .01$.

Solution a. The computed test statistic for this large-sample test about $(\mu_1 - \mu_2)$ was given as $z_c = 3.22$. Since the test is two-tailed, the associated p-value is

$$2 \cdot P(z \geq |z_c|) = 2 \cdot P(z \geq |3.22|)$$
$$= 2 \cdot P(z \geq 3.22)$$
$$\approx 2(.5 - .4999) = 2(.0001) = .0002$$

Thus, the approximate attained significance level of the test is .0002.

b. Since the attained significance level of .0002 is less than the maximum tolerable Type I error probability of $\alpha = .01$, we will reject H_0 and conclude that a significant difference exists between the mean amounts of carbon monoxide in smoke emitted from the two brands of cigarette (Marlboro and Kool). In fact, we could choose a Type I error probability as small as $\alpha = .0002$ and still have sufficient evidence to reject H_0. Thus, the result is highly significant.

EXAMPLE 10.17 Refer to Example 10.6. Compute the attained significance level of the test and interpret your result.

Solution The computed test statistic for this small-sample test concerning μ was given as $t_c = -2.0$. Since the t-test is one-tailed, the associated p-value is found by calculating

$$P(t \geq |t_c|) = P(t \geq |-2.0|) = P(t > 2.0)$$

where the distribution of t is based on $(n - 1) = 15$ degrees of freedom. To find the p-value from the table of critical t-values provided in Table 4, Appendix E, search for the value 2.0 in the row corresponding to 15 df. You can see that 2.0 does not appear in this row but falls between the values 1.753 (in the $t_{.05}$ column) and 2.131 (in the $t_{.025}$ column). The p-value associated with 1.753 is .05, and the p-value associated with 2.131 is .025. Thus, the p-value associated with $t_c = -2.0$ is somewhere between .025 and .05. Since the exact p-value is unknown, we take the conservative approach and report the approximate p-value as the larger of the two endpoints, namely .05. This (approximate) p-value indicates that the null hypothesis H_0: $\mu = 45$ will be rejected in favor of H_a: $\mu < 45$ for any fixed significance level α larger than or equal to .05.

The above will be the usual procedure when computing a p-value from Table 4, Appendix E. That is, the exact value of the test statistic t_c will usually not appear in the row corresponding to the appropriate number of df. You will need to locate the tabled t-values between which t_c falls and determine the p-values associated with these tabled values (by noting the subscripts (α) in the respective column headings). Report the approximate attained significance level of the test as the larger of the two p-values.

Whether we conduct a test using p-values or the rejection region approach, our choice of a maximum tolerable Type I error probability becomes critical to the decision concerning H_0 and should not be hastily made. In either case, care should be taken to weigh the seriousness of committing a Type I error in the context of the problem.

EXERCISES

10.29 For a large-sample test of

H_0: $(\mu_1 - \mu_2) = 0$

H_a: $(\mu_1 - \mu_2) > 0$

compute the p-value associated with each of the following computed test statistic values:

a. $z_c = 1.96$ **b.** $z_c = 1.645$ **c.** $z_c = 2.67$ **d.** $z_c = 1.25$

10.30 For a large-sample test of

H_0: $(p_1 - p_2) = 0$

H_a: $(p_1 - p_2) \neq 0$

compute the p-value associated with each of the following computed test statistic values:

a. $z_c = -1.01$ **b.** $z_c = -2.37$ **c.** $z_c = 4.66$ **d.** $z_c = -1.45$

10.31 Refer to Exercise 10.1. Compute the attained significance level (p-value) of the test. Interpret the result.

10.32 Refer to Example 10.7. Compute the attained significance level (p-value) of the test. What is your decision regarding H_0 if you are willing to risk a maximum Type I error probability of only $\alpha = .01$?

10.33 Refer to Exercise 10.20. Compute the attained significance level (p-value) of the test. Compare to the fixed significance level, α, and make an appropriate conclusion.

10.34 Refer to Exercise 10.28. Compute the attained significance level (p-value) of the test. What is your decision regarding H_0 if you are willing to increase the probability of a Type I error to $\alpha = .10$?

10.9 Summary

In this chapter, we have summarized the procedures for testing hypotheses about various population parameters. As we noted with the estimation techniques of Chapter 8, fewer assumptions about the sampled populations are required when the sample sizes are large. We also wish to emphasize that *statistical* significance differs from *practical* significance, and the two must not be confused. A reasonable approach to hypothesis testing blends a valid application of the formal statistical procedures with the researcher's knowledge of the subject matter.

KEY WORDS

Statistical significance
Practical significance
Attained significance level
p-Value

SUPPLEMENTARY EXERCISES

[*Note:* For each of these exercises, carefully define any notation used, perform all steps of the relevant hypothesis test, state a conclusion in terms of the problem, and specify any assumptions required for the validity of the procedure.]

10.35 The percentage of body fat and skinfold thickness are good indicators of the energy metabolic status and general health of humans. A study of the percentage of body fat of college students in India was recently undertaken (*American Journal of Physical Anthropology,* January 1981). Two groups of healthy male college students, urban and rural, from different colleges in eastern India were independently and randomly selected and the percentage of body fat in each was measured. The data are summarized in Table 10.16. Does the sample information provide sufficient evidence to conclude that the mean percentage of body fat in healthy male college students residing in urban areas of India differs from the corresponding mean for students residing in rural areas? Use a significance level of $\alpha = .05$.

TABLE 10.16
Percent Body Fat Data, Exercise 10.35

URBAN STUDENTS	RURAL STUDENTS
$n_1 = 193$	$n_2 = 188$
$\bar{x}_1 = 12.07$	$\bar{x}_2 = 11.04$
$s_1 = 3.04$	$s_2 = 2.63$

10.36 In a particular state, it is known that the mean number of prior convictions for youthful offenders is 2. A juvenile probation officer has recently instituted a rehabilitation program which he believes will reduce the number of prior convictions for future offenders. Six months after inception of the program, he randomly sampled 100 recent juvenile offenders, and recorded the number of prior convictions for each. The results are summarized as follows:

$\bar{x} = 1.90$

$s = .52$

Is there evidence (at significance level $\alpha = .02$) that the mean number of prior convictions for youthful offenders has dropped since institution of the rehabilitation program?

10.37 In order to stock their various departments with the type and style of goods that appeal to their potential group of customers, a downtown Los Angeles department store is interested in determining whether the average age of downtown Los Angeles shoppers is less than 35 years. A random sample of 20 downtown shoppers revealed the following summary statistics on their ages:

$\bar{x} = 34.6$ years

$s = 10.8$ years

Is there sufficient evidence to conclude that the average age of downtown Los Angeles shoppers is less than 35 years? Test at $\alpha = .10$.

10.38 As part of its energy conservation plan, the city of Portland, Oregon, has built an efficient new bus system. It is hoped that the new bus system will fight fuel waste by discouraging the use of automobiles. Suppose that the proportion of the city's residents who rode the old bus system regularly was known to be 14%. Three months after installation of the new bus system, a survey of 100 city residents indicated that 22 now ride the buses regularly. Is there evidence that the true proportion of the city's residents who ride the new bus system regularly is greater than .14? Test at $\alpha = .05$.

10.39 The rising price of oil has prompted Brazil to use alcohol, distilled mainly from its bumper crops of sugar cane, as a power source. Many of the automobiles traveling along Brazil's roads are powered by pure alcohol instead of gasoline. Suppose that it is desired to compare the proportion of cars powered by pure alcohol in Brazil to the proportion of cars powered by pure alcohol in the United States. Random samples of automobiles are selected independently in Brazil and the United States and the number using alcohol as a motor fuel (gasohol) determined. The data are shown in Table 10.17. Test (at significance level $\alpha = .01$) whether there is a significant difference between the proportion of Brazil's cars powered by alcohol and the corresponding proportion in the United States.

TABLE 10.17 Gasohol Data, Exercise 10.39

	BRAZIL	UNITED STATES
Number of automobiles sampled	70	108
Number of sampled automobiles which are using alcohol as a motor fuel	15	6

10.40 A field experiment was conducted to ascertain the impact of desert granivores (seed-eaters) on the density and distribution of seeds in the soil (*Ecology,* December 1979). Since some desert rodents are known to hoard seeds in surface caches, the study was specifically designed to determine if these caches eventually produce more seedlings, on the average, than an adjacent control area. Forty small areas excavated by rodents were located and covered with plastic cages to prevent rodents from reusing the caches. A caged control area was set up adjacent to each of the caged caches. The number of seedlings germinating from the caches and from the control areas was then observed. A summary of the data is provided in Table 10.18. Is there sufficient evidence (at $\alpha = .05$) to indicate that the average number of seedlings germinating from the seed caches of desert rodents is significantly higher than the corresponding average for the control (non-cache) areas?

TABLE 10.18 Seedlings Germinating, Exercise 10.40

CACHES	CONTROL AREAS (NON-CACHE)
$n_1 = 40$	$n_2 = 40$
$\bar{x}_1 = 5.3$	$\bar{x}_2 = 2.7$
$s_1 = 1.3$	$s_2 = 0.7$

Source: *Ecology,* December 1979, *60,* 1089–1092. Copyright 1979, the Ecological Society of America.

10.41 The following was reported in the Gainesville *Sun* (May 7, 1981): "Medical tests show that Haitian refugees arriving in this country from their impoverished Caribbean island homeland have a much higher incidence of tuberculosis than any other segment of the Dade County (Miami) population." The incidence of tuberculosis among Dade County residents is known to be .0002 (i.e., 2 cases per 10,000 people). Twenty-five cases of tuberculosis were found in the 3,107 Haitians screened at a refugee camp in Dade County. Does this sample evidence support the *Sun's* claim? Test at a significance level of $\alpha = .01$. [*Hint:* Let p = true proportion of all Haitian refugees with tuberculosis and test H_0: $p = .0002$.]

10.42 Suppose that Ford Motor Company is involved in a multi-million dollar court case concerning their LTD model luxury cars. To support their case, Ford needs to show that the mean dollar amount of damage done to an LTD as a result of a 35-mile-per-hour crash into the rear bumper of a parked car is less than $400. Ford test-crashed 32 of their LTDs into parked cars at 35 miles per hour and then had an independent claims adjuster appraise the dollar amount of damage done to each. The results are:

$\bar{x} = \$366$

$s = \$54$

Perform the appropriate test of hypothesis for Ford Motor Company. Base your conclusion on the attained significance level (p-value) of the test.

10.43 The manager of a fast-food hamburger chain is concerned with the unusually high proportion of crushed hamburger buns being delivered by a supplier. If the next shipment of 2,000 buns contains more than 10% which are crushed, the manager will return the entire shipment to the supplier. Because it is physically impossible to check every hamburger bun in such a large shipment, a random sample of 100 buns is selected for inspection. Based on this sample, the manager will decide whether to accept or reject the shipment. Suppose that on the next shipment, 14 of the 100 buns sampled were crushed. Is there sufficient evidence for the manager to reject the entire shipment, i.e., is the true proportion of crushed hamburger buns in the shipment larger than 10%? Test at $\alpha = .05$.

10.44 A school official wishes to compare a new method of teaching reading to "slow learners" to the current standard method. The official will base this comparison upon the results of a reading test given at the end of a learning period of 6 months. Of a random sample of 20 slow learners, 8 are taught by the new method and 12 are taught by the standard method. All 20 children are taught by qualified instructors under similar conditions for a 6-month period. The results of the reading test at the end of this period are summarized in Table 10.19. Does the sample evidence indicate (at significance level $\alpha = .01$) that the mean reading test score of children taught by the new method exceeds the mean reading test score of children taught by the standard method?

TABLE 10.19
Reading Test Scores, Exercise 10.44

NEW METHOD	STANDARD METHOD
$n_1 = 8$	$n_2 = 12$
$\bar{x}_1 = 76.9$	$\bar{x}_2 = 72.7$
$s_1 = 4.85$	$s_2 = 6.35$

10.45 The ion balance of our atmosphere has a significant effect on human health. A high concentration of positive ions in a room can induce fatigue, stress, and respiratory problems in the room's occupants. However, research has shown that introduction of additional negative ions into the room's atmosphere (through a negative ion generator), in combination with constant ventilation, regains the natural balance of ions which is conducive to human health. One experiment was conducted as follows: One hundred employees of a large factory were randomly selected and divided into two groups of 50 each. Both groups were told that they would be working in an atmosphere with an ion balance (controlled through negative ion generators). However, unknown to the employees, the generators were switched on only in the experimental group's work area. At the end of the day, the number of employees reporting migraine, nausea, fatigue, faintness, or some other physical discomfort was recorded for each group. The results are summarized in Table 10.20. Perform a test of hypothesis to determine if the proportion of employees in the experimental group who experience some type of physical discomfort at the end of the day is significantly less than the corresponding proportion for the control group. Use a significance level of $\alpha = .03$.

TABLE 10.20
Ion Balance Experiment, Exercise 10.45

	EXPERIMENTAL GROUP (ION GENERATORS ON)	CONTROL GROUP (ION GENERATORS OFF)
Number in sample	$n_1 = 50$	$n_2 = 50$
Number in sample who experience some type of physical discomfort	3	12

10.46 In July, 1977, the United States Congress was considering a bill that would make it illegal for an employer to force an employee to retire at age sixty-five. The bill was not passed, due basically to the lack of support from elderly and retired workers. Suppose that one Congressman believes that the majority of today's elderly and retired workers would have supported the bill. From a random sample of 300 workers who were asked to retire at age sixty-five, the Congressman found that 162 would have preferred to stay on the job. Is there evidence to support the Congressman's belief, i.e., is the true proportion of retired workers who would have preferred to stay on the job past age sixty-five larger than .50? Base your conclusion on the attained significance level (p-value) of the test.

10.47 To what extent, if any, can we influence local weather conditions? Some Texas farmers have hired a meteorologist to investigate the effectiveness of cloud seeding in the artificial production of rainfall. Two farming areas in Texas with similar past meteorological records were selected for the experiment. One is seeded regularly throughout the year, while the other is left unseeded. The monthly precipitation at the farms for the first six months of the year is monitored and the difference in precipitation between the seeded and unseeded areas recorded each month. These six measurements resulted in a sample mean of $\bar{x} = .18$ inch and a sample standard deviation of $s = .40$ inch. Using a significance level of $\alpha = .02$, test whether μ, the true mean difference between the average monthly precipitations in the seeded and unseeded farm areas, is greater than 0.

10.48 The advertising department of a major western Pennsylvania automobile dealership, which has branches in the cities of Pittsburgh and Erie, wished to determine whether advertising is really effective in influencing prospective customers. The dealership decided to heavily advertise its new diesel model car in the city of Pittsburgh for one week, while maintaining a low profile on the car in the city of Erie. One month after the campaign ended, 28 of the 215 cars sold by the dealer in Pittsburgh were diesels, while 2 of the 37 cars sold by the dealer in Erie were diesels.

Do the results of the study indicate that the proportion of diesels sold in a city exposed to heavy advertising is higher than the proportion of diesels sold in a city where no advertising is planned? Base your decision on the attained significance level (p-value) of the test. Compare to your choice of a fixed significance level, α.

10.49 A breeder and seller of live hogs wishes to determine if there is a difference between the average weekly selling prices of hogs at two different Florida markets: Live Oak and Gainesville. The weekly selling prices at the two markets were obtained for independent random samples of $n_1 = 4$ and $n_2 = 5$ weeks. These results are summarized in Table 10.21. Conduct an hypothesis test to determine whether there is a difference between the average weekly selling prices of live hogs at the two Florida markets. Use a significance level of $\alpha = .05$.

TABLE 10.21
Selling Price of Live Hogs (¢ per Lb.), Exercise 10.49

LIVE OAK	GAINESVILLE
$n_1 = 4$	$n_2 = 5$
$\bar{x}_1 = 53.15$	$\bar{x}_2 = 54.20$
$s_1 = 3.57$	$s_2 = 3.32$

10.50 The amount of milk ordered daily at a large grocery store is very carefully monitored by the store owner. If not enough milk is ordered, the demand will not be met. If too much milk is ordered, there will be an excess supply and thus spoilage, costing the store money. Based on past experience, the store owner ordered, on the average, 1,000 gallons of milk per day to meet demand. However, the owner believes that the milk demand per day has changed. A check of the amount of milk (in gallons) purchased per day for a random sample of 50 days revealed the following summary information:

$$\bar{x} = 977 \text{ gallons}$$
$$s = 121 \text{ gallons}$$

Is there evidence that the average number of gallons of milk purchased per day is different from 1,000? Test at a significance level of $\alpha = .10$.

10.51 Last year's severe winter and late, cool spring played a major role in Wisconsin's small commercial apple crop production. The U.S. Department of Agriculture (USDA) will investigate the progress of this winter's crop. The USDA randomly selects 10 commercial orchards in the state and determines the number of pounds of McIntosh apples produced at each so far this year. A second random sample (independent of the first) of 20 commercial orchards is selected and the number of pounds of Red Delicious apples produced at each so far this year is recorded. (Each orchard selected contained 100 or more trees of bearing age.) The data are summarized in Table 10.22. At a level of significance of $\alpha = .05$, is there evidence of a difference between the average commercial productions so far this year of Wisconsin's two winter variety apples, McIntosh and Red Delicious?

TABLE 10.22
Wisconsin Apple Production (Thousands of Pounds), Exercise 10.51

McINTOSH	RED DELICIOUS
$n_1 = 10$	$n_2 = 20$
$\overline{x}_1 = 10.5$	$\overline{x}_2 = 9.7$
$s_1 = 4.1$	$s_2 = 6.6$

10.52 A new insecticide is advertised to kill more than 95% of roaches upon contact. In a laboratory test, the insecticide was applied to 400 roaches and, although all 400 eventually died, only 384 died immediately after contact. Is this sufficient evidence to support the advertised claim? Base your decision upon the computed *p*-value.

10.53 The following experiment was performed for a producer of 3-minute egg timers (i.e., timers that, if they are operating correctly, will track time for a period of 180 seconds). A random sample of 9 egg timers were timed in a vertical (upright) position, while a second random sample (independent of the first) of 9 egg timers were timed in a position 20 degrees from vertical. The results are given in Table 10.23. Conduct an hypothesis test to determine if there is a difference between the mean operating times of 3-minute egg timers in the two different positions. Test at $\alpha = .01$.

TABLE 10.23
Egg Timer Operation Times, Exercise 10.53

VERTICAL	20° FROM VERTICAL
$n_1 = 9$	$n_2 = 9$
$\overline{x}_1 = 187$ seconds	$\overline{x}_2 = 183$ seconds
$s_1 = 9$ seconds	$s_2 = 6$ seconds

10.54 A dietitian has developed a diet that is low in fats, carbohydrates, and cholesterol. Although the diet was initially intended to be used by people with heart disease, the dietitian wishes to examine the effect this diet has on the weight of obese people. Two independent random samples of 100 obese people are selected, and one group of 100 is placed on the low-fat diet. The other 100 are placed on a diet that contains approximately the same quantity of food, but is not as low in fats, carbohydrates, and cholesterol. For each person, the amount of weight lost (or gained) in a 3-week period is recorded. The data are summarized in Table 10.24. The dietitian wishes to test whether there is a difference between the mean weight loss (or gain) for people on the two diets. Perform the test (at significance level $\alpha = .05$) of interest to the dietitian.

TABLE 10.24
Weight Losses or Gains (In Pounds), Exercise 10.54

LOW-FAT DIET	OTHER DIET
$n_1 = 100$	$n_2 = 100$
$\overline{x}_1 = 9.31$	$\overline{x}_2 = 3.75$
$s_1 = 4.73$	$s_2 = 4.04$

10.55 In a nationwide Gallup Poll conducted in January 1981, 65% of the respondents favored the death penalty as a punishment for those convicted of murder. A Florida opinion poll conducted in May 1981 indicated that 415 of the 568 respondents (all residents of Florida) were in favor of the death penalty (Gainesville *Sun*, May 21, 1981). Test the hypothesis that the true proportion of Florida residents who favor the death penalty is larger than .65, the estimated proportion nationwide. Use a significance level of $\alpha = .02$.

10.56 Most supermarket chains give each store manager a detailed plan showing exactly where each product belongs on each shelf. Since more products are picked up from shelves at eye level than from any others, the most profitable and fastest-moving items are placed at eye level to make them easy for shoppers to reach. Traditionally, the eye-level shelf has been slightly under 5 feet from the floor—just the right height for the average female shopper, 5-feet 4-inches tall. But nowadays, more men are shopping than ever before. Since the average male shopper is 5-feet 10-inches tall, how will this affect the eye-level shelf? To investigate this, a random sample of 100 supermarkets from across the country were selected and the height of the eye-level shelf (i.e., the shelf with the most popular products) was recorded for each. The results were:

$$\bar{x} = 62 \text{ inches}$$
$$s = 3 \text{ inches}$$

Is there evidence that the average height of the eye-level shelf at supermarkets is now higher than 60 inches from the floor (i.e., higher than the traditional height of 5 feet)? Base your decision on the attained significance level (p-value) of the test.

10.57 Many "solar" homes waste the sun's valuable energy. To be efficient, a solar house must be specially designed to trap solar radiation and use it effectively. Basically, homes with energy-efficient solar heating systems can be categorized into two groups, *passive* solar heating systems and *active* solar heating systems. In a passive solar heating system, the house itself is a solar energy collector, while in an active solar heating system, elaborate mechanical equipment is used to convert the sun's rays into heat. Suppose that we wish to determine whether there is a difference between the proportions of passive solar and active solar heating systems which require less than 200 gallons of oil per year in fuel consumption. Independent random samples of 50 passive and 50 active solar-heated homes are selected and the number which required less than 200 gallons of oil last year is noted. The results are given in Table 10.25. Is there evidence of a difference between the proportions of passive and active solar-heated homes which require less than 200 gallons of oil in fuel consumption last year? Test at a level of significance of $\alpha = .02$.

TABLE 10.25
Solar-Heated Home Data, Exercise 10.57

	PASSIVE SOLAR	ACTIVE SOLAR
Number of homes	50	50
Number which required less than 200 gallons of oil last year	37	46

10.58 An advertisement for the Chrysler Corporation in a national magazine reads: "In an independent test, 50 mid-specialty car owners compared the 1981 Chrysler Cordoba to the 1981 Oldsmobile Cutlass Supreme. They compared each car 39 different ways, from the quality of their paint jobs to the quality of their seat seams. From the way they felt on the road to the way they looked standing still. Out of the 39 categories, Cordoba beat Cutlass 34 times." A footnote at the bottom of the advertisement informs us that the results are "based on responses preferring Cordoba over Cutlass 27 out of 50 to 46 out of 50 participants, depending on the category." Let us consider the unspecified category in which 27 out of 50 respondents favored Cordoba over Cutlass. Test the hypothesis that the true proportion of mid-specialty car owners who favor the Cordoba over the Cutlass in this particular category is greater than .5. Use a significance level of $\alpha = .05$.

10.59 The *Statistical Bulletin* (April–June, 1978) reports that "with motorcycle registrations exceeding five million in 1977, there is now, on the average, one motorcycle for every 43 persons in the United States. Increasing motorcycle use has been accompanied by a mounting death toll from motorcycle accidents. Inexperience and lack of skill are contributing factors to the (high) fatality rates among motorcyclists."

Another major contributor to the high motorcycle fatality rate is that many motorcyclists fail to wear helmets. Studies have shown that the number of fatal or serious head injuries is significantly reduced when properly constructed and fitted protective headgear is worn. The *Statistical Bulletin* reported on a recent study conducted in Sacramento County, California, which determined that serious head injuries were 50% lower for those who wore helmets than for those who did not. At present, laws in 23 states and the District of Columbia require that helmets be worn by all motorcycle riders. In addition, 16 states require helmets only for motorcyclists under 18 years of age, one state for those only under age 17, and one state for those only under 16. This leaves 9 states with absolutely no motorcycle helmet requirements.

Suppose we wish to test whether the proportion of motorcycle accidents in Illinois (which has no helmet law) that result in fatal or serious head injuries is at least .50 higher than the corresponding proportion in Michigan (which has a helmet law). Set up the null and alternative hypotheses for this test.

10.60 Refer to Exercise 10.59. A random sample of 55 motorcyclists who received medical treatment in Illinois last year—as identified from police reports, death certificates, and the admission and emergency records of hospitals—showed that 41 received fatal or serious head injuries. In Michigan, 8 of the 31 sampled motorcycle accidents resulted in fatal or serious head injuries to the motorcyclist. Use this information to test the hypotheses of Exercise 10.59 at significance level $\alpha = .05$.

CASE STUDY 10.1 The Invisible Birds

If you have ever lost a contact lens, a key, or some other small object of value, you have probably enlisted one or two of your friends to help you find it. Common sense tells you that two people will have a better chance of locating the lost object than one. But what if ten people offered you assistance? Would you gain in utilizing the services of ten searchers, or would too many searchers cause confusion and actually hinder your chances of finding the object?

Bird watchers and observers of asteroids, mineral deposits, caterpillars, or other objects of natural history or geology have struggled with a similar question: "What is the optimum number of observers to use in a search party, to be reasonably sure of seeing 95% of the objects (e.g., birds, caterpillars, etc.) that are available to be seen?" Frank Preston (1979)* has these thoughts on the matter:

> Two observers in one unit (of time), or one observer in two units of time, will in most cases see more than one observer in one unit (of time). This has been known for many decades, presumably for ages. . . . [However], if the observers are fairly skilled and reasonably lucky, there is little to be gained in having more than three or at most four observers in a single party. . . .

Preston considered the problem of "bird spotting" from a probabilistic point of view. [As does Preston, this case study uses the term "bird" to cover any object of search (e.g., asteroids, caterpillars, etc.).]

> Whether an observer sees a bird . . . , or does not see it, is a matter of chance; whether he is looking in the right direction at the right instant, for example. Or perhaps it depends on the lighting, or the degree of screening by foliage or other obstacles, or on the bird making a movement that attracts attention.

Assuming p, the probability that a single observer sees a particular bird, is .5, and that this probability is the same for all observers in the party, Preston used a binomial probability model to construct the following table.

TABLE 10.26 Proportion of Birds Missed of Those Birds Available To Be Seen

NUMBER OF OBSERVERS	PROPORTION MISSED
1	.50
2	.25
3	.13
4	.06
5	.03
6	.02
7	.01
8	.00

The table indicates that 1 observer will miss half the total number of birds available to be seen, 2 observers will miss 25% of the total number of birds available to be seen, but that 4 observers will miss only 6% of the total number. Preston calculated similar proportions for differing values of the probability that a single observer sees a bird. If p is as low as .2, 2 observers will miss 64% of the birds present (i.e., available to be seen). This latter result did not surprise a veteran bird watcher who, along with

*Frank W. Preston. "The invisible birds." *Ecology,* June 1979, *60,* 451–454. Copyright 1979, the Ecological Society of America.

another competent observer, had tried to conduct a census of the birds in the Buckeye Lake region of Ohio some 50 years ago. Relates Preston:

> He (the veteran bird watcher) said that on one Christmas foray, when many observers were in the field, he and his companion saw a line of blackbirds extending from horizon to horizon, and accounting for more than half the birds seen by the whole group of observers; yet only he and his companion saw this flock.

Preston's results, of course, are all theoretical. He admits: "If all these approximations or rough estimates were in fact exactly correct, we should still have the general 'noise' or statistical fluctuations." We can use the methods of this chapter to test Preston's theory. Let us consider the population parameter p, the true proportion of times that a search party of one competent observer spots a particular bird in a long series of forays. According to Preston, the percentage of birds missed (undetected) depends upon the value of p. If the observer is "fairly skilled and reasonably lucky," then p will exceed .5. Consider the following experiment: A single (competent) observer is placed at a feeding station located near a lake for the purpose of conducting a census of the birds in the region. The observer will count the number of birds seen during a 1-hour period each day for a total of 25 days. On each of these 25 forays, a trained bird tagged with a bright red deflector is released among the other birds at the lake region, but within view of the observer. However, the observer does not know the time or direction of the release. We are interested in the proportion of forays in which the observer spots the marked bird.

a. Suppose the observer detects the marked bird on 14 of the 25 days. Test the hypothesis that p, the true proportion of times the observer spots the marked bird, exceeds .5. Use a significance level of $\alpha = .05$.
b. Compute the attained significance level (p-value) of the test and interpret its value.
c. Would you be willing to risk a Type I error probability as large as the p-value you computed in part b? Explain.

CASE STUDY 10.2 Florida's "New Look" Jury System

> Greetings, You are hereby requested to report for jury duty at 9 AM Monday in the courthouse. Please do not be upset about the inconvenience of being away from your job, sitting idle for several hours while attorneys plea bargain, and possibly reporting every day of the week for a trial that might be either postponed or cancelled.

With this facetious description of a jury summons, Frank Dorman, staff writer for the Gainesville *Sun,* began his article on Florida's jury selection system.* To most Floridians, the system is simply a waste of a lot of people's time that costs a lot of the taxpayers' money. Nevertheless, serving on jury duty is one of our most important civic responsibilities and should not be taken lightly.

Dorman reports that "serious attempts" at making "permanent improvements in the state's jury systems" have been underway since 1978. It was at that time that the Office of the (Florida) State Courts Administrator (OSCA) began a two-year,

---

*Gainesville *Sun,* February 22, 1981.

$300,000 pilot study of the new jury procedures in seven Florida counties. After the study, OSCA estimated that if all 67 Florida counties tried the new system, more than $500,000 in jury selection costs could be saved per year, a reduction of 50%. The study also indicates that the new system has markedly improved Floridians' outlook on jury service. Some of the highlights of the new juror system are:

1. Single-day juror empanelment (i.e., all judges use only one day to select jurors for trials during the week) which eliminates the need for having jurors continue to report to the court each day of the week and perhaps never being selected
2. Twenty-four-hour toll-free telephone call-in service which enables jurors to call the night prior to their service date to learn from a recorded message whether they should report the next day
3. The use of first class mail for summons distribution rather than registered mail or hand delivery
4. The use of computers to randomly choose venires, the large group (usually 100 or more) of people from which juries are selected

a. Jurors in Florida are paid $10 a day and 10¢ a mile for travel expenses. Under the old system, a jury trial in Alachua County involved an average juror cost of $600. With the new procedural changes described above, it is believed that the average juror cost will be less than 50% of the old cost, i.e., less than $300. Suppose that a random sample of 8 Alachua County jury trials operating under the new system had an average cost of $\overline{x} = \$292$ and a standard deviation of $s = \$83$. Test the hypothesis that the average juror cost under the new system in Alachua County is less than $300. Use a significance level of $\alpha = .05$.

b. Suppose that OSCA wishes to determine if the average length of actual service on jury duty (in days) is less for jurors selected under the new system than under the old system. What are the null and alternative hypotheses of interest to OSCA?

c. OSCA claims that only 7% of the "new system" jurors have unfavorable impressions of jury service. If a random sample of 250 jurors selected under the new system included 18 who had unfavorable impressions of jury service, test the OSCA claim. [*Hint:* Test H_0: $p = .07$ against H_a: $p > .07$, using a significance level of $\alpha = .05$.]

CASE STUDY 10.3 Drug Prescriptions Linked to Medical Advertising

Have you ever purchased a product solely on the basis of an appealing advertisement? If you have, you are no exception, for almost all consumers have, at one time or another, been influenced by an eye-catching advertisement in a magazine, newspaper, or on television. Most well-planned advertising campaigns are successful because the advertiser (or manufacturer) of the product is able to correspond, on a one-to-one basis, with the consumer (or reader) through the media. With the majority of consumers' likes and dislikes already known in advance (through extensive market studies), the successful advertiser is able to conjure in the consumer's mind an appealing picture of the product being advertised.

This approach to boosting a product's sales may not be so effective when the direct line between the advertiser and consumer is broken. For example, in the marketing of prescription drugs, the sales success of a certain product depends almost entirely upon the physician who writes the drug prescription. Although the consumer is the ultimate user of the product, it is the prescribing physician at whom medical (drug) advertisements are aimed. Harold Walton writes* that "those charged with responsibilities for advertising to physicians through the printed media rarely, if ever, expect a 'sales miracle.'" Medical advertisers do, however, "expect ongoing advertising to contribute to or to maintain a [physician's] awareness of a therapeutic message. His recollection . . . will be vivid and accurate enough for him to consider prescribing the advertised product at the appropriate time." Nevertheless, says Walton, almost all physicians "deny that advertising shapes their professional decisions." Physicians are very skeptical "about therapeutic concepts elaborated within a commercial frame," and this skepticism has many medical advertisers pursuing their "programs with little enthusiasm, not fully convinced that medical advertising can deliver results. . . ."

Despite this outward negativity on the part of the physician, Walton conducted research for the purpose of showing that "advertising appearing in medical journals . . . has a positive effect on the prescribing behavior of physicians, an effect that is far from coincidental." A sample of 100 physicians involved in patient care was randomly selected from the circulation list of a medical journal which is among the most widely read and reaches approximately 90% of the physicians in private practice in the United States. Each physician was first asked to examine an ad for which all references to product name, generic name, company name, ingredients, and logo were blotted out, and then asked the question, "Have you seen this ad before?" After recording the response, the product in the ad was revealed and the physician was asked if he or she had prescribed or recommended the product in the month prior to the interview. The resulting data were then used in an hypothesis test to decide whether H_a: $(p_1 - p_2) > 0$ is true, that is, to determine if the proportion of physicians who prescribe a certain product is greater for those aware of the ad (p_1) than for those not aware of the ad (p_2). The results of the analysis for six such ads (i.e., six different products) are given in Table 10.27 on page 362.

a. What assumptions are necessary for the hypothesis-testing procedure to be valid? Are they satisfied in each of the six cases (brands)?

b. For each case in which the assumptions are satisfied, compute the value of the appropriate test statistic.

c. Calculate the corresponding attained significance levels, i.e., p-values, for the computed test statistics, part b.

d. Based on the attained significance levels of the tests, for which of the products is the proportion of physicians who prescribe the product significantly greater for those aware of the ad than for those unaware of the ad (i.e., for which of the products does the sample evidence indicate that $(p_1 - p_2) > 0$)? Assume that you are willing to tolerate a Type I error probability of $\alpha = .05$ in each case.

---

*H. Walton. "Ad recognition and prescribing by physicians." *Journal of Advertising Research*, June 1980, *20*, 39–48. Reprinted from the *Journal of Advertising Research*. © Copyright 1980, by the Advertising Research Foundation.

TABLE 10.27
Physicians' Responses to Ad Survey, Case Study 10.3

PRODUCT ADVERTISED		PHYSICIANS' RESPONSE Ad Recalled	Ad Not Recalled
BRAND 1	Number in Sample	25	75
	Number Who Prescribed Product	17	30
BRAND 2	Number in Sample	46	54
	Number Who Prescribed Product	34	33
BRAND 3	Number in Sample	34	66
	Number Who Prescribed Product	1	2
BRAND 4	Number in Sample	40	60
	Number Who Prescribed Product	26	39
BRAND 5	Number in Sample	40	60
	Number Who Prescribed Product	7	0
BRAND 6	Number in Sample	34	66
	Number Who Prescribed Product	13	20

REFERENCES

Bandyopadhyay, B., & Chattopadhyay, H. "Body fat in urban and rural male college students of eastern India." *American Journal of Physical Anthropology,* January 1981, *54,* 119–122.

Dorman, F. "'New look' in jury selection saving time and money." Gainesville *Sun,* February 22, 1981, 1a, 12a.

Johnson, R. *Elementary statistics.* 3d ed. North Scituate, Mass.: Duxbury, 1980. Chapters 9, 10 and 11.

McClave, J. T., & Dietrich, F. H. *Statistics.* 2d ed. San Francisco: Dellen, 1982. Chapters 7 and 8.

Mendenhall, W. *Introduction to probability and statistics.* 5th ed. North Scituate, Mass.: Duxbury, 1979. Chapters 8 and 9.

Metropolitan Life Insurance Company. "Motorcycle accident fatalities." *Statistical Bulletin,* April–June 1978, 7–9.

Preston, F. W. "The invisible birds." *Ecology,* June 1979, *60,* 451–454.

Reichman, O. J. "Desert granivore foraging and its impact on seed densities and distributions." *Ecology,* December 1979, *60,* 1085–1092.

Snedecor, G. W., & Cochran, W. G. *Statistical Methods.* 6th ed. Ames, Iowa: Iowa State University Press, 1967. Chapters 2, 3, 10, and 11.

Walton, H. "Ad recognition and prescribing by physicians." *Journal of Advertising Research,* June 1980, *20,* 39–48.

Eleven

Comparing More Than Two Population Means: An Analysis of Variance

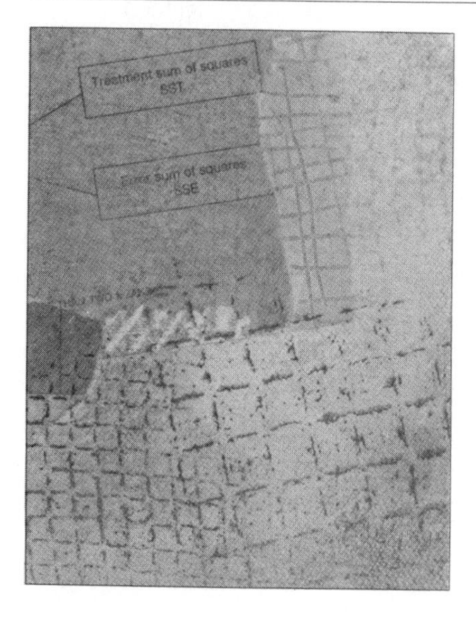

Is the moral upbringing of juvenile delinquents any different from that of lower-class and middle-class nondelinquents? Suppose you were to present adolescents selected from each of the three sociological groups with a moral dilemma (e.g., damaging of property), record their responses, and assign each a moral development score. How would you decide whether the sample data indicate a difference among the mean moral development scores of the three sociological groups? In this chapter, we will consider the general problem of comparing more than two population means and will examine the comparison of moral developments of juvenile delinquents, lower-class nondelinquents, and middle-class nondelinquents in greater detail in Case Study 11.1.

Contents

11.1 Introduction

Sociologists often conduct experiments to investigate the relationship between socioeconomic status and college performance. Generally, socioeconomic status is partitioned into three groups: (1) lower class, (2) middle class, and (3) upper class. Consider the problem of comparing the mean grade-point averages of those college freshmen associated with the lower class, those associated with the middle class, and those associated with the upper class. We can think of the totality of the grade-point averages for all college freshmen in a particular socioeconomic group as constituting a population that characterizes the performance of college freshmen in that socioeconomic group. Thus, we wish to compare the means of the three grade-point-average populations corresponding to the three socioeconomic groups.

The grade-point averages for random samples of 7 college freshmen associated with the lower socioeconomic class, 7 with the middle class, and 6 with the upper class* were selected from a university's files at the end of the academic year. The data are recorded in Table 11.1. Scan the data. If the freshmen in each socioeconomic group tend to be similar in ability and motivation, we would expect the mean grade-point averages in the populations to be nearly equal. If the freshmen differ substantially in these characteristics, we would expect the population mean grade-point averages to be correspondingly different. Based on your scanning of the data, do you think that this sample information presents sufficient evidence to indicate a difference among the three population (socioeconomic class) means?

TABLE 11.1
Grade-Point Averages for Three Socioeconomic Groups

LOWER CLASS	MIDDLE CLASS	UPPER CLASS
2.87	3.23	2.25
2.16	3.45	3.13
3.14	2.78	2.44
2.51	3.77	3.27
1.80	2.97	2.81
3.01	3.53	1.36
2.16	3.01	

We have explained in earlier sections that the reliability of inferences based upon intuition cannot be measured. Consequently, we will demonstrate a statistical procedure for deciding whether differences exist among two or more population means. The method we will use is known as ***analysis of variance,*** or ***ANOVA.***

11.2 One-Way Analysis of Variance and the *F* Statistic

We learned an hypothesis-testing technique for comparing the means of two populations in Chapters 9 and 10. The null and alternative hypotheses in the analysis of variance test for comparing more than two population means take a familiar form.

*The grade-point average of one freshman from the upper socioeconomic class had to be eliminated since the freshman was forced to withdraw from the university in mid-semester because of medical problems.

EXAMPLE 11.1 Refer to the grade-point-average–socioeconomic-class discussion in Section 11.1. State the null and alternative hypotheses for a test to determine whether there is a difference among the true mean grade-point averages for freshmen in the three socioeconomic classes.

Solution Our objective is to determine whether differences exist among the three population (socioeconomic class) grade-point average means. Consequently, we will select the null hypothesis "All three population means, μ_1, μ_2, μ_3, are equal." That is,

$$H_0: \quad \mu_1 = \mu_2 = \mu_3$$

where μ_1 is the mean grade-point average for those college freshmen in the lower socioeconomic class, μ_2 is the mean grade-point average for freshmen in the middle class, and μ_3 is the mean grade-point average for freshmen in the upper class.

The alternative hypothesis of interest is

H_a: At least two of the population means differ

If the differences among the sample means are large enough to indicate differences among the corresponding population means, we will reject the null hypothesis H_0 in favor of the alternative hypothesis H_a; otherwise, we will fail to reject H_0 and conclude that there is insufficient evidence to indicate differences among the population means.

HYPOTHESES FOR A TEST OF k POPULATION MEANS

$$H_0: \quad \mu_1 = \mu_2 = \cdots = \mu_k$$

H_a: At least two population means differ

Experiments for the comparison of more than two population means can be designed using a variety of sampling schemes. However, in this chapter we shall consider only those experiments which use an ***independent sampling design.***

DEFINITION 11.1

An ***independent sampling design*** is one in which independent random samples are drawn from each of the target populations.

When a test for the equality of more than two population means is conducted using the data obtained from an independent sampling design, the procedure is often referred to as a ***one-way analysis of variance.*** The prefix ***one-way*** is used because each of the sampling units is classified in only ***one*** direction. For example, for the grade-point-average–socioeconomic-class problem of Section 11.1, each college freshmen (sampling unit) included in the sample was classified only according to socioeconomic status (lower class, middle class, upper class). In contrast, if the freshmen were also classified according to sex (male or female), then the classification would be ***two-dimensional*** and, under the appropriate sampling design, a subsequent test for equality of means would be called a ***two-way*** analysis of variance. Throughout this chapter, when we use the terminology *analysis of variance,* we are referring to a one-way analysis of variance, i.e., a technique for comparing more than two population means with an independent sampling design.

EXAMPLE 11.2 To see how the principle behind the analysis of variance method works, let us consider the following simple example. The means (μ_1 and μ_2) of two populations are to be compared using independent random samples of size 5 from each of the populations. The sample observations and the sample means are shown in Table 11.2.

a. Do you think these data provide sufficient evidence to indicate a difference between the population means μ_1 and μ_2?

TABLE 11.2
Data for Example 11.2a

SAMPLE FROM POPULATION 1	SAMPLE FROM POPULATION 2
6	5
−1	1
0	3
3	2
2	4
$\bar{x}_1 = 2$	$\bar{x}_2 = 3$

b. Now look at two more samples of $n_1 = n_2 = 5$ measurements from the populations, as shown in Table 11.3. Do these data appear to provide evidence of a difference between μ_1 and μ_2?

TABLE 11.3
Data for Example 11.2b

SAMPLE FROM POPULATION 1	SAMPLE FROM POPULATION 2
2	3
2	3
2	3
2	3
2	3
$\bar{x}_1 = 2$	$\bar{x}_2 = 3$

Solution **a.** One way to determine whether a difference exists between the population means μ_1 and μ_2 is to examine the spread (or variation) ***between*** the sample means $\bar{x}_1$ and $\bar{x}_2$, and to compare it to a measure of variability ***within*** the samples. The greater the difference in the variations, the greater will be the evidence to indicate a difference between μ_1 and μ_2.

You can see from the data of Table 11.2 that *the difference between the sample means is small in relation to the variability within the sample obser-*

vations. Thus, we think you will agree that the difference between $\bar{x}_1$ and $\bar{x}_2$ is not large enough to indicate a difference between μ_1 and μ_2.

b. Notice that the difference between the sample means for the data of Table 11.3 is identical to the difference shown in Table 11.2. However, since there is now no variability within the sample observations, *the difference between the sample means is large in comparison with the variability within the sample observations.* Thus, the data appear to give clear evidence of a difference between μ_1 and μ_2.

We can apply this principle to the general problem of comparing *k* population means. If the *variability among the k sample means* is large in relation to the *variability within the k samples,* then there is evidence to indicate that a difference exists among the *k* population means. The term *analysis of variance* is thus quite descriptive of the procedure. We give and illustrate the use of the computational formulas for these two measures of variability in the following examples.

EXAMPLE 11.3 Refer to Example 11.1, and suppose the data of Table 11.1 represent random samples selected ***independently*** from the three socioeconomic groups. Compute the test statistic appropriate for testing

H_0: $\mu_1 = \mu_2 = \mu_3$

H_a: At least two population means are different

where μ_1 is the true mean grade-point average for freshmen classified as lower class, μ_2 is the true mean grade-point average for those classified as middle class, and μ_3 is the true mean grade-point average for those classified as upper class.

Solution As stated in Example 11.2, the criterion for testing the equality of means involves a comparison of two measures of variability: (1) the variation among the *k* sample means; and (2) the variation within the *k* samples. The first step in obtaining a measure of the variation among the sample means is to compute a weighted sum of squares of deviations of the sample means $\bar{x}_1, \bar{x}_2, \ldots, \bar{x}_k$ about the overall mean. In analysis of variance, this quantity is often called the ***sum of squares for treatments*** (SST).* The second step is to divide SST by the quantity $(k - 1)$ to obtain the ***mean square for treatments*** (MST). The mean square for treatments, MST, is then used to measure the variation among the *k* sample means.

When the data are given in the column form of Table 11.1, the computational formulas for SST and MST are simplified. Table 11.4 gives the column totals necessary for computing SST and MST in this example.

---

*The term *treatment* evolves from agricultural experiments in which the analysis of variance technique first achieved importance. The sampling units (for example, fields, animals, etc.) in the experiment were "treated" in two or more different ways, and then a comparison of the means of the populations of measurements corresponding to the different treatments made.

STEPS IN COMPUTING A MEASURE OF THE VARIATION AMONG k SAMPLE MEANS

1. $$\text{SST} = \frac{T_1^2}{n_1} + \frac{T_2^2}{n_2} + \cdots + \frac{T_k^2}{n_k} - \text{CM}$$

where T_i = Total of all observations in column i

n_i = Number of observations in column i

CM = Correction for the mean

$$= \frac{(\text{Total of all } n \text{ observations})^2}{n}$$

$$n = n_1 + n_2 + \cdots + n_k$$

2. $$\text{MST} = \frac{\text{SST}}{k-1}$$

where k = Number of means to be compared

TABLE 11.4 Column Totals for Data of Table 11.1

LOWER CLASS	MIDDLE CLASS	UPPER CLASS
2.87	3.23	2.25
2.16	3.45	3.13
3.14	2.78	2.44
2.51	3.77	3.27
1.80	2.97	2.81
3.01	3.53	1.36
2.16	3.01	
$T_1 = 17.65$	$T_2 = 22.74$	$T_3 = 15.26$
$n_1 = 7$	$n_2 = 7$	$n_3 = 6$
Total of all observations = 55.65		

To obtain a measure of the variation among the three sample means, $\bar{x}_1$, $\bar{x}_2$, $\bar{x}_3$, in this example, we first compute

$$\text{SST} = \frac{T_1^2}{n_1} + \frac{T_2^2}{n_2} + \frac{T_3^2}{n_3} - \text{CM}$$

$$= \frac{(17.65)^2}{7} + \frac{(22.74)^2}{7} + \frac{(15.26)^2}{7} - \frac{(\text{Total of all } n \text{ observations})^2}{n}$$

$$= 44.50 + 73.87 + 38.81 - \frac{(55.65)^2}{(7+7+6)}$$

$$= 157.18 - 154.85 = 2.33$$

Dividing SST by $(k-1) = 2$, since we are comparing $k = 3$ means, we obtain

$$\text{MST} = \frac{\text{SST}}{k-1} = \frac{2.33}{2} = 1.165$$

To determine how large the value of MST must be before we reject the null hypothesis H_0, we must compare its value, MST = 1.165, to the variability within the sample observations themselves. A measure of this within-sample variability, called the ***mean square for error*** (MSE), is obtained by first computing the ***sum of squared errors*** (SSE) and then dividing SSE by $(n-k)$.

It can be shown (proof omitted) that the SSE is the pooled sum of squares of deviations of the x-values about their respective sample means:

$$\text{SSE} = \sum_{i=1}^{n_1} (x_{1i} - \bar{x}_1)^2 + \sum_{i=1}^{n_2} (x_{2i} - \bar{x}_2)^2 + \cdots + \sum_{i=1}^{n_k} (x_{ki} - \bar{x}_k)^2$$

Thus, the MSE is a pooled measure of the variability within the k samples (an extension of the pooled estimator of σ^2 first discussed in Chapter 8).

STEPS IN COMPUTING A MEASURE OF THE WITHIN-SAMPLE VARIABILITY

1. SSE = SS(Total) − SST

where SS(Total) = Total sum of squares

= (Sum of squares of all observations) − CM

SST = Sum of squares for treatments (see formula in previous box)

CM = Correction for mean (see formula in previous box)

2. $\text{MSE} = \dfrac{\text{SSE}}{n-k}$

where n = Total number of observations

k = Number of means to be compared

To compute SSE and MSE for this example we need, in addition to the previously computed quantities CM = 154.85 and SST = 2.33,

$$\begin{aligned}\text{SS(Total)} &= \text{(Sum of squares of all observations)} - \text{CM} \\ &= (2.87)^2 + (2.16)^2 + (3.14)^2 + \cdots + (1.36)^2 - 154.85 \\ &= 161.89 - 154.85 = 7.04\end{aligned}$$

Then

$$\begin{aligned}\text{SSE} &= \text{SS(Total)} - \text{SST} \\ &= 7.04 - 2.33 = 4.71\end{aligned}$$

and

$$\text{MSE} = \frac{\text{SSE}}{n-k} = \frac{4.71}{(20-3)} = \frac{4.71}{17} = 0.277$$

TEST STATISTIC FOR COMPARING k MEANS (INDEPENDENT RANDOM SAMPLES)

Test statistic: $F = \frac{\text{MST}}{\text{MSE}}$

where MST = Mean square for treatments

$$= \frac{\text{SST}}{k-1}$$

MSE = Mean square for error

$$= \frac{\text{SSE}}{n-k}$$

n = Total number of observations

k = Number of means to be compared

SUMMARY OF STEPS IN COMPUTING THE TEST STATISTIC FOR COMPARING k POPULATION MEANS

1. CM = Correction for mean

$$= \frac{(\text{Total of all observations})^2}{\text{Total number of observations}} = \frac{(\Sigma x_i)^2}{n}$$

2. SS(Total) = Total sum of squares

= (Sum of squares of all observations) − CM

$$= \Sigma x_i^2 - \text{CM}$$

3. SST = Sum of squares for treatments

$$= \begin{pmatrix} \text{Sum of squares of column totals} \\ \text{with each square divided by the} \\ \text{number of observations in that column} \end{pmatrix} - \text{CM}$$

$$= \frac{T_1^2}{n_1} + \frac{T_2^2}{n_2} + \cdots + \frac{T_k^2}{n_k} - \text{CM}$$

4. SSE = Sum of squares for error

= SS(Total) − SST

5. MST = Mean square for treatments $= \frac{\text{SST}}{k-1}$

6. MSE = Mean square for error $= \frac{\text{SSE}}{n-k}$

7. F = Test statistic $= \frac{\text{MST}}{\text{MSE}}$

Once we have computed measures of the two sources of variability—a measure of the variability due to differences among the sample means (MST = 1.165) and a measure of the variability due to within-sample differences among the observations (MSE = 0.277)—we are ready to form the test statistic appropriate for testing H_0: $\mu_1 = \mu_2 = \mu_3$. The statistic, called an ***F statistic,*** is simply the ratio of the mean squares (see box at top of page 370):

$$F = \frac{\text{MST}}{\text{MSE}}$$

The appropriate test statistic for this problem is then

$$F = \frac{\text{MST}}{\text{MSE}} = \frac{1.165}{0.277} = 4.21$$

Large values of the F statistic indicate that differences between the sample means are large, and therefore support the alternative hypothesis that the population means differ. We discuss the appropriate rejection region for the analysis of variance F test in Example 11.4. The steps necessary for the computation of the F statistic are summarized in the box at the bottom of page 370.

EXAMPLE 11.4 Refer to Example 11.3. Specify the rejection region for testing H_0: $\mu_1 = \mu_2 = \mu_3$ using a significance level of $\alpha = .05$. Is there evidence that the mean grade-point averages of freshmen differ among the three socioeconomic classes?

Solution Under certain conditions (see Section 11.5), the F statistic has a repeated sampling distribution known as the ***F distribution.*** The shape of the F distribution will depend upon two quantities: ***numerator degrees of freedom*** and ***denominator degrees of freedom.*** In a one-way analysis of variance procedure, the F distribution has $(k-1)$ numerator degrees of freedom and $(n-k)$ denominator degrees of freedom. Note that the numerator degrees of freedom, $(k-1)$, was divided into SST to obtain MST (the quantity which appears in the ***numerator*** of the F statistic). Similarly, the denominator degrees of freedom, $(n-k)$, was divided into SSE to obtain MSE (the quantity which appears in the ***denominator*** of the F statistic). An F distribution with 7 and 9 df (numerator and denominator df, respectively) is shown in Figure 11.1 on page 372. As you can see, the distribution is skewed to the right.

In order to establish the rejection region for our test of hypothesis, we need to be able to find F values corresponding to the tail areas of this distribution. We need to find only upper-tail F-values, however, because we will reject H_0 if the value of the computed F statistic is too large. The upper-tail F-values can be found in Tables 5, 6, and 7 of Appendix E. Table 6, partially reproduced in Figure 11.2 (page 372), gives F-values that correspond to $\alpha = .05$ upper-tail areas for different pairs of degrees of freedom. (Tables 5 and 7 give F-values that correspond to $\alpha = .10$ and $\alpha = .025$, respectively.) The columns of the table correspond to various numerator degrees of freedom, while the rows correspond to various denominator degrees of freedom. Thus, if the numerator degrees of freedom is 7 and the denominator degrees of freedom is 9, we look in the seventh column and ninth row to find the F-value

$$F_{.05} = 3.29$$

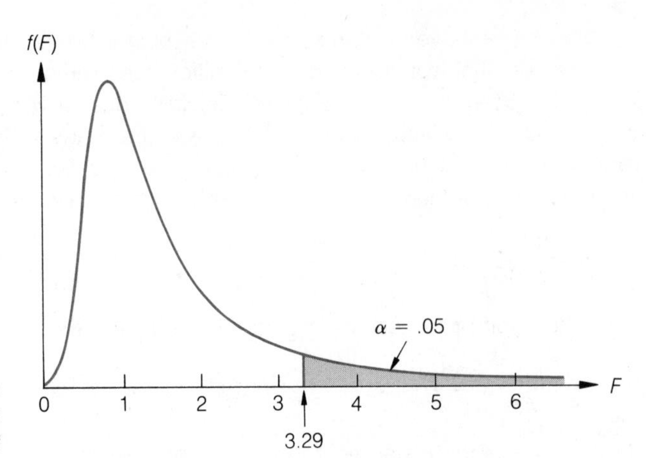

FIGURE 11.1 An F Distribution with 7 Numerator df and 9 Denominator df

As shown in Figure 11.1, $\alpha = .05$ is the tail area to the right of 3.29 in the F distribution with 7 numerator df and 9 denominator df, i.e., the probability that the F statistic will exceed 3.29 is $\alpha = .05$.

FIGURE 11.2 Reproduction of Part of Table 6, Appendix E: $\alpha = .05$

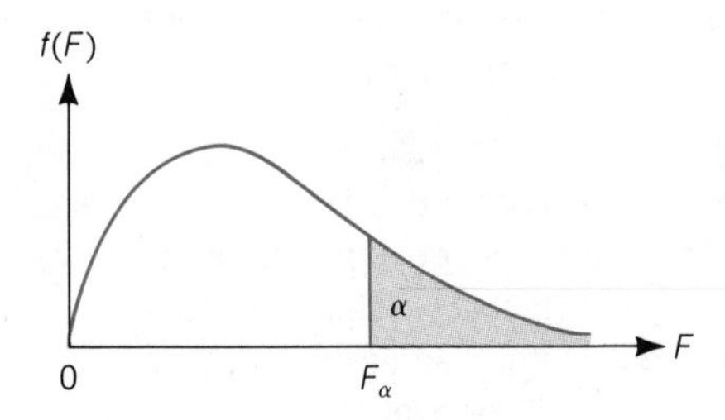

DENOMINATOR DEGREES OF FREEDOM ν_2 \ ν_1 NUMERATOR DEGREES OF FREEDOM	1	2	3	4	5	6	7	8	9
1	161.4	199.5	215.7	224.6	230.2	234.0	236.8	238.9	240.5
2	18.51	19.00	19.16	19.25	19.30	19.33	19.35	19.37	19.38
3	10.13	9.55	9.28	9.12	9.01	8.94	8.89	8.85	8.81
4	7.71	6.94	6.59	6.39	6.26	6.16	6.09	6.04	6.00
5	6.61	5.79	5.41	5.19	5.05	4.95	4.88	4.82	4.77
6	5.99	5.14	4.76	4.53	4.39	4.28	4.21	4.15	4.10
7	5.59	4.74	4.35	4.12	3.97	3.87	3.79	3.73	3.68
8	5.32	4.46	4.07	3.84	3.69	3.58	3.50	3.44	3.39
9	5.12	4.26	3.86	3.63	3.48	3.37	3.29	3.23	3.18
10	4.96	4.10	3.71	3.48	3.33	3.22	3.14	3.07	3.02
11	4.84	3.98	3.59	3.36	3.20	3.09	3.01	2.95	2.90
12	4.75	3.89	3.49	3.25	3.11	3.00	2.91	2.85	2.80
13	4.67	3.81	3.41	3.18	3.03	2.92	2.83	2.77	2.71
14	4.60	3.74	3.34	3.11	2.96	2.85	2.76	2.70	2.65

Given this information on the *F* distribution, we are now able to find the rejection region for the analysis of variance *F* test.

REJECTION REGION FOR A TEST TO COMPARE *k* POPULATION MEANS

Rejection region: $F > F_\alpha$

where the distribution of *F* is based on $(k-1)$ numerator df and $(n-k)$ denominator df, and F_α is the *F*-value such that $P(F > F_\alpha) = \alpha$

Since we are comparing $k = 3$ means in Example 11.3, the numerator degrees of freedom is $(k-1) = (3-1) = 2$. There are $n = 20$ measurements in the combined samples, so the denominator degrees of freedom is $(n-k) = (20-3) = 17$. Using $\alpha = .05$, we will reject the null hypothesis that the three means are equal if

$$F > F_{.05}$$

where from Table 6, Appendix E, the *F*-value associated with 2 numerator df and 17 denominator df is $F_{.05} = 3.59$. This rejection region is shown in Figure 11.3. The next step is to compare the computed value, $F = 4.21$, of the test statistic (obtained in Example 11.3) with the tabulated value, $F_{.05} = 3.59$. Since this calculated value exceeds the tabulated value of *F*, it lies in the rejection region (see Figure 11.3). Consequently, we have sufficient evidence (at significance level $\alpha = .05$) to conclude that the true mean freshmen grade-point averages differ for at least two of the three socioeconomic groups. The chance that this procedure will result in a Type I error (conclude that there are differences if in fact no differences among the means exist) is, at most, $\alpha = 05$.

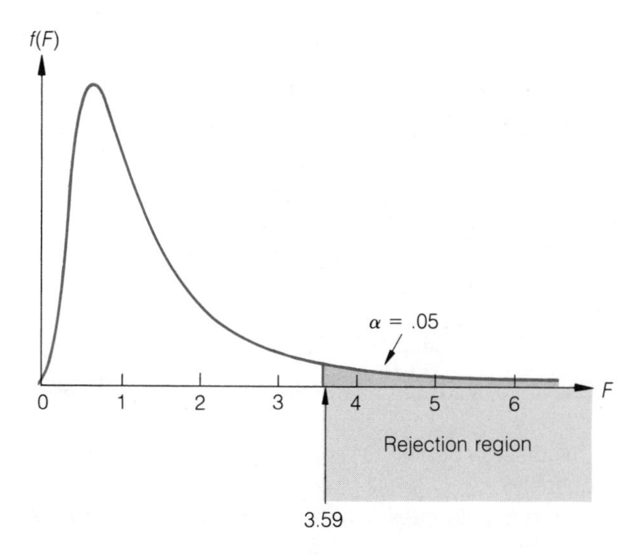

FIGURE 11.3 Rejection Region for Example 11.4: Numerator df = 2, Denominator df = 17

The key elements in a test to compare more than two population means using independent random samples are summarized in the box.

TEST TO COMPARE *k* POPULATION MEANS FOR AN INDEPENDENT SAMPLING DESIGN

H_0: $\mu_1 = \mu_2 = \cdots = \mu_k$

H_a: At least two treatment means differ

Test statistic: $F = \dfrac{\text{MST}}{\text{MSE}}$

Rejection region: $F > F_\alpha$

where the distribution of F is based on $(k-1)$ numerator df and $(n-k)$ denominator df, and F_α is the F-value such that $P(F > F_\alpha) = \alpha$

Assumptions: See Section 11.5.

EXERCISES

11.1 Find $F_{.05}$ for an F distribution with:

a. Numerator df = 7, denominator df = 25
b. Numerator df = 10, denominator df = 8
c. Numerator df = 30, denominator df = 60
d. Numerator df = 15, denominator df = 4

Show the location of $F_{.05}$ on a sketch of the F distribution for each part of the exercise.

11.2 Find F_α for an F distribution with 15 numerator df and 12 denominator df for the following values of α:

a. $\alpha = .025$ **b.** $\alpha = .05$ **c.** $\alpha = .10$

Show the location of F_α on a sketch of the F distribution for each part of the exercise.

11.3 Independent random samples were selected from three populations. The data are shown in Table 11.5.

TABLE 11.5 Data for Exercise 11.3

SAMPLE 1	SAMPLE 2	SAMPLE 3
2.1	4.4	1.1
3.3	2.6	0.2
0.2	3.0	2.0
	1.9	

a. Calculate MST for the data. What type of variability is measured by this quantity?
b. Calculate MSE for the data. What type of variability is measured by this quantity?
c. How many degrees of freedom are associated with MST?

d. How many degrees of freedom are associated with MSE?

e. Compute the test statistic appropriate for testing H_0: $\mu_1 = \mu_2 = \mu_3$ against the alternative that at least one population mean is different from the other two.

f. Specify the rejection region, using a significance level of $\alpha = .05$.

g. Make the proper conclusion, based on your answers to parts e and f.

11.4 Most new products are test marketed in several locations, frequently using different advertising techniques. A new type of bourbon, called California Brandy, is test marketed at each of four locations in the state, San Francisco, Los Angeles, San Diego, and Sacramento. The number of sales for the product at each city during each of 5 randomly selected days last month is recorded in Table 11.6.

TABLE 11.6 California Brandy Sales (Bottles), Exercise 11.4

SAN FRANCISCO	LOS ANGELES	SAN DIEGO	SACRAMENTO
37	40	22	28
15	42	10	49
30	36	8	20
21	12	26	42
53	29	15	41

a. Assuming the samples were independently selected, specify the hypotheses for testing whether there is a difference among the mean daily sales of California Brandy in the four cities.

b. Calculate a measure of the variability among the four sample means (i.e., compute MST).

c. Calculate a measure of the within-sample variability (i.e., compute MSE).

d. Calculate the appropriate statistic for testing the hypothesis of part a.

e. How many degrees of freedom are associated with MST, the numerator of the F statistic?

f. How many degrees of freedom are associated with MSE, the denominator of the F statistic?

g. Using a significance level of $\alpha = .025$, specify the rejection region.

h. Is there evidence of a difference among the mean daily sales of California Brandy in the four cities?

11.5 As oil drilling costs rise at unprecedented rates, the task of measuring drilling performance becomes essential to a successful oil company. One method of lowering drilling costs is to increase drilling speed. Researchers at Cities Service Co. have developed a drill bit, called the PD-1, which they believe penetrates rock at a faster rate than any other bit on the market. It is decided to compare the speed of the PD-1 with the two fastest drill bits known, the IADC 1-2-6 and the IADC 5-1-7, at 12 drilling locations in Texas. Four drilling sites were randomly assigned to each bit, and the rate of penetration (RoP) in feet per hour (fph) was recorded after drilling 3,000 feet at each site. The data are given in Table 11.7 on page 376. Based on this information, can Cities Service Co. conclude that the mean RoP differs for at least two of the three drill bits? Test at the $\alpha = .05$ level of significance.

TABLE 11.7
Rate of Penetration (fph), Exercise 11.5

PD-1	IADC 1-2-6	IADC 5-1-7
35.2	25.8	14.7
30.1	29.7	28.9
37.6	26.6	23.3
34.3	30.1	16.2

11.6 The guinea pig has the highest adrenal-to-body weight ratio of any mammal. In addition, the adrenal gland of the guinea pig is similar to the human adrenal. Thus, experiments on guinea pig adrenals have been of interest to scientists for some time. One such experiment was conducted for the purpose of comparing the mean weights of guinea pigs injected with different growth hormones. Twenty-four adult male English short-haired guinea pigs weighing 450–600 grams were divided into three groups of eight pigs each. Each pig was injected with a growth hormone twice daily for a period of two weeks. However, one group received the hormone DEX, another group received the hormone ACTH, and the third group (the control group) received no hormone (sterile water). At the end of the two-week period, the adrenals of the pigs were removed and weighed. The results are listed in Table 11.8. Test (at significance level $\alpha = .10$) for a difference among the mean adrenal weights of guinea pigs in the three groups.

TABLE 11.8
Adrenal Weight (Milligrams), Exercise 11.6

DEX	ACTH	CONTROL
147	151	130
150	183	146
157	156	121
126	105	110
133	167	148
117	155	132
146	146	109
156	170	142

11.7 Studies conducted at the University of Melbourne (Australia) indicate that there may be a difference in the pain thresholds of blonds and brunettes. Men and women of various ages were divided into four categories according to hair color: light blond, dark blond, light brunette, and dark brunette. The purpose of the experiment was to determine whether hair color is related to the amount of pain produced by common types of mishaps and assorted types of trauma. Each person in the experiment was given a pain threshold score based upon his or her performance in a pain sensitivity test (the higher the score, the lower the person's pain

TABLE 11.9
Pain Sensitivity Scores, Exercise 11.7

LIGHT BLOND	DARK BLOND	LIGHT BRUNETTE	DARK BRUNETTE
62	63	42	32
60	57	50	39
71	52	41	51
55		37	30
48			35

tolerance). Consider the results shown in Table 11.9. Is there evidence of a difference among the mean pain thresholds for people possessing the four types of hair color? Test at a significance level of $\alpha = .025$.

11.8 Psychologists generally agree that natural languages are learned systematically. One method of studying the effects of language stimuli on the systematic acquisition of language rules is through the use of miniature linguistic systems (MLS). An MLS consists of monosyllabic nonsense words denoting colors, shapes, actions, etc. In one sequence of experiments, an MLS of nine three-letter words was created; three words were designated as subjects, three as objects, and three as verbs. Participants in the study were randomly assigned to one of the word classes and shown the corresponding list of words. The number of trials required for each subject to learn the meanings of the words was recorded in Table 11.10. A psychologist is concerned about the possibility that the three word classes (subjects, verbs, and objects) are acquired at different rates. Conduct a test of hypothesis to determine if the mean number of trials required to learn the words differs among the three word classes. Use $\alpha = .05$.

TABLE 11.10
Trials to Learn, Exercise 11.8

WORD CLASS		
Subjects	Objects	Verbs
3	4	2
6	7	1
8	2	4
12	9	6
2	5	3
5	3	3

11.9 In U.S. business, two basic types of management attitudes prevail: Theory-X bosses believe that workers are basically lazy and untrustworthy, and Theory-Y managers hold that employees are hard-working, dependable individuals (*Time*, March 2, 1981). Japanese firms take a third approach: Theory-Z companies emphasize long-range planning, consensus decision-making, and strong, mutual worker-employer loyalty. Suppose we wish to compare the hourly wage rates of workers at Theory-X-, -Y-, and -Z-style corporations. Independent random samples of six engineering firms of each managerial philosophy were selected, and the starting hourly wage rates for laborers at each recorded in Table 11.11. Is there evidence of a difference among the mean starting hourly wages of engineers at Theory-X-, -Y-, and -Z-style firms? Test at a significance level of $\alpha = .025$.

TABLE 11.11
Hourly Labor Rates (Dollars), Exercise 11.9

MANAGERIAL ATTITUDE		
Theory X	Theory Y	Theory Z
5.20	6.25	5.50
5.20	6.80	5.75
6.10	6.87	4.60
6.00	7.10	5.36
5.75	6.30	5.85
5.60	6.35	5.90

11.3 Confidence Intervals for Means

An analysis of variance for an independent sampling design may include the construction of confidence intervals for a single mean or for the difference between two means. Because the independent sampling design involves the selection of independent random samples, we can find a confidence interval for a single mean using the method of Section 8.3 and for the difference between two population means using the method of Section 8.6. The only modification we will make in these two procedures is that we will use an estimate of σ^2 based on the information contained in all k samples, namely, the pooled measure of variability within the k samples; that is,

$$\text{MSE} = s^2 = \frac{\text{SSE}}{n-k}$$

Note that this estimate of σ^2 is based upon $(n-k)$ degrees of freedom and that it is the same quantity used in the denominator for the analysis of variance F test. The formulas for the confidence intervals of Chapter 8 are reproduced in the box.

CONFIDENCE INTERVALS FOR MEANS

Single population mean μ_i: $\quad \bar{x}_i \pm t_{\alpha/2}\frac{s}{\sqrt{n_i}}$

Difference $(\mu_i - \mu_j)$ between two population means: $\quad (\bar{x}_i - \bar{x}_j) \pm t_{\alpha/2}s\sqrt{\frac{1}{n_i}+\frac{1}{n_j}}$

where $s = \sqrt{\text{MSE}}$; the distribution of t is based on $(n-k)$ degrees of freedom (the denominator degrees of freedom in the ANOVA and the degrees of freedom associated with s^2); and $t_{\alpha/2}$ is the t-value such that $P(t > t_{\alpha/2}) = \alpha/2$

EXAMPLE 11.5 Refer to Example 11.3. Construct a 95% confidence interval for μ_1, the true mean grade-point average of all college freshmen in the lower socioeconomic class.

Solution From Example 11.3,

$$\text{MSE} = 0.277$$

Then

$$s = \sqrt{\text{MSE}} = \sqrt{0.277} = 0.526$$

The sample mean grade-point average for those freshmen in the lower class is

$$\bar{x}_1 = \frac{T_1}{n_1} = \frac{17.65}{7} = 2.521$$

and the tabulated value, $t_{.025}$, for 17 df (the same as the denominator df in the ANOVA) is (from Table 4, Appendix E)

$$t_{.025} = 2.110$$

Therefore, a 95% confidence interval for μ_1, the mean grade-point average for all freshmen in the lower socioeconomic class, is

$$\bar{x}_1 \pm t_{\alpha/2}\frac{s}{\sqrt{n_1}} = 2.521 \pm 2.110\frac{.526}{\sqrt{7}}$$

$$= 2.521 \pm 0.419$$

or (2.102, 2.940). We say that the interval 2.102 to 2.940 encloses the true mean, μ_1, with 95% confidence.

EXAMPLE 11.6 Refer to Example 11.3. Find a 95% confidence interval for $(\mu_2 - \mu_1)$, the difference between mean grade-point averages for freshmen in the middle socioeconomic class and those in the lower class.

Solution The mean grade-point average of the sample of freshmen in the middle class is

$$\bar{x}_2 = \frac{T_2}{n_2} = \frac{22.74}{7} = 3.249$$

and from Example 11.5, $\bar{x}_1 = 2.521$. The tabulated t-value, $t_{.025}$, is the same as for Example 11.5, namely 2.110. Then the 95% confidence interval for $(\mu_2 - \mu_1)$ is

$$(\bar{x}_2 - \bar{x}_1) \pm t_{\alpha/2}s\sqrt{\frac{1}{n_2} + \frac{1}{n_1}} = (3.249 - 2.521) \pm (2.110)(0.526)\sqrt{\frac{1}{7} + \frac{1}{7}}$$

$$= 0.728 \pm 0.593$$

or (0.135, 1.321).

Since the interval includes only positive numbers, we can conclude, with 95% confidence, that the mean grade-point average (μ_2) of freshmen in the middle socioeconomic class exceeds the mean grade-point average (μ_1) of those in the lower class. The difference could be as large as 1.321 grade points or as small as 0.135 grade point.

To obtain narrower confidence intervals than those constructed in Examples 11.5 and 11.6, we would have to select larger samples of college freshmen from within each socioeconomic group.

EXERCISES

11.10 Refer to Exercise 11.4 and the data of Table 11.6. Calculate a 99% confidence interval for $(\mu_3 - \mu_1)$, the difference between mean daily sales of California Brandy in San Diego and San Francisco. Interpret the interval.

11.11 Refer to Exercise 11.5 and the data of Table 11.7.

a. Find a 95% confidence interval for μ_1, the mean RoP for the new PD-1 drill bit. Interpret the interval.

b. Find a 95% confidence interval for $(\mu_1 - \mu_2)$, the difference between the mean RoPs for the PD-1 and the IADC 1-2-6 drill bits. Which of the two drill bits appears to have the faster mean RoP? Explain.

11.12 Refer to Exercise 11.6 and the data of Table 11.8.

a. Construct 90% confidence intervals for each of the three population means tested.

b. Find a 90% confidence interval for $(\mu_2 - \mu_1)$, the difference between the mean adrenal weights for guinea pigs injected with ACTH and for those injected with DEX.

11.13 Refer to Exercise 11.7 and the data of Table 11.9. Construct a 95% confidence interval for $(\mu_4 - \mu_1)$, the true difference between mean pain sensitivity scores for dark brunettes and light blonds. Interpret the interval.

11.14 Refer to Exercise 11.8 and the data of Table 11.10. Construct a 90% confidence interval for the true mean number of trials required by participants to learn MLS verbs. Is there evidence that the true mean is greater than 3?

11.15 Refer to Exercise 11.9 and the data of Table 11.11.

a. Calculate 99% confidence intervals for each of the three differences, $(\mu_1 - \mu_2)$, $(\mu_1 - \mu_3)$, and $(\mu_2 - \mu_3)$.

b. Based on your intervals, part a, which type of engineering firm appears to pay its employees a higher average starting hourly wage rate, Theory-X-, Theory-Y-, or Theory-Z-style firms? [*Hint:* Compare the means in a pairwise fashion. For example, first compare μ_1 to μ_2 by checking the interval for the difference $(\mu_1 - \mu_2)$. If the interval indicates that μ_1 is greater than μ_2, then compare μ_1 to μ_3 by checking the interval for the difference $(\mu_1 - \mu_3)$. If this interval indicates that μ_1 is greater than μ_3, then there is evidence that μ_1 is the largest of the three means, i.e., greater than both μ_2 and μ_3.]

11.4 An Example of a Computer Printout

As the sizes of the samples in the independent samples design are increased, the computations in an analysis of variance become tedious. This difficulty can easily be circumvented (if you have access to a computer) by using one of the many statistical computer packages designed to carry out a complete analysis of variance. In this section, we discuss the location and interpretation of the key ANOVA elements on a computer printout from one of the more popular computer packages, called the Statistical Analysis System (SAS). At your institution, other computer packages such as Biomed, Minitab, and SPSS may be installed. These packages also have ANOVA procedures available, and their respective printouts are similar to that of the SAS. Whichever computer package you have access to, you should be able to understand the ANOVA results after reading the discussion that follows.

The results of an analysis of variance are often summarized in tabular form. The general form of an ***ANOVA table*** for an independent samples design is shown in Table 11.12. ***Source*** refers to the source of variation (***Treatments*** for the variability among the sample means, and ***Error*** for the within-sample variability), and for each source, **df** refers to the degrees of freedom, **SS** to the sum of squares, **MS** to the mean square, and ***F*** to the *F* statistic. Most computer packages present the analysis of variance results in the form of an ANOVA table.

TABLE 11.12 ANOVA Summary Table for an Independent Samples Design

SOURCE	df	SS	MS	F
Treatments	$k-1$	SST	MST	MST/MSE
Error	$n-k$	SSE	MSE	
Total	$n-1$	SS(Total)		

EXAMPLE 11.7 Consider the problem of comparing the mean starting salaries of June 1980–March 1981 University of Florida bachelor's degree graduates in the five colleges (Business Administration, Education, Engineering, Liberal Arts, Sciences) discussed in Section 1.1. Many of these graduates did not return the CRC questionnaire, so we will never know their starting salaries, but we can still imagine that a starting salary exists for each graduate of a college and that the totality of these salaries constitutes a population that characterizes the first-year financial compensations of graduates of that college. Thus, we wish to compare the mean starting salaries corresponding to the starting salary populations of the five colleges, i.e., we wish to test

H_0: $\mu_1 = \mu_2 = \mu_3 = \mu_4 = \mu_5$

H_a: At least two means are different

where

μ_1 = Mean starting salary of graduates of the College of Business Administration

μ_2 = Mean starting salary of graduates of the College of Education

μ_3 = Mean starting salary of graduates of the College of Engineering

μ_4 = Mean starting salary of graduates of the College of Liberal Arts

μ_5 = Mean starting salary of graduates of the College of Sciences

In order to perform the comparison, independent random samples of starting salaries were selected from the actual starting salaries for the five colleges listed in Appendix B (see Table 11.13) and the data subjected to an analysis of variance using the SAS. A portion of the SAS printout is reproduced in Figure 11.4. Interpret these results.

TABLE 11.13 Starting Salaries of Graduates of the Five Colleges

BUSINESS ADMINISTRATION	EDUCATION	ENGINEERING	LIBERAL ARTS	SCIENCES
$ 5,500	$15,500	$21,800	$13,800	$15,000
11,500	13,100	25,100	14,000	15,400
16,600	5,000	14,800	14,500	18,300
15,400	7,800	13,700	10,600	14,700
14,800	9,300	21,500	20,700	13,900
17,100	6,800	22,500	9,700	10,600
14,900	7,300	12,000	22,700	11,300
13,800	18,600	22,100	11,600	19,300

FIGURE 11.4 Portion of the SAS Printout of the ANOVA, Example 11.7

```
DEPENDENT VARIABLE: SALARY
SOURCE              DF      SUM OF SQUARES          MEAN SQUARE       F VALUE     PR > F
MODEL                4   314678500.00000000   78669625.00000000        4.29     0.0063
ERROR               35   641812500.00000000   18337500.00000000
CORRECTED TOTAL     39   956491000.00000000

SOURCE              DF             ANOVA SS     F VALUE    PR > F

COLLEGE              4   314678500.00000000        4.29    0.0063
```

Solution You can see from Figure 11.4 that the SAS printout presents the results in the form of an ANOVA table. The source of variation attributable to treatments, i.e., to the variability among the sample means for the five colleges, is labelled MODEL and the source of variation attributable to error, i.e., to the within-sample variability, is labelled ERROR. Their corresponding sums of squares and mean squares are

$$\text{SST} = 314{,}678{,}500.0$$
$$\text{SSE} = 641{,}812{,}500.0$$
$$\text{MST} = 78{,}669{,}625.0$$
$$\text{MSE} = 18{,}337{,}500.0$$

The computed value of the test statistic, given under the column heading F VALUE, is

$$F = 4.29$$

To determine whether to reject the null hypothesis

$$H_0: \quad \mu_1 = \mu_2 = \cdots = \mu_5$$

in favor of the alternative

H_a: At least two population means are different

we may consult Appendix E for tabulated values of the F distribution corresponding to an appropriately chosen significance level α. However, since the SAS printout gives the attained significance level (p-value) of the test, we will use this quantity to assist us in reaching a conclusion.

Under the column headed PR > F is the attained significance level of the test. This value, .0063, implies that H_0 will be rejected at any chosen level of α larger than .0063. Thus, there is very strong evidence of a difference among the mean starting salaries of graduates of the five colleges. The probability that this procedure will lead to a Type I error (conclude that there is a difference among the means if in fact they are all equal) is .0063.

EXERCISES **11.16** Refer to Exercise 11.4. Give the results of the ANOVA in the form of a summary table, Table 11.12.

11.17 Refer to Exercise 11.7. Give the results of the ANOVA in the form of a summary table, Table 11.12.

11.18 Refer to Exercise 11.8. Give the results of the ANOVA in the form of a summary table, Table 11.12.

11.19 A drug company synthesized three new drugs that should alleviate pain due to ulcers. To determine whether the drugs will be absorbed by the stomach (and hence have a possibility of being effective), 24 pigs were randomly assigned, 8 to each drug, to receive oral doses. After a given amount of time, the concentration of the drug in the stomach lining of each pig was determined. The data are shown in Table 11.14. In order to determine if there is a difference among the mean concentrations for the three drugs, the data of Table 11.14 were subjected to an ANOVA using the SAS. A portion of the SAS printout is shown in Figure 11.5. Locate the key elements of the ANOVA on the printout and interpret their values. At a significance level of $\alpha = .025$, is there evidence of a difference among the mean concentrations for the three drugs after the fixed period of time?

TABLE 11.14 Drug Concentration (Cubic Centimeters), Exercise 11.19

DRUG		
1	2	3
1.70	1.73	1.67
1.72	1.79	1.63
1.81	1.76	1.60
1.79	1.75	1.55
1.75	1.70	1.63
1.66	1.80	1.61
1.83	1.81	1.67
1.75	1.73	1.59

FIGURE 11.5 Output from the SAS for the Drug Concentration Data, Exercise 11.19

```
DEPENDENT VARIABLE: CONCENTRATION
SOURCE              DF    SUM OF SQUARES    MEAN SQUARE    F VALUE    PR > F
MODEL                2        0.09923333     0.04961667      23.12    0.0001
ERROR               21        0.04506250     0.00214583
CORRECTED TOTAL     23        0.14429583

SOURCE              DF         ANOVA SS    F VALUE    PR > F
DRUGS                2       0.09923333      23.12    0.0001
```

11.20 Refer to Case Study 2.3 and the data of Table 2.6. Most tobacco companies now market light or extra light cigarettes which are low in tar and nicotine. Are these light cigarettes also significantly lower than the regular brands in terms of carbon monoxide ranking (as determined by the Federal Trade Commission)? Suppose we wish to compare the mean carbon monoxide (CO) rankings of regular, light, and extra light cigarettes.

a. Specify the appropriate null and alternative hypotheses.
b. The data of Table 2.6 were subjected to an analysis of variance using the SAS. A portion of the SAS printout is shown in Figure 11.6. Locate the key elements of the ANOVA on the printout and interpret their values.
c. Is there evidence of a difference among the mean CO rankings for the three types (regular, light, extra light) of cigarettes?

FIGURE 11.6 Output from the SAS for the Carbon Monoxide Data, Exercise 11.20

DEPENDENT VARIABLE: CO

SOURCE	DF	SUM OF SQUARES	MEAN SQUARE	F VALUE	PR > F
MODEL	2	574.89341231	287.44670616	10.70	0.0001
ERROR	184	4944.26669464	26.87101464		
CORRECTED TOTAL	186	5519.16010695			

SOURCE	DF	ANOVA SS	F VALUE	PR > F
LIGHT TYPE	2	574.89341231	10.70	0.0001

11.21 Before purchasing a $10,000 life insurance policy, a young executive wishes to determine if there are significant differences among the mean monthly premiums of four types of policies that he is considering: (1) individual convertible term, (2) individual renewable term, (3) group renewable term, and (4) government group insurance. For each of these policies he obtains price quotations from six randomly selected insurance companies (thus, 24 insurance companies, 6 for each type of policy, are included in the total sample). The data were analyzed for differences among the mean prices for the four policy types by using the ANOVA procedure of the SAS. The results are given in Figure 11.7.

a. Locate the following quantities on the SAS printout and interpret their values: SST, SSE, MST, MSE, F.
b. Are there significant differences among the mean monthly premiums of the four types of insurance policies?

FIGURE 11.7 SAS ANOVA Procedure, Exercise 11.21

DEPENDENT VARIABLE: PREMIUM

SOURCE	DF	SUM OF SQUARES	MEAN SQUARE	F VALUE	PR > F
MODEL	3	7.01500000	2.33833333	24.83	0.0001
ERROR	20	1.88333333	0.09416667		
CORRECTED TOTAL	23	8.89833333			

SOURCE	DF	ANOVA SS	F VALUE	PR > F
POLICIES	3	7.01500000	24.83	0.0001

11.22 A clinical psychologist wished to compare the mean hostility levels (as measured by a certain psychological test) of four groups of patients: schizophrenics, manic depressives, psychosomatics, and heroin addicts. Independent random samples of patients in each group were selected and administered the hostility test. Use the information provided by the SAS printout, Figure 11.8, to test whether there is a difference among the mean hostility levels for the four groups of patients.

FIGURE 11.8 SAS ANOVA Procedure, Exercise 11.22

DEPENDENT VARIABLE: HOSTILITY LEVEL

SOURCE	DF	SUM OF SQUARES	MEAN SQUARE	F VALUE	PR > F
MODEL	3	712.58643892	237.52881297	3.77	0.0280
ERROR	19	1196.63095238	62.98057644		
CORRECTED TOTAL	22	1909.21739130			

SOURCE	DF	ANOVA SS	F VALUE	PR > F
GROUPS	3	712.58643892	3.77	0.0280

11.5 Assumptions: When the Analysis of Variance *F* Test Is Appropriate

Most of the statistical procedures which we have discussed throughout Chapters 8–10 require that certain assumptions be satisfied in order for their inferences to be valid. The assumptions necessary for a valid analysis of variance F test for comparing more than two population means are extensions of those assumptions required for comparing two means (Chapters 8–10). These assumptions, which must be satisfied regardless of the sample sizes employed in the sampling design, are summarized in the box.

ASSUMPTIONS FOR A TEST TO COMPARE *k* POPULATION MEANS: INDEPENDENT SAMPLES DESIGN

1. All k population probability distributions are normal.
2. The k population variances are equal.

In most real-world applications the assumptions will not be satisfied exactly. However, the analysis of variance procedure is flexible in the sense that slight departures from the assumptions will not significantly affect the analysis or the validity of the resulting inferences.

11.6 Summary

This chapter presents an extension of the independent sampling experiment to allow for the comparison of more than two means. The independent sampling design uses independent random samples selected from each of k populations. The comparison of the population means is made by comparing the variation among the sample means, as measured by the mean square for treatments (MST), to the variation attributable to differences within the samples, as measured by the mean square for error (MSE). If the ratio of MST to MSE is large, we conclude that a difference exists between the means of at least two of the *k* populations.

We point out that there are various methods of collecting data and designing experiments for the purpose of comparing more than two population means; an independent sampling design is the simplest of these preconceived plans. If you would like to study other types of experimental designs and the ANOVA technique associated with each, consult the references at the end of this chapter.

KEY WORDS

Analysis of variance
Independent sampling design
Variability among the sample means
Within-sample variability
F distribution

KEY SYMBOLS

Analysis of variance: ANOVA
Sum of squares for treatments: SST
Sum of squared errors: SSE
Mean square for treatments: MST
Mean square for error: MSE
Ratio of mean squares: *F*

SUPPLEMENTARY EXERCISES

[*Note:* List the assumptions necessary to ensure the validity of the procedure you use to solve these problems.]

11.23 Suppose that you wish to compare the means of three treatments using independent random samples of size $n_1 = n_2 = n_3 = 15$.

a. How many degrees of freedom are associated with the numerator of the *F* statistic?

b. How many degrees of freedom are associated with the denominator of the *F* statistic?

11.24 Suppose that you wish to compare the means of eight treatments using independent random samples of size $n_1 = n_2 = \cdots = n_8 = 4$.

a. How many degrees of freedom are associated with the numerator of the *F* statistic?

b. How many degrees of freedom are associated with the denominator of the F statistic?

11.25 Complete the ANOVA summary table shown below.

SOURCE	df	SS	MS	F
Treatments	9		15.2	
Error		200.7		
Total	39			

11.26 Complete the ANOVA summary table shown below.

SOURCE	df	SS	MS	F
Treatments				
Error	16	32		
Total	17	60		

11.27 In hopes of attracting more riders, a city transit company plans to have express bus service from a suburban terminal to the downtown business district. These buses will travel along a major city street where there are numerous traffic lights that will affect travel time. The city decides to perform a study on the effect of four different plans (a special bus lane, traffic signal progression, etc.) on the travel times for the buses. Travel times (in minutes) are measured for several weekdays during a morning rush-hour trip while each plan is in effect. The results are recorded in Table 11.15.

TABLE 11.15
Bus Travel Times, Exercise 11.27

PLAN			
1	2	3	4
27	25	34	30
25	28	29	33
29	30	32	31
26	27	31	
	24	36	

a. Construct an ANOVA summary table for this experiment.

b. Is there evidence of a difference among the mean travel times for the four plans? Use $\alpha = .025$.

c. Form a 95% confidence interval for the difference between the mean travel times of buses under plan 1 (express lane) and plan 3 (a control—no special travel arrangements).

11.28 The oxygen supply of coastal marine sediments is of great importance to floral (plant) and faunal (animal) communities in the ocean. A scientist wishes to compare the mean depths of penetration of oxygen into coastal marine sediments at five different water depths: 5, 10, 15, 20, and 40 meters. Independent samples of marine sediments at each water depth were randomly selected and measured for depth of oxygen penetration by polarograhic oxygen microelectrodes. The data are recorded (in millimeters) in Table 11.16.

TABLE 11.16
Depth of Penetration of Oxygen, Exercise 11.28

WATER DEPTH (METERS)				
5	10	15	20	40
1.2	3.7	1.7	2.8	4.4
3.6	4.6	2.2	4.6	5.3
1.3	3.1	1.6	3.7	3.6
1.8	2.0	1.5	3.2	4.8

a. Construct an ANOVA summary table for this experiment.
b. Is there evidence of a difference among the mean depths of penetration of oxygen into marine sediments for the five water depths? Use $\alpha = .05$.
c. Find a 95% confidence interval for $(\mu_4 - \mu_2)$, the difference between the mean depths of penetration of oxygen at water depths of 20 and 10 meters. Interpret the interval.

11.29 Great Britain has experimented with different 40-hour work weeks to maximize production and minimize expenses. A factory tested a 5-day week (8 hours per day), a 4-day week (10 hours per day), and a 3⅓-day week (12 hours per day). The weekly production results are shown in Table 11.17 (in thousands of dollars worth of items produced).

TABLE 11.17
Weekly Production, Exercise 11.29

8-HOUR DAY	10-HOUR DAY	12-HOUR DAY
87	75	95
96	82	76
75	90	87
90	80	82
72	73	65
86	87	71

a. Construct an ANOVA summary table for this experiment.
b. Is there evidence of a difference among the mean weekly productivity levels for the three lengths of workdays? Test using a significance level of $\alpha = .10$.
c. Using a 90% confidence interval, estimate the mean weekly productivity level when 12-hour workdays are used.
d. Estimate the difference between mean weekly productivity levels for 8-hour workdays and 10-hour workdays, using a 90% confidence interval.

11.30 The possibility of a genetic component to the severity of stuttering was recently investigated. A random sample of 184 adult stutterers (as diagnosed by professionals) was selected and each subject was classified according to one of ten groups according to age, sex, and whether or not a relative ever stuttered. Each subject was required to read a 500-word standard passage to determine the severity of stuttering. (The severity measure was defined as the percent of words on which at least one disfluency—repetition of sounds, prolongation of sounds, audible struggle behavior, or phonatory blockage—was observed.) A partial ANOVA summary table for the data is shown.

SOURCE	df	SS	MS	F
Groups	9	37.101		
Error	174	157.023		
Total	183	194.124		

a. Complete the ANOVA summary table.

b. Is there evidence of a difference among the mean severities of stuttering for the ten groups? Test at $\alpha = .05$.

11.31 A plastics company hypothesizes that treatment, after casting, of a plastic used in optic lenses will improve wear. Six different treatments are to be tested. To determine whether any differences in mean wear exist among the treatments, 18 castings from a single formulation of the plastic were made, and 3 castings were randomly assigned to each of the treatments. Wear was determined by measuring the increase in "haze" after 200 cycles of abrasion (better wear being indicated by smaller increases). These results are shown in Table 11.18.

TABLE 11.18 Plastic "Haze" Measurements, Exercise 11.31

TREATMENT					
A	B	C	D	E	F
13.29	15.15	9.54	10.00	8.73	14.86
12.07	11.95	11.47	12.45	9.75	15.03
11.97	14.75	11.26	12.38	8.01	11.18

a. Construct an ANOVA summary table for this experiment.

b. Is there evidence of a difference in mean wear among the six treatments? Test using $\alpha = .05$.

c. Estimate the difference between the mean increases in haze for treatments B and E, using a 95% confidence interval.

d. Estimate the mean increase in haze for optic lenses receiving treatment A, using a 95% confidence interval.

11.32 Kent (1981) conducted a study of the lifespan of a certain species of predatory, ciliate protozoan.* One hundred sixteen individuals of this species were captured from the Sacramento River (California) and randomly divided into five groups. All individuals were fed on a predetermined schedule; however, the food level varied among groups. The first group was fed 3 paramecia per day, the second group 1 paramecium per day, the third group ½ paramecium per day, the fourth ¼ paramecium per day, and the fifth group was starved. The lifespan, in days, was

*Ecology, April 1981, 62, 296–302. Copyright © 1981, the Ecological Society of America.

recorded for each individual. An analysis of variance was performed with the following results:

SOURCE	df	SS	MS	F
Food Levels	4		196.14	
Error	111		59.12	
Total	115			

a. Complete the ANOVA summary table.

b. Is there evidence of a difference (at significance level $\alpha = .025$) among the mean lifespans for the five food levels?

11.33 A Japanese watchmaking firm manufactures three types of watches: digital, mechanical, and quartz analog. The firm conducted the following experiment to compare the performances of the watches. Five watches of each type were timed for a period of twenty-four hours and the time gains (or losses) were recorded for each. The data are recorded in Table 11.19. [*Note:* A negative gain denotes a loss in time.]

TABLE 11.19 Watch Time Gains (Seconds), Exercise 11.33

DIGITAL	MECHANICAL	QUARTZ ANALOG
0.10	1.33	0.01
0.08	−1.20	−0.12
0.22	−2.17	0.00
−0.67	0.03	0.45
0.91	−1.55	−0.28

a. Construct an ANOVA summary table for the experiment.

b. Is there evidence of a difference among the mean time gains for the three types of watches? Test using $\alpha = .025$.

c. Find a 90% confidence interval for the difference between the mean time gains for digital and quartz analog watches.

11.34 Refer to the data on DDT measurements in fish, Appendix D. Suppose we wish to compare the mean DDT concentrations of the three species of fish (channel catfish, large mouth bass, and small mouth buffalo) collected from the Tennessee River and its creek tributaries. An analysis of variance on DDT concentrations of the three species was carried out using the SAS. A portion of the SAS printout is given in Figure 11.9.

a. Locate the following quantities on the SAS printout and interpret their values: SST, SSE, MST, MSE, F.

b. Interpret the value of the quantity PR > F. Are there significant differences among the true mean DDT concentrations of the three species of fish inhabiting the Tennessee River?

FIGURE 11.9 SAS ANOVA Procedure, Exercise 11.34

DEPENDENT VARIABLE: DDT

SOURCE	DF	SUM OF SQUARES	MEAN SQUARE	F VALUE	PR > F
MODEL	2	23454.45013750	11727.22506875	1.22	0.2997
ERROR	141	1360549.03766250	9649.28395505		
CORRECTED TOTAL	143	1384003.48780000			

SOURCE	DF	ANOVA SS	F VALUE	PR > F
SPECIES	2	23454.45013750	1.22	0.2997

CASE STUDY 11.1 Moral Development of Teenagers

Recent years have seen an increase in psychological studies of the moral development of adolescents. However, almost all research has been carried out with children or teenagers attending regular schools; very few experiments have used juvenile delinquents as subjects. Sagi and Eisikovits addressed the issue in their article "Juvenile delinquency and moral development" (*Criminal Justice and Behavior,* March 1981). Their purpose was to "compare delinquent and nondelinquent populations, using concepts and measures taken from the area of moral development."

Sagi and Eisikovits randomly sampled 249 adolescents, representing both males and females between the ages of 13 and 17, drawn from three "sociological affiliations"—middle-class nondelinquents, lower-class nondelinquents, and lower-class delinquents. The study employed a version of MOTEC, a morality test for children which has been widely acclaimed for its high degree of reliability and validity. The test consisted of seven situations presented in the form of a booklet. Sagi and Eisikovits* explain:

> Each situation represented a young protagonist facing a moral dilemma, and the subject was asked to indicate how the protagonist would solve the dilemma. The dilemmas represented a range of seven immoral acts, as follows: stealing, cheating, expropriation, minor violence, major violence, damaging of property, and lying.
>
> The following situation (damaging of property) is an example: "The picture you see represents two young persons lighting a fire for a barbecue party. As they are running short of wood, one of them is looking for more in the area and discovers a wooden shack. No one else is around." Upon exposure to the dilemma, the subject was asked to indicate whether the protagonist would or would not resist the temptation. The answer was either "Yes" or "No." A response of "Yes" (the subject did not resist temptation) resulted in assignment of 0 points to the subject. A "No" response (the subject did resist temptation) granted one point.
>
> The subject was then instructed to turn to the appropriate page in the booklet, depending on the answer. If the answer was "Yes," the subject was asked to turn to a

*Reprinted from "Juvenile Delinquency and Moral Development" by A. Sagi and Z. Eisikovits, in *Criminal Justice and Behavior* 8: 79–93, © 1981 American Association of Correctional Psychologists, with permission of the publisher, Sage Publications, Beverly Hills.

page where four possible reasons for not resisting temptation were provided. If the answer was "No," the subject was asked to turn to a page where there were four possible reasons for resisting temptation. The subject was then asked to indicate which one of the four reasons best explained the protagonist's behavior."

The four reasons (with points in parentheses) for each answer are as follows:

Yes: Because he likes to eat barbecued meat very much (1)
Because no one sees him (2)
Because he promised to bring wood to his friends (3)
Because there were clear signs the shack was abandoned (4)

No: Because he may get hurt (1)
Because he is afraid of cops hanging out in the area (2)
His friends may get punished because of him (3)
Because one shouldn't damage property belonging to others (4)

Test scores were tabulated for five moral dimensions. These dimensions (with range of scores in parentheses) are as follows: (1) resistance to temptation (0–7); (2) moral reasoning (7–28); (3) feelings after offense (7–21); (4) severity of punishment (7–28); (5) confession (0–7). The mean moral dimension scores for the three groups of teenagers are reported in Table 11.20.

TABLE 11.20
Mean Moral Dimension Scores, Case Study 11.1

MORAL DIMENSION	SOCIOLOGICAL AFFILIATION		
	Delinquents	Lower-class Nondelinquents	Middle-class Nondelinquents
(1) Resistance to temptation	3.39	4.82	4.48
(2) Moral development	18.74	22.85	21.71
(3) Feelings after offense	13.60	15.95	14.85
(4) Severity of punishment	16.34	17.10	15.56
(5) Confession	2.51	3.47	3.70

a. Consider the moral dimension, Resistance to temptation. Do you think it is possible, based on the sample means, to determine whether differences exist among the true mean MOTEC test scores for the three sociological affiliations? Why is a statistical test needed?

b. Set up the appropriate null and alternative hypotheses to determine if the mean MOTEC "resistance to temptation" test scores differ for at least two of the three sociological affiliations.

c. Sagi and Eisikovits conducted a one-way analysis of variance on MOTEC test scores for each of the five moral dimensions.* The corresponding test statistics and attained significance levels (p-values) are given in Table 11.21. Interpret these results.

*The actual analysis was conducted as a three-way ANOVA, but for the purpose of this case study, we have chosen to discuss it as a one-way ANOVA.

TABLE 11.21 Test Statistics and *p*-Values

MORAL DIMENSION	*F*	*p*-VALUE
(1) Resistance to temptation	15.80	.001
(2) Moral development	22.47	.001
(3) Feelings after offense	13.91	.001
(4) Severity of punishment	2.79	.070
(5) Confession	10.55	.001

CASE STUDY 11.2 A Comparison of the Shopping Habits of Four Types of Grocery Shoppers

Classifying shoppers, particularly grocery shoppers, into various groups based upon certain market characteristics has long been a research tradition. Williams, Painter, and Nicholas ("A policy-oriented typology of grocery shoppers," *Journal of Retailing,* 1978) present a grouping of grocery shoppers oriented toward store "policy which not only offers insight into consumer behavior, but also provides suggestions for the choice of marketing tools to influence particular market segments."

Williams et al. derive their "policy-oriented" typology from a competition angle. They first identify the two most important areas in which grocery retailers compete for buyers: (1) pricing policies, and (2) customer service policies.

> The price images customers maintain about different stores may be the one most important factor differentiating these stores. Also, customer service, defined here to include location and shopping convenience, is also a major dimension along which customers select stores to patronize. Given that store managers and marketing executives can design and implement strategy along at least these two dimensions, customers likewise maintain differing degrees of concern or interest in the pricing practices and customer service policies of the stores where they shop. Specifically, customers can be highly involved with the pricing or customer service practices or rather uninvolved or apathetic about these procedures.

By cross-classifying shoppers on the basis of either high or low involvement along both price and customer service dimensions, a typology of grocery shoppers results. The four groups identified by the authors (and shown in Table 11.22) are (1) Involved shoppers, (2) Convenience shoppers, (3) Price shoppers, and (4) Apathetic shoppers.

TABLE 11.22 Grocery Shopping Orientations or Buying Styles

		STORE OR CHAIN'S PRICING PRACTICES	
		High Customer Involvement	**Low Customer Involvement**
STORE OR CHAIN'S CUSTOMER SERVICE PRACTICES	**High Customer Involvement**	Involved shopper	Convenience shopper
	Low Customer Involvement	Price shopper	Apathetic (uninvolved) shopper

Williams et al. point out that the four types of shoppers "actually represent rough groupings or aggregations of shoppers which in an empirical sense would only be expected to display a general or dominant orientation. Consequently within each shopper group a researcher would expect to find some diversity. The important question, however, is not whether some diversity exists [within the groups] but whether the aggregation process leads to a useful market segmentation." That is, if shoppers in each group are subjected to various marketing mix variables, will significant differences among the groups' responses exist? If so, a market segmentation strategy based upon this "policy-oriented" typology should be employed.

In order to investigate this phenomenon, Williams et al. gathered data from a representative sample of households located in metropolitan Salt Lake City during April and May of 1974. The principal grocery buyer in each household was interviewed, and the data collected included information on a respondent's favorite store, shopping habits, store loyalty, and various socioeconomic and demographic variables. The responses for a total of 298 households were deemed usable in the analysis.

The first phase of the study involved an empirical derivation of the four groups of grocery shoppers illustrated in Table 11.22. Using a statistical technique called ***cluster analysis,*** the authors were able to classify each of the 298 respondents into one of the four groups (clusterings): Apathetic shoppers (59 respondents), Convenience shoppers (125), Price shoppers (81), and Involved shoppers (33). For the purposes of this case study, we will assume that the data were obtained using an independent sampling design—random samples of $n_1 = 59$ Apathetic shoppers, $n_2 = 125$ Convenience shoppers, $n_3 = 81$ Price shoppers, and $n_4 = 33$ Involved shoppers, independently selected from among the four shopping groups.

For the second phase of the study, of particular interest in this case study, Williams et al. subjected the data to a one-way analysis of variance in order to detect significant differences among the mean responses to questions concerning socioeconomic, demographic, and shopping habit variables. A listing of the variables, along with the sample mean responses* of the shoppers in each group, is given in Table 11.23. The extreme right column of the table gives the F-values, corresponding to a one-way analysis of variance test for equality of mean responses among the four groups for each of the five variables.

TABLE 11.23 Mean Response and F-Values for Socioeconomic and Demographic Variables

	MEAN RESPONSES				
SOCIOECONOMIC AND DEMOGRAPHIC VARIABLES	Apathetic Shoppers ($n_1 = 59$)	Convenience Shoppers ($n_2 = 125$)	Price Shoppers ($n_3 = 81$)	Involved Shoppers ($n_4 = 33$)	F-VALUE
1. Total weekly grocery* expenditure ($)	2.10	2.12	2.25	2.12	.95
2. Household size (number)	3.22	3.18	3.57	3.52	1.28
3. Age (years)	37.36	40.60	43.90	49.85	4.69
4. Income* ($)	3.98	3.94	4.16	4.06	.30
5. Life-cycle stage*	3.59	3.92	4.38	4.73	4.56

*Coded variables

*Some of the responses were coded for the convenience of the shopper who filled out the questionnaire. For example, weekly expenditures were coded 1 for expenditures under $20 per week, 2 for between $20 and $50, 3 for between $50 and $100, and 4 for expenditures greater than $100 per week.

The detection of significant differences among the mean responses (for a particular variable) for the four shopping groups could be of invaluable help to a grocery store manager or supermarket chain executive who is planning a market strategy. In summarizing their results, Williams et al. write:

> Each [of the four major types of shoppers identified in this research] represents a unique shopper orientation or buying style, and thus represents varying propensities for being influenced by different marketing methods. Store managers and supermarket chain store executives must consciously choose which group or groups of customers to satisfy and then implement a planned marketing program which has been especially designed for the target segment or segments the decision-maker has chosen.

a. Use the results of Table 11.23 to test the hypothesis of a difference among the true mean total weekly grocery expenditures for the four shopping groups. Use a significance level of $\alpha = .05$.

b. Repeat part a for the variables Household size, Age, Income, and Life-cycle stage.

c. What assumptions are necessary for the valid implementation of the one-way analysis of variance F test used in parts a and b?

CASE STUDY 11.3 A Child's-Eye View of the President

How do children develop their opinions concerning the president? Numerous studies have dealt with children's perceptions of political authority figures, notably the president. Differences in perceptions have been found to be related to age, education, region, race, social class, and family partisanship. Maddox and Handberg (1980) investigated the "role of television in the process of political socialization."* Specifically, they considered the question: "Does reliance on television as the major source of political information significantly affect how children perceive a new president in the first weeks of his administration?"

Maddox and Handberg suggest that "children who rely primarily on television for their political information will report more positive attitudes about the new president than those reported by children who rely on other sources of information." They base their belief on the fact that, in the early days of an administration—the so-called "honeymoon" period—children and adults alike have relatively little concrete information about the president. The pageantry and ceremony of the new president's inauguration and initial organization of the new administration portrayed on television may tend to favorably influence children's views. In contrast, children who rely on newspapers (which do not offer as dramatic a visual presentation) or their parents for political opinions may receive more divergent views of the new president.

In order to test their hypothesis, Maddox and Handberg randomly sampled 759 sixth grade students in Seminole County, Florida, in February 1977 (the first full month of the Carter presidency) and administered to each a questionnaire. Information source was measured by the question, "From what one source do you get the

---

*Reprinted from "Children view the new president" by W. S. Maddox and R. Handberg, in *Youth and Society*, *12*, 3–16, © 1980 Sage Publications, Beverly Hills, with permission of the publisher.

most information about current events and politics?" Responses to the question were divided into two groups according to source of information: television (412 children) and other sources of information (347). The main objective of the study was to compare the overall attitudes toward the president (measured on a six-point scale) of the two groups of children. However, data pertaining to political interest ("How interested in politics are you?"), party differences, partisanship, and chauvinism ("To what extent do you approve of expressing criticism of America?") were also collected and differences among attitudes toward the president for children grouped according to their responses to these questions also explored. The ANOVA summary table for the experiment is reproduced in Table 11.24.

TABLE 11.24 ANOVA Summary Table for Children's "Attitudes Toward the President" Scores

SOURCE	df	SS	MS	F	p-Value
Information source	1	5.298	5.298	1.039	.3080
Political interest	3	21.753	7.251	1.423	.2350
Partisanship	2	127.790	63.895	12.536	.0001
Party differences	4	31.731	7.933	1.556	.1840
Chauvinism	15	174.477	11.632	2.282	.0040
Error	733	3736.834			
Corrected total	758	4097.883			

Notice that the table is not in the form of a one-way ANOVA summary table. This is because the children were classified not only according to information source (television or other) but also according to degree of political interest, degree of partisanship, etc. An analysis of variance for this type of design requires additional sum of squares calculations; however, interpretations of the quantities SS, MS, F, and p-value are identical to the one-way ANOVA. For example, to determine whether a difference exists between the mean presidential attitude scores for the two categories of information source, check the value of F and the p-value in the row corresponding to "Information source." Similarly, to determine whether differences exist among the mean presidential attitude scores for the three categories of partisanship, check the value of F and the p-value in the row corresponding to "Partisanship."

Consider the following comments by Maddox and Handberg:

> In this period of presidential transition (the initial period of an administration) when there are few major cues to guide children in their evaluations, *partisan identification* and general *chauvinistic attitudes* serve as primary points of reference. The *source of political information,* however, had no influence [on their presidential evaluations.]

a. Point out the key quantities in the ANOVA summary table which led Maddox and Handberg to their conclusions.

b. Interpret the p-values corresponding to the sources of variability "Political interest" and "Party difference."

[*Note:* In an attempt to answer the question, "Why is the source of information not a crucial variable in this particular situation?" Maddox and Handberg write:

(Although) the television coverage (of presidential transition) may be more dramatic, it is not perceived as being any more or less positive than the presentation received from other sources. One can plausibly argue that the child who (tends) to be either critical or supportive of the president is still operating in the environment which generated the initial predisposition.

In addition, . . . the initial media coverage of the president tends to be generally supportive or positive in tone. What occurs during this period is a meeting of public partisan differences. (Without the view of the media), children may in effect return to the cues that have served them well in the past: partisanship and chauvinism. Like adults, children selectively perceive what they wish to see in the political world.]

REFERENCES

Barr, A., Goodnight, J., Sall, J., Blair, W., & Chilko, D. *SAS user's guide.* 1979 ed. SAS Institute, P.O. Box 10066, Raleigh, N. C. 27605.

Cochran, W. G., & Cox, G. M. *Experimental designs.* 2d ed. New York: Wiley, 1957.

Kent, E. B. "Life history responses to resource variation in a sessile predator, the ciliate protozoan *Tokophrya Lemnarum* Stein." *Ecology,* April 1981, *62,* 296–302.

Maddox, W. S., & Handberg, R. "Children view the new president." *Youth and Society,* September 1980, *12,* 3–16.

McClave, J. T., & Dietrich, F. H. *Statistics.* 2d ed. San Francisco: Dellen, 1982. Chapter 9.

Mendenhall, W. *Introduction to linear models and the design and analysis of experiments.* Belmont, Ca.: Wadsworth, 1968. Chapter 8.

Sagi, A., & Eisikovits, Z. "Juvenile delinquency and moral development." *Criminal Justice and Behavior,* March 1981, *8,* 79–93.

Scheffé, H. *The analysis of variance.* New York: Wiley, 1959.

Williams, R. H., Painter, J. J., & Nicholas, H. R. "A policy-oriented typology of grocery shoppers." *Journal of Retailing,* 1978, *54,* 27–41.

Twelve

Simple Linear Regression and Correlation

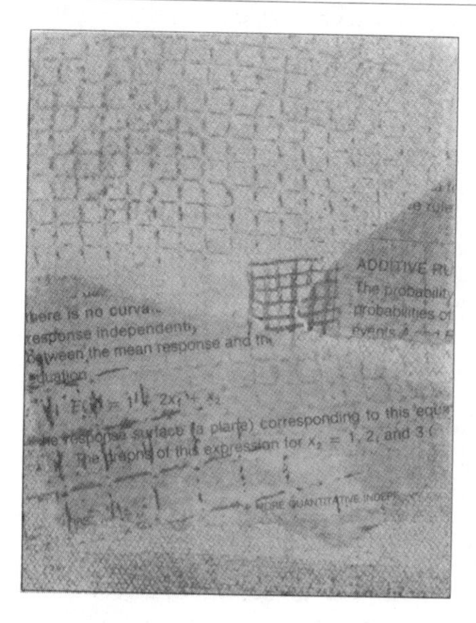

Are you a pill taker? If you are, some would contend that this behavior was learned in childhood, spawned by the flood of TV advertising for over-the-counter drugs. Are children's attitudes toward these drugs linked to the amount of TV advertising they receive? Does an increase in drug advertising imply an increase in child acceptance of over-the-counter drugs? And can you use the level of advertising to actually predict a level of acceptance? These and other questions concerning the relationship between two variables will be discussed in this chapter, and you will learn more about television, children, and drugs in Case Study 12.1.

Contents

12.1 Introduction: Bivariate Relationships

The procedures discussed in the previous four chapters are most useful in cases where we are interested in testing hypotheses about or estimating the values of one or more population parameters based on random sampling. However, a more important concern may be the relationship between two different random variables, x and y, known as a ***bivariate*** relationship. For example, a farmer may be interested in the relationship between the level of fertilizer, x, and the yield of potatoes, y; or, a psychologist may be interested in the bivariate relationship between a child's creativity score, x, and flexibility score, y; or, a medical researcher may be interested in the bivariate relationship between a patient's blood pressure, x, and heart rate, y; etc. In each case, the object of this interest is not merely academic. The farmer wants to know the level of fertilizer which gives the maximum yield of potatoes; the psychologist would like to know if a child's creativity score is a reliable predictor of the child's flexibility score; and the medical researcher wishes to determine if blood pressure is a good indicator of a patient's heart rate. How can we determine if one variable, x, is a reliable predictor of another variable, y? In order to answer this question, we must be able to model the bivariate relationship, that is, describe how the two variables, x and y, are related. In this chapter we present a method useful for modeling the (straight-line) relationship between two variables—a method called *simple linear regression analysis.*

12.2 Simple Linear Correlation

If two variables are related in such a way that the value of one is indicative of the value of the other, we sometimes say that the variables are ***correlated.*** For example, the claim is often made that the crime rate and the unemployment rate are "highly correlated." Another popular belief is that the GNP and the rate of inflation are "correlated." In the following example we show how to calculate a numerical descriptive measure of the correlation between two variables, x and y.

EXAMPLE 12.1 In recent years, physicians have used the so-called "diving reflex" to reduce abnormally rapid heartbeats in humans by submerging the patient's face in cold water. (The reflex, triggered by cold water temperatures, is an involuntary neural response that shuts off circulation to the skin, muscles, and internal organs, and diverts extra oxygen-carrying blood to the heart, lungs, and brain.) A research physician conducted an experiment to investigate the effects of various cold temperatures on the pulse rates of 10 small children; the results are presented in Table 12.1. Calculate a numerical descriptive measure of the correlation between temperature of water, x, and reduction in pulse rate, y.

Solution The first step in computing a measure of correlation between x and y for the $n = 10$ pairs of observations is to find the sums of the x-values (Σx) and y-values (Σy), the squares of the x-values (Σx^2), the squares of the y-values (Σy^2), and the cross products of the corresponding x- and y-values (Σxy). As an aid in finding these

TABLE 12.1
Temperature of Water–Pulse Rate Data, Example 12.1

CHILD	TEMPERATURE OF WATER x, °F	REDUCTION IN PULSE RATE y, beats/minute
1	68	2
2	65	5
3	70	1
4	62	10
5	60	9
6	55	13
7	58	10
8	65	3
9	69	4
10	63	6

quantities, construct a "sums of squares" table of the type shown in Table 12.2. Notice that the quantities Σx, Σy, Σx^2, Σy^2, and Σxy appear in the bottom row of the table.

TABLE 12.2
Sums of Squares for Data of Table 12.1

	x	y	x^2	y^2	xy
	68	2	4,624	4	136
	65	5	4,225	25	325
	70	1	4,900	1	70
	62	10	3,844	100	620
	60	9	3,600	81	540
	55	13	3,025	169	715
	58	10	3,364	100	580
	65	3	4,225	9	195
	69	4	4,761	16	276
	63	6	3,969	36	378
TOTALS	$\Sigma x = 635$	$\Sigma y = 63$	$\Sigma x^2 = 40{,}537$	$\Sigma y^2 = 541$	$\Sigma xy = 3{,}835$

The second step is to calculate the quantities SS_{xy}, SS_{xx}, and SS_{yy}, as shown below:

$$SS_{xy} = \Sigma xy - \frac{(\Sigma x)(\Sigma y)}{n} = 3{,}835 - \frac{(635)(63)}{10} = -165.5$$

$$SS_{xx} = \Sigma x^2 - \frac{(\Sigma x)^2}{n} = 40{,}537 - \frac{(635)^2}{10} = 214.5$$

$$SS_{yy} = \Sigma y^2 - \frac{(\Sigma y)^2}{n} = 541 - \frac{(63)^2}{10} = 144.1$$

Finally, compute the measure of correlation, denoted by the symbol r, as follows:

$$r = \frac{SS_{xy}}{\sqrt{SS_{xx}SS_{yy}}} = \frac{-165.5}{\sqrt{(214.5)(144.1)}} = \frac{-165.5}{175.8} = -.94$$

The formal name given to r is the ***Pearson product moment coefficient of correlation,*** and its formula is given in the box. It can be shown (proof omitted) that r is scaleless, i.e., it is not measured in dollars, pounds, etc., but always assumes a value between -1 and $+1$ regardless of the units of measurement of the variables x and y. In the examples that follow, we will learn that the correlation coefficient r is a measure of the strength of the linear relationship between x and y.

DEFINITION 12.1

The ***Pearson product moment coefficient of correlation,*** r, is computed as follows for a sample of n measurements on x and y:

$$r = \frac{SS_{xy}}{\sqrt{SS_{xx}SS_{yy}}}$$

where

$$SS_{xy} = \Sigma xy - \frac{(\Sigma x)(\Sigma y)}{n}$$

$$SS_{xx} = \Sigma x^2 - \frac{(\Sigma x)^2}{n}$$

$$SS_{yy} = \Sigma y^2 - \frac{(\Sigma y)^2}{n}$$

It is a measure of the strength of the linear relationship between two random variables x and y.

EXAMPLE 12.2 What are the implications of various possible values of the correlation coefficient r?

Solution We have suggested that the coefficient of correlation r is a measure of the strength of the linear relationship between two variables. We can gain insight into this interpretation by observing the plots of typical sample data presented in Figure 12.1. These plots, called ***scattergrams,*** are constructed by plotting the pairs of sample observations (x, y) on a piece of graph paper (x-values on the horizontal axis and y-values on the vertical axis).

Consider first the scattergram in Figure 12.1(b). The correlation coefficient for this set of points is near zero. We can see that a value of r near or equal to zero implies little or no linear relationship between x and y. That is, as x increases (or decreases), there is no definite trend in the values of y. In contrast, Figure 12.1(a) shows that positive values of r imply a positive linear relationship between y and x, i.e., y tends to increase as x increases. Similarly, a negative value of r implies a negative linear relationship between y and x, i.e., y tends to decrease as x increases (see Figure 12.1(c)).

A perfect linear relationship exists when all the (x, y) points fall exactly along a straight line, as shown in Figures 12.1(d) and 12.1(e). We can see that a value of $r = +1$ implies a perfect positive linear relationship between y and x (Figure 12.1(d)),

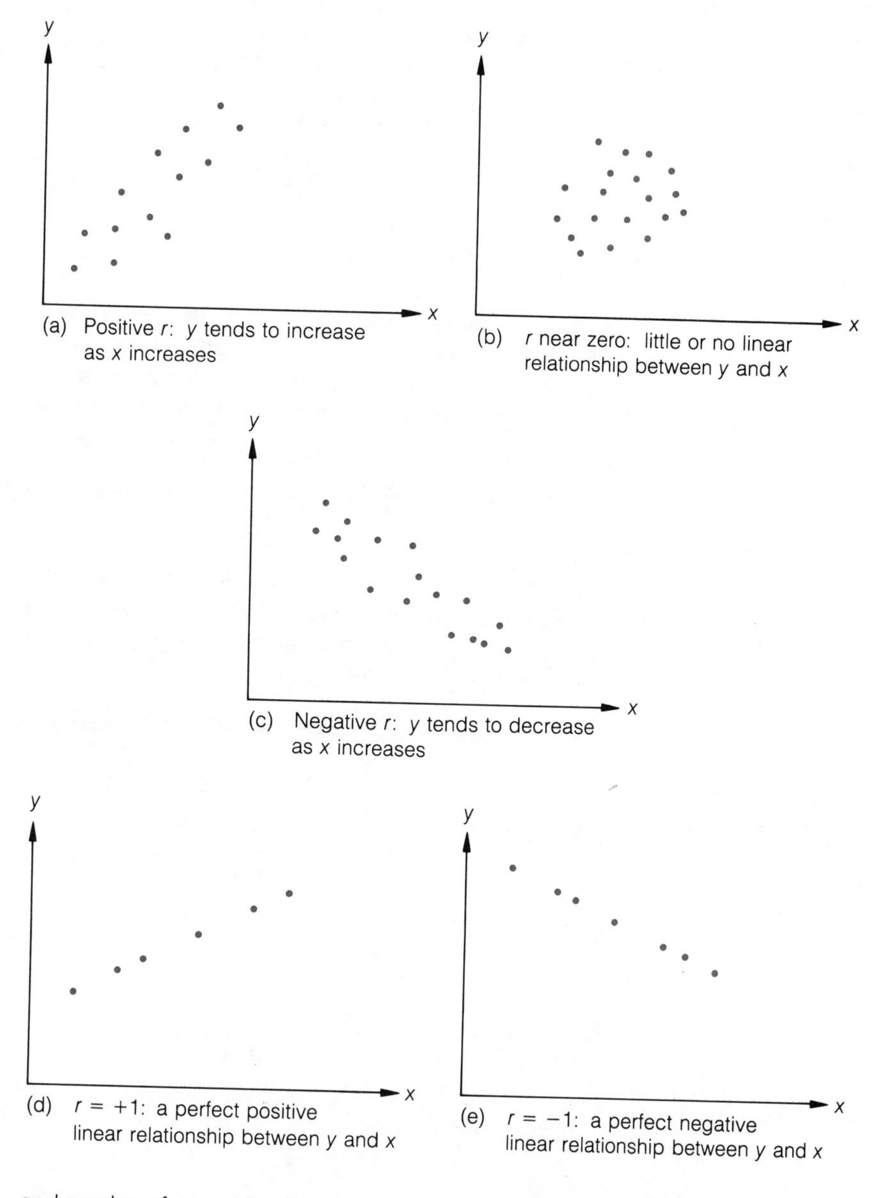
y
x
(a) Positive r: y tends to increase as x increases
(b) r near zero: little or no linear relationship between y and x
(c) Negative r: y tends to decrease as x increases
(d) r = +1: a perfect positive linear relationship between y and x
(e) r = −1: a perfect negative linear relationship between y and x

FIGURE 12.1 Values of r and Their Implications

and a value of $r = -1$ implies a perfect negative linear relationship between y and x (Figure 12.1(e)).

EXAMPLE 12.3 Interpret the value of r obtained for the temperature of water–reduction in pulse rate data of Table 12.1, Example 12.1.

Solution From Example 12.1, the coefficient of correlation was calculated as $r = -.94$. How should the research physician interpret this large negative value of r? From our previous discussion, the implication is that a strong negative linear relationship between temperature of water and reduction in pulse rate exists for the 10 sampled children, i.e., the reduction in pulse rate tends to decrease as the temperature of the water increases, *for this sample of 10 children.* We can see this nearly perfect negative linear relationship clearly in the scattergram, Figure 12.2, for the data of Table 12.1. However, the research physician should not use this result to conclude that the best way to reduce a child's abnormally rapid heartbeat is to submerge the child's face in extremely cold water; it is incorrect to assume that there is a ***causal relationship*** between the two variables. In this example, there are probably many variables that have contributed to the children's reductions in heart rate, e.g., the length of time the children are submerged, the conscious or unconscious state of the children, the children's general physiological conditions, etc. The only appropriate conclusion to be made by the research physician based on the high negative correlation in the sample data is that a negative linear trend may exist between the temperature of water, x, and the reduction in pulse rate, y.

WARNING

High correlation does not imply causality. If a large positive or negative value of the sample correlation coefficient r is observed, it is incorrect to conclude that a change in x causes a change in y. The only valid conclusion is that a linear trend ***may*** exist between x and y.

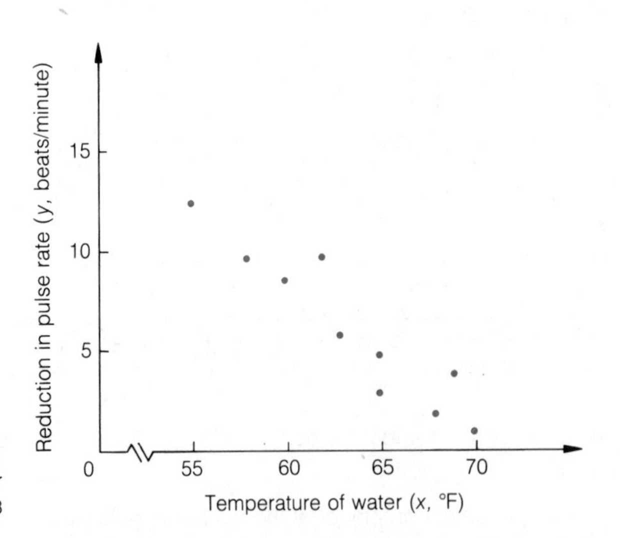

FIGURE 12.2 Scattergram for Example 12.3

We pause here to remind you that the correlation coefficient r is defined as a measure of the linear correlation between x-values and y-values ***in the sample;*** thus, r is a sample statistic. Similarly, we define the ***population correlation coefficient*** to be a measure of the linear correlation for the population from which the sample of (x, y)

data points was selected.* Our interpretation of the value of the population correlation coefficient is analogous to that of r: the population coefficient of correlation measures the strength of the linear relationship between x-values and y-values in the entire population.

EXAMPLE 12.4 Does linear correlation in a sample imply correlation in the population? That is, if the calculated value of r is nonzero, does this imply that the population correlation coefficient is nonzero?

Solution Sometimes, but not always. For example, we think you will agree that a sample correlation coefficient of $r = -.97$ provides strong evidence that the x-values and y-values in the population from which the sample was obtained are negatively linearly correlated. Likewise, a sample correlation coefficient of $r = .01$ indicates that little or no correlation between x and y exists in the population. Consider though, the value $r = .36$. Here the decision is not clear-cut. This sample value may be due to actual linear correlation in the population, or it may be due simply to random variation in the sample even though no linear correlation exists in the population. To help us decide, we employ the statistical decision-making tools of Chapters 9 and 10: an hypothesis test.

The null hypothesis which we wish to test is

H_0: There is no linear correlation between the variables x and y (i.e., the population correlation coefficient equals zero)

As our test statistic, we use the sample coefficient of correlation, r. The method outlined in Chapter 9 for determining the form of the rejection region requires that we choose an appropriate significance level α and then find the critical value based on the sampling distribution of the test statistic r. Like the distribution of the Student's t statistic, the sampling distribution of r under the null hypothesis is symmetric with a mean of 0. Table 10, Appendix E, gives critical values of r for various values of α and the sample size n. A portion of this table is reproduced in Table 12.3. Notice that the table is constructed similar to the table of critical t-values, Table 4, Appendix E. The

TABLE 12.3 Reproduction of a Portion of Table 10, Appendix E: Critical Values of the Sample Correlation Coefficient, r

SAMPLE SIZE n	$r_{.050}$	$r_{.025}$	$r_{.010}$	$r_{.005}$
3	.988	.970	.951	.988
4	.900	.950	.980	.900
5	.805	.878	.934	.959
6	.729	.811	.882	.917
7	.669	.754	.833	.875
8	.621	.707	.789	.834
9	.582	.666	.750	.798
10	.549	.632	.715	.765

*The population correlation coefficient is often represented by the greek letter ρ (rho).

critical value, r_α, has the same interpretation as t_α. That is, we define r_α to be the value such that $P(r > r_\alpha) = \alpha$. We illustrate the use of these critical values in the following example.

EXAMPLE 12.5

a. Test (at significance level $\alpha = .05$) for evidence of linear correlation between the variables x and y if the sample size (number of measurements in the sample) is $n = 8$ and the value of r computed from the sample is $r = -.83$.

b. Test (at significance level $\alpha = .05$) for evidence that the variables x and y are positively correlated if the sample size is $n = 10$ and the value of r computed from the sample is $r = .37$.

Solution

a. Since we are interested in detecting whether the variables x and y are linearly correlated (either positively or negatively), we wish to test the hypotheses

H_0: The population correlation coefficient equals zero

H_a: The population correlation coefficient is nonzero

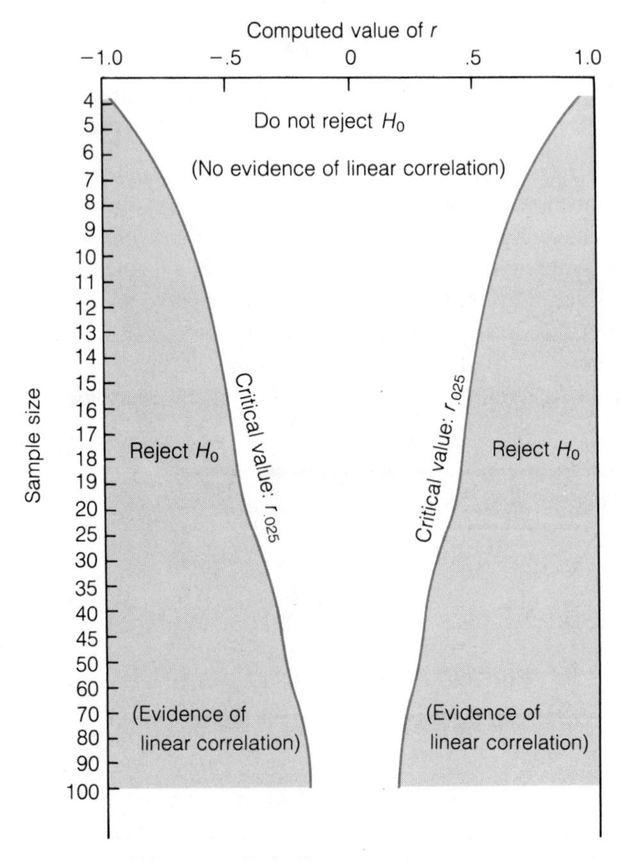

FIGURE 12.3
Rejection Region for Testing H_0: No Linear Correlation (Two-Tailed Test, $\alpha = .05$)

Hence, our test is two-sided and we must consider both large and small values of r as possible areas of rejection. For a two-tailed test conducted at significance level α, we will reject the null hypothesis of no linear correlation if the computed value of the test statistic r is either greater than $r_{\alpha/2}$ or less than $-r_{\alpha/2}$, i.e., our rejection region takes the form

$$r > r_{\alpha/2} \quad \text{or} \quad r < -r_{\alpha/2}$$

(This decision-making procedure is illustrated in Figure 12.3 for $\alpha = .05$.)

For $\alpha = .05$ and $n = 8$, the critical value of r obtained from Table 12.3 is .707 (the entry in the row corresponding to $n = 8$ and the column corresponding to $r_{\alpha/2} = r_{.025}$). Thus, we will reject H_0 if $r > .707$ or $r < -.707$. Since the computed value $r = -.83$ falls below the lower-tailed critical value of $-.707$, we reject H_0 and conclude (at the $\alpha = .05$ level of significance) that linear correlation exists between x and y. This procedure yields an $\alpha = .05$ chance of committing a Type I error, i.e., concluding that x and y are linearly correlated, if in fact no correlation exists.

b. Now we want to determine whether the two variables x and y are positively correlated. That is, we wish to test

H_0: The population coefficient of correlation equals zero

H_a: The population coefficient of correlation is positive (greater than zero)

This test is one-tailed (an upper-tailed test), and we will reject H_0 if the computed value of r is too large, specifically if $r > r_\alpha$, where r_α depends on the

---

TEST OF HYPOTHESIS FOR LINEAR CORRELATION

a. One-tailed test

H_0: There is no linear correlation between x and y

H_a: The variables x and y are positively correlated

(or H_a: The variables x and y are negatively correlated)

Test statistic: r

Rejection region:

$r > r_\alpha$ (or $r < -r_\alpha$)

b. Two-tailed test

H_0: There is no linear correlation between x and y

H_a: The variables x and y are linearly correlated

Test statistic: r

Rejection region:

$r > r_{\alpha/2}$ or $r < -r_{\alpha/2}$

where the distribution of r depends on the sample size n, and r_α and $r_{\alpha/2}$ are the critical values obtained from Table 10, Appendix E, such that

$$P(r > r_\alpha) = \alpha \quad \text{and} \quad P(r > r_{\alpha/2}) = \alpha/2$$

---

sample size n. From Table 12.3, the critical value corresponding to $n = 10$ is $r_\alpha = r_{.05} = .549$. Thus, our rejection region takes the form

$$r > r_{.05} = .549$$

Since the computed value $r = .37$ does not exceed .549, we fail to reject H_0. There is insufficient evidence (at $\alpha = .05$) to conclude that the variables x and y are positively correlated in the population.

A test for negative correlation between x and y, i.e., a test to determine whether the population correlation coefficient is negative (less than zero) is conducted similarly. Both the one-tailed and two-tailed tests for linear correlation in the population are summarized in the box on page 407.

Recall that our ultimate objective is to predict the value of one variable, y, from the value of another variable, x. While the correlation coefficient r indicates the strength of the linear relationship between the variables, it does not tell us what the exact relationship is. In Section 12.3, we introduce a model which will enable us to make predictions.

EXERCISES

12.1 The number of buyers of new automobiles in a specific month, say, February, is likely to be correlated with the prime interest rate. Will this correlation be positive or negative? Explain.

12.2 Research by law enforcement agencies has shown that the crime rate is correlated with the United States population. Would you expect the correlation to be positive or negative? Explain.

12.3 Is the demand for a product, say, hamburgers, correlated with its price? If it is, would you expect the correlation to be positive or negative? Explain.

12.4 Do you believe the grade-point average (GPA) of a college student is correlated with the student's intelligence quotient (IQ)? If so, will the correlation be positive or negative? Explain.

12.5 Give an example of two variables in your field of study that are:

a. Positively correlated **b.** Negatively correlated

12.6 Consider the five data points:

x	−1	0	1	2	3
y	−1	1	1	2.5	3.5

a. Construct a scattergram for the data. After examining the scattergram, do you think that x and y are correlated? If correlation is present, is it positive or negative?

b. Find the correlation coefficient r and interpret its value.

c. Do the data provide sufficient evidence to indicate that x and y are linearly correlated? Test using $\alpha = .05$.

12.7 Consider the seven data points:

x	-5	-3	-1	0	1	3	5
y	.8	1.1	2.5	3.1	5.0	4.7	6.2

a. Construct a scattergram for the data. After examining the scattergram, do you think that x and y are correlated? If correlation is present, is it positive or negative?

b. Find the correlation coefficient r and interpret its value.

c. Do the data provide sufficient evidence to indicate that x and y are linearly correlated? Test using $\alpha = .05$.

12.8 A company conducted a survey of its customers to investigate the relationship between the demand (in numbers of units per month) for the company's products and the price per unit. Five prices were selected for the study and two customers were randomly selected for each price and asked to estimate their monthly purchase rate. The data are shown in the table.

PRICE PER UNIT x, dollars	PURCHASE RATE y, units per month
450	32
450	36
475	34
475	28
500	26
500	23
525	25
525	17
550	12
550	16

a. Construct a scattergram for the data. After examining the scattergram, do you think that x and y are correlated? If correlation is present, is it positive or negative?

b. Find the correlation coefficient r and interpret its value.

c. Do the data provide sufficient evidence to indicate that x and y are linearly correlated? Test using $\alpha = .05$.

12.9 A breeder of thoroughbred horses wished to determine the correlation between the gestation period and the length of life of a horse. The breeder obtained information (recorded in the table on page 410) on the gestation period and length of life of seven horses from various thoroughbred stables across the state.

a. Construct a scattergram for the data. After examining the scattergram, do you think that x and y are correlated? If correlation is present, is it positive or negative?

b. Find the correlation coefficient r and interpret its value.

HORSE	LENGTH OF LIFE y, years	GESTATION PERIOD x, days
1	24	416
2	25.5	279
3	20	298
4	21.5	307
5	22	356
6	23.5	403
7	21	265

c. Do the data provide sufficient evidence to indicate that x and y are linearly correlated? Test using $\alpha = .05$.

12.10 Is the maximal oxygen uptake, a measure often used by physiologists to indicate an individual's cardiovascular fitness, related to the performance of distance runners? Six long-distance runners submitted to treadmill tests for determination of their maximal oxygen uptake. These results, along with each runner's best time for the mile run, are shown in the table.

ATHLETE	MAXIMAL OXYGEN UPTAKE y, milliliters/kilogram	MILE TIME x, seconds
1	63.3	241.5
2	60.1	249.8
3	53.6	246.1
4	58.8	232.4
5	67.5	237.2
6	62.6	238.4

a. Construct a scattergram for the data. After examining the scattergram, do you think that x and y are correlated? If correlation is present, is it positive or negative?

b. Find the correlation coefficient r and interpret its value.

c. Do the data provide sufficient evidence to indicate that x and y are linearly correlated? Test using $\alpha = .05$.

12.11 The manager of a clothing store decided to investigate the relationship between the number of sales clerks on duty and the amount (in hundreds of dollars) of merchandise lost (called *shrinkage*) due to shoplifting or other causes. The dollar volume lost due to shrinkage was recorded for each of twelve weeks. The number of sales clerks was held constant within a given week but was varied from one week to another. The data are shown in the table on page 411.

a. Would you expect the coefficient of correlation between shrinkage and number of sales clerks on duty to be positive or negative? Explain.

b. Find the coefficient of correlation, r, and interpret its value.

c. Do the data provide sufficient evidence to indicate a linear correlation between x and y? Test using $\alpha = .05$.

WEEK	SHRINKAGE y, hundreds of dollars	NUMBER OF SALES CLERKS x
1	20	20
2	23	21
3	18	22
4	17	23
5	17	24
6	12	25
7	17	25
8	26	20
9	21	22
10	11	24
11	10	26
12	4	28

12.3 Straight-Line Probabilistic Models

Consider the Federal Trade Commission's tar, nicotine, and carbon monoxide rankings of American cigarette brands given in Table 2.6, Case Study 2.3. Suppose you wish to model the relationship between the carbon monoxide ranking and the nicotine content of the cigarette brands.

EXAMPLE 12.6 Do you believe that an exact relationship exists between the two variables, Carbon monoxide ranking and Nicotine content? That is, would it be possible to state the exact carbon monoxide ranking of an American cigarette brand if you knew the nicotine content of the brand?

Solution In reality, the answer is a very definite no! The amount of carbon monoxide in the smoke of any given cigarette will depend not only on the nicotine content of the brand but also on such variables as the tar content, length, filter-type, light-type, and menthol flavor of the cigarette. You can probably think of additional variables which play an important role in determining the FTC's carbon monoxide ranking of a cigarette. How can we construct a model, then, for two variables for which no exact relationship exists? We illustrate with an example.

EXAMPLE 12.7 Given in Table 12.4 (on page 412) are the nicotine content and carbon monoxide ranking of a sample of five different cigarette brands obtained from Table 2.6. Hypothesize a reasonable model for the relationship between carbon monoxide (CO) ranking and nicotine content.

Solution Upon examination of the data, we see that y, the carbon monoxide ranking of a cigarette brand, is approximately 15 times x, the brand's nicotine content. In Figure 12.4 we have plotted the actual CO rankings against the respective nicotine contents in a scattergram. The graph of the straight line $y = 15x$ has been superimposed on the figure. Notice that CO ranking appears to vary randomly about the straight line

TABLE 12.4
Carbon Monoxide–Nicotine Data, Example 12.7

BRAND	NICOTINE CONTENT x, milligrams	CO RANKING y, milligrams
Multi-filter	.78	12.5
Belair	.82	11.3
Salem	1.46	20.0
Kool	1.20	18.6
Carlton	.13	1.9

$y = 15x$. A reasonable model, then, is one which will allow for unexplained variation in CO ranking caused by important variables not included in the model (such as those discussed in Example 12.6) or simply by random phenomena which cannot be modeled or explained. Models that account for this ***random error*** are called ***probabilistic models.***

The probabilistic model relating the CO ranking y to the nicotine content x of a cigarette is written

$$y = 15x + \text{Random error}$$

We note that probabilistic models include two components: a ***deterministic component*** and a ***random error component.*** For this model, the deterministic component is $15x$. If y could always be determined exactly when x is known, then a deterministic relationship, such as $y = 15x$, would hold true. By including a random error component in our model, we allow for the random variation of CO ranking shown in Figure 12.4.

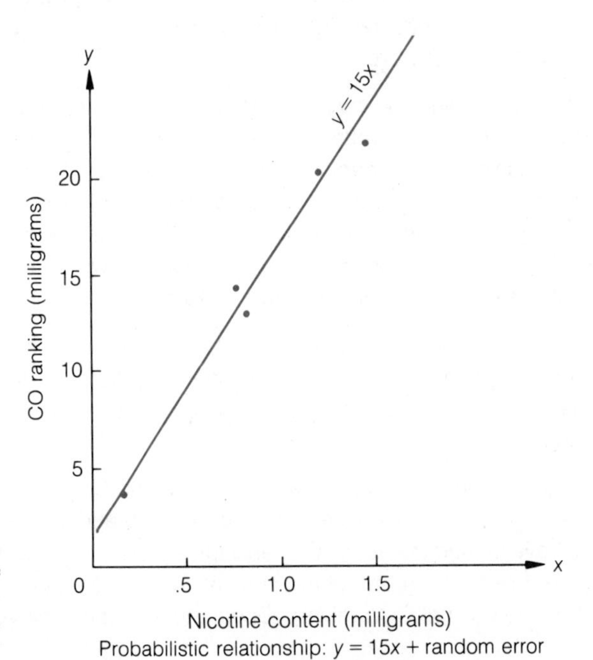

FIGURE 12.4
Scattergram of CO Ranking Versus Nicotine Content, Example 12.7

GENERAL FORM OF PROBABILISTIC MODELS

y = Deterministic component + Random error

where y is the variable to be predicted.

Assumption: The mean value of the random error equals 0. This is equivalent to assuming that the mean value of y, $E(y)$, equals the deterministic component of the model, i.e.,

$$E(y) = \text{Deterministic component}$$

In this chapter, we consider only the simplest of probabilistic models—***the straight-line model***—which derives its name from the fact that the deterministic portion of the model graphs as a straight line. The elements of the straight-line model are summarized in the box.

THE STRAIGHT-LINE PROBABILISTIC MODEL

$$y = \beta_0 + \beta_1 x + \text{Random error}$$

where y = Variable to be predicted, called the ***dependent variable***

x = Variable to be used as a predictor of y, called the ***independent variable***

$E(y) = \beta_0 + \beta_1 x$ is the deterministic portion of the model (the equation of a straight line)

β_0 (beta zero) = y-intercept of the line, i.e., the point at which the line intercepts or cuts through the y-axis (see Figure 12.5)

β_1 (beta one) = Slope of the line, i.e., amount of increase (or decrease) in the deterministic component of y for every 1 unit increase in x (see Figure 12.5)

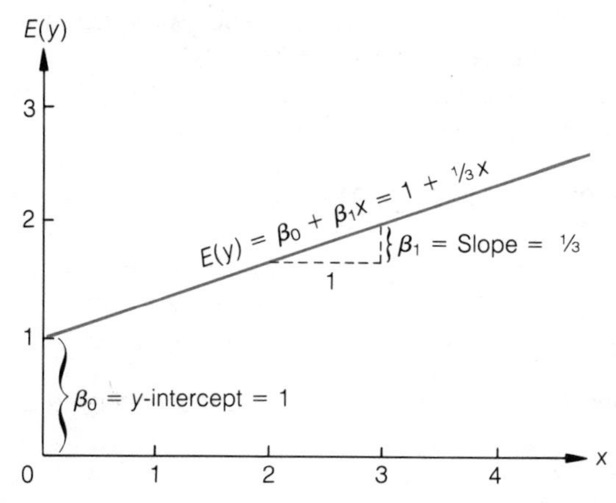

FIGURE 12.5
The Straight-Line Model

We use the Greek symbols, β_0 and β_1, to represent the y-intercept and the slope of the model, respectively. They are population parameters which will be known only if we have access to the entire population of (x, y) measurements. In the section that follows, we will use the sample data to estimate the slope (β_1) and the y-intercept (β_0) of the deterministic portion of our straight-line model.

EXERCISES

12.12 Suppose that y is exactly related to x by the equation

$$y = 1.5 + 2x$$

a. Find the value of y that corresponds to $x = 1$.
b. Find the value of y that corresponds to $x = 2$.
c. Plot the two (x, y) points found in parts a and b on graph paper and draw a line through the points. This line corresponds to the equation, $y = 1.5 + 2x$.
d. Find the value of y that corresponds to $x = 1.5$. Plot this point on the graph, part c, and confirm that it falls on the line that passes through the points found in parts a and b.
e. Part d illustrates an important relationship between graphs and equations. All the points that satisfy the equation, $y = 1.5 + 2x$, possess a common property. What is it?

12.13 Refer to Exercise 12.12.
a. Find the y-intercept for the line and interpret its value.
b. Find the slope of the line and interpret its value.
c. If you increase x by one unit, how much will y increase or decrease?
d. If you decrease x by one unit, how much will y increase or decrease?
e. What is the value of y when $x = 0$?

12.14 Answer the questions contained in Exercise 12.12, using the line, $y = 1.5 - 2x$.

12.15 Refer to Exercise 12.14.
a. Find the y-intercept for the line and interpret its value.
b. Find the slope of the line and interpret its value.
c. If you increase x by one unit, how much will y increase or decrease?
d. What is the value of y when $x = 0$?
e. What do the two lines, Exercises 12.12 and 12.14, have in common? How do they differ?

12.16 Graph the lines corresponding to the following equations:
a. $y = 1 + 3x$
b. $y = 1 - 3x$
c. $y = -1 + (1/2)x$
d. $y = -1 - 3x$
e. $y = 2 - (1/2)x$
f. $y = -1.5 + x$
g. $y = 3x$
h. $y = -2x$

12.17 Give the values of β_0 and β_1 corresponding to each of the lines, Exercise 12.16.

12.4 How to Fit the Model: The Least Squares Approach

The following example illustrates the technique we will use to ***fit*** the straight-line model to the data, i.e., to estimate the slope and y-intercept of the line using information provided by the sample data.

EXAMPLE 12.8 We return to modeling the relationship between the CO ranking, y, and the nicotine content, x, of an American-made cigarette (as determined by the Federal Trade Commission). We have hypothesized the deterministic component of the probabilistic model as

$$E(y) = \beta_0 + \beta_1 x$$

If we were able to obtain the nicotine content and CO ranking of all American-made cigarettes, i.e., the entire population of (x, y) measurements, then the values of the population parameters β_0 and β_1 could be determined exactly. Of course, we will never have access to the entire population of (x, y) measurements, since the FTC tested only one cigarette of each of the 187 American-made brands. The problem, then, is to estimate the unknown population parameters based upon the information contained in a sample of (x, y) measurements. Suppose that we randomly sample five cigarettes that were tested by the FTC. The CO rankings and nicotine content values are given in Table 12.5. (The numbers are simplified for computational ease.) How can we best use the sample information to estimate the unknown y-intercept β_0 and the slope β_1?

TABLE 12.5 CO Ranking–Nicotine Content Data, Example 12.8

CIGARETTE	NICOTINE CONTENT x, milligrams	CO RANKING y, milligrams
1	.2	2
2	.4	10
3	.6	13
4	.8	15
5	1.0	20

Solution Estimates of the unknown parameters β_0 and β_1 are obtained by finding the best-fitting straight line through the sample data points of Table 12.5. (These points are plotted in Figure 12.6.) The procedure we will use to find the best fit is known as the ***method of least squares,*** and the best-fitting line, called the ***least squares line,*** is written

$$\hat{y} = \hat{\beta}_0 + \hat{\beta}_1 x$$

The first step in finding the least squares line is to construct a sums of squares table in order to find the sums of the x-values (Σx), y-values (Σy), the squares of the x-values (Σx^2), and the cross products of the corresponding x- and y-values (Σxy). The sums of squares table for the CO ranking–nicotine content data is given in Table 12.6.

TABLE 12.6
Sums of Squares for Data of Table 12.5

	x	y	x^2	xy
	.2	2	.04	.4
	.4	10	.16	4.0
	.6	13	.36	7.8
	.8	15	.64	12.0
	1.0	20	1.00	20.0
TOTALS	$\Sigma x = 3.0$	$\Sigma y = 60$	$\Sigma x^2 = 2.20$	$\Sigma xy = 44.2$

The second step is to substitute the values of Σx, Σy, Σx^2, and Σxy into the formulas for SS_{xy} and SS_{xx} given in Section 12.2:

$$SS_{xy} = \Sigma xy - \frac{(\Sigma x)(\Sigma y)}{n} = 44.2 - \frac{(3.0)(60)}{5}$$

$$= 44.2 - 36.0 = 8.2$$

$$SS_{xx} = \Sigma x^2 - \frac{(\Sigma x)^2}{n} = 2.20 - \frac{(3.0)^2}{5}$$

$$= 2.20 - 1.80 = .4$$

Next, use these values of SS_{xy} and SS_{xx} to compute the estimate $\hat{\beta}_1$, as shown in the box.

SLOPE OF THE LEAST SQUARES LINE

$$\hat{\beta}_1 = \frac{SS_{xy}}{SS_{xx}}$$

Substituting, we find $\hat{\beta}_1$, the slope of the least squares line, to be

$$\hat{\beta}_1 = \frac{SS_{xy}}{SS_{xx}} = \frac{8.2}{.4} = 20.5$$

Finally, calculate $\hat{\beta}_0$, the y-intercept of the least squares line, as follows:

***y*-INTERCEPT OF THE LEAST SQUARES LINE**

$$\hat{\beta}_0 = \bar{y} - \hat{\beta}_1\bar{x}$$

$$\hat{\beta}_0 = \bar{y} - \hat{\beta}_1\bar{x} = \frac{\Sigma y}{5} - \hat{\beta}_1\left(\frac{\Sigma x}{5}\right)$$

$$= \frac{60}{5} - (20.5)\left(\frac{3.0}{5}\right)$$

$$= 12 - (20.5)(.6) = 12 - 12.3 = -.3$$

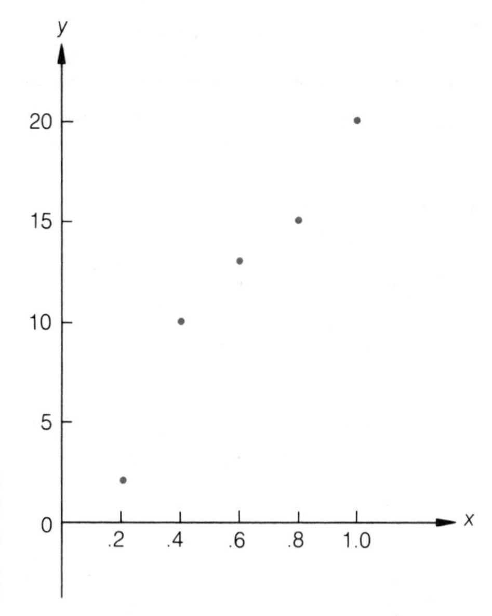

FIGURE 12.6 Scattergram of CO Ranking–Nicotine Content Data, Table 12.5

Therefore, $\hat{\beta}_0 = -.3$, $\hat{\beta}_1 = 20.5$, and the least squares line is

$$\hat{y} = -.3 + (20.5)x$$

A graph of this line is shown in Figure 12.7.

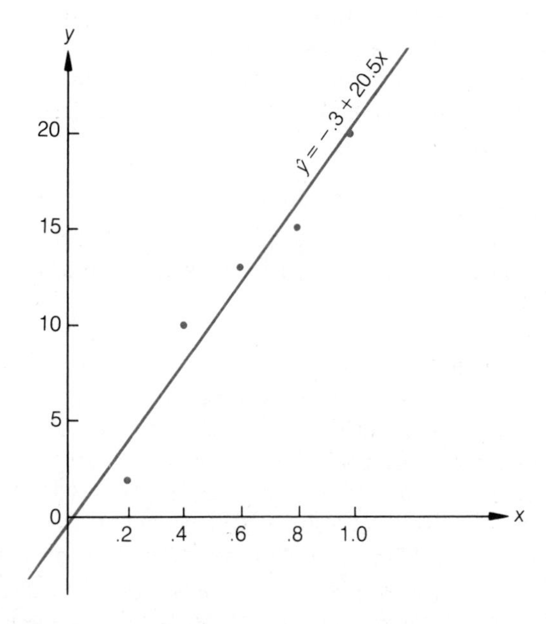

FIGURE 12.7 Least Squares Line, Example 12.8

The four-step least squares procedure is summarized in the box.

STEPS TO FOLLOW IN FITTING A LEAST SQUARES LINE TO A SET OF DATA

1. Construct a table (similar to Table 12.6) to find Σx, Σy, Σx^2, and Σxy.
2. Substitute the values into the formulas for SS_{xy} and SS_{xx}:

$$SS_{xy} = \Sigma xy - \frac{(\Sigma x)(\Sigma y)}{n}$$

$$SS_{xx} = \Sigma x^2 - \frac{(\Sigma x)^2}{n}$$

   where n = sample size (number of pairs of observations).
3. Substitute the values into the formula for $\hat{\beta}_1$. Then find $\hat{\beta}_0$.

$$\hat{\beta}_1 = \frac{SS_{xy}}{SS_{xx}}$$

$$\hat{\beta}_0 = \bar{y} - \hat{\beta}_1\bar{x}$$

4. Use the computed values of $\hat{\beta}_0$ and $\hat{\beta}_1$ to form the equation of the least squares line, i.e.,

$$\hat{y} = \hat{\beta}_0 + \hat{\beta}_1 x$$

EXAMPLE 12.9 In what sense is the least squares line the "best-fitting" straight line to a set of data?

Solution In deciding whether a line provides a good fit to a set of data, we examine the vertical distances, or ***deviations,*** between the data points and the fitted line. (Since we are attempting to predict y, a measure of fit will involve the difference between the observed value y and the predicted value $\hat{y}$—a quantity which is represented by the vertical deviation between the data point and the fitted line.) The deviations for the least squares line

$$\hat{y} = -.3 + 20.5x$$

are shown in Figure 12.8(a). Let us compare the deviations of the least squares line with the deviations of another fitted line (one fitted visually), given by the equation

$$\hat{y} = -2.5 + 22.5x$$

The deviations of the visually fitted line are shown in Figure 12.8(b). Notice first that some of the deviations are positive, some are negative, and that even though two of the five data points fall exactly on the visually fitted line, the individual deviations tend to be smaller for the least squares line than for the visually fitted line. Second, note that the sum of squares of deviations ($SSE = \Sigma(y - \hat{y})^2$) is smaller for the least squares line. (The values of SSE for the least squares and visually fitted lines are given at the bottom of Figures 12.8(a) and 12.8(b), respectively.) *In fact, it can be shown that there is one and only one line which will minimize the sum of squares of deviations of the points about the fitted line. It is the least squares line.*

FIGURE 12.8 Deviations about the Fitted Line, Example 12.9

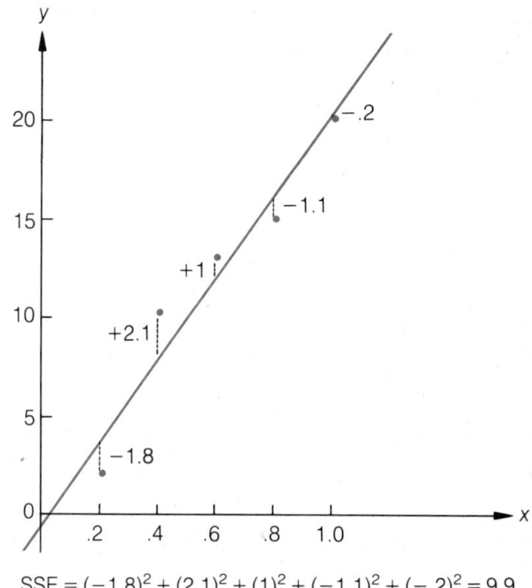

(a) Least squares line: $\hat{y} = -.3 + 20.5x$

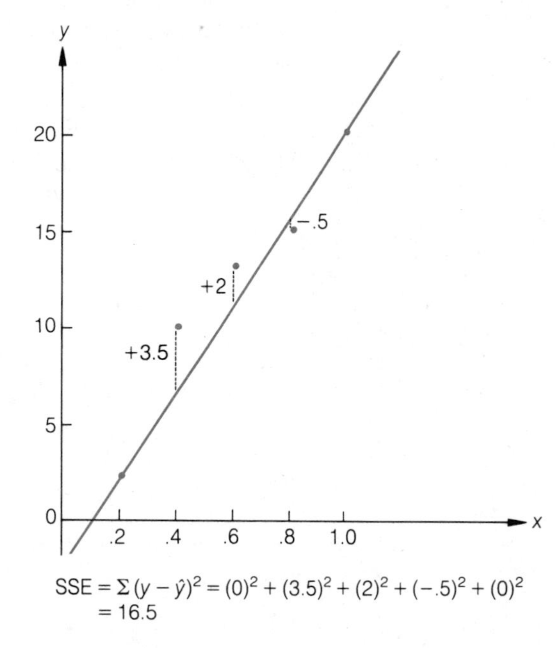

(b) Visually fitted line: $\hat{y} = -2.5 + 22.5x$

LEAST SQUARES CRITERION FOR FINDING THE "BEST-FITTING" LINE

Choose the line that minimizes the sum of squared deviations, SSE. This is called the ***least squares line,*** or alternatively, the ***least squares prediction equation.***

The fact that the least squares line is the one that minimizes the sum of squares of deviations does not guarantee that it is the "best" line to fit the data. However, intuitively, it would seem that this is a desirable property for a good fitting line.* A second reason for liking the method of least squares is that we know the sampling distributions of the estimates of β_0 and β_1, something that would be unknown for lines fitted intuitively or visually. A third and final reason is that, under certain conditions, the least squares estimators of β_0 and β_1 will have smaller standard errors than other types of estimators.

EXAMPLE 12.10 A psychologist is interested in modeling the relationship between the grade-point average (GPA) at the end of the freshman year and the college entrance exam score

*The sum of the deviations from the least squares line will also always equal 0. Since there are many other fitted lines which also have this property, we do not use this as a criterion for choosing the "best-fitting" line.

for college freshmen. The psychologist believes that the relationship between the two variables can best be described by a straight line. The sample data shown in Table 12.7 were supplied by the university's office of the registrar.

TABLE 12.7
GPA–Entrance Exam Data, Example 12.10

FRESHMAN	GRADE-POINT AVERAGE y	COLLEGE ENTRANCE EXAM SCORE x
1	3.0	1000
2	3.2	1300
3	2.1	800
4	3.5	1200
5	2.5	900

Use the method of least squares to estimate the y-intercept and slope of the line. According to the least squares line, approximately what GPA should a college student expect to earn at the end of his freshman year if he scored 1150 on the college entrance exam?

Solution The straight-line model describing the relationship between GPA and college entrance exam score is

$$E(y) = \beta_0 + \beta_1 x$$

To use the method of least squares to estimate β_0 and β_1, we must first (step 1) make the preliminary computations shown in Table 12.8.

TABLE 12.8
Preliminary Computations, Example 12.10

	x	y	x^2	xy
	1000	3.0	1,000,000	3,000
	1300	3.2	1,690,000	4,160
	800	2.1	640,000	1,680
	1200	3.5	1,440,000	4,200
	900	2.5	810,000	2,250
TOTALS	$\Sigma x = 5200$	$\Sigma y = 14.3$	$\Sigma x^2 = 5{,}580{,}000$	$\Sigma xy = 15{,}290$

We now calculate (step 2)

$$SS_{xy} = 15{,}290 - \frac{(5200)(14.3)}{5} = 15{,}290 - 14{,}872 = 418$$

$$SS_{xx} = 5{,}580{,}000 - \frac{(5200)^2}{5} = 5{,}580{,}000 - 5{,}408{,}000 = 172{,}000$$

The slope of the least squares line is then (step 3)

$$\hat{\beta}_1 = \frac{SS_{xy}}{SS_{xx}} = \frac{418}{172{,}000} = .00243$$

and the y-intercept is

$$\hat{\beta}_0 = \bar{y} - \hat{\beta}_1(\bar{x}) = \frac{14.3}{5} - (.00243)\frac{(5200)}{5} = 2.86 - 2.53 = .33$$

The least squares line can then be written (step 4)

$$\hat{y} = .33 + (.00243)x$$

To answer the second question, we need to obtain the predicted value for y, the GPA for a student at the end of his freshman year, if he scored $x = 1150$ on the college entrance exam. Substituting $x = 1150$ into the least squares equation, we obtain

$$\hat{y} = .33 + (.00243)(1150) = .33 + 2.79 = 3.12$$

On the average then, a student can expect to earn a GPA of 3.12 at the end of his freshman year if he scored 1150 on the college entrance exam. We will obtain a measure of reliability for a prediction such as this in Section 12.8.

EXERCISES

12.18 The data for Exercise 12.6 are reproduced below:

x	−1	0	1	2	3
y	−1	1	1	2.5	3.5

a. Construct a scattergram for the data.
b. Find the least squares prediction equation.
c. Graph the least squares line on the scattergram and visually confirm that it provides a good fit to the data points.

12.19 Consider the four data points:

x	1	1.5	1.9	2.5
y	3.1	2.2	1.0	.3

a. Construct a scattergram for the data.
b. Find the least squares prediction equation.
c. Graph the least squares line on the scattergram and visually confirm that it provides a good fit to the data points.

12.20 The data for Exercise 12.7 are reproduced below:

x	−5	−3	−1	0	1	3	5
y	.8	1.1	2.5	3.1	5.0	4.7	6.2

a. Construct a scattergram for the data.
b. Find the least squares prediction equation.
c. Graph the least squares line on the scattergram and visually confirm that it provides a good fit to the data points.

12.21 Consider the four data points in the table on page 422.
a. Construct a scattergram for the data.
b. Find the least squares prediction equation.

x	−3.0	2.4	−1.1	2.0
y	2.7	.4	1.3	.5

c. Graph the least squares line on the scattergram and visually confirm that it provides a good fit to the data points.

12.22 The data for Exercise 12.8 are reproduced in the table.

PRICE PER UNIT x, dollars	PURCHASE RATE y, units per month
450	32
450	36
475	34
475	28
500	26
500	23
525	25
525	17
550	12
550	16

a. Construct a scattergram for the data.
b. Find the least squares prediction equation.
c. Graph the least squares line on the scattergram.
d. Use the least squares prediction equation to predict the purchase rate y when the price per unit is $x = \$480$. [*Note:* We will find a measure of the reliability of this prediction in Section 12.8.]

12.23 The data for Exercise 12.9 are reproduced in the table.

HORSE	LENGTH OF LIFE y, years	GESTATION PERIOD x, days
1	24	416
2	25.5	279
3	20	298
4	21.5	307
5	22	356
6	23.5	403
7	21	265

a. Construct a scattergram for the data.
b. Find the least squares prediction equation.
c. Graph the least squares line on the scattergram.

d. Use the least squares prediction equation to predict the length of life y of a horse if the horse's gestation period was $x = 320$ days. [*Note:* We will find a measure of reliability of this prediction in Section 12.8.]

12.24 The data for Exercise 12.10 are reproduced in the table.

ATHLETE	MAXIMAL OXYGEN UPTAKE y, milliliters/kilogram	MILE TIME x, seconds
1	63.3	241.5
2	60.1	249.8
3	53.6	246.1
4	58.8	232.4
5	67.5	237.2
6	62.6	238.4

a. Construct a scattergram for the data.
b. Find the least squares prediction equation.
c. Graph the least squares line on the scattergram.
d. Use the least squares prediction equation to predict an athlete's maximal oxygen uptake y when the athlete's best time in the mile run is $x = 240$ seconds. [*Note:* We will find a measure of reliability of this prediction in Section 12.8.]

12.25 The data for Exercise 12.11 are reproduced in the table.

WEEK	SHRINKAGE y, hundreds of dollars	NUMBER OF SALES CLERKS x
1	20	20
2	23	21
3	18	22
4	17	23
5	17	24
6	12	25
7	17	25
8	26	20
9	21	22
10	11	24
11	10	26
12	4	28

a. Construct a scattergram for the data.
b. Find the least squares prediction equation.
c. Graph the least squares line on the scattergram.
d. Use the least squares prediction equation to estimate the mean weekly shrinkage y when $x = 23$ clerks are on duty. [*Note:* We will find a measure of reliability for this estimate in Section 12.8.]

12.5 Estimating σ^2

Is the carbon monoxide ranking y of Example 12.8 really related to a cigarette's nicotine content x, or is the linear relation that we seem to see a figment of our imagination? That is, could it be the case that x and y are completely unrelated, and that the apparent linear configuration of the data points in the scattergram, Figure 12.6, is due to random variation? We could obtain an answer to this question by testing for the existence of correlation in the population as outlined in Section 12.2. A second method, one which will also enable us to predict the value of y from a given x and attach a measure of reliability to our predictions, requires that we know how much y will vary for a given value of x. That is, we need to know the value of the quantity, called σ^2, which measures the variability of the y-values about the least squares line. Since the variance σ^2 will rarely be known, we will estimate its value using the sum of the squared deviations (sum of the squared errors, SSE) and the procedure shown in the box.

ESTIMATION OF σ^2, A MEASURE OF THE VARIABILITY OF THE y-VALUES ABOUT THE LEAST SQUARES LINE

An estimate of σ^2 is given by

$$s^2 = \frac{\text{SSE}}{(n-2)}$$

where

$$\text{SSE} = \Sigma(y - \hat{y})^2 = \text{SS}_{yy} - \hat{\beta}_1 \text{SS}_{xy}$$

$$\text{SS}_{yy} = \Sigma(y - \bar{y})^2 = \Sigma y^2 - \frac{(\Sigma y)^2}{n}$$

Warning: When performing these calculations, you may be tempted to round the calculated values of SS_{yy}, $\hat{\beta}_1$, and SS_{xy}. We recommend carrying at least six significant figures for each of these quantities to avoid substantial rounding errors in the calculation of SSE.

[*Note:* The denominator of s^2 is termed *the number of degrees of freedom for error variance estimation.*]

EXAMPLE 12.11 Refer to Example 12.8. Estimate the value of σ^2 for the data of Table 12.5.

Solution According to the formulas given in the box, the first step is to compute SS_{yy}. We have

$$\text{SS}_{yy} = \Sigma y^2 - \frac{(\Sigma y)^2}{n}$$

$$= (2)^2 + (10)^2 + (13)^2 + (15)^2 + (20)^2 - \frac{(60)^2}{5}$$

$$= 898 - 720 = 178$$

Recall from Example 12.8 that $\hat{\beta}_1 = 20.5$ and $SS_{xy} = 8.2$. Thus, we compute

$$\begin{aligned} SSE &= SS_{yy} - \hat{\beta}_1 SS_{xy} \\ &= 178 - (20.5)(8.2) = 178 - 168.1 = 9.9 \end{aligned}$$

Notice that this value of SSE agrees with the value previously given in Figure 12.8(a). Our estimate of σ^2 is therefore

$$s^2 = \frac{SSE}{n-2} = \frac{9.9}{5-2} = \frac{9.9}{3} = 3.3$$

We could also compute the estimated standard deviation s by taking the square root of s^2. In this example, we have

$$s = \sqrt{s^2} = \sqrt{3.3} = 1.82$$

Since s measures the spread of the distribution of the y-values about the least squares line, we should not be surprised to find that most of the observations lie within $2s$ or $2(1.82) = 3.64$ of the least squares line. From Figure 12.9 we see that, for this example, all five data points have y-values which lie within $2s$ of $\hat{y}$, the least squares predicted value.

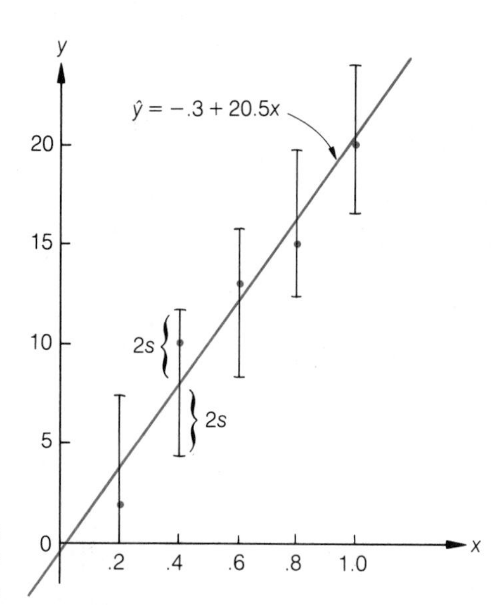

FIGURE 12.9 Observations within $2s$ of the Least Squares Line

12.6 Making Inferences about the Slope β_1

After fitting the model to the data and computing an estimate of σ^2, our next task is to statistically check the usefulness of the model. That is, use a statistical procedure (a test of hypothesis or confidence interval) to determine whether the least squares straight-line (linear) model is a reliable tool for predicting y for a given value of x.

EXAMPLE 12.12 Consider the probabilistic model

$$y = \beta_0 + \beta_1 x + \text{Random error}$$

How do we determine, statistically, whether this model is useful for prediction purposes, i.e., how could we test whether x provides useful information for the prediction of y?

Solution Suppose that x is ***completely unrelated*** to y. What could we say about the values of β_0 and β_1 in the probabilistic model if in fact x contributes no information for the prediction of y? We think you will agree that for y to be independent of x, the true slope of the line, β_1, must be equal to zero. Therefore, to test the null hypothesis that x contributes no information for the prediction of y against the alternative that these variables are linearly related with a slope differing from zero, we test

$$H_0: \quad \beta_1 = 0$$
$$H_a: \quad \beta_1 \neq 0$$

If the data support the alternative hypothesis, we will conclude that x does contribute information for the prediction of y using the straight-line model (although the true relationship between $E(y)$ and x could be more complex than a straight line). [*Note:* This test is equivalent to the test for correlation discussed in Section 12.2. If x and y are in fact linearly correlated, then the slope β_1 of the straight-line model will be different from 0.]

Using the hypothesis-testing techniques developed in Chapters 9 and 10, we set up the test for the predictive ability of the model as shown in the box.

TEST OF HYPOTHESIS FOR DETERMINING WHETHER THE STRAIGHT-LINE MODEL IS USEFUL FOR PREDICTING y FROM x

a. One-tailed test

$H_0: \quad \beta_1 = 0$

$H_a: \quad \beta_1 > 0$

(or $H_a: \quad \beta_1 < 0$)

Test statistic:

$$t = \frac{\hat{\beta}_1}{s/\sqrt{SS_{xx}}}$$

Rejection region:

$t > t_\alpha$ (or $t < -t_\alpha$)

b. Two-tailed test

$H_0: \quad \beta_1 = 0$

$H_a: \quad \beta_1 \neq 0$

Test statistic:

$$t = \frac{\hat{\beta}_1}{s/\sqrt{SS_{xx}}}$$

Rejection region:

$t < -t_{\alpha/2}$ or $t > t_{\alpha/2}$

where the distribution of t is based on $(n - 2)$ degrees of freedom, t_α is the t-value such that $P(t > t_\alpha) = \alpha$, and $t_{\alpha/2}$ is the t-value such that $P(t > t_{\alpha/2}) = \alpha/2$.

Assumptions: See Section 12.10

[*Note:* The test statistic is derived from the sampling distribution of the least squares estimator of the slope, $\hat{\beta}_1$.]

Inferences based upon this hypothesis test require certain assumptions about the random error term (see Section 12.10). However, the test statistic has a sampling distribution that remains relatively stable for minor departures from the assumptions. That is, our inferences remain valid for practical cases in which the assumptions are nearly, but not completely, satisfied.

We illustrate the test for determining whether the model is useful for predicting y from x with a practical example.

EXAMPLE 12.13 Let us return to the nicotine content–carbon monoxide ranking problem of Example 12.8. At significance level $\alpha = .05$, test the hypothesis that the nicotine content x of a cigarette contributes useful information for the prediction of carbon monoxide ranking y, i.e., test the predictive ability of the least squares straight-line model

$$\hat{y} = -.3 + 20.5x$$

Solution Testing the usefulness of the model requires testing the hypotheses

$$H_0: \quad \beta_1 = 0$$

$$H_a: \quad \beta_1 \neq 0$$

With $n = 5$ and $\alpha = .05$, the critical value based on $(5 - 2) = 3$ df is obtained from Table 4, Appendix E:

$$t_{\alpha/2} = t_{.025} = 3.182$$

Thus, we will reject H_0 if $t < -3.182$ or $t > 3.182$.

In order to compute the test statistic, we need the values of $\hat{\beta}_1$, s, and SS_{xx}. In previous examples, we computed $\hat{\beta}_1 = 20.5$, $s = 1.82$, and $SS_{xx} = .4$. Hence, our test statistic is

$$t = \frac{\hat{\beta}_1}{s/\sqrt{SS_{xx}}} = \frac{20.5}{1.82/\sqrt{.4}} = 7.12$$

Since this calculated t-value falls in the upper-tail rejection region (see Figure 12.10), we reject the null hypothesis and conclude that the slope β_1 is not zero. At the $\alpha = .05$ level of significance then, the sample data provide sufficient evidence to conclude that nicotine content does contribute useful information for prediction of carbon monoxide ranking via the linear model.

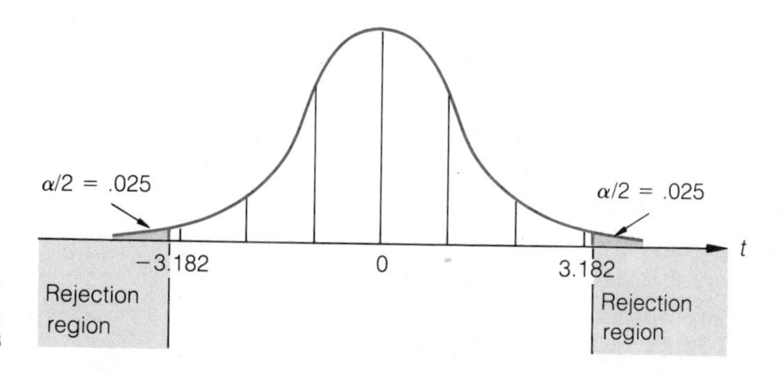

FIGURE 12.10 Rejection Region, Example 12.13

If the test statistic had not fallen in the rejection region, would we have concluded that $\beta_1 = 0$? The answer to this question is "no" (recall the discussion in Chapter 9). Rather, we acknowledge that additional data might indicate that β_1 differs from zero, or that a more complex relationship (other than a straight line) may exist between y and x. We may also wish to examine the attained significance level of the test.

EXAMPLE 12.14 A consumer investigator obtained the following least squares straight-line model relating the yearly food cost y for a family of four and annual income x:

$$\hat{y} = 467 + .26x$$

based on a sample of $n = 100$ families. In addition, the investigator computed the quantities $s = 1.1$ and $SS_{xx} = 26$. Compute the attained significance level (p-value) for a test to determine whether mean yearly food cost y increases as annual income x increases, i.e., whether the slope of the line, β_1, is positive. Interpret this value.

Solution The consumer investigator wishes to test

$$H_0\colon \quad \beta_1 = 0$$

$$H_a\colon \quad \beta_1 > 0$$

To compute the attained significance level of the test we must first find the calculated value of the test statistic, t_c. Since $\hat{\beta}_1 = .26$, $s = 1.1$, and $SS_{xx} = 26$, we have

$$t_c = \frac{\hat{\beta}_1}{s/\sqrt{SS_{xx}}} = \frac{.26}{1.1/\sqrt{26}} = 1.21$$

From Section 10.8, the attained significance level or p-value is given by

$$P(t > t_c) = P(t > 1.21)$$

where the distribution of t is based on $(n - 2) = (100 - 2) = 98$ degrees of freedom. Since df is greater than 30, we can approximate the t distribution with the z distribution. Thus,

$$\begin{aligned} p\text{-value} &= P(t > 1.21) \approx P(z > 1.21) \\ &= .5 - .3869 = .1131 \end{aligned}$$

In order to conclude that the mean yearly food cost increases as annual income increases (i.e., $\beta_1 > 0$), the investigator would have to be willing to tolerate a Type I error probability, α, of .1131 or larger. Since it is very doubtful that the investigator would be willing to take this risk, we consider the sample result to be statistically insignificant.

In addition to testing whether the slope β_1 is zero, we may also be interested in estimating its value with a confidence interval.

EXAMPLE 12.15 Using the information supplied in Example 12.13, construct a 95% confidence interval for the slope β_1 in the straight-line model relating carbon monoxide ranking to nicotine content of a cigarette.

Solution The methods of Chapter 8 are used to construct a confidence interval for β_1. The interval, derived from the sampling distribution of $\hat{\beta}_1$, is given in the box.

A 100(1 − α) PERCENT CONFIDENCE INTERVAL FOR THE SLOPE β_1

$$\hat{\beta}_1 \pm t_{\alpha/2}\left(\frac{s}{\sqrt{SS_{xx}}}\right)$$

where the distribution of t is based on $(n-2)$ degrees of freedom and $t_{\alpha/2}$ is the value of t such that $P(t > t_{\alpha/2}) = \alpha/2$.

For a 95% confidence interval, $\alpha = .05$. Therefore we need to find the value of $t_{.025}$ based on $(n-2) = (5-2) = 3$ df. In Example 12.13, we found that $t_{.025} = 3.182$. Also we have $\hat{\beta}_1 = 20.5$, $s = 1.82$, and $SS_{xx} = .4$. Thus, a 95% confidence interval for the slope in the model relating carbon monoxide to nicotine content is

$$\hat{\beta}_1 \pm (t_{.025})\frac{s}{\sqrt{SS_{xx}}} = 20.5 \pm (3.182)\frac{1.82}{\sqrt{.4}} = 20.5 \pm 9.16$$

Our interval estimate of the slope parameter β_1 is then 11.34 to 29.66.

EXAMPLE 12.16 Interpret the interval estimate of β_1 derived in Example 12.15.

Solution Since all the values in the interval (11.34, 29.66) are positive, we say that we are 95% confident that the slope β_1 is positive. That is, we are 95% confident that the mean carbon monoxide ranking, $E(y)$, increases as nicotine content, x, increases. In addition, we can say that for every 1 milligram increase in the nicotine content x of a cigarette, the increase in mean carbon monoxide ranking $E(y)$ of the cigarette could be as small as 11.34 milligrams or as large as 29.66 milligrams. However, the rather large width of the interval reflects the small number of data points (and, consequently, a lack of information) in the experiment. We would expect a narrower interval if the sample size were increased.

EXERCISES

12.26 The data for Exercise 12.6 are reproduced below:

x	−1	0	1	2	3
y	−1	1	1	2.5	3.5

a. Find SSE and s^2 for the data.
b. How many degrees of freedom are associated with s^2?
c. Test the null hypothesis that the slope β_1 of the line equals 0 against the alternative hypothesis that β_1 is not equal to 0. Use $\alpha = .10$.
d. Compute the approximate attained significance level of the test.
e. Find a 90% confidence interval for the slope β_1.

12.27 The data for Exercise 12.7 are reproduced below:

x	−5	−3	−1	0	1	3	5
y	.8	1.1	2.5	3.1	5.0	4.7	6.2

a. Find SSE and s^2 for the data.
b. How many degrees of freedom are associated with s^2?
c. Test the null hypothesis that the slope β_1 of the line equals 0 against the alternative hypothesis that β_1 is not equal to 0. Use $\alpha = .10$.
d. Compute the approximate attained significance level of the test.
e. Find a 90% confidence interval for the slope β_1.

12.28 The data for Exercise 12.8 are reproduced in the table.

PRICE PER UNIT x, dollars	PURCHASE RATE y, units per month
450	32
450	36
475	34
475	28
500	26
500	23
525	25
525	17
550	12
550	16

a. Find SSE and s^2 for the data.
b. How many degrees of freedom are associated with s^2?
c. Do the data provide sufficient evidence to indicate that price per unit x contributes information for the prediction of purchase rate y? Test using $\alpha = .05$.

12.29 The data for Exercise 12.10 are reproduced in the table.

ATHLETE	MAXIMAL OXYGEN UPTAKE y, milliliters/kilogram	MILE TIME x, seconds
1	63.3	241.5
2	60.1	249.8
3	53.6	246.1
4	58.8	232.4
5	67.5	237.2
6	62.6	238.4

a. Find SSE and s^2 for the data.
b. How many degrees of freedom are associated with s^2?

c. Do the data provide sufficient evidence to indicate that the best mile time x of an athlete contributes information for the prediction of the athlete's maximal oxygen uptake y? Use $\alpha = .01$.

12.30 A large car rental agency sells its cars after using them for a year. Among the records kept for each car are mileage and maintenance costs for the year. To evaluate the performance of a particular car model in terms of maintenance costs, the agency wants to use a 95% confidence interval to estimate the mean increase in maintenance cost for each additional 1,000 miles driven. Assume the relationship between maintenance cost and miles driven is linear.

CAR	MILES DRIVEN x, thousands	MAINTENANCE COST y, dollars
1	54	326
2	27	159
3	29	202
4	32	200
5	28	181
6	36	217

a. Use the six data points in the table to find the least squares prediction equation.
b. Find SSE and s^2.
c. Find a 95% confidence interval for the mean increase in maintenance cost per additional 1,000 miles driven. [*Hint:* Find a 95% confidence interval for the slope β_1.]

12.31 Suppose a fire insurance company wants to relate the amount of fire damage in major residential fires to the distance between the residence and the nearest fire station. A random sample of eight recent fires in a large suburb of a major city is selected and the amount of damage, y, and the distance, x, between the fire and the nearest fire station are recorded for each. The data are listed in the table below. Do the data provide sufficient evidence to indicate that mean fire damage increases as the distance between the fire and the nearest fire station increases? Use $\alpha = .05$. [*Note:* You wish to determine whether the slope β_1 is greater than 0. Therefore, you will conduct a one-tailed test of hypothesis.]

FIRE	DISTANCE FROM FIRE STATION x, miles	FIRE DAMAGE y, thousands of dollars
1	4.6	31.3
2	6.1	43.2
3	3.0	22.3
4	4.8	36.4
5	5.5	36.0
6	3.8	26.1
7	2.3	23.1
8	4.3	31.3

12.7 How Well Does the Least Squares Line Fit the Data?

So far, we have discussed a numerical descriptive measure of the correlation between two variables and a method of evaluating the usefulness of the straight-line model. The correlation coefficient r measures the strength of the straight-line (linear) relationship between two variables x and y. An inference about the slope β_1 of a straight-line model (either an hypothesis test or confidence interval) leads to a determination of whether the independent variable x in the model contributes information for the prediction of the dependent variable y. In this section, we define an alternative numerical descriptive measure of how well the least squares line fits the sample data. This measure, called the ***coefficient of determination,*** is very useful for assessing how much the errors of prediction of y can be reduced by using the information provided by x.

EXAMPLE 12.17 Refer to the carbon monoxide ranking–nicotine content examples. Suppose that you do not use x, nicotine content of a cigarette, to predict y, carbon monoxide ranking. If you have access to a sample of carbon monoxide rankings of cigarettes only, what quantity would you use as the best predictor for any y-value?

Solution If we have no information on the relative frequency distribution of the y-values other than that provided by the sample, then $\bar{y}$, the sample average CO ranking, would be the best predictor for *any* y-value. Using $\bar{y}$ as our predictor, the sum of the squared prediction errors would be $\Sigma(\text{actual } y - \text{predicted } y)^2 = \Sigma(y - \bar{y})^2$ which is the familiar quantity SS_{yy}. The magnitude of SS_{yy} is an indicator of how well $\bar{y}$ behaves as a predictor of y.

EXAMPLE 12.18 Refer to Example 12.17. Suppose now that you use the information on nicotine content, x, to predict CO ranking, y. How do we measure the additional information provided by using the value of x in the least squares prediction equation rather than $\bar{y}$ to predict y?

Solution If we use the information on x to predict y, the sum of squares of the deviations of the y-values about the predicted values obtained from the least squares equation $\hat{y} = \hat{\beta}_0 + \hat{\beta}_1 x$ is

$$SSE = \Sigma(y - \hat{y})^2$$

A convenient way of measuring how well the least squares equation performs as a predictor of y is to compute the reduction in the sum of squares of deviations that can be attributed to x, expressed as a proportion of SS_{yy}. This quantity, called the ***coefficient of determination,*** is

$$\frac{SS_{yy} - SSE}{SS_{yy}}$$

It can be shown that this proportion is equal to the square of the simple linear coefficient of correlation r.

Note that r^2 is always between 0 and 1 since r is between -1 and $+1$. Thus, $r^2 = .75$ means that 75% of the sum of squares of deviations of the y-values about their mean is attributable to the linear relationship between y and x. In other words, the error of prediction can be reduced by 75% when the least squares equation, rather than $\bar{y}$, is used to predict y.

DEFINITION 12.2

The *coefficient of determination* is

$$r^2 = \frac{SS_{yy} - SSE}{SS_{yy}} = 1 - \frac{SSE}{SS_{yy}}$$

It represents the proportion of the sum of squares of deviations of the y-values about their mean that can be attributed to a linear relationship between y and x. (It may also be computed as the square of the coefficient of correlation.)

EXAMPLE 12.19 Calculate the coefficient of determination for the nicotine content–carbon monoxide ranking data of Example 12.8 and interpret its value. (The data are repeated in Table 12.9 for convenience.)

TABLE 12.9

CIGARETTE	NICOTINE CONTENT x, milligrams	CARBON MONOXIDE RANKING y, milligrams
1	.2	2
2	.4	10
3	.6	13
4	.8	15
5	1.0	20

Solution We will use the formula given in the box of Definition 12.2 to compute r^2. From previous calculations, $SS_{yy} = 178$ and $SSE = 9.9$. Therefore,

$$r^2 = \frac{SS_{yy} - SSE}{SS_{yy}} = \frac{178 - 9.9}{178} = \frac{168.1}{178} = .9444$$

We interpret this value as follows: The use of nicotine content, x, to predict carbon monoxide ranking, y, with the least squares line

$$\hat{y} = -.3 + 20.5x$$

accounts for approximately 94% of the total sum of squares of deviations of the five sample CO rankings about their mean. That is, we can reduce the total sum of squares of our prediction errors by more than 94% by using the least squares equation $\hat{y} = -.3 + 20.5x$, instead of $\bar{y}$, to predict y.

Since the two numerical descriptive measures r and r^2 are very closely related, there may be some confusion as to when each should be used. Our recommendations are as follows: If you are only interested in measuring the strength of the linear relationship between two variables x and y, use the coefficient of correlation r. However, if you wish to determine how well the least squares straight-line model fits the data, use the coefficient of determination r^2.

EXERCISES

12.32 Suppose you were to fit a straight line to model the relationship between the annual United States crime rate y, and the population x. Suppose also that the data were collected over a period of ten years. The value of r^2 for the least squares prediction equation was computed to be .73. Interpret this value.

12.33 Refer to the data of Exercises 12.6, 12.18, and 12.26. Find the coefficient of determination r^2 and interpret its value.

12.34 Refer to the data of Exercises 12.7, 12.20, and 12.27. Find the coefficient of determination r^2 and interpret its value.

12.35 Refer to the data of Exercises 12.8, 12.22, and 12.28. Find the coefficient of determination r^2 and interpret its value.

12.36 Refer to the data of Exercises 12.10, 12.24, and 12.29. Find the coefficient of determination r^2 and interpret its value.

12.8 Using the Model for Estimation and Prediction

After we have statistically checked the usefulness of our straight-line model and are satisfied that x contributes information for the prediction of y, we are ready to accomplish our original objective—using the model for prediction, estimation, etc.

The most common uses of a probabilistic model for making inferences can be divided into two categories and are listed in the box.

USES OF THE PROBABILISTIC MODEL FOR MAKING INFERENCES

1. Use the model for estimating the mean value of y, $E(y)$, for a specific value of x.
2. Use the model for predicting a particular y-value for a given value of x.

In the first case, we wish to estimate the mean value of y for a very large number of experiments at a given x-value. For example, we may want to estimate the mean carbon monoxide ranking for all cigarettes which have a nicotine content of .4 milligram. In the second case, we wish to predict the outcome of a single experiment (predict an individual value of y) at the given x-value. For example, we may want to predict the carbon monoxide ranking of a particular American-made cigarette which has been tested by the Federal Trade Commission and found to have a nicotine content of .4 milligram.

We will use the least squares model

$$\hat{y} = \hat{\beta}_0 + \hat{\beta}_1 x$$

both to estimate the mean value of y, $E(y)$, and to predict a particular value of y for a given value of x.

EXAMPLE 12.20 Refer to Example 12.8. We found the least squares model relating carbon monoxide ranking, y, to nicotine content, x, to be

$$\hat{y} = -.3 + 20.5x$$

Give a point estimate for the mean carbon monoxide ranking of all cigarettes which have a nicotine content of .4 milligram.

Solution We need to find an estimate of $E(y)$. Using the least squares model, our estimate is simply $\hat{y}$. Then, when $x = .4$,

$$\hat{y} = -.3 + (20.5)(.4) = -.3 + 8.2 = 7.9$$

Thus, the estimated mean carbon monoxide ranking for all cigarettes with a nicotine content of .4 milligram is 7.9 milligrams.

EXAMPLE 12.21 Refer to Example 12.20. Use the least squares model to predict the carbon monoxide ranking of a particular cigarette whose nicotine content is .4 milligram.

Solution Just as we may use $\hat{y}$ from the least squares model to estimate $E(y)$, we also use $\hat{y}$ to predict a particular value of y for a given value of x. Again, when $x = .4$, we obtain $\hat{y} = 9.7$. Thus, we predict that an American-made cigarette with a nicotine content of .4 milligram will have a carbon monoxide ranking of 9.7 milligrams.

Since the least squares model is used to obtain both the estimator of $E(y)$ and the predictor of y, how do the two model uses differ? The difference lies in the accuracies with which the estimate and the prediction are made. These accuracies are best measured by the repeated sampling errors of the least squares line when it is used as an estimator and predictor, respectively. These errors are given in the box.

SAMPLING ERRORS FOR THE ESTIMATOR OF THE MEAN OF y AND THE PREDICTOR OF AN INDIVIDUAL y

1. The standard deviation of the sampling distribution of the estimator $\hat{y}$ of the mean value of y at a fixed x is

$$\sigma_{\hat{y}} = \sigma\sqrt{\frac{1}{n} + \frac{(x - \bar{x})^2}{SS_{xx}}}$$

where σ is the square root of σ^2, the measure of variability discussed in Section 12.5.

2. The standard deviation of the prediction error for the predictor $\hat{y}$ of an individual y-value at a fixed x is

$$\sigma_{(y-\hat{y})} = \sigma\sqrt{1 + \frac{1}{n} + \frac{(x - \bar{x})^2}{SS_{xx}}}$$

where σ is the square root of σ^2, the measure of variability discussed in Section 12.5.

Since the true value of σ will rarely be known, we estimate σ by s. The sampling errors are then used in estimation and prediction intervals as shown in the box.

A 100(1 − α) PERCENT CONFIDENCE INTERVAL FOR THE MEAN VALUE OF *y* AT A FIXED *x*

$$\hat{y} \pm (t_{\alpha/2})s\sqrt{\frac{1}{n} + \frac{(x - \bar{x})^2}{SS_{xx}}}$$

A 100(1 − α) PERCENT PREDICTION INTERVAL FOR AN INDIVIDUAL *y* AT A FIXED *x*

$$\hat{y} \pm (t_{\alpha/2})s\sqrt{1 + \frac{1}{n} + \frac{(x - \bar{x})^2}{SS_{xx}}}$$

EXAMPLE 12.22 Find a 95% confidence interval for the mean carbon monoxide ranking of cigarettes which have nicotine contents of .4 milligram.

Solution For a nicotine content of .4 milligram, $x = .4$ and the confidence interval for the mean of y is

$$\hat{y} \pm (t_{\alpha/2})s\sqrt{\frac{1}{n} + \frac{(x - \bar{x})^2}{SS_{xx}}} = \hat{y} \pm (t_{.025})s\sqrt{\frac{1}{5} + \frac{(.4 - \bar{x})^2}{SS_{xx}}}$$

where the distribution of t is based on $(n - 2) = 3$ degrees of freedom. Recall from previous examples that $\hat{y} = 7.9$, $s = 1.82$, $\bar{x} = .6$, and $SS_{xx} = .4$. From Table 4, Appendix E, $t_{.025} = 3.182$. Thus, we have

$$7.9 \pm (3.182)(1.82)\sqrt{\frac{1}{5} + \frac{(.4 - .6)^2}{.4}} = 7.9 \pm (3.182)(1.82)(.548)$$

$$= 7.9 \pm 3.17 = (4.73,\ 11.07)$$

Hence, the 95% confidence interval for the mean carbon monoxide ranking of all American-made cigarettes with a nicotine content of .4 milligram is 4.73 milligrams to 11.07 milligrams. Note that the small sample size ($n = 5$ cigarettes) is reflected in the large width of the confidence interval.

EXAMPLE 12.23 Using a 95% prediction interval, predict the carbon monoxide ranking of a cigarette if its nicotine content (as determined by the FTC) is .4 milligram.

Solution For $x = .4$, the 95% prediction interval for y is computed as

$$\hat{y} \pm (t_{\alpha/2})s\sqrt{1 + \frac{1}{n} + \frac{(x - \bar{x})^2}{SS_{xx}}} = 7.9 \pm (3.182)(1.82)\sqrt{1 + \frac{1}{5} + \frac{(.4 - .6)^2}{.4}}$$

$$= 7.9 \pm (3.182)(1.82)(1.140)$$

$$= 7.9 \pm 6.60 = (1.30,\ 14.50)$$

Thus, we predict that the carbon monoxide ranking for this particular cigarette will fall within the interval from 1.30 milligrams to 14.50 milligrams. Again, the large width of this interval can be attributed to the unusually small number of data points (only five) used to fit the least squares line. The width of the prediction interval could be reduced by using a larger number of data points.

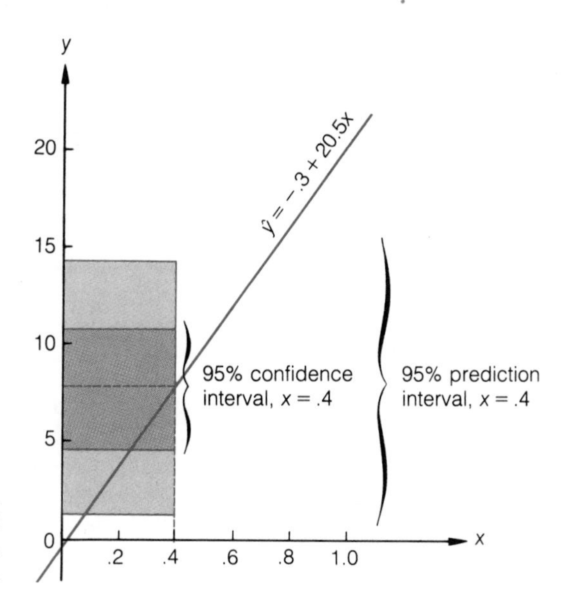

FIGURE 12.11
95% Confidence Interval for $E(y)$ and Prediction Interval for y When $x = .4$

In comparing the results of Examples 12.22 and 12.23, it is important to note that the prediction interval for an individual cigarette carbon monoxide ranking is wider than the corresponding confidence interval for the mean carbon monoxide ranking (see Figure 12.11). By examining the formulas for the two intervals, you can see that this will always be true.

WARNING

Using the least squares prediction equation to estimate the mean value of y or to predict a particular value of y for values of x that fall ***outside the range*** of the values of x contained in your sample data may lead to errors of estimation or prediction that are much larger than expected. Although the least squares model may provide a very good fit to the data over the range of x-values contained in the sample, it could give a poor representation of the true model for values of x outside this region.

EXERCISES

12.37 The data for Exercises 12.6, 12.18, and 12.26 are reproduced below:

x	−1	0	1	2	3
y	−1	1	1	2.5	3.5

a. Estimate the mean value of y when $x = 1$, using a 90% confidence interval. Interpret the interval.

b. Suppose that you plan to observe the value of y for a particular experimental unit with $x = 1$. Find a 90% prediction interval for the value of y that you will observe. Interpret the interval.

c. Which of the two intervals, parts a and b, is wider?

12.38 The data for Exercises 12.7, 12.20, and 12.27 are reproduced below:

x	−5	−3	−1	0	1	3	5
y	.8	1.1	2.5	3.1	5.0	4.7	6.2

a. Estimate the mean value of y when $x = -1$, using a 90% confidence interval. Interpret the interval.

b. Suppose that you plan to observe the value of y for a particular experimental unit with $x = -1$. Find a 90% prediction interval for the value of y that you will observe. Interpret the interval.

c. Which of the two intervals, parts a and b, is wider?

12.39 In Exercise 12.23, you found the least squares prediction equation relating the length of life y of a thoroughbred horse to the gestation period x of the horse and used it to predict the length of life of a horse when $x = 320$. The data are reproduced in the table.

HORSE	LENGTH OF LIFE y, years	GESTATION PERIOD x, days
1	24	416
2	25.5	279
3	20	298
4	21.5	307
5	22	356
6	23.5	403
7	21	265

a. Find a 90% prediction interval for this value of y and interpret it.

b. Suppose you were to estimate the mean length of life of all thoroughbred horses with gestation periods of $x = 320$ days. Find a 90% confidence interval for this mean and interpret it.

12.40 An engineer conducted a study to determine whether there is a linear relationship between the breaking strength, y, of wooden beams and the specific gravity, x, of the wood. Ten randomly selected beams of the same cross-sectional dimensions were stressed until they broke. The breaking strength and the specific gravity of the wood are shown in the table for each of the ten beams.

BEAM	BREAKING STRENGTH y	SPECIFIC GRAVITY x
1	11.14	.499
2	12.74	.558
3	13.13	.604
4	11.51	.441
5	12.38	.550
6	12.60	.528
7	11.13	.418
8	11.70	.480
9	11.02	.406
10	11.41	.467

a. Find the least squares prediction equation relating the breaking strength y of a wooden beam to the beam's specific gravity x.

b. Find a 95% confidence interval for the mean breaking strength of beams with specific gravity .590. Interpret this interval.

c. Predict the breaking strength of a beam if its specific gravity is .590. Use a 95% prediction interval. Interpret this interval.

12.9 Simple Linear Regression: An Example of a Computer Printout

In the previous sections we have presented the basic elements necessary to fit and use the straight-line regression model $E(y) = \beta_0 + \beta_1 x$. Throughout, we have illustrated the numerical techniques through an example relating the carbon monoxide ranking of a cigarette to its nicotine content. Even with a small number of measurements (five data points), the required computations, if performed without the aid of a pocket or desk calculator, can become tedious and cumbersome. Many institutions have installed computer packages which fit a straight-line regression model by the method of least squares. These packages enable the user to greatly decrease the burden of calculation. In this section, we locate, discuss, and interpret the elements of a simple linear regression on a computer printout.

We will again present output from the Statistical Analysis System (SAS), first introduced in Section 11.4. Since the linear regression output of the SAS is similar to that of most other package regression programs (such as Minitab and SPSS), you should have little trouble interpreting output from other packages. We relate all the examples in this section to the SAS output for the carbon monoxide ranking–nicotine content data in Table 12.5. A portion of the SAS printout is given in Figure 12.12. Again, we let y be the carbon monoxide ranking of an American-made cigarette and x the nicotine content of the cigarette.

DEPENDENT VARIABLE: Y (CO)

SOURCE	DF	SUM OF SQUARES	MEAN SQUARE	F VALUE
MODEL	1	168.10000000	168.10000000	50.94
ERROR	3	9.90000000	3.30000000	PR > F
CORRECTED TOTAL	4	178.00000000		0.0057

R-SQUARE	C.V.	STD DEV	Y MEAN
0.944382	15.1383	1.81659021	12.00000000

PARAMETER	ESTIMATE	T FOR H0: PARAMETER=0	PR > |T|	STD ERROR OF ESTIMATE
INTERCEPT	-0.30000000	-0.16	0.8849	1.90525589
X (NICOTINE)	20.50000000	7.14	0.0057	2.87228132

FIGURE 12.12 Portion of the SAS Printout for the Carbon Monoxide Ranking–Nicotine Content Data

EXAMPLE 12.24 Locate on the SAS printout in Figure 12.12 the least squares estimates of the y-intercept β_0 and the slope β_1 for the straight-line model relating carbon monoxide ranking to nicotine content.

Solution The least squares estimates of the y-intercept and slope are shaded in Figure 12.13. Note that the estimate of the y-intercept, i.e., $\hat{\beta}_0 = -0.30000000$, and the estimate of the slope, i.e., $\hat{\beta}_1 = 20.50000000$, given in the printout agree with our previous calculations made by hand. The least squares model relating CO ranking to nicotine content can thus be written

$$\hat{y} = -.3 + 20.5x$$

DEPENDENT VARIABLE: Y (CO)

SOURCE	DF	SUM OF SQUARES	MEAN SQUARE	F VALUE
MODEL	1	168.10000000	168.10000000	50.94
ERROR	3	9.90000000	3.30000000	PR > F
CORRECTED TOTAL	4	178.00000000		0.0057

R-SQUARE	C.V.	STD DEV	Y MEAN
0.944382	15.1383	1.81659021	12.00000000

PARAMETER	ESTIMATE	T FOR H0: PARAMETER=0	PR > |T|	STD ERROR OF ESTIMATE
INTERCEPT	-0.30000000	-0.16	0.8849	1.90525589
X (NICOTINE)	20.50000000	7.14	0.0057	2.87228132

FIGURE 12.13 SAS Printout with the Least Squares Estimates Shaded

EXAMPLE 12.25 Locate the SSE on the SAS printout in Figure 12.12. Also, find s^2 and s, the estimates of σ^2 and σ, respectively.

Solution The SSE is found by locating the entry under the column heading SUM OF SQUARES in the row labelled ERROR. This quantity, shaded in Figure 12.14, is SSE = 9.90000000. A check of our previous calculations confirms its validity. The estimate of σ^2 is given in the figure (shaded) as MEAN SQUARE for ERROR and is located to the immediate right of SSE. Thus, we have $s^2 = 3.30000000$. The least squares estimate of σ, shaded in Figure 12.14 under the heading STD DEV, is $s = 1.81659021$. Except for rounding errors, our computed values of s^2 and s agree with the figures given in the printout.

```
DEPENDENT VARIABLE: Y (CO)
SOURCE               DF    SUM OF SQUARES     MEAN SQUARE      F VALUE
MODEL                 1      168.10000000    168.10000000        50.94
ERROR                 3        9.90000000      3.30000000       PR > F
CORRECTED TOTAL       4      178.00000000                       0.0057

R-SQUARE           C.V.           STD DEV          Y MEAN
0.944382        15.1383        1.81659021     12.00000000

                                  T FOR H0:    PR > |T|     STD ERROR OF
PARAMETER           ESTIMATE   PARAMETER=0                     ESTIMATE

INTERCEPT        -0.30000000         -0.16      0.8849       1.90525589
X (NICOTINE)     20.50000000          7.14      0.0057       2.87228132
```

FIGURE 12.14
SAS Printout with SSE, s^2, and s Shaded

EXAMPLE 12.26 Use the SAS output to test the null hypothesis that x contributes no information for the prediction of y against the alternative hypothesis that x and y are linearly related.

Solution We desire a test of the hypotheses

$$H_0\colon\ \beta_1 = 0$$
$$H_a\colon\ \beta_1 \neq 0$$

```
DEPENDENT VARIABLE: Y (CO)
SOURCE               DF    SUM OF SQUARES     MEAN SQUARE      F VALUE
MODEL                 1      168.10000000    168.10000000        50.94
ERROR                 3        9.90000000      3.30000000       PR > F
CORRECTED TOTAL       4      178.00000000                       0.0057

R-SQUARE           C.V.           STD DEV          Y MEAN
0.944382        15.1383        1.81659021     12.00000000

                                  T FOR H0:    PR > |T|     STD ERROR OF
PARAMETER           ESTIMATE   PARAMETER=0                     ESTIMATE

INTERCEPT        -0.30000000         -0.16      0.8849       1.90525589
X (NICOTINE)     20.50000000          7.14      0.0057       2.87228132
```

FIGURE 12.15
SAS Printout with r^2 and Value of the Test Statistic Shaded

The value of the test statistic for this test is shaded in Figure 12.15 under the column heading T FOR H0: PARAMETER = 0 in the lower portion of the printout. The value shown here is $t = 7.14$, which agrees with our computed test statistic. To determine whether this value falls within the rejection region, check the shaded quantity to the immediate right under PR > |T|. This quantity is the attained significance level or p-value of the test. Generally, if the p-value is smaller than .05, there is evidence to reject the null hypothesis that the slope is zero in favor of the alternative hypothesis that the slope differs from zero. In this example, an attained significance level of .0057 indicates that nicotine content x of a cigarette and carbon monoxide ranking y are linearly related (a result consistent with the test conducted in Example 12.13).

EXAMPLE 12.27 To determine how well the least squares straight-line model fits the data, find the value of the coefficient of determination r^2.

Solution The value of r^2, shaded in Figure 12.15, is given as $r^2 = .944382$. The result that we computed in a previous example is accurate to four decimal places. We say, then, that by using the least squares equation, instead of $\bar{y}$, to predict y, we can reduce the total sum of squares of our prediction errors by approximately 94%. Notice that the coefficient of correlation r is not given in the SAS printout.

EXAMPLE 12.28 Construct a 95% confidence interval for $E(y)$, the mean carbon monoxide ranking of all American-made cigarettes that have a nicotine content of .4 milligram.

Solution A portion of the SAS printout not previously shown is given in Figure 12.16. For $x = .4$ (i.e., nicotine content of .4 milligram), we see that the least squares predicted value of y is $\hat{y} = 7.9$, and that a 95% confidence interval for the mean, $E(y)$, is 4.73345260 to 11.06654740. Thus, the mean carbon monoxide ranking for cigarettes with a nicotine content of .4 milligram falls between 4.7335 milligrams and 11.0665 milligrams, with 95% confidence. In Example 12.22, we computed the interval to be 4.73 milligrams to 11.07 milligrams. The differences in the intervals are due to rounding error.

FIGURE 12.16 Portion of SAS Printout for the Carbon Monoxide Ranking–Nicotine Content Data

X	PREDICTED VALUE	LOWER 95% CL FOR MEAN	UPPER 95% CL FOR MEAN
.4	7.90000000	4.73345260	11.06654740

EXAMPLE 12.29 Find a 95% prediction interval for the carbon monoxide ranking of a cigarette with a nicotine content of .4 milligram.

Solution The 95% prediction interval for an individual value of y with $x = .4$ is given in Figure 12.17 as (1.30830593, 14.49169407). We predict that the carbon monoxide ranking for a particular cigarette with a nicotine content of .4 milligram will fall between 1.3083 milligrams and 14.4917 milligrams. The difference between this result and the interval we computed in Example 12.23, 1.30 milligrams to 14.50 milligrams, is again due to rounding error.

FIGURE 12.17
Portion of SAS Printout for the Carbon Monoxide Ranking–Nicotine Content Data

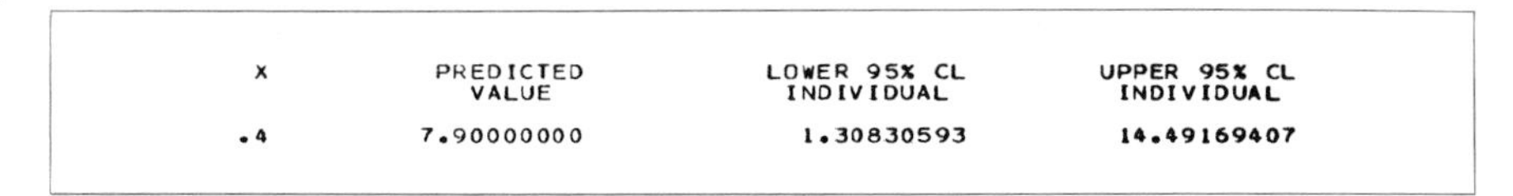

X	PREDICTED VALUE	LOWER 95% CL INDIVIDUAL	UPPER 95% CL INDIVIDUAL
.4	7.90000000	1.30830593	14.49169407

In Chapter 13 we will discuss the interpretation of those portions of the SAS printout which were not shaded or mentioned here. However, the important elements of a linear regression analysis have been located, and you should be able to use this example as a guide to interpreting linear regression computer printouts.

12.10 Assumptions Required for a Linear Regression Analysis

As with most statistical procedures, the validity of the confidence intervals, prediction intervals, and statistical tests associated with a simple linear regression analysis depend on certain assumptions being satisfied. These assumptions are made about the random error term in the straight-line probabilistic model.

ASSUMPTIONS REQUIRED FOR A LINEAR REGRESSION ANALYSIS

1. The mean of the probability distribution of the random error is 0. That is, for each setting of the independent variable x, the average of the errors over an infinitely long series of experiments is 0.
2. The variance of the probability distribution of the random error is constant for all settings of the independent variable x and is equal to σ^2, i.e., the variance of the random error is equal to σ^2 for all values of x.
3. The probability distribution of the random error is normal.
4. The errors associated with any two observations are independent. That is, the error associated with one value of y has no effect on the errors associated with other y-values.

Figure 12.18 (page 444) shows a pictorial representation of the assumptions given in the box. For each value of x shown in the figure, the relative frequency distribution of the errors is normal with mean zero, and with a constant variance (all the distributions shown have the same amount of spread or variability) equal to σ^2.

In practice, you will never know whether your data satisfy the four assumptions listed above. Fortunately, the estimators and test statistics used in a simple linear regression have sampling distributions that remain relatively stable for minor departures from the assumptions.

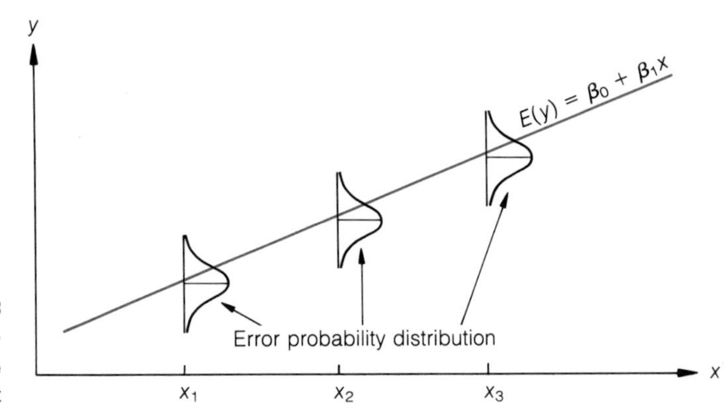

FIGURE 12.18 The Probability Distribution of the Random Error Component

12.11 Summary

In this chapter we introduced bivariate relationships and showed how to compute the coefficient of correlation, r, a measure of the strength of the linear relationship between two variables. We also introduced an extremely useful tool—the method of least squares—for fitting a straight-line model to a set of data. This procedure, along with associated statistical tests and estimations, is called a regression analysis.

After hypothesizing the straight-line probabilistic model

$$y = \beta_0 + \beta_1 x + \text{Random error}$$

perform the following steps:

1. Use the method of least squares to estimate the unknown parameters in the deterministic component, $\beta_0 + \beta_1 x$. You may obtain the estimates by either applying the computational formulas given in this chapter or, if you have access to a computer package, from a computer printout. The least squares estimates will yield a model $\hat{y} = \hat{\beta}_0 + \hat{\beta}_1 x$ with a sum of squared errors (SSE) that is smaller than that produced by any other straight-line model.

KEY WORDS

Bivariate relationships
Coefficient of correlation
Linear regression analysis
Method of least squares
Probabilistic models
Deterministic component
Random error component
Least squares line (prediction equation)
Coefficient of determination

2. Check that the assumptions about the random error component (outlined in the box in Section 12.10) are satisfied. You should also determine s^2 (either by hand calculation or from a computer printout), an estimate of σ^2, the variance of the random error component.
3. Assess the usefulness of the hypothesized model, i.e., determine how well x performs as a predictor of y. Included here are making inferences about the slope β_1 and computing the coefficient of determination, r^2.
4. Finally, if you are satisfied with the model, use it to estimate the mean y-value, $E(y)$, for a given x-value and/or to predict an individual y-value for a specific x.

KEY SYMBOLS

y-intercept of the line: β_0
Slope of the line: β_1
Coefficient of correlation: r
Coefficient of determination: r^2

SUPPLEMENTARY EXERCISES

12.41 Use the method of least squares to fit a straight line to the following six data points:

x	1	2	3	4	5	6
y	1	2	2	3	5	5

a. What are the least squares estimates of β_0 and β_1?
b. Plot the data points and graph the least squares line. Does the line pass through the data points?
c. Do the data provide sufficient evidence to indicate that x contributes information for the prediction of y? Test at $\alpha = .05$.

12.42 Refer to Exercise 12.41.
a. If y and x are linearly correlated, will the correlation be positive or negative? Explain.
b. Calculate the coefficient of correlation r. Interpret its value.
c. Calculate the coefficient of determination r^2 and interpret its value.
d. Based on the test, Exercise 12.41(c), can you conclude that y and x are linearly correlated? Explain.
e. Test for linear correlation between y and x, using the method of Section 12.2. Does your result agree with that of Exercise 12.41(c)?

12.43 Refer to Exercise 12.41.
a. Find a 90% confidence interval for the mean value of y when $x = 2$.
b. Find a 90% prediction interval for y when $x = 2$.
c. Interpret the intervals, parts a and b.

12.44 Use the method of least squares to fit a straight line to the following five data points:

x	−2	−1	0	1	2
y	4	3	3	1	−1

a. What are the least squares estimates of β_0 and β_1?
b. Plot the data points and graph the least squares line. Does the line pass through the data points?
c. Do the data provide sufficient evidence to indicate that x contributes information for the prediction of y? Test at $\alpha = .05$.

12.45 Refer to Exercise 12.44.
a. If y and x are linearly correlated, will the correlation be positive or negative? Explain.
b. Calculate the coefficient of correlation r. Interpret its value.
c. Calculate the coefficient of determination r^2 and interpret its value.
d. Based on the test, Exercise 12.44(c), can you conclude that y and x are linearly correlated? Explain.
e. Test for linear correlation between y and x, using the method of Section 12.2. Does your result agree with that of Exercise 12.44(c)?

12.46 Refer to Exercise 12.44.
a. Find a 90% confidence interval for the mean value of y when $x = 2$.
b. Find a 90% prediction interval for y when $x = 2$.
c. Interpret the intervals, parts a and b.

12.47 As a result of the increase in the number of suburban shopping centers, many center-city stores are suffering financially. A downtown department store thinks that increased advertising might help lure more shoppers into the area. To study the relationship between sales and advertising, records were obtained for several mid-year months during which the store varied advertising expenditures.

SALES y, thousands of dollars	ADVERTISING EXPENSE x, thousands of dollars
30	0.9
34	1.1
32	0.8
37	1.2
31	0.7

a. Estimate the coefficient of correlation between sales and advertising expenditures.
b. Do the data provide sufficient evidence of a correlation between sales, y, and advertising expense, x?

12.48 At temperatures approaching absolute zero (273 degrees below zero Celsius), helium exhibits traits that defy many laws of conventional physics. An experiment has been conducted with helium in solid form at various temperatures near absolute zero. The solid helium is placed in a dilution refrigerator along with a solid impure substance, and the proportion (by weight) of the impurity passing through the solid helium is recorded. (This phenomenon of solids passing directly through solids is known as "quantum tunnelling.") The data are given in the table.

PROPORTION OF IMPURITY PASSING THROUGH HELIUM y	TEMPERATURE x, °C
.315	−262
.202	−265
.204	−256
.620	−267
.715	−270
.935	−272
.957	−272
.906	−272
.985	−273
.987	−273

a. Fit a least squares line to the data.
b. Plot the data and graph the line as a check on your calculations.
c. Calculate r and r^2. Interpret these values.
d. Estimate the mean proportion of impurity passing through helium when the temperature is set at −270°C. Use a 99% confidence interval.
e. Predict the proportion of impurity passing through helium at a temperature of −270°C using a 99% prediction interval.

12.49 A group of children, ranging from 10 to 12 years of age, were administered a verbal test in order to study the relationship between the number of words used and the silence interval before response. A psychologist believes a linear relationship may exist between the two variables. Each subject was asked a series of questions, and the total number of words used in answering and the time before the subject responded were recorded. The data are given in the table.

SUBJECT	TOTAL SILENCE TIME x, seconds	TOTAL WORDS y
1	23	61
2	37	70
3	38	42
4	25	52
5	17	91
6	21	63
7	42	71
8	16	55

a. Estimate the coefficient of correlation between total silence time and total words.

b. Do the data provide sufficient evidence to indicate a correlation between total words, y, and total silence time, x? Use $\alpha = .05$.

c. Fit a least squares line to the data.

d. Do the data provide sufficient evidence to indicate that total silence time contributes information for the prediction of total words used in answering? [*Hint:* This test is identical to that of part b.]

12.50 The management of a manufacturing firm is considering the possibility of setting up its own market research department rather than continuing to use the services of a market research firm. The management wants to know what salary should be paid to a market researcher, based on years of experience. An independent consultant checks with several other firms in the area and obtains the information on market researchers shown in the table.

ANNUAL SALARY y, thousands of dollars	EXPERIENCE x, years
20.3	2
20.2	1.5
32.0	11
37.4	15
29.5	9
25.3	6

a. Fit a least squares line to the data.

b. Plot the data and graph the line as a check on your calculations.

c. Calculate r and r^2. Interpret these values.

d. Estimate the mean annual salary of market researchers with 8 years of experience. Use a 90% confidence interval.

e. Predict the salary of a market researcher with 7 years of experience, using a 90% prediction interval.

12.51 The importance of islands as sampling units for flora and fauna population studies has been widely recognized by biogeographers and evolutionists; the theory of equilibrium island biology states that larger islands should have more species than smaller islands. Does such a relationship exist among the species of flora found in the vernal pools (i.e., pools of water formed in low-lying areas) of the Central Valley of California? To investigate this phenomenon, Holland and Jain (*The American Naturalist,* January 1981) selected six sites in California where vernal pools are in abundance. At each site, 10 to 20 pools were surveyed for species richness (i.e., number of different flora species inhabiting the pool) and pool surface area. A linear model relating species richness y and surface area x (in square feet) of the pools was fit to the data using the method of least squares with the following result: $\hat{y} = 18.4 + .04x$.

a. Give the null and alternative hypotheses for a test to determine whether surface area x is useful for predicting species richness y in a linear model.

b. The reported p-value for the test of part a is greater than .05. Interpret this result in terms of the problem.

c. The coefficient of determination for the simple linear regression is $r^2 = .06$. Interpret this value.

12.52 "In the analysis of urban transportation systems it is important to be able to estimate expected travel time between locations." Cook and Russell (*Transportation Research*, June 1980) collected data in the city of Tulsa on the urban travel times and distances between locations for two types of vehicles, large hoist compactor trucks and passenger cars. A simple linear regression analysis was conducted for both sets of data (y = urban travel time in minutes, x = distance between locations in miles) with the following results:

PASSENGER CARS	TRUCKS
$\hat{y} = 2.50 + 1.93x$	$\hat{y} = 1.85 + 3.86x$
$r^2 = .676$; p-value $< .05$	$r^2 = .758$; p-value $< .01$

a. Is there sufficient evidence to indicate that distance between locations is linearly related to urban travel time for passenger cars? Test at $\alpha = .05$.

b. Is there sufficient evidence to indicate that distance between locations is linearly related to urban travel time for trucks? Test at $\alpha = .01$.

c. Interpret the values of r^2 for the two prediction equations.

d. Estimate the mean urban travel time for all passenger cars traveling a distance of 3 miles on Tulsa's highways.

e. Predict the urban travel time for a particular truck traveling a distance of 5 miles on Tulsa's highways.

f. Explain how we could attach a measure of reliability to the inferences derived in parts d and e.

CASE STUDY 12.1 Television, Children, and Drugs

Concern has been expressed that Americans are becoming a nation of pill takers. Many critics contend that this behavior is learned in childhood and that it is encouraged by TV advertising for proprietary (advertised nonprescription or "over-the-counter") drugs.

This is how J. R. Rossiter and T. S. Robertson began their article* on children's attitudes toward drugs advertised on television. Despite this belief, lack of information on the potential relationship of TV drug advertising to children's attitudes caused the Federal Communications Commission (FCC) to deny a petition seeking to prohibit TV advertising for proprietary drugs until after 9:00 PM when fewer children would be in the audience. At present, proprietary drugs cannot be advertised on children's programs, but they can be advertised during adult and family programs which constitute 85% of children's viewing.

*J.R. Rossiter and T.S. Robertson. "Children's dispositions toward proprietary drugs and the role of television drug advertising." *Public Opinion Quarterly, 44*, 317–329, Fall 1980. Reprinted by permission. Copyright 1980 by Elsevier North Holland, Inc.

Previous studies have shown that "children's and teenagers' dispositions toward proprietary drugs are only weakly related to their exposure to TV drug advertising." One researcher "found correlations in the .07 to .22 range with children's exposure to TV drug advertising, with only about half of the correlations reaching the .05 level of significance." However, Rossiter and Robertson criticize these studies because they "focused on a limited set of proprietary drugs, including cold and stomachache remedies but excluding cough and headache remedies." Thus, in March 1977 the authors conducted their own study of the relationship between children's attitudes toward proprietary drugs and TV advertising. Their study included a sample of 668 children drawn from inner-city and suburban areas of Philadelphia. With permission from their respective parents, each child was asked to fill out a questionnaire of the type shown below:

Sample Questions

Belief variable: (1) "When I have a cold, cold medicine can make me feel better."
Scoring: 1 = strongly disagree to 4 = strongly agree.

Affect variable: (2) "When I have a cold, I like to take cold medicine."
Scoring: 1 = strongly disagree to 4 = strongly agree.

Intention variable: (3) "When I have a cold, I want to take cold medicine."
Scoring: 1 = never to 4 = always.

Request variable: (4) "When I have a cold, I ask my parents for cold medicines."
Scoring: 1 = never to 4 = always.

Usage variable: (5) "Since school started in September, how many times have your parents given you cold medicines?"
Scoring: 1 = never to 5 = five or more times.

TV drug advertising exposure variable: Number of proprietary drug commercials viewed per program.

Rossiter and Robertson calculated the Pearson product moment coefficient of correlation (r) of each of the five variables (Belief, Affect, Intention, Request, and Usage) with TV drug ad exposure. Results are given in the table for two subsamples of the 668 children: (1) $n = 132$ children from educationally "disadvantaged" families who receive little parental instruction about proprietary drugs and the way they are advertised; and (2) $n = 55$ children who are never allowed to take proprietary drugs without parental supervision. Does it appear that children's attitudes toward proprietary drugs are linearly related to TV drug advertising exposure? Use Table 12.3

VARIABLE	LOW PARENT EDUCATION ($n = 132$) r	NO SELF-ADMINISTRATION ($n = 55$) r
Belief	.06	.29
Affect	.13	.26
Intention	.44	.23
Request	.39	.34
Usage	.12	.23

to determine which, if any, of the sample correlations are large enough to indicate (at $\alpha = .05$) that linear correlation exists between TV drug ad exposure and the variable measured.

CASE STUDY 12.2 Evaluating the Protein Efficiency of a New Food Product

One aspect of the preliminary evaluation of a new food product is determination of the nutritive quality of the product. This is often accomplished by feeding the food product to animals whose metabolic processes are very similar to our own. In a 1978 article,* Elizabeth Street and Mavis Carroll discussed such an evaluation of a new product developed by General Foods Corporation.

General Foods wished to compare the protein efficiency of two forms of a product (known by the pseudonym H), one solid and the other liquid. Because previous experience had shown that 28-day feeding of 10 to 15 rats on a diet gives a fairly reliable estimate of the diet's protein efficiency, General Foods conducted a rat-feeding study. Thirty male rats, all newly weaned, were used in the experiment. Ten rats were randomly assigned a diet of solid H, ten a diet of liquid H, and ten a standard (control) diet. During the feeding period, each rat was permitted to eat as much as it wished. At the end of the 28 days, the total protein intake x (in grams) and the weight gain y (in grams) were recorded for each of the 30 rats.

Using a computer program, General Foods fit a straight line to the 10 (x, y) data points for each diet by the method of least squares. The three least squares prediction equations were

Liquid H: $\hat{y} = 109.3 + 3.72x$

Solid H: $\hat{y} = 106.7 + 3.66x$

Control: $\hat{y} = 50.6 + 2.91x$

a. Graph the three lines on the same piece of graph paper. For a given increase in protein intake, which of the three diets resulted in the greatest increase in weight gain? [*Hint:* Which line has the greatest slope?]
b. Estimate the mean weight gain in rats with a 28-day protein intake of 40 grams for each of the three diets. Compare these values.
c. Predict the weight gain for a rat on each of the three diets if the rat's 28-day protein intake is 40 grams. Compare these values.
d. How would the width of a confidence interval for any of the estimates, part b, compare with the width of a prediction interval for the corresponding prediction, part c? Explain.

CASE STUDY 12.3 Mental Imagery and the Third Dimension

How do people form a mental image of an object in three-dimensional space? The question is an interesting one when we consider that our visual senses lack a three-dimensional structure. When we view a scene, for example, a person water-skiing, our eyesight does not process the three-dimensional layout of the scene

*Elizabeth Street and Mavis B. Carroll. "Preliminary evaluation of a new food product." In *Statistics: A guide to the unknown.* Tanur et al., eds. San Francisco: Holden-Day, 1978, 269–278.

directly. The scene is reconstructed in our brain from the two-dimensional projections of the scene onto the retina of our eye. Consequently, certain perspective effects result, e.g., as the water-skier recedes from our view, he appears to shrink in size or if a large sailboat interrupts our line of sight, the scene of the water-skier is blocked. However, suppose we could explore the scene with our sense of touch. A more distant object (the water-skier) would not feel smaller, nor would the sailboat prevent us from locating the water-skier with our touch. The three-dimensional structure remains free of the perspective effects which exist in vision.

Psychologists have just begun to research this phenomenon of mental imagery in three dimensions. In this case study, we focus on one of a series of experiments conducted by Steven Pinker (1980)* of Harvard University. Pinker wished to answer the following question: "Do mental images preserve interval information concerning the distances between objects in three dimensions?" In this experiment, the time that subjects require to scan from one object in a three-dimensional image to another is used as a measure of the "distance" between those objects in the image. Pinker hypothesizes that if three-dimensional distances are preserved in the image, the scanning times should be highly positively linearly correlated with these distances, i.e., scanning times should increase linearly with increasing distances in three dimensions.

Ten volunteers for the study, all affiliated with Harvard University, were asked to view a gray box, open at the top and front, in which five small toys (a hat, an apple, a teddy bear, a tire, and a sea shell) were suspended by clear nylon thread. The objects' positions were chosen so that the interobject distances in three dimensions correlated poorly with the corresponding distances in the two-dimensional plane. Each subject was asked to form a mental image of the box and its contents, making sure each object was imagined at its proper location. After the layout was memorized (this was verified through extensive testing), the front of the box was covered with an opaque screen. The subject was then asked to shut his or her eyes, and, upon hearing the name of one of the objects in the box, to mentally focus on that object. Four seconds later another object was named, and the subject was to "scan" to it by moving in a straight line as quickly as possible from the first to the second object. When the subject "arrived" at the destination object, he or she was to press a button indicating the end of the trial. Trials continued until all possible pairs of objects were scanned. The response times (in milliseconds) for scanning between the members of each pair of objects and the three-dimensional distance (in centimeters) between the objects were recorded for each pair of objects for each subject.

a. Pinker plotted the ten response time means (one for each pair of stimulus objects, averaged over subjects) against the distance between objects. The result is reproduced in Figure 12.19. Does it appear that response time is linearly related to distance between objects?

b. The correlation between response (scanning) time and distance between objects is $r = .92$. Interpret this value.

---

*Steven Pinker. "Mental imagery and the third dimension." *Journal of Experimental Psychology: General*, September 1980, *109*, 354–371. Copyright by the American Psychological Association. Adapted by permission of the publisher and author.

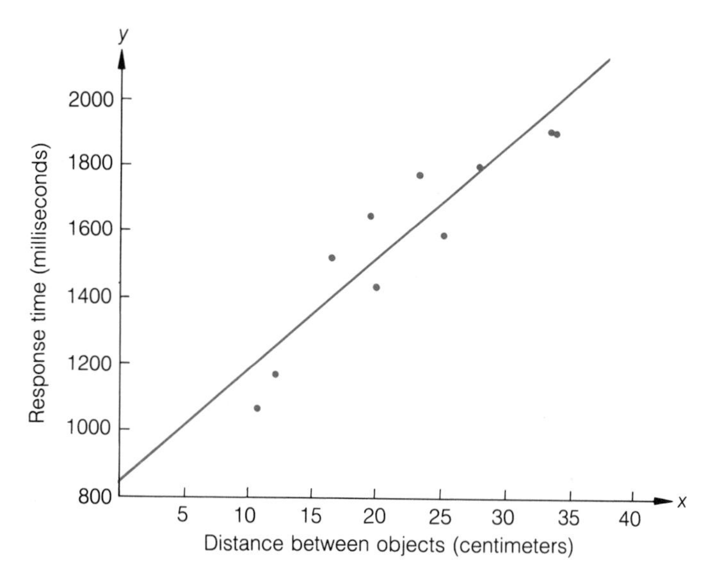

FIGURE 12.19
Mean Response Times for Scanning Mentally in Three Dimensions between Imagined Objects Separated by Different Three-Dimensional Distances

c. The least squares line relating response time y to distance x between objects was found to be $\hat{y} = 833 + 33.8x$. Set up the null and alternative hypotheses for testing whether response time increases linearly as the distance between objects increases.

d. The test statistic for testing the hypothesis of part c was found to be $t = 8.28$, corresponding to df = 81. Is there sufficient evidence (at $\alpha = .01$) to indicate that response time increases linearly as distance between objects increases?

e. What assumptions are necessary for the validity of the test, part d?

REFERENCES

Barr, A., Goodnight, J., Sall, J., Blair, W., & Chilko, D. *SAS user's guide.* 1979 ed. SAS Institute, P. O. Box 10066, Raleigh, N. C. 27605.

Cook, T. M., & Russell, R. A. "Estimating urban travel times: A comparative study." *Transportation Research,* June 1980, *14A,* 173–175.

Draper, N., & Smith, H. *Applied regression analysis.* New York: Wiley, 1966, Chapter 1.

Holland, R. F., & Jain, S. "Insular biogeography of vernal pools in the Central Valley of California." *The American Naturalist,* January 1981, *117,* 24–37.

Johnson, R. *Elementary statistics.* 3d ed. North Scituate, Mass.: Duxbury, 1980. Chapter 11.

Mendenhall, W., & McClave, J. T. *A second course in business statistics: Regression analysis.* San Francisco: Dellen, 1981. Chapter 3.

Neter, J., & Wasserman, W. *Applied linear statistical models.* Homewood, Ill.: Richard Irwin, 1974. Chapters 2–6.

Pinker, S. "Mental imagery and the third dimension." *Journal of Experimental Psychology: General,* September 1980, *109,* 354–371.

Rossiter, J. R., & Robertson, T. S. "Children's dispositions toward proprietary drugs and the role of television drug advertising." *Public Opinion Quarterly,* Fall 1980, *44,* 316–329.

Tanur, J. M., Mosteller, F., Kruskal, W. H., Link, R. F., Pieters, R. S., & Rising, G. R. *Statistics: A guide to the unknown.* San Francisco: Holden-Day, 1978.

Thirteen Multiple Regression

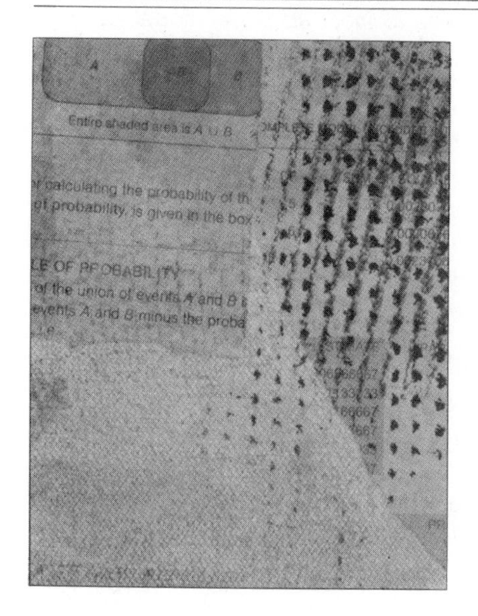

"Other than food and sex, nothing is quite as universally interesting as the size of our paychecks." So say N. L. Preston and E. R. Fiedler in an article that examines the relationship between the compensation of an economist and the economist's qualifications (see Case Study 13.2). In this chapter, we will extend the ideas of Chapter 12 and build a regression model to relate a variable y to two or more independent variables. We will see that Preston and Fiedler use this technique to develop a model for determining "fair compensation" for an economist.

Contents

13.1 Introduction: Linear Statistical Models and Multiple Regression Analysis

Most practical applications of regression analysis require models that are more complex than the simple straight-line model. For example, a realistic probabilistic model for the carbon monoxide ranking y of an American-made cigarette would include more variables than the nicotine content x of the cigarette (discussed in Chapter 12). Additional variables such as tar content, length, filter-type, and menthol flavor might also be related to carbon monoxide ranking. Thus, we would want to incorporate these and other potentially important independent variables into the model if we needed to make accurate predictions of the carbon monoxide ranking y. This more complex probabilistic model relating y to these various independent variables, say $x_1, x_2, x_3, \ldots$, is called a ***general linear statistical model,*** or more simply, a ***linear model.***

EXAMPLE 13.1 How does a general linear model differ from the following simple straight-line model?

$$y = \beta_0 + \beta_1 x + \text{Random error}$$

Solution General linear models are more flexible than straight-line models in the sense that they may include more than one independent variable. A linear model for y, the carbon monoxide ranking of a cigarette, could be written

$$y = \beta_0 + \beta_1 x_1 + \beta_2 x_2 + \beta_3 x_3 + \text{Random error}$$

In addition to the independent variable x_1, the nicotine content of a cigarette, the model includes two other independent variables: the tar content, x_2, and the length of a cigarette, x_3. (Note how the data for fitting general linear models would be collected: For each experimental unit—in our example, a cigarette—we measure the dependent variable y and record the values of the independent variables x_1, x_2, and x_3.)

A general linear model may include some independent variables which appear at higher orders, e.g., terms such as x_1^2, $x_1 x_2$, x_3^3, etc. For example,

$$y = \beta_0 + \beta_1 x_1 + \beta_2 x_2 + \beta_3 x_1 x_2 + \beta_4 x_1^2 + \beta_5 x_2^2 + \text{Random error}$$

is a linear model. You may wonder why the model is referred to as a linear model if these higher order terms are present. The model is called linear because it represents a linear function of the unknown parameters, $\beta_0, \beta_1, \beta_2, \ldots$. That is, each term contains only one of the β parameters and each β is a coefficient of the remaining portion of the term. For example, the term $\beta_1 x_1 x_2^2$ satisfies this requirement, but the term $\beta_1 x_1^{\beta_2}$ does not because it contains two unknown parameters (β_1 and β_2) and, secondly, because β_2 appears as an exponent rather than a multiplicative coefficient.

This chapter will employ the method of least squares to fit a general linear model to a set of data. This process, along with the estimation and test procedures associated with it, is called a ***multiple regression analysis.*** Because the computations involved in a multiple regression analysis are very complex, most regression analyses are performed on a computer. In the following sections we will present an example of a multiple regression analysis and will examine and interpret the printouts for one of the standard multiple regression computer program packages.

13.2 The Quadratic Model

The formula for a general linear model is given in the box.

THE GENERAL LINEAR MODEL

$$y = \beta_0 + \beta_1 x_1 + \beta_2 x_2 + \cdots + \beta_k x_k + \text{Random error}$$

where y is the dependent variable (variable to be predicted) and $x_1, x_2, \ldots, x_k$ are the independent variables.

[*Note:* Remember that the symbols $x_1, x_2, \ldots, x_k$ may represent higher order terms. For example, x_1 might represent the length of a cigarette, x_2 might represent x_1^2, etc.]

$E(y) = \beta_0 + \beta_1 x_1 + \beta_2 x_2 + \cdots + \beta_k x_k$ is the deterministic portion of the model

β_0 is the y-intercept

β_i determines the contribution of the independent variable x_i

In this chapter we consider a special case of the general linear model, a case in which the model includes only two independent variables, one of which is a higher order term. The form for this model, called the ***quadratic*** (or ***second-order) model,*** is

$$y = \beta_0 + \beta_1 x + \beta_2 x^2 + \text{Random error}$$

Technically, the quadratic model includes only one independent variable, x, but we can think of the model as a general linear model in two independent variables with

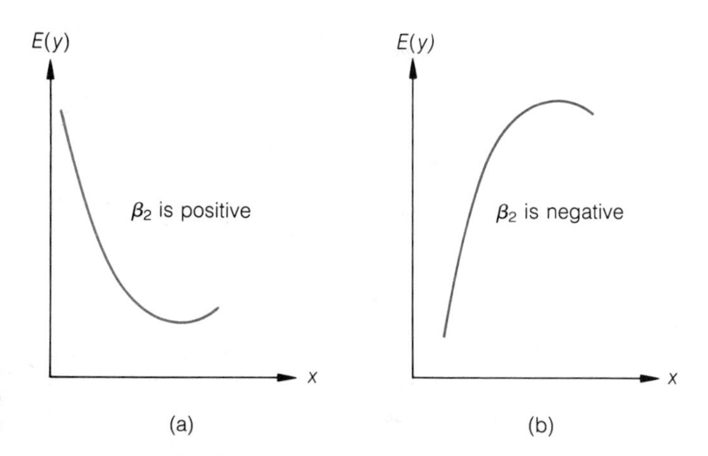

FIGURE 13.1 Graphs for Two Quadratic Models

$x_1 = x$ and $x_2 = x^2$. The term involving x^2, called the ***quadratic term,*** enables us to hypothesize curvature in the graph of the response model relating y to x. Graphs of the quadratic model for two different values of β_2 are shown in Figure 13.1. When the curve opens upward (i.e., the curve "holds water"), the sign of β_2 is positive (see Figure 13.1(a)); when the curve opens downward (i.e., the curve "spills water"), the sign of β_2 is negative (see Figure 13.1(b)).

EXAMPLE 13.2 Of interest to real estate investors, city tax assessors, and real estate appraisers is the relationship between the appraised improvements value of a property and its eventual selling price. Is appraised improvements value a good indicator of the price at which the property will be sold? Suppose a real estate investor wishes to model the relationship between the sale price of a residential property located in a particular neighborhood of a city and the corresponding appraised improvements value of the property. A random sample of 10 properties in the neighborhood which were sold last year was selected for the analysis. The resulting data are given in Table 13.1.

TABLE 13.1
Sale Price–Appraised Improvements Value for 10 Properties in the Neighborhood

SALE PRICE **y, dollars**	APPRAISED IMPROVEMENTS **x, dollars**
22,500	13,140
27,000	15,890
31,900	18,700
33,000	20,730
29,000	16,510
46,600	24,180
51,500	26,150
54,700	35,870
55,000	40,100
54,900	43,460

a. Construct a scattergram for the sale price–appraised improvements value data of Table 13.1.

b. Hypothesize a probabilistic model relating sale price to appraised improvements value for all properties in the neighborhood which were sold last year.

Solution **a.** A plot of the data of Table 13.1 is given in Figure 13.2. You can see that the sale price of properties appears to increase in a curvilinear manner with appraised improvements value.

b. The apparent curvature in the graph relating appraised improvements value, x, to sale price, y, provides some support for the inclusion of a quadratic term, x^2, in the response model. Thus, we think that the quadratic model

$$y = \beta_0 + \beta_1 x + \beta_2 x^2 + \text{Random error}$$

might yield a better prediction equation than one based on the straight-line model of Chapter 12. To determine if this model gives an adequate representation of the relationship between y and x, we follow the same steps that we applied in developing the straight-line model.

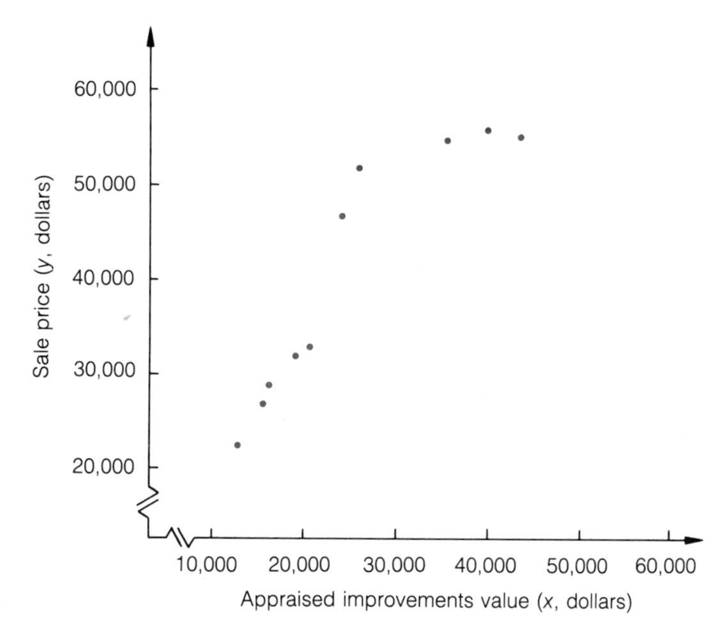

FIGURE 13.2 Scattergram of the Sale Price–Appraised Improvements Data

STEPS TO FOLLOW IN A MULTIPLE REGRESSION ANALYSIS

1. Hypothesize the form of the linear model.
2. Estimate the unknown parameters $\beta_0, \beta_1, \beta_2, \ldots$.
3. Check whether the fitted model is useful for predicting y.
4. If we decide that the model is useful, use it to estimate the mean value of y or to predict a particular value of y for given values of the independent variables.

13.3 Fitting the Model

To fit a general linear model by the method of least squares, we choose the estimated model

$$\hat{y} = \hat{\beta}_0 + \hat{\beta}_1 x_1 + \hat{\beta}_2 x_2 + \cdots + \hat{\beta}_k x_k$$

that minimizes $SSE = \Sigma(y_i - \hat{y}_i)^2$. We will use the Statistical Analysis System (SAS), referred to in Chapters 11 and 12, to fit the linear model and to illustrate each of the remaining steps in a multiple regression analysis.

EXAMPLE 13.3 Refer to Example 13.2. Use the method of least squares to estimate the unknown parameters β_0, β_1, β_2 in the quadratic model relating sale price, y, to appraised improvements value, x, for properties located in a particular neighborhood of the city.

Solution Part of the output of the SAS multiple regression routine for the data of Table 13.1 is reproduced in Figure 13.3. The least squares estimates of the β parameters appear (shaded) in the column labelled ESTIMATE. You can see that $\hat{\beta}_0 = -26{,}391.07$, $\hat{\beta}_1 = 4.2624$, and $\hat{\beta}_2 = -.00005522$. Therefore, the equation that minimizes SSE for this data is

$$\hat{y} = -26{,}391.07 + 4.2624x - .00005522x^2$$

From Figure 13.4 we see that the graph of the quadratic regression model provides a good fit to the data of Table 13.1. We note here that the small value of $\hat{\beta}_2$ does ***not*** imply that the curvature is insignificant, since the numerical scale of $\hat{\beta}_2$ is dependent upon the scale of measurements. We will test the contribution of the quadratic coefficient in Section 13.4.

DEPENDENT VARIABLE: Y

SOURCE	DF	SUM OF SQUARES	MEAN SQUARE	F VALUE
MODEL	2	1484489413.91039800	742244706.95519900	84.40
ERROR	7	61559586.08960193	8794226.58422885	PR > F
CORRECTED TOTAL	9	1546049000.00000000		0.0001

R_SQUARE	C.V.	STD DEV	Y MEAN
.960183	7.3024	2965.50612615	40610.00000000

PARAMETER	ESTIMATE	T FOR H0: PARAMETER=0	PR > |T|	STD ERROR OF ESTIMATE
INTERCEPT	-26391.06854166	-2.95	0.0214	8945.31427948
X	4.26240155	6.03	0.0005	0.70732695
X*X	-5.52192855E-05	-4.48	0.0029	0.00001233

FIGURE 13.3 Output from the SAS for the Sale Price–Appraised Improvements Data

EXAMPLE 13.4 Locate the minimum value of SSE on the SAS printout reproduced in Figure 13.3. Also, obtain an estimate of σ^2, the variance of the random error term in the probabilistic model.

Solution The minimum value of SSE, 61,559,586.09, is shaded in the row labelled ERROR under the column labelled SUM OF SQUARES in the printout shown in Figure 13.3. Recall from Section 12.5 that we can use this quantity to estimate σ^2. The estimator for the straight-line model was $s^2 = \text{SSE}/(n-2)$. Note that the denominator is $n -$ (Number of estimated β parameters) which, in the case of the straight-line model, is equal to $n - 2$. Since we must estimate one more parameter, β_2, for the quadratic model, $y = \beta_0 + \beta_1 x + \beta_2 x^2 + \text{Random error}$, the estimator of σ^2 is $s^2 = \text{SSE}/(n-3)$. That is, the denominator becomes $(n-3)$ because there are now three β parameters in the model.

The numerical estimate of σ^2 for this example is

$$s^2 = \frac{\text{SSE}}{10-3} = \frac{61{,}559{,}586.09}{7} = 8{,}794{,}226.584$$

Note that this estimate appears on the printout as the MEAN SQUARE for ERROR. Similarly, the standard deviation $s = 2{,}965.506$ appears in the column headed STD DEV (see Figure 13.3).

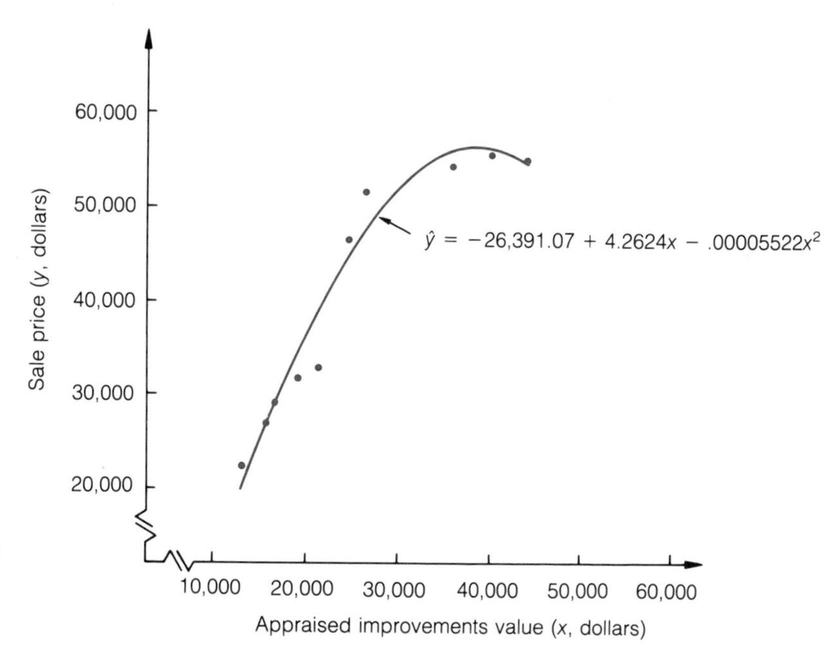

FIGURE 13.4
Least Squares Model for the Sale Price–Appraised Improvements Data

The importance of the estimator of σ^2 is that we use its numerical value both to check the predictive ability of the model (Sections 13.4, 13.5, and 13.6) and to provide a measure of the reliability of predictions and estimates when the model is used for those purposes (Section 13.7).

ESTIMATOR OF σ^2

$$s^2 = \frac{\text{SSE}}{n - (\text{Number of estimated } \beta \text{ parameters})}$$

where n is the number of data points.

13.4 Estimating and Testing Hypotheses about the β Parameters

Recall from our discussion in Chapter 12 of the straight-line model

$$y = \beta_0 + \beta_1 x + \text{Random error}$$

that β_1 has a practical interpretation. It is the mean change in y for every 1 unit increase in x. One method of determining whether the mean value of y changes as x increases is to test the null hypothesis H_0: $\beta_1 = 0$.

Consider now the quadratic model

$$y = \beta_0 + \beta_1 x + \beta_2 x^2 + \text{Random error}$$

What is the practical interpretation of β_2? As noted earlier, β_2 measures the amount of curvature in the response curve (see Figure 13.1). Thus, one method of determining whether curvature exists in the population is to test the null hypothesis H_0: $\beta_2 = 0$. A test of hypothesis about β_2 is illustrated in the following example.

EXAMPLE 13.5 Refer to Example 13.3. Test the hypothesis that the quadratic term in the model contributes significantly to the prediction of sale price, y (i.e., test the null hypothesis H_0: $\beta_2 = 0$). Use $\alpha = .05$.

Solution We require a test of the hypotheses

H_0: $\beta_2 = 0$ (No curvature in the response curve)

H_a: $\beta_2 \neq 0$ (Curvature exists in the response curve)

This test, a t test, is quite similar to the test of the slope of the simple straight-line model (Section 12.6). The details for a test about any β parameter in the general linear model are given in the box.

TEST ABOUT AN INDIVIDUAL PARAMETER COEFFICIENT IN THE GENERAL LINEAR MODEL (TWO-TAILED)

$$y = \beta_0 + \beta_1 x_1 + \beta_2 x_2 + \cdots + \beta_k x_k + \text{Random error}$$

H_0: $\beta_i = 0$

H_a: $\beta_i \neq 0$

Test statistic: $t = \hat{\beta}_i / s_{\hat{\beta}_i}$

Rejection region: $t > t_{\alpha/2}$ or $t < -t_{\alpha/2}$

where

n = Number of observations

$s_{\hat{\beta}_i}$ = Estimated standard deviation of the repeated sampling distribution of $\hat{\beta}_i$

and the distribution of t is based on (n − Number of β parameters in the model) degrees of freedom, and $t_{\alpha/2}$ is the t-value such that $P(t > t_{\alpha/2}) = \alpha/2$.

Assumptions: See Section 13.9

We use the symbol $s_{\hat{\beta}_2}$ to represent the estimated standard deviation of $\hat{\beta}_2$. Since the formula for $s_{\hat{\beta}_2}$ is very complex, we will not present it here. However, this will not cause difficulty since the printouts for most computer packages list the estimated

DEPENDENT VARIABLE: Y

SOURCE	DF	SUM OF SQUARES	MEAN SQUARE	F VALUE
MODEL	2	1484489413.91039800	742244706.95519900	84.40
ERROR	7	61559586.08960193	8794226.58422885	PR > F
CORRECTED TOTAL	9	1546049000.00000000		0.0001

R_SQUARE	C.V.	STD DEV	Y MEAN
.960183	7.3024	2965.50612615	40610.00000000

PARAMETER	ESTIMATE	T FOR H0: PARAMETER=0	PR > |T|	STD ERROR OF ESTIMATE
INTERCEPT	-26391.06854166	-2.95	0.0214	8945.31427948
X	4.26240155	6.03	0.0005	0.70732695
X*X	-5.5219285E-05	-4.48	0.0029	0.00001233

FIGURE 13.5
Output from the SAS

standard deviation $s_{\hat{\beta}_i}$ for each of the estimated model coefficients $\hat{\beta}_i$ in the linear model as well as the corresponding calculated t-values.

In order to test the null hypothesis that $\beta_2 = 0$, we again consult the SAS printout for the sale price–appraised improvements value example. From Figure 13.5, we see that the computed value of the test statistic corresponding to the test of H_0: $\beta_2 = 0$ (shaded under the column headed T FOR H0: PARAMETER = 0) is $t = -4.48$. The appropriate rejection region is obtained by consulting Table 4, Appendix E. For $\alpha = .05$ and $(n-3) = 7$ degrees of freedom, we have $t_{\alpha/2} = t_{.025} = 2.365$. Note that the critical t-value used to specify the rejection region depends upon $n-3$ degrees of freedom because the quadratic model contains three parameters (β_0, β_1, β_2). Then the rejection region (shown in Figure 13.6) is

$$t > 2.365 \quad \text{or} \quad t < -2.365$$

Since $t = -4.48$ falls into the lower tail of the rejection region, we conclude that the quadratic term $\beta_2 x^2$ makes an important contribution to the prediction model of the sale price of properties located in the neighborhood last year.

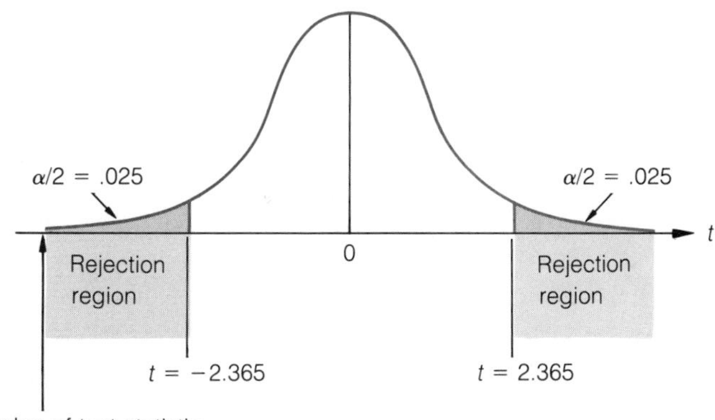

FIGURE 13.6
Rejection Region for Test about β_2

This result could be obtained directly from the SAS printout, which lists the two-tailed attained significance levels (p-values) for each t-value under the column headed PR > |T|. The significance level .0029 corresponds to the quadratic term, and this implies that we would reject H_0: $\beta_2 = 0$ in favor of H_a: $\beta_2 \neq 0$ at any α level larger than .0029. Thus, there is very strong evidence of curvature in the response model relating sale price to appraised improvements value for properties located in the neighborhood.

EXAMPLE 13.6 Refer to Example 13.3. Form a 95% confidence interval for the parameter β_2 in the quadratic model.

Solution A confidence interval for any β parameter in a general linear model is given in the box. From Figure 13.5, $\hat{\beta}_2 = -.00005522$. The estimated standard deviations of the model coefficients appear in the SAS printout under the column labelled STD ERROR OF ESTIMATE. The value, $s_{\hat{\beta}_2} = .00001233$, is shaded in Figure 13.5. Substituting the values of $\hat{\beta}_2$, $s_{\hat{\beta}_2}$, and $t_{.025} = 2.365$ (based on $n - 3 = 7$ degrees of freedom) into the formula for a confidence interval, we find the 95% confidence interval for β_2 to be

$$\hat{\beta}_2 \pm t_{\alpha/2} s_{\hat{\beta}_2} = -.00005522 \pm (2.365)(.00001233)$$

or $(-.00008438, -.00002606)$. This interval can be used to estimate the rate of curvature in mean sale price as appraised improvements value is increased. Note that all values in the interval are negative, reconfirming our test conclusion that β_2 is nonzero.

A 100(1 − α) PERCENT CONFIDENCE INTERVAL FOR AN INDIVIDUAL PARAMETER COEFFICIENT IN THE GENERAL LINEAR MODEL

$$\hat{\beta}_i \pm t_{\alpha/2} s_{\hat{\beta}_i}$$

where

n = Number of observations

$s_{\hat{\beta}_i}$ = Estimated standard deviation of the repeated sampling distribution of $\hat{\beta}_i$

and the distribution of t is based on (n − Number of β parameters in the model) degrees of freedom, and $t_{\alpha/2}$ is the t-value such that $P(t > t_{\alpha/2}) = \alpha/2$.

13.5 Measuring How Well the Model Fits the Data

Recall that the coefficient of determination, r^2, is a measure of how well a straight-line model fits a set of data (Chapter 12). To measure how well a general linear model (e.g., a quadratic model) fits a set of data, we compute the multiple regression equivalent of r^2, called the ***multiple coefficient of determination,*** denoted by the symbol R^2.

DEFINITION 13.1

The ***multiple coefficient of determination,*** R^2, is defined as

$$R^2 = 1 - \frac{\text{SSE}}{\text{SS}_{yy}}$$

where $\text{SSE} = \Sigma(y_i - \hat{y}_i)^2$, $\text{SS}_{yy} = \Sigma(y_i - \bar{y})^2$, and $\hat{y}_i$ is the predicted value of y_i for the multiple regression model.

Just as for the simple linear model, R^2 represents the proportion of the sum of squares of deviations (SS_{yy}) of the y-values about $\bar{y}$ that can be attributed to the regression model. Thus, $R^2 = 0$ implies a complete lack of fit of the model to the data and $R^2 = 1$ implies a perfect fit, with the model passing through every data point. In general, the larger the value of R^2, the better the model fits the data.

EXAMPLE 13.7 Refer to Example 13.3. Locate the value of R^2 on the SAS printout and interpret its value. Does the quadratic model appear to provide a good fit to the sale price data for the properties in the neighborhood?

Solution The SAS printout for the sale price–appraised improvements data is reproduced in Figure 13.7. The value of R^2 (shaded) is shown to be $R^2 = .960$. This very high value implies that, by using the independent variable Appraised improvements value in a quadratic model instead of $\bar{y}$ to predict y, we can reduce the sum of squared prediction errors by 96%. Thus, this large value of R^2 indicates that the quadratic model provides a good fit to the $n = 10$ sample data points.

```
DEPENDENT VARIABLE: Y
SOURCE              DF     SUM OF SQJARES          MEAN SQUARE        F VALUE
MODEL                2    1484489413.91039800    742244706.95519900      84.40
ERROR                7      61559586.08960193      8794226.58422885     PR > F
CORRECTED TOTAL      9    1546049000.00000000                            0.0001

R_SQUARE          C.V.            STD DEV                Y MEAN
.960183         7.3024       2965.50612615        40610.00000000

                                   T FOR H0:      PR > |T|       STD ERROR OF
PARAMETER          ESTIMATE      PARAMETER=0                       ESTIMATE

INTERCEPT     -26391.06854166         -2.95       0.0214        8945.31427948
X                  4.26240155          6.03       0.0005           0.70732695
X*X           -5.5219285E-05          -4.48       0.0029           0.00001233
```

FIGURE 13.7
Output from the SAS

A large value of R^2 computed from the ***sample*** data does not necessarily mean that the model provides a good fit to all of the data points in the ***population.*** For example, a quadratic model which contains 3 parameters will provide a perfect fit to a sample of 3 data points and R^2 will equal 1. Likewise, you will always obtain a perfect fit ($R^2 = 1$) to a set of n data points if your model contains exactly n parameters. Consequently, if you wish to use the value of R^2 as a measure of how useful the model will be for predicting y, it should be based on a sample that contains substantially more data points than the number of parameters in the model.

WARNING

In a multiple regression analysis, use the value of R^2 as a measure of how useful a linear model will be for predicting y only if the sample contains substantially more data points than the number of β parameters in the model.

We discuss a more formal method of checking the predictive ability of a general linear model, a statistical test of hypothesis, in the following section.

13.6 Testing Whether the Model Is Useful for Predicting y

A test of the utility of a general linear model, i.e., a test to determine whether the model is really useful for predicting y, involves testing all the β parameters in the model simultaneously. Conducting individual t tests on each β parameter in a model (Section 13.4) is generally not a good way to determine whether a model is contributing information for the prediction of y. Even if all the β parameters (except β_0) in the model are in fact equal to zero, $100(\alpha)\%$ of the time you will incorrectly reject the null hypothesis and conclude that some β parameter differs from zero. A better way to test the overall utility of a linear model is to conduct a test involving all the β parameters (except β_0). The null and alternative hypotheses for this test of model utility are given in the box.

HYPOTHESES FOR TESTING WHETHER A GENERAL LINEAR MODEL IS USEFUL FOR PREDICTING y

H_0: $\beta_1 = \beta_2 = \cdots = \beta_k = 0$

H_a: At least one of the β parameters in H_0 is nonzero

Practically speaking, this test for model utility is a comparison of the predictive ability of the estimated general linear model (which uses the predictor $\hat{y} = \hat{\beta}_0 + \hat{\beta}_1 x_1 + \hat{\beta}_2 x_2 + \cdots + \hat{\beta}_k x_k$) with a model that contains no x's (which uses the predictor $\hat{y} = \bar{y}$). If the test shows that at least one of the β's is nonzero, then the value of $\hat{y}$ obtained from the estimated linear model will, generally speaking, more accurately predict a future value of y than the sample mean $\bar{y}$. We illustrate a test of model utility in the following example.

EXAMPLE 13.8 Refer to Example 13.3. Test (using $\alpha = .05$) whether the quadratic model is useful for predicting y by testing the null hypothesis

$$H_0: \quad \beta_1 = \beta_2 = 0$$

Solution The test statistic used in the test for model utility is an F statistic. The formula for computing the F statistic is given in the box. However, most computer regression analysis packages give this F-value. The F-value for the sale price example is shaded in the SAS computer printout shown in Figure 13.7. The value of the test statistic is $F = 84.40$. To determine whether this F-value is statistically significant, we read the value of the attained significance level given in the SAS printout. (For details on the appropriate rejection region, see Section 13.8.) The attained significance level for this test, .0001, is shaded in Figure 13.7 in the column headed PR > F. This implies that we would reject the null hypothesis for any α-value larger than .0001. Thus, we have strong evidence to reject H_0 and to conclude that at least one of the model coefficients, β_1 and β_2, is nonzero. Since the attained significance level is so small, there is ample evidence to indicate that $\hat{y} = \hat{\beta}_0 + \hat{\beta}_1 x + \hat{\beta}_2 x^2$ is useful for predicting the sale price of properties located in the neighborhood.

TEST STATISTIC FOR TESTING WHETHER A GENERAL LINEAR MODEL IS USEFUL FOR PREDICTING y

$$F = \frac{R^2/k}{(1 - R^2)/[n - (k + 1)]}$$

where

R^2 = Multiple coefficient of determination
n = Number of observations
k = Number of β parameters in the model (excluding β_0)

13.7 Using the Model for Estimation and Prediction

After checking the utility of the linear model and finding it to be useful for prediction and estimation, we are ready to use it for those purposes. Our methods for prediction and estimation using the quadratic or any general linear model are identical to those discussed for the simple straight-line model (Section 12.8). We will use the model to form a confidence interval for the mean $E(y)$ for a given value of x, or a prediction interval for a future value of y for a given x.

EXAMPLE 13.9 Refer to Example 13.3. Using the least squares quadratic model

$$\hat{y} = -26{,}391.07 + 4.2624x - .00005522x^2$$

estimate the mean sale price, $E(y)$, of a property with an appraised improvements value of $x = \$18{,}700$. Use a 95% confidence interval.

Solution Substituting $x = 18{,}700$ into the least squares prediction equation, the estimate of $E(y)$ is

$$\hat{y} = -26{,}391.07 + 4.2624(18{,}700) - .00005522(18{,}700)^2$$
$$= 34{,}006.21$$

To form a confidence interval for the mean, we need to know the standard deviation of the sampling distribution for the estimator $\hat{y}$. For general linear models, the form of this standard deviation is very complex. However, the SAS regression package allows us to obtain the confidence intervals for mean values of y at any given setting of the independent variables. This portion of the SAS output for the sale price example is shown in Figure 13.8. The 95% confidence interval for $E(y)$, the mean sale price for all properties with an appraised improvements value of $x = \$18{,}700$, is shown to be \$31,293.41 to \$36,719.00 (see Figure 13.9).

FIGURE 13.8 SAS Printout for Estimated Mean and Corresponding Confidence Interval for $x = 18{,}700$

X	PREDICTED VALUE	LOWER 95% CL FOR MEAN	UPPER 95% CL FOR MEAN
18700	34006.20866573	31293.41293639	36719.00439507

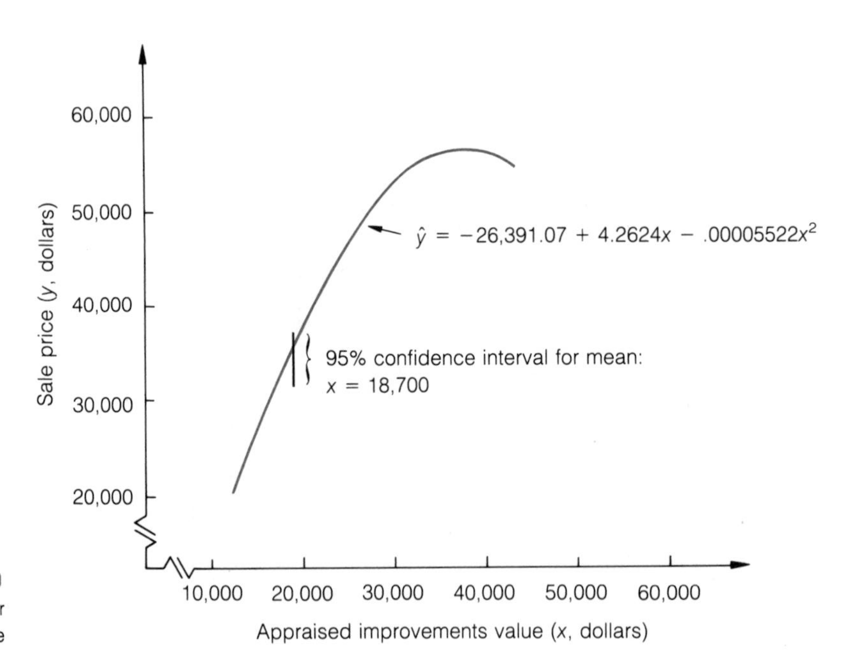

FIGURE 13.9 Confidence Interval for Mean Sale Price

EXAMPLE 13.10 Refer to Example 13.3. Construct a 95% prediction interval for y, the sale price of a particular property with an appraised improvements value of $x = \$18{,}700$.

Solution When $x = 18{,}700$, the predicted value for y is again $\hat{y} = 34{,}006.21$. However, the prediction interval for a particular value of y will be wider than the confidence interval for the mean value. This is reflected in the SAS printout shown in Figure 13.10. The prediction interval extends from \$26,487.40 to \$41,525.02 (see Figure 13.11).

FIGURE 13.10
SAS Printout for Predicted Value and Corresponding Prediction Interval for $x = 18{,}700$

X	PREDICTED VALUE	LOWER 95% CL INDIVIDUAL	UPPER 95% CL INDIVIDUAL
18700	34006.20866573	26487.40120860	41525.01612285

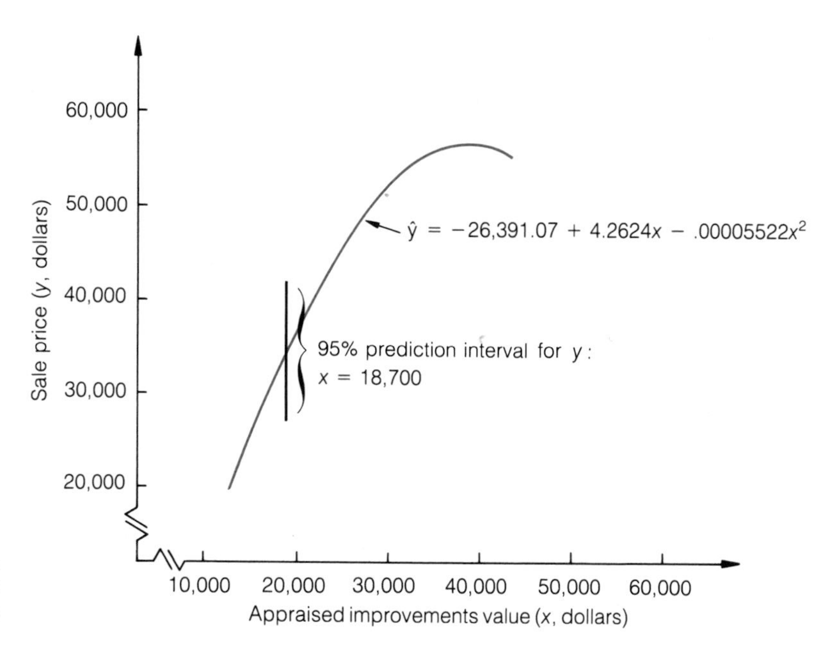

FIGURE 13.11
Prediction Interval for Sale Price

Just as in simple linear regression, it is dangerous to use the quadratic or any general linear prediction model for making predictions outside the region in which the sample data fall. (In our example, we would not use the estimated model to make estimates or predictions for properties with appraised improvements values less than \$13,140 or greater than \$43,460.) In general, the fitted model might not provide a good model for the relationship between the mean of y and the value of x when stretched over a wider range of x-values.

WARNING

Do not use the least squares model to predict a value of y outside the region in which the sample data fall, i.e., do not predict y for values of the independent variables $x_1, x_2, \ldots, x_k$ which are not within the range of the sample data.

13.8 Other Computer Printouts

We have highlighted the key elements of a multiple regression analysis as they appear on the SAS printout. However, there are a number of other different statistical program packages, such as Biomed, Minitab, and SPSS, to which you may have access at your computer center. The multiple regression computer programs for these packages may differ in programming methodology and in the appearance of their computer printouts, but all of them print sufficient information for conducting a multiple regression analysis. To illustrate, we compare the SAS regression analysis computer printout with the Minitab and SPSS printouts.

EXAMPLE 13.11 In Example 13.3, we used the SAS to fit the quadratic model

$$y = \beta_0 + \beta_1 x + \beta_2 x^2 + \text{Random error}$$

to $n = 10$ data points, where

y = Sale price of a residential property in the neighborhood

x = Appraised improvements value of the property

The Minitab and SPSS printouts for this example are given in Figure 13.12. (For convenience, we also reprint the SAS output originally given in Figure 13.3.) Compare the Minitab and SPSS printouts to the SAS.

Solution In any multiple regression analysis, nine key elements are utilized. They are: (1) estimates of the β parameters; (2) the standard deviation, $s_{\hat{\beta}_i}$, of each least squares estimate $\hat{\beta}_i$; (3) computed values of t, the test statistics for testing H_0: $\beta_i = 0$; (4) attained significance levels of the t tests; (5) SSE; (6) s^2; (7) R^2; (8) computed value of the F statistic for testing whether the overall model is useful for predicting y; and (9) the attained significance level of the F test. With the aid of the arrows in Figure 13.12, you can see that all nine of these elements are located on the SAS printout (Figure 13.12(a)). However, some of these elements are missing on the Minitab (Figure 13.12(b)) and SPSS (Figure 13.12(c)) printouts.

Although it gives the computed t-values for testing H_0: $\beta_i = 0$, the Minitab printout (Figure 13.12(b)) does not give the attained significance levels of the tests. Thus, in order to draw conclusions about the individual β parameters from the Minitab printout, we must compare the computed values of t with the critical values given in a t table (Table 4, Appendix E). Also, the Minitab printout does not give the F statistic for testing the overall utility of the model, i.e., testing the null hypothesis that all model parameters (except β_0) simultaneously equal zero. We show you how to obtain this F-value from the Minitab in an example that follows.

The elements missing on the SPSS printout (Figure 13.12(c)) also involve t and F tests. The SPSS does not print the computed t-values for testing the individual β parameters, nor does it give the corresponding attained significance levels. While the SPSS printout does report the value of the F statistic for testing whether the overall model is useful for predicting y, it fails to give the attained significance level of the F test. We learn how to obtain these elements missing from SPSS in the examples that follow.

FIGURE 13.12 Computer Printouts for Example 13.11

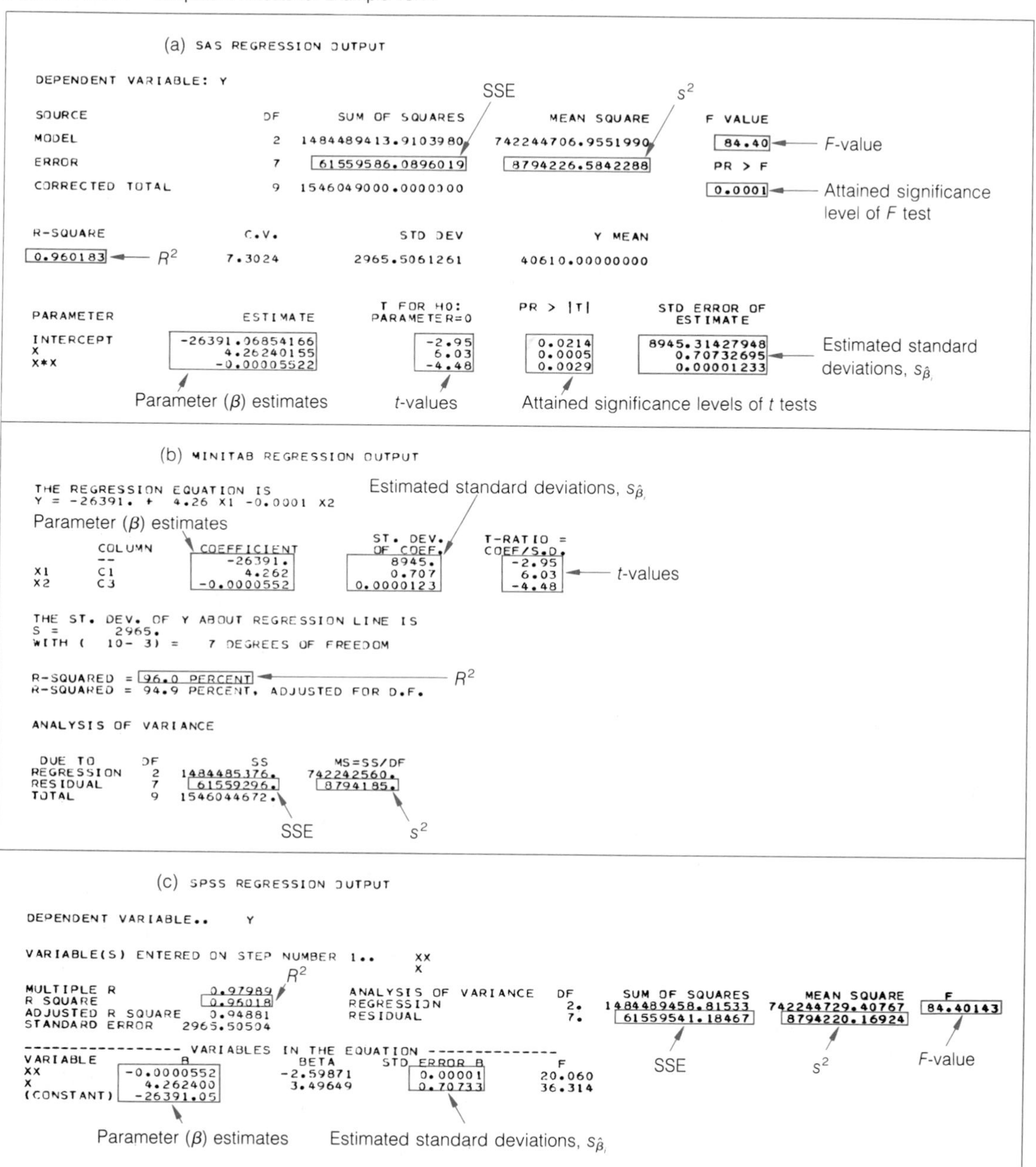

(a) SAS REGRESSION OUTPUT

DEPENDENT VARIABLE: Y

SOURCE	DF	SUM OF SQUARES	MEAN SQUARE	F VALUE
MODEL	2	1484489413.9103980	742244706.9551990	84.40
ERROR	7	61559586.0896019	8794226.5842288	PR > F
CORRECTED TOTAL	9	1546049000.0000000		0.0001

R-SQUARE	C.V.	STD DEV	Y MEAN
0.960183	7.3024	2965.5061261	40610.00000000

PARAMETER	ESTIMATE	T FOR H0: PARAMETER=0	PR > |T|	STD ERROR OF ESTIMATE
INTERCEPT	-26391.06854166	-2.95	0.0214	8945.31427948
X	4.26240155	6.03	0.0005	0.70732695
X*X	-0.00005522	-4.48	0.0029	0.00001233

(b) MINITAB REGRESSION OUTPUT

THE REGRESSION EQUATION IS
Y = -26391. + 4.26 X1 -0.0001 X2

	COLUMN	COEFFICIENT	ST. DEV. OF COEF.	T-RATIO = COEF/S.D.
	--	-26391.	8945.	-2.95
X1	C1	4.262	0.707	6.03
X2	C3	-0.0000552	0.0000123	-4.48

THE ST. DEV. OF Y ABOUT REGRESSION LINE IS
S = 2965.
WITH (10- 3) = 7 DEGREES OF FREEDOM

R-SQUARED = 96.0 PERCENT
R-SQUARED = 94.9 PERCENT, ADJUSTED FOR D.F.

ANALYSIS OF VARIANCE

DUE TO	DF	SS	MS=SS/DF
REGRESSION	2	1484485376.	742242560.
RESIDUAL	7	61559296.	8794185.
TOTAL	9	1546044672.	

(c) SPSS REGRESSION OUTPUT

DEPENDENT VARIABLE.. Y

VARIABLE(S) ENTERED ON STEP NUMBER 1.. XX
X

		ANALYSIS OF VARIANCE	DF	SUM OF SQUARES	MEAN SQUARE	F
MULTIPLE R	0.97989	REGRESSION	2.	1484489458.81533	742244729.40767	84.40143
R SQUARE	0.96018	RESIDUAL	7.	61559541.18467	8794220.16924	
ADJUSTED R SQUARE	0.94881					
STANDARD ERROR	2965.50504					

------------------ VARIABLES IN THE EQUATION ------------------

VARIABLE	B	BETA	STD ERROR B	F
XX	-0.0000552	-2.59871	0.00001	20.060
X	4.262400	3.49649	0.70733	36.314
(CONSTANT)	-26391.05			

For those elements of a multiple regression analysis which appear in all three printouts (SSE, s^2, R^2, etc.) you may have noticed the slight differences in their values. These differences are due to rounding errors inherent in the package programs.

EXAMPLE 13.12 How can we conduct tests about the individual model parameters, i.e., tests of H_0: $\beta_i = 0$, from the information provided by the SPSS printout, Figure 13.12(c)?

Solution In contrast to the *t* tests of the SAS and Minitab packaged programs, SPSS conducts tests of H_0: $\beta_i = 0$ using an *F* statistic. The computed *F*-values are given in the SPSS printout in the column titled F in the bottom portion of the printout labelled VARIABLES IN THE EQUATION (see Figure 13.12(c)). It can be shown (proof omitted) that the square root of the *F* statistic appropriate for testing an individual β parameter in the model is equivalent to the absolute value of the familiar Student's *t* statistic, i.e.,

$$|t| = \sqrt{F}$$

For example, to conduct a test of H_0: $\beta_2 = 0$ using the SPSS printout, first locate the quantity $F = 20.060$ corresponding to the variable XX in Figure 13.12(c). Then compute

$$|t| = \sqrt{F} = \sqrt{20.060} = 4.48$$

To determine whether the actual computed *t*-value is positive or negative, observe the sign (positive or negative) of the corresponding parameter estimate, in this case, $\hat{\beta}_2$. The sign of the computed *t*-value will be identical to the sign of the parameter estimate. Since $\hat{\beta}_2$ is negative, the appropriate test statistic is $t = -4.48$. This computed value is then compared to the critical value of *t*, obtained from Table 4, Appendix E, as outlined in Section 13.4.

EXAMPLE 13.13 How can we compute the value of the *F* statistic for testing whether the overall model is useful for predicting *y*, using the information provided by the Minitab printout, Figure 13.12(b)?

Solution In order to compute the value of the *F* statistic when using Minitab, we must use the formula given in Section 13.6,

$$F = \frac{R^2/k}{(1 - R^2)/[n - (k + 1)]}$$

Using the Minitab value $R^2 = .96$ from Figure 13.12(b), and the values $n = 10$ and $k = 2$, we have

$$F = \frac{.96/2}{(1 - .96)/[10 - (3)]} = \frac{.48}{(.04)/7} = 84.0$$

This value agrees with the values given in the SAS (Figure 13.12(a)) and SPSS (Figures 13.12(c)) printouts (except for rounding errors). We outline the method for finding the critical region for the test in Example 13.14.

EXAMPLE 13.14 The *F* statistic for testing whether the overall model is useful for predicting *y*, i.e., testing the null hypothesis that all model parameters (except β_0) equal zero, is shown in the SPSS printout and must be computed if Minitab is used. But in neither case is the attained significance level of the test given. Thus, if we use the SPSS or Minitab regression analysis package, we must compare the printed or computed F-value with those tabulated in the F tables, Tables 5, 6, and 7, Appendix E. What is the appropriate rejection region of the test?

Solution The method for determining the appropriate critical value is given in the box. In Example 13.3, there are $n = 10$ data points and $k = 2$ parameters (β_1 and β_2) in the model other than β_0; hence, the critical value is based on an F distribution with $k = 2$ numerator degrees of freedom and $[10 - (2 + 1)] = 7$ denominator degrees of freedom. Using $\alpha = .05$, the critical value (obtained from Table 6) is $F_{.05} = 4.74$; thus, we will reject H_0 if

$$F > 4.74$$

Since the computed F-value given in the SPSS printout (Figure 13.12(c)) and computed from the Minitab (Figure 13.12(b)) falls within the rejection region, we reach the same conclusion that we reached in Section 13.6: Reject H_0 and conclude that at least one of the model parameters (β_1 and β_2) is nonzero; i.e., the model appears to be useful for predicting y, sale price.

We will not comment on the merits or drawbacks of the various computer program packages because you will have to use the package(s) available at your computer center and will have to become familiar with its output. As we have seen, most of the computer printouts are similar, and it is relatively easy to learn how to read one output after you have become familiar with another.

REJECTION REGION FOR A TEST OF WHETHER THE OVERALL MODEL IS USEFUL FOR PREDICTING y

H_0: $\beta_1 = \beta_2 = \cdots = \beta_k = 0$

Rejection region: $F > F_\alpha$

where the distribution of F depends on k numerator degrees of freedom and $[n - (k + 1)]$ denominator degrees of freedom, F_α is the value such that $P(F > F_\alpha) = \alpha$, and

n = Number of observations

k = Number of parameters in the model (excluding β_0)

13.9 Model Assumptions

In order for the statistical tests, confidence intervals, and prediction intervals of the preceding sections to be valid, certain assumptions must be satisfied. These assumptions, made about the random error term in the general linear model, follow the same general pattern as for the straight-line model. The assumptions are given in the box at the top of page 474.

ASSUMPTIONS ABOUT THE RANDOM ERROR TERM IN THE GENERAL LINEAR MODEL

1. The mean of the probability distribution of the random error is 0.
2. The variance σ^2 of the probability distribution of the random error is constant for all settings of the independent variables in the model.
3. The probability distribution of the random error is normal.
4. The errors associated with any two observations are independent.

Various statistical techniques exist for checking the validity of these assumptions, and there are remedies to be applied when they appear invalid. But these are beyond the scope of this text. However, just as we stated in Chapter 12, these assumptions need not hold exactly in order for the results of a multiple regression analysis to be valid. In fact, in many practical applications they will be adequately satisfied.

13.10 Summary

In this chapter we have discussed some of the methodology of multiple regression analysis. To illustrate the procedure, we utilized the quadratic model

$$y = \beta_0 + \beta_1 x + \beta_2 x^2 + \text{Random error}$$

to explain the curvilinear relationship between a dependent variable, y, and an independent variable, x. The quadratic model is just one type of linear model that can be fitted to a set of data using the method of least squares.

The steps employed in a multiple regression analysis (fitting, testing, and using the prediction equation) are identical to those employed in a simple linear regression analysis (Chapter 12):

1. The form of the probabilistic model is hypothesized (and the appropriate model assumptions are made).
2. The model coefficients are estimated using the method of least squares.
3. The utility of the model is checked.
4. If the model is deemed useful, it may be used to make estimates and to predict values of y to be observed in the future.

We stress that this is not intended to be a complete coverage of multiple regression analysis. Whole texts have been devoted to this topic. However, we have presented the core necessary for a basic understanding of multiple regression and general linear models. If you are interested in a more extensive coverage, you may wish to consult the references at the end of this chapter.

KEY WORDS

General linear model
Multiple regression analysis
Quadratic model
Multiple coefficient of determination: R^2
F test for determining whether the model is useful for predicting y

EXERCISES

13.1 Underinflated or overinflated tires can increase tire wear and decrease gas mileage. A new tire was tested for wear at different pressures with the results shown in Table 13.2. Suppose you are interested in modeling the relationship between y and x.

TABLE 13.2

PRESSURE x, pounds per square inch	MILEAGE y, thousands
30	29
31	32
32	36
33	38
34	37
35	33
36	26

a. Plot the data on a scattergram.
b. If you were given only the information for $x = 30, 31, 32, 33$, what kind of model, linear or quadratic, would you suggest? For $x = 33, 34, 35, 36$? For all the data?

13.2 A study was conducted to compare the effects of several factors on teachers' attitudes toward handicapped students. One factor, number of years of teaching experience, was found to be at least linearly related to teachers' attitudes toward the handicapped.

a. Hypothesize a linear model relating y, Attitude toward handicapped students (as measured with a standardized attitude scale), to x, Number of years of teaching experience.
b. A researcher believes that curvature may be present in the graph of the response model relating y to x. Give the form of the quadratic model relating teachers' attitudes toward the handicapped to years of teaching experience.
c. Twenty teachers were randomly selected and the values of y and x measured for each. The researcher fit the model $y = \beta_0 + \beta_1 x + \beta_2 x^2 + \text{Random error}$ to the data with the following results:

$$\hat{y} = 50 + 5x - .1x^2$$

$$s_{\hat{\beta}_2} = .03$$

Is there sufficient evidence to indicate that the quadratic term in years of experience, x^2, is useful for predicting attitude score? Use $\alpha = .05$.

13.3 The length of a mosquito's proboscis plays a large role in determining its feeding habits. An entomologist has proposed the following model for predicting the length of a mosquito's proboscis:

$$y = \beta_0 + \beta_1 x_1 + \beta_2 x_2 + \beta_3 x_3 + \text{Random error}$$

where

y = Length of proboscis (millimeters)

x_1 = Dry weight (milligrams)

x_2 = Length of wing (millimeters)

x_3 = Width of wing (millimeters)

An analysis of data obtained from a sample of 44 mosquitos of a particular species produced the following least squares model:

$$\hat{y} = .968 + .292x_1 + .614x_2 - .201x_3$$

Also, $s_{\hat{\beta}_1} = .248$, $s_{\hat{\beta}_2} = .131$, $s_{\hat{\beta}_3} = .267$, $R^2 = .536$

a. Do these statistics indicate that the overall model is useful in predicting proboscis length? Use a significance level of $\alpha = .10$. [*Hint:* Use the value of R^2.]

b. Do the data provide sufficient evidence (at $\alpha = .05$) to indicate that x_3 (width of wing) is an important variable for predicting y using the first-order model specified above? [*Hint:* Test H_0: $\beta_3 = 0$.]

13.4 Before accepting a job, a computer at a major university estimates the cost of running the job in order to see if the user's account contains enough money to cover the cost. As part of the job submission, the user must specify estimated values for two variables—Central processing unit (CPU) time and Lines printed. While the CPU time required and the lines printed do not account for the complete cost of the run, it is thought that knowledge of their values should allow a good prediction of job cost. The following model is proposed to explain the relationship of lines printed and CPU time to job cost:

$$E(y) = \beta_0 + \beta_1 x_1 + \beta_2 x_2 + \beta_3 x_1 x_2$$

where

y = Job cost

x_1 = Lines printed

x_2 = CPU time (tenths of a second)

Records from twenty previous runs were used to fit this model. A portion of the SAS printout is shown in Figure 13.13.

a. Identify the least squares model that was fitted to the data.

b. What are the values of SSE and s^2 (estimate of σ^2) for the data?

c. What do we mean by the statement: This value of SSE (see part b) is minimum?

FIGURE 13.13 Portion of the SAS Printout for Exercise 13.4

SOURCE	DF	SUM OF SQUARES	MEAN SQUARE	F VALUE	PR > F
MODEL	3	43.25090461	14.41696820	84.96	0.0001
ERROR	16	2.71515039	0.16969690	R-SQUARE	STD DEV
CORRECTED TOTAL	19	45.96605500		0.940931	0.41194283

PARAMETER	ESTIMATE	T FOR HO: PARAMETER = 0	PR > |T|	STD ERROR OF ESTIMATE
INTERCEPT	0.04564705	0.22	0.8313	0.21082636
X1	0.00078505	5.80	0.0001	0.00013537
X2	0.23737262	7.50	0.0001	0.03163301
X1*X2	−0.00003809	−2.99	0.0086	0.00001273

X1	X2	PREDICTED VALUE	LOWER 95% CL FOR MEAN	UPPER 95% CL FOR MEAN
2000	42	8.38574865	7.32284845	9.44864885

13.5 Refer to Exercise 13.4 and the portion of the SAS printout shown.

a. Is there evidence that the overall model is useful for predicting job cost? Test at $\alpha = .05$.

b. What assumptions are necessary for the validity of the test conducted in part a?

13.6 Refer to Exercise 13.4 and the portion of the SAS printout shown. Use a 95% confidence interval to estimate the mean cost of computer jobs that print 2,000 lines and require 4.2 seconds of CPU time.

13.7 Refer to Exercise 12.23. The breeder of thoroughbred horses has been advised that the prediction model could probably be improved if a quadratic term were added. The following model is therefore proposed:

$$y = \beta_0 + \beta_1 x + \beta_2 x^2 + \text{Random error}$$

where, as before,

y = Length of life of horse (years)

x = Gestation period of horse (days)

The least squares model was found to be

$$\hat{y} = 20.70 - .0145x + .00005x^2$$

with $F = 26.25$. Test the overall utility of the quadratic model (recall that $n = 7$) at significance level $\alpha = .05$.

13.8 Many colleges and universities develop regression models for predicting the grade-point average (GPA) of incoming freshmen. This predicted GPA can then be used to make admission decisions. Consider the proposed model

$$y = \beta_0 + \beta_1 x_1 + \beta_2 x_2 + \text{Random error}$$

where

y = Freshman college GPA

x_1 = Verbal score on college entrance exam (percentile)

x_2 = Mathematics score on college entrance exam (percentile)

Data collected on a random sample of 100 college freshmen were used to fit the model, with the following results:

$$\hat{y} = -1.5705 + .02573x_1 + .033614x_2$$

$$F = 39.51;\ R^2 = .4489$$

a. Interpret the value of R^2.

b. Is there evidence that the overall model is useful for predicting freshman GPA? Use $\alpha = .05$.

c. Predict the GPA of an incoming college freshman with college entrance exam scores of $x_1 = 75$ and $x_2 = 90$.

13.9 Most companies institute rigorous safety programs in order to assure employee safety. Suppose that sixty reports of accidents over the last year at a company are randomly selected, and that the number of hours the employee had worked before the accident occurred, x, and the amount of time the employee lost from work, y, are recorded. A quadratic model is proposed to investigate a fatigue hypothesis that more serious accidents occur near the end of a day than near the beginning. Thus the proposed model

$$E(y) = \beta_0 + \beta_1 x + \beta_2 x^2$$

was fitted to the data, and part of the computer printout appears in Figure 13.14.

FIGURE 13.14
Portion of the Computer Printout for Exercise 13.9

SOURCE	DF	SUM OF SQUARES	MEAN SQUARE	F VALUE
MODEL	2	112.110	56.055	1.28
ERROR	57	2496.201	43.793	R-SQUARE
TOTAL	59	2608.311		.0430

a. Do the data support the fatigue hypothesis? Use $\alpha = .05$ to test whether the proposed model is useful in predicting the lost work time, y.

b. Does the result of the test in part a necessarily mean that no fatigue factor exists? Explain. [*Hint:* The true model of the relationship between y and x may include higher order or other terms.]

13.10 Refer to Exercise 13.9. Suppose the company persists in using the quadratic model, despite its apparent lack of utility. The fitted model is

$$\hat{y} = 12.3 + .25x - .0033x^2$$

where $\hat{y}$ is the predicted time lost (days) and x is the number of hours worked prior to an accident.

a. Use the model to estimate the mean number of days missed by all employees who have an accident after 6 hours of work.

b. Suppose the 95% confidence interval for the estimated mean in part a is (1.35, 26.01). What is the interpretation of this interval? Does this interval reconfirm your conclusion about this model in Exercise 13.4?

13.11 Mendenhall and McClave (1981) present a case study that involves fitting the linear model

$$E(y) = \beta_0 + \beta_1 x_1 + \beta_2 x_2$$

where

y = Demand for a product

x_1 = Price of a product

$x_2 = \frac{1}{a}$, where a is the advertising expenditure employed to market the product.

The data and an SPSS computer printout for the regression analysis are shown in Table 13.3 and Figure 13.15 (see next page), respectively.

TABLE 13.3
Annual Demand, Price, and Advertising Expenditures for Processed Grapefruit

OBSERVATION	y, million gallons	x_1, dollars per gallon	a, million dollars
1. 1972–1973	53.52	1.294	1.837
2. 1971–1972	51.34	1.344	1.053
3. 1970–1971	49.31	1.332	.905
4. 1969–1970	45.93	1.274	.462
5. 1968–1969	51.65	1.056	.576
6. 1967–1968	38.26	1.102	.260
7. 1966–1967	44.29	.930	.363

a. Write the least squares prediction equation.

b. Find R^2 on the computer printout, Figure 13.15, and interpret its value.

c. Do the data provide sufficient evidence to indicate that the overall model contributes information for the prediction of product demand? Explain.

d. Conduct tests about the individual β parameters, i.e., test $H_0\colon\ \beta_1 = 0$ and $H_0\colon\ \beta_2 = 0$. Use $\alpha = .05$ in each case. Do both the price, x_1, and advertising expenditure, x_2, contribute information for the prediction of product demand, y?

13.12 Refer to Exercise 12.52. In an attempt to improve upon the ability of the model to predict urban travel times, Cook and Russell (1980) added a second

FIGURE 13.15 SPSS Computer Printout for Exercise 13.11

```
DEPENDENT VARIABLE..     Y              ANNUAL DEMAND
VARIABLE(S) ENTERED ON STEP NUMBER  1..    X1          ANNUAL AVERAGE PRICE
                                           X2          RECIPROCAL OF A


MULTIPLE R            0.97937                ANALYSIS OF VARIANCE    DF       SUM OF SQUARES        MEAN SQUARE           F
R SQUARE              0.95917                REGRESSION               2.           162.26148           81.13074      45.98577
ADJUSTED R SQUARE     0.93876                RESIDUAL                 4.             6.90683            1.72671
STANDARD ERROR        1.31404

----------------- VARIABLES IN THE EQUATION ------------------
VARIABLE              B              BETA       STD ERROR B        F
X1             -10.09196         -0.30553        4.51577        4.994
X2              -5.334766        -1.15883        0.62937       71.849
(CONSTANT)      69.75354

               OBSERVED            PREDICTED
SEQNUM            Y                    Y             RESIDUAL
     1         53.52000             53.79047       -0.2704874
     2         51.34000             51.12367        0.2163125
     3         49.31000             50.41628       -1.106280
     4         45.93000             45.34926        0.5807204
     5         51.65000             49.83467        1.815310
     6         38.26000             38.11386        0.1461202
     7         44.29000             45.67168       -1.381695
```

independent variable—weighted average speed limit between the two urban locations. The proposed model takes the form

$$y = \beta_0 + \beta_1 x_1 + \beta_2 x_2 + \text{Random error}$$

where

y = Urban travel time (minutes)

x_1 = Distances between locations (miles)

x_2 = Weighted speed limit between locations (miles per hour)

This model was fitted to the data pertaining to passenger cars and to the data pertaining to trucks, with the following results:

PASSENGER CARS	TRUCKS
$\hat{y} = 5.46 + 2.15x_1 - .09x_2$	$\hat{y} = 4.84 + 3.92x_1 - .09x_2$
$R^2 = .687$; $n = 567$	$R^2 = .771$; $n = 918$

a. Is the model useful for predicting the urban travel times of passenger cars? Use $\alpha = .05$.

b. Is the model useful for predicting the urban travel times of trucks? Use $\alpha = .05$.

13.13 A psychologist conducted an experiment to investigate the effects of various factors on an individual's level of arousal. The following model was proposed:

$$y = \beta_0 + \beta_1 x_1 + \beta_2 x_2 + \beta_3 x_3 + \beta_4 x_4 + \text{Random error}$$

where

y = Level of arousal (standardized score)

x_1 = Oral temperature (°F)

x_2 = Time of day (hours)

x_3 = Level of extraversion (standardized score)

x_4 = Administration of caffeine (yes = 1, no = 0)

The study included a random sample of 69 subjects; data collected on the subjects were subjected to a multiple regression analysis, with the following results:

$$\hat{y} = 78 - .1x_1 - .05x_2 + .3x_3 + 17x_4$$

$$s_{\hat{\beta}_1} = .0611,\ s_{\hat{\beta}_2} = .0052,\ s_{\hat{\beta}_3} = .0537,\ s_{\hat{\beta}_4} = 4.73$$

$$F = 117.33,\ R^2 = .88$$

a. Interpret the value of R^2.

b. Is the overall model useful for predicting level of arousal? Use $\alpha = .01$.

c. Conduct independent t tests on the individual β's (i.e., test H_0: $\beta_i = 0, i = 1, 2, 3, 4$). Use a significance level of $\alpha = .01$ in each test. Which of the independent variables are useful for predicting y?

13.14 A supermarket chain conducted an experiment to investigate the effect of price p (in dollars) on the weekly demand y (in pounds) for a house brand of coffee. Eight supermarkets that had nearly equal past records of demand for the product were used in the experiment. Eight prices were randomly assigned to the stores and were advertised using the same procedures. The number of pounds of coffee sold during the following week was recorded for each of the stores and is shown in Table 13.4.

TABLE 13.4 Coffee Sales, Exercise 13.14

DEMAND y, pounds	PRICE p, dollars
1,120	3.00
999	3.10
932	3.20
884	3.30
807	3.40
760	3.50
701	3.60
688	3.70

a. Find the least squares prediction equation for fitting the model

$$E(y) = \beta_0 + \beta_1 x$$

to the data, letting $x = 1/p$. The Minitab computer printout for fitting this model to the data is shown in Figure 13.16.

b. Do the data provide sufficient evidence to indicate that the model contributes information for the prediction of demand?

c. Find the value of the coefficient of determination and interpret its value.

```
THE REGRESSION EQUATION IS
Y = - 1180. + 6808. X

                                          ST. DEV.     T-RATIO =
          COLUMN      COEFFICIENT         OF COEF.     COEF/S.D.
          --             -1180.             108.         -10.96
X         C3              6808.             358.          19.00

R-SQUARED = 98.4 PERCENT

ANALYSIS OF VARIANCE

 DUE TO        DF          SS          MS=SS/DF
 REGRESSION     1       157717.        157717.
 RESIDUAL       6         2622.           437.
 TOTAL          7       160339.
```

FIGURE 13.16 Minitab Printout for Exercise 13.14

13.15 A physiologist wished to investigate the relationship between the physical characteristics of preadolescent boys and their maximal oxygen uptake (measured in milliliters of oxygen per kilogram of body weight). The data shown in Table 13.5

were collected on a random sample of ten preadolescent boys. As a first step in the data analysis, the researcher fit the first-order model

$$y = \beta_0 + \beta_1 x_1 + \beta_2 x_2 + \beta_3 x_3 + \beta_4 x_4 + \text{Random error}$$

to the data. The output for a SAS regression analysis is shown in Figure 13.17.

a. Write the least squares prediction equation.
b. Give the value of R^2 and interpret its value.
c. Test the hypothesis that chest depth contributes significantly to the prediction of maximal oxygen uptake, i.e., test H_0: $\beta_4 = 0$. Use $\alpha = .05$.
d. If you have access to the appropriate computer package, find a 95% prediction interval for maximal oxygen uptake for a boy with Age = 8.8, Height = 128.1, Weight = 28.0, and Chest depth = 13.7.

FIGURE 13.17 SAS Printout for Exercise 13.15

SOURCE	DF	SUM OF SQUARES	MEAN SQUARE	F VALUE	PR>F
MODEL	4	0.20206274	0.05051568	23.18	0.0020
ERROR	5	0.01089726	0.00217945	R-SQUARE	STD DEV
CORRECTED TOTAL	9	0.21296000		0.948830	0.04668461

PARAMETER	ESTIMATE	T FOR H0: PARAMETER = 0	PR>|T|	STD ERROR OF ESTIMATE
INTERCEPT	−3.42634621	−4.08	0.0096	0.84078488
AGE	−0.03999447	−1.94	0.1105	0.02065180
HEIGHT	0.04814593	6.34	0.0014	0.00759695
WEIGHT	0.00179116	0.33	0.7556	0.00544641
CHEST	−0.07709906	−0.92	0.3979	0.08343441

TABLE 13.5 Maximal Oxygen Uptake Data, Exercise 13.15

MAXIMAL OXYGEN UPTAKE y	AGE x_1, years	HEIGHT x_2, centimeters	WEIGHT x_3, kilograms	CHEST DEPTH x_4, centimeters
1.54	8.4	132.0	29.1	14.4
1.74	8.7	135.5	29.7	14.5
1.32	8.9	127.7	28.4	14.0
1.50	9.9	131.1	28.8	14.2
1.46	9.0	130.0	25.9	13.6
1.35	7.7	127.6	27.6	13.9
1.53	7.3	129.9	29.0	14.0
1.71	9.9	138.1	33.6	14.6
1.27	9.3	126.6	27.7	13.9
1.50	8.1	131.8	30.8	14.5

13.16 [*Note:* This exercise is for students who have access to a multiple regression computer routine.] Literacy rate is a reflection of the educational facilities and quality of education available in a country, and mass communication plays a large part in the educational process. In an effort to relate the literacy rate of a country to various mass communication outlets, a demographer proposed to relate the response

$$y = \text{Literacy rate}$$

to the independent variables

x_1 = Number of daily newspaper copies (per 1,000 population)

x_2 = Number of radios (per 1,000) population

x_3 = Number of television sets (per 1,000 population)

Use a multiple regression computer routine to fit the model

$$y = \beta_0 + \beta_1 x_1 + \beta_2 x_2 + \beta_3 x_3 + \text{Random error}$$

to the data of Table 13.6.

TABLE 13.6 Literacy Rates, Exercise 13.16

COUNTRY	NEWSPAPER COPIES per 1,000	RADIOS per 1,000	TELEVISION SETS per 1,000	LITERACY RATE
Czechoslovakia	280	266	228	.98
Italy	142	230	201	.93
Kenya	10	114	2	.25
Norway	391	313	227	.99
Panama	86	329	82	.79
Philippines	17	42	11	.72
Tunisia	21	49	16	.32
USA	314	1,695	472	.99
USSR	333	430	185	.99
Venezuela	91	182	89	.82

CASE STUDY 13.1 Tree Clones and Their Water Balance

Poplar trees are known by ecologists and forestry experts for their rapid growth rate and cross-breeding capacity. In fact, because of this high growth potential, clones of certain hybrid poplars are being considered as potential sources of fiber and fuel. Associated with this rapid growth rate is the poplar's need for an abundance of water—the transpiration rate of poplars (on the basis of leaf area) may be twice as high as that of other deciduous trees. Pallardy and Kozlowski (1981) conducted an investigation of water relations in poplars.* They write: "The substantial water requirements of poplars make them interesting subjects for studies of water balance, and particularly of adaptive responses by which they avoid severe water deficits."

*S. G. Pallardy and T. T. Kozlowski. "Water relations of *Populus* clones." *Ecology,* February 1981, 62, 159–169. Copyright 1981, the Ecological Society of America.

Two plants from each of two poplar clones (identified by the numbers 5263 and 5271) and three plants from each of two other poplar clones (identified by the numbers 5331 and 5319) were selected for the study from a group of cuttings planted at the Hugo Sauer Nursery near Rhinelander, Wisconsin. (The four clones were chosen because of their differing growth rates.) Since clones of poplar trees may differ in root penetration, and thus in their capacity to extract water from the soil, the researchers decided to examine the relationship between the soil water potential of the clones and their transpiration rates. For a period of approximately two weeks, the soil water potential y (in megapascals) and transpirational flux density x (in micrograms per centimeter squared per second) were measured each day for individually selected leaves of the plants. The data for each of the four clones were to be analyzed separately.

Previous studies had shown that the slope of the relationship between y and x estimates the resistance to water flow in the soil and plant. Pallardy and Kozlowski, wishing to obtain an estimate of this slope, first fit the linear model

$$y = \beta_0 + \beta_1 x + \text{Random error}$$

and then the quadratic model

$$y = \beta_0 + \beta_1 x + \beta_2 x^2 + \text{Random error}$$

to the four data sets using multiple regression. Since the quadratic relationship produced higher coefficients of determination (R^2), they report only those results.

TABLE 13.7 Regression Results, Case Study 13.1

CLONE	LEAST SQUARES PREDICTION EQUATION	$s_{\hat{\beta}_1}$	$s_{\hat{\beta}_2}$	R^2	n
5263	$\hat{y} = -1.47 - 2.53x + .12x^2$	.14	.01	.62	418
5319	$\hat{y} = -1.48 - 1.90x + .08x^2$	.14	.01	.59	417
5331	$\hat{y} = -1.11 - 2.43x + .14x^2$	.19	.02	.53	315
5271	$\hat{y} = -1.67 - 1.89x + .07x^2$	.10	.01	.68	315

For each of the four clones:

a. Interpret the value of R^2.
b. Test the hypothesis that the overall model is useful for predicting the soil water potential of leaves from the poplar clone. Use $\alpha = .05$.
c. Is there evidence of curvature in the response model relating soil water potential y to transpirational flux density x? Use $\alpha = .05$. [*Hint:* Test H_0: $\beta_2 = 0$.]
d. List any assumptions required for the validity of the tests conducted in parts b and c.

CASE STUDY 13.2 Business Economists: Overworked and Underpaid?

Other than food and sex, nothing is quite as universally interesting as the size of our paychecks. Workers everywhere are preoccupied with the subject, and by the nature of their profession, this is no doubt especially true for economists. Why am I paid so little? Why does my friend in Chicago get a bigger salary for a less demanding job? Where does my boss get the crazy notion that my pay is so generous? And what can I do to justify a raise?

In an attempt to answer these questions, put forth in the introductory paragraph to their article,* Preston and Fiedler developed a model for determining "fair" compensation for an economist, based upon his or her contribution to a firm's total output. The authors use the method of multiple regression analysis to examine the relationship between an economist's pay and productivity. "Linear regression methods," they write, "are the mainstay of economic research, but all too frequently economists find themselves using such procedures for information about sales, new orders, or steel production when they are thinking about their own compensation and working conditions. This [research] uses the familiar statistical tools [i.e., multiple regression], applied to the NABE salary survey data, to isolate significant findings."

The NABE is the National Association of Business Economists, an organization which conducts periodic "salary characteristic" surveys of its members. These surveys provide a wealth of information that allows business economists to compare their own salaries with others in the profession. However, Preston and Fiedler point out that the survey's content allows salary comparisons only within single categories of classification. For example, "an NABE member can compare his salary with the average for his industry, the average for his level of education, and the average for his location. But he cannot make a comparison for all three characteristics simultaneously. That is, he cannot find out how he stacks up against other NABE members who, e.g., have a Master's degree and work for a bank in New York City." To make the required comparisons, the authors conducted a multiple regression analysis on the data provided by the $n = 1{,}393$ responses to the 1978 NABE survey. The results are given below.

$$\hat{y} = 9.393 + .224x_1 + .019x_2 + .049x_3 + .190x_4 + .245x_5 - .180x_6 - .281x_7 - .266x_8 + .067x_9 + .078x_{10} + .122x_{11}$$

where

y = Natural logarithm of *total annual compensation* ($)

x_1 = Natural logarithm of *experience* (years)

x_2 = Number of *persons supervised*

x_3 = Level of *education* (no degree = 0, bachelor's = 4, master's = 6, all but Ph.D. = 8, Ph.D. = 10)

x_4 = *Sex* (male = 1, female = 0)

x_5 = Employed in *investments industry* (yes = 1, no = 0)

x_6 = Employed in *nonprofit research organization* (yes = 1, no = 0)

x_7 = Employed in *government* (yes = 1, no = 0)

x_8 = Employed by *academic institution* (yes = 1, no = 0)

x_9 = Located in *New York City* (yes = 1, no = 0)

x_{10} = *Economic advisor* (yes = 1, no = 0)

x_{11} = *General administration-economist* (yes = 1, no = 0)

---

*N. L. Preston and E. R. Fiedler, "Overworked and underpaid?" *Business Economics*, January 1980. Reprinted by permission of the National Association of Business Economists.

Also, $R^2 = .54$ and all individual tests on the β coefficients are significant at $\alpha = .01$ (i.e., all tests have attained significance levels less than .01).

Preston and Fiedler illustrate the use of the least squares prediction equation for determining "fair" compensation for an economist with an example, shown here in the box.

USING THE REGRESSION EQUATION

For those not familiar with the process, here is an illustration of how to compare your compensation with the profession as a whole by plugging your own characteristics into the estimating equation. Our example is a business economist with 10 years of professional experience, who supervises 3 other employees, has earned a Master's degree, is male, is employed in banking in New York City, and whose responsibilities are in corporate planning.

y = Natural log of total compensation = Sum of the following:

Constant term		=		=	9.393
x_1	(experience)	=	$.224 \times$ natural log of 10		
			$.224 \times 2.3025$	=	.516
x_2	(no. supervised)	=	$.019 \times 3$	=	.057
x_3	(education)	=	$.049 \times 6$	=	.294
x_4	(sex)	=	$.190 \times 1$	=	.190
x_5	(securities and investments)	=	$.245 \times 0$	=	0
x_6	(nonprofit res.)	=	$-.180 \times 0$	=	0
x_7	(government)	=	$-.281 \times 0$	=	0
x_8	(academic)	=	$-.266 \times 0$	=	0
x_9	(New York City)	=	$.067 \times 1$	=	.067
x_{10}	(economic advisor)	=	$.078 \times 0$	=	0
x_{11}	(general administration-economist)	=	$.122 \times 0$	=	0
			SUM	=	10.517

Natural antilog of 10.517 = $36,938, which is the estimated total compensation for a business economist with this set of characteristics.

a. Interpret the value of R^2.

b. Test whether the overall model is useful for predicting total compensation for business economists. Use a significance level of $\alpha = .05$. [*Hint:* Use the value of R^2 to calculate the F statistic.]

c. Predict the total compensation for a business economist with 6 years of experience, who supervises 25 other employees, has earned a Ph.D. degree, is female, is employed in government in Washington, D.C., and whose responsibilities are as economic advisor to the president.

We conclude this case study with some final remarks by Preston and Fiedler.

> Every economist can use this regression to gauge where he or she stands competitively in the salary derby. Are you ahead of the pack, or behind? Plug your own characteristics into the equation and it will yield an estimate of your total compensation—a normative value of your services. If your actual salary is higher, you are a 'winner' in the derby! Alternatively, an actual pay level below the estimate would provide some justification for the feeling that you are overworked and underpaid. At the same time, however, it marks you a 'loser'! In the latter case you can . . . use the regression [equation] to demand higher pay from your boss! But if you're a winner, you may want to quickly discard this issue of *Business Economics* before your boss happens upon it!

(The authors' interpretations of the multiple regression results in terms of "winners" and "losers" is, of course, facetious. They note that the prediction equation is far from complete (refer to your answer to part a), for the model fails to include such unmeasurable factors as innate ability, communications skills, "horse sense," and just plain luck.)

CASE STUDY 13.3 It Never Rains in California

The advent of multiple regression analysis has brought to geography a more precise, numerical approach to the age-old problem of describing the interrelationships we observe around us.*

Taylor (1980) sought to describe the method of multiple regression to the research geographer "in a completely nontechnical manner." For the purposes of illustration, he chose to investigate the variation in average annual precipitation in California—"a typical research problem which would be tackled using multiple regression analysis."

Data on average annual precipitation (y), altitude (x_1), latitude (x_2), and distance from the Pacific coast (x_3) were obtained for thirty meteorological stations scattered throughout the state. The data set is reproduced in Table 13.8. As a first attempt at explaining the average annual precipitation in California, Taylor proposed the model

$$y = \beta_0 + \beta_1 x_1 + \beta_2 x_2 + \beta_3 x_3 + \text{Random error}$$

Taylor's interpretations of the β parameters in the model are as follows:

> [β_0] is the base constant and is an estimate of the value of the dependent variable when all the independent variables are zero. In our example it is level of precipitation associated with zero altitude, latitude, and distance from the coast. [β_1, β_2, β_3, and β_4] are regression coefficients which relate each independent variable to the dependent variable. . . . They tell how much change in the dependent variable is associated with a change of one unit of an independent variable. In our example the regression coefficient (β_1) that relates altitude to precipitation is an estimate of how much precipitation increased in inches for an increase of one foot of altitude.

*P. J. Taylor. "A pedagogic application of multiple regression analysis." *Geography*, July 1980, *65*, 203–212.

TABLE 13.8

STATION	AVERAGE ANNUAL PRECIPITATION y, inches	ALTITUDE x_1, feet	LATITUDE x_2, degrees	DISTANCE FROM COAST x_3, miles
1. Eureka	39.57	43	40.8	1
2. Red Bluff	23.27	341	40.2	97
3. Thermal	18.20	4,152	33.8	70
4. Fort Bragg	37.48	74	39.4	1
5. Soda Springs	49.26	6,752	39.3	150
6. San Francisco	21.82	52	37.8	5
7. Sacramento	18.07	25	38.5	80
8. San Jose	14.17	95	37.4	28
9. Giant Forest	42.63	6,360	36.6	145
10. Salinas	13.85	74	36.7	12
11. Fresno	9.44	331	36.7	114
12. Pt. Piedras	19.33	57	35.7	1
13. Pasa Robles	15.67	740	35.7	31
14. Bakersfield	6.00	489	35.4	75
15. Bishop	5.73	4,108	37.3	198
16. Mineral	47.82	4,850	40.4	142
17. Santa Barbara	17.95	120	34.4	1
18. Susanville	18.20	4,152	40.3	198
19. Tule Lake	10.03	4,036	41.9	140
20. Needles	4.63	913	34.8	192
21. Burbank	14.74	699	34.2	47
22. Los Angeles	15.02	312	34.1	16
23. Long Beach	12.36	50	33.8	12
24. Los Banos	8.26	125	37.8	74
25. Blythe	4.05	268	33.6	155
26. San Diego	9.94	19	32.7	5
27. Daggett	4.25	2,105	34.09	85
28. Death Valley	1.66	−178	36.5	194
29. Crescent City	74.87	35	41.7	1
30. Colusa	15.95	60	39.2	91

The least squares model relating y to x_1, x_2, and x_3 was found to be

$$\hat{y} = -102.5314 + .004x_1 + 3.4536x_2 - .1426x_3$$

with a multiple coefficient of determination of $R^2 = .5942$.

a. Interpret the value of R^2.

b. Predict the value of y at the Eureka station ($x_1 = 43$, $x_2 = 40.8$, $x_3 = 1$).

c. The difference between the observed value y (in this example, the actual precipitation) and the predicted value $\hat{y}$, i.e., $(y - \hat{y})$, is the error of prediction or *residual*. Use your answer to part b and the data in Table 13.8 to compute the residual for the Eureka station.

Taylor used the residuals for all 30 stations to detect the presence of an additional independent variable. He found that the residuals exhibited a fairly consistent pattern. Stations located on the westward-facing slopes of the California mountains invariably had residuals which were positive (i.e., the least squares model under-predicted the level of precipitation), while stations on the leeward of the mountains had residuals which were negative (i.e., the least squares model over-predicted the level of precipitation). In Taylor's words, "This suggests a very clear shadow effect of the mountains, for which California is known. We can add this to the model by incorporating a further variable (x_4) which we will term shadow effect. This will be what statisticians refer to as a 'dummy variable' taking only the values 0 and 1. All stations in the lee of mountains will score 1, other stations score 0." Stations 1, 4, 5, 6, 9, 12, 16, 17, 21, 22, 23, 26, and 29 were assigned a value of $x_4 = 0$, the remaining stations a value of $x_4 = 1$.

The model with shadow effect takes the form

$$y = \beta_0 + \beta_1 x_1 + \beta_2 x_2 + \beta_3 x_3 + \beta_4 x_4 + \text{Random error}$$

Taylor fit the shadow effect model to the data and obtained the least squares equation

$$\hat{y} = -99.1909 + .0021x_1 + 3.4893x_2 - .6518x_3 - 16.1660x_4$$

d. The multiple coefficient of determination for the shadow effect model was found to be $R^2 = .7374$. Compare this value to the value of R^2 for the first model. Interpret your findings.

e. Use the value of R^2 to test whether the overall model (with shadow effect) is useful for predicting average annual precipitation y. Use $\alpha = .05$.

f. Predict the value of y at the Eureka station using the shadow effect model. (At the Eureka station, $x_1 = 43$, $x_2 = 40.8$, $x_3 = 1$, and $x_4 = 0$.) Compute the residual for this prediction and interpret its value.

g. List any assumptions required for the validity of the multiple regression technique used in this case study.

REFERENCES

Barr, A., Goodnight, J., Sall, J., Blair, W., & Chilko, D. *SAS user's guide.* 1979 ed. SAS Institute, P. O. Box 10066, Raleigh, N. C. 27605.

Mendenhall, W., & McClave, J. T. *A second course in business statistics: Regression analysis.* San Francisco: Dellen, 1981. Chapter 4.

Neter, J., & Wasserman, W. *Applied linear statistical models.* Homewood, Ill.: Richard Irwin, 1974. Chapter 7.

Pallardy, S. G., & Kozlowski, T. T. "Water relations of *Populus* clones." *Ecology,* February 1981, *62,* 159–169.

Preston, N. L., & Fiedler, E. R. "Overworked and underpaid?" *Business Economics,* January 1980, *15,* 9–15.

Ryan, T. A., Joiner, B. L., & Ryan, B. F. *Minitab student handbook.* North Scituate, Mass.: Duxbury, 1979.

Taylor, P. J. "A pedagogic application of multiple regression analysis." *Geography,* July 1980, *65,* 203–212.

Fourteen

Categorical Data and the Chi-Square Distribution

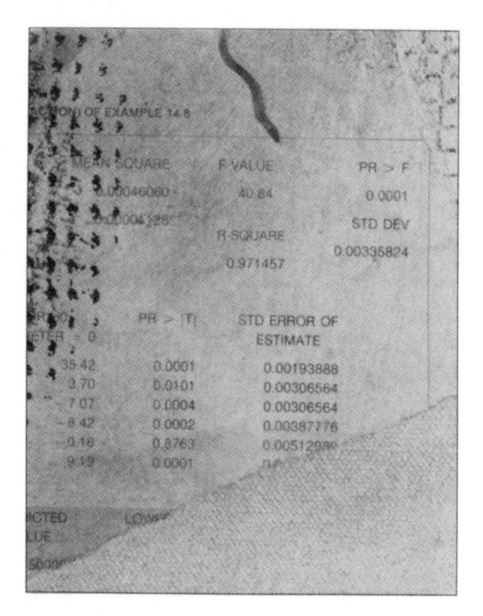

If you shop for groceries, are you an apathetic shopper, a convenience shopper, an involved shopper, or a price shopper? But more important to a grocery chain is whether the perceived price level of a store depends upon your shopping type. To answer this question, we need to be able to test for a dependence between two qualitative variables, Type of store and Type of shopper. In this chapter we will show you how to conduct this test, and we will deal more thoroughly with the grocery shopping problem in Case Study 14.3.

Contents

14.1 An Example of Categorical Data: The Case of the Herring Gulls

In Chapter 2 we stated that data could be one of two types, quantitative or qualitative. Qualitative (or ***categorical***) data are responses that can be classified or categorized, each into exactly one of two or more categories. Opinion polls in which each response could fall into one of three categories (favor, do not favor, and no opinion) produce categorical data, as do many other experiments in the biological, social, and physical sciences. We learned in Chapters 8–10 how to analyze categorical data for the special case where the data were collected in a binomial experiment (an experiment with two categories). In this chapter we will learn how to analyze categorical data for the general case where the number of categories is two or more.

To illustrate the types of questions that arise in the analysis of categorical data, we will examine one portion of data collected by Joanna Burger (1981). Burger's paper, "On becoming independent in herring gulls: Parent-young conflict" (*The American Naturalist,* April 1981*), deals with the competition for survival between parents and newly born offspring. Specifically, she notes that parental care of offspring, the amount of care and the length of time it is administered, affects parents and offspring in different ways. The greater the amount of care and the longer it is received by the offspring, the better is the quality of their lives and the greater are their chances of survival. In contrast, increasing the amount of parental care and providing it over a longer period of time produces a physical drain on the parents and reduces their prospects of long-term survival. Thus, a gain-loss relationship exists between offspring and parents. As the survival prospects for the offspring increase, the long-term survival prospects for the parents decrease (and vice versa). Fortunately, the amount of parental care provided by most species of animals, birds, etc., strikes a reasonable balance between these survival rates but it does vary from species to species.

Burger examined the amount of parental care (primarily feeding) provided by parent herring gulls to their young during the months that followed the hatching of their chick offspring. (The gulls were observed on Clam Island, a salt marsh island in Barnegat Bay, New Jersey, during 1978.) One aspect of her study dealt with the length of time that parental care was provided to the offspring for this species. Parents were observed to feed their young on a continual and steady basis until fledging (when the chicks began to fly), an event that took place approximately 45 days after birth during the month of July. Parents and offspring returned to the nesting area for parental feeding after fledging but the amount of feeding dropped off and appeared to stop approximately 3 to 4 months after birth (in September).

To support this theory, Burger gives the number of dead herring gull young found per month in the vicinity of Clam Island during the months July to December and she provides data from similar studies made on the same species at two other locations.** These data, expressed as percentages of the total number of deaths per

*Paraphrased and reprinted by permission of The University of Chicago Press. © 1981 by The University of Chicago.

**Burger gives data collected at five locations. To simplify our analysis, we present the data at only three locations.

location, are shown in Table 14.1. Note how the percentage of dead herring gull young tends to rise and peak in September and October, two to three months after fledging, at all three locations.

TABLE 14.1
Percentage of Deaths in the Six Months Following Fledging for Herring Gulls

TIME PERIOD	LOCATION		
	New Jersey	Netherlands	England
July	9%	1%	4%
August	16	7	25
September	46	32	37
October	22	37	16
November	5	15	13
December	2	8	5
TOTAL NUMBER OF DEATHS	42	406	241

Burger notes that the dead chicks (in September) seemed to have died from starvation, thus suggesting that parental feeding was withdrawn approximately two months after July fledging. Similar results were obtained in England. In contrast, the percentage of dead young gulls in the Netherlands study seems to peak a month later, in October. Can we draw the conclusion that the length of time parental feeding is provided to herring gull young varies from one location to another, or can the differences among the three distributions of percentages be attributed to sampling variation? To answer this question, we will test the null hypothesis that the distributions of percentages are identical for the three locations, New Jersey, the Netherlands, and England. If we are able to reject this hypothesis, we will conclude that the differences seen in Table 14.1 are not due to chance but are, instead, likely due to differences in environmental conditions, the availability of food, etc., at the three locations in the years that the data were collected.

14.2 The Chi-Square Statistic

The data of Table 14.1 represent the sample percentages of the number of dead herring gull young who died at each of the three locations during the months July to December. If there is evidence of a significant difference in the distributions of these sample percentages from one location to another, we will conclude that the distributions differ for the populations of all young herring gulls who might have been observed at the three locations during years in which the data were collected.

Unfortunately, we cannot compare the sample percentages because the sample sizes (and hence the reliability of the data) vary from one location to another. Consequently, we must convert the sample percentages to numbers of dead gulls and we must make certain that the number of birds in each cell is moderately large.* To satisfy this latter requirement of our method, we will combine the data for July and August and also the data for November and December, thus producing four time

*We will state precisely how large in Section 14.3.

periods. The resulting number of birds who died during these time periods, at each of the three locations, is shown in Table 14.2. This table, showing the number of dead gulls in each of the twelve categories (or ***cells***) of the table, as well as the row and column totals, is called a ***contingency table.*** As you will subsequently see, the row and column totals, as well as the total number of dead gulls observed at the three locations, are used in calculating the value of the appropriate test statistic. [*Note:* In all of the examples and exercises that follow, the data will be presented in the form of a contingency table, i.e., as cell counts rather than as percentages.]

TABLE 14.2
A Contingency Table for the Dead Herring Gull Young (Entries Are Numbers of Dead Gulls)

TIME PERIOD	LOCATION			TOTALS
	New Jersey	Netherlands	England	
July–August	11	32	70	113
September	19	130	89	238
October	9	150	39	198
November–December	3	94	43	140
TOTALS	42	406	241	689

EXAMPLE 14.1 Let us suppose the distributions of percentages of dead herring gull young observed in the four time period categories are identical for the three locations, i.e., the distribution of percentages of dead gulls is *independent* of location. Now examine the observed counts in Table 14.2. How can we determine whether these observed counts contradict our assumption of identical distributions of percentages for the three locations?

Solution One method of detecting a difference in the distributions of percentages of responses in a category (say, July–August) is to compare the observed number of responses in each of the three cells *corresponding to location* to the number of responses we would expect to see if in fact the percentages were identical. The larger the deviations between the observed numbers and their respective expected numbers, the more evidence there is that the true percentages are different.

The expected number of responses in the cells, i.e., the expected cell counts, although unknown, can be estimated using the row and column totals of the contingency table. An estimate of the expected number of responses for any cell in the table is computed by multiplying the row total and column total corresponding to the row and column in which the cell is located, and then dividing this product by the total sample size (which appears in the lower right-hand corner of the table). For example, the estimated number of responses for the cell in the upper left-hand corner of the table, denoted e_1, is

$$e_1 = \frac{(113)(42)}{689} = 6.888$$

Thus, if there is no difference in the distributions of percentages of responses for the three locations, we would expect to observe approximately seven dead herring gull young at the New Jersey location during the July–August time period. Similarly, the estimated expected number of responses for the cells in the first row corresponding to the Netherlands and England are, respectively,

$$e_2 = \frac{(113)(406)}{689} = 66.586$$

and

$$e_3 = \frac{(113)(241)}{689} = 39.525$$

Compare these expected counts with the observed counts in the table.

In Example 14.2, we will compute a statistic which will help us determine if the differences between the observed cell counts and expected cell counts are large enough for us to conclude that the distributions of percentages are different.

In the box we give the general formula for computing an estimate of the expected number of responses in any cell of the table.

GENERAL FORMULA FOR COMPUTING EXPECTED CELL COUNTS

$$e_i = \frac{(R_i)(C_i)}{n}$$

where

e_i = Estimated expected count for cell i

R_i = Row total corresponding to the row in which cell i appears

C_i = Column total corresponding to the column in which cell i appears

n = Sample size

EXAMPLE 14.2 What is the appropriate test statistic in a test of hypothesis to determine whether the distributions of percentages of responses in the categories corresponding to time periods differ from location to location?

Solution From Example 14.1, we learned that the differences between the observed number of responses in the cells of Table 14.2 and their respective expected number of responses are used to detect differences in the response percentages. How should this information be combined into a single statistic? Let o_i denote the observed number of responses in cell i and e_i denote the (estimated) expected number of responses in cell i. Then the appropriate statistic which measures the disagreement between the data and the assumption that the distributions of percentages are identical is denoted by the symbol χ^2 and is computed as

$$\chi^2 = \frac{(o_1 - e_1)^2}{e_1} + \frac{(o_2 - e_2)^2}{e_2} + \cdots + \frac{(o_k - e_k)^2}{e_k}$$

where k is the number of cells in the table (i.e., the number of possible categories into which the responses may be classified). For the data of Table 14.2, we have 4 row categories (time periods) and 3 column categories (location); hence, $k = (4)(2) = 8$ cells for which we need to compute the expected counts. These expected cell counts and the observed cell counts are then substituted into the formula, and the χ^2 statistic computed.

HOW TO COMPUTE THE χ^2 STATISTIC IN A TEST FOR DIFFERENCES OF PERCENTAGES OF CATEGORICAL RESPONSES

1. Compute the (estimated) expected cell counts, $e_1, e_2, \ldots, e_k$, for each of the k cells of the table.
2. Compute the differences $(o_1 - e_1), (o_2 - e_2), \ldots, (o_k - e_k)$ for each of the k cells of the table, where o_i denotes the observed number of responses in cell i.
3. Compute the χ^2 test statistic as follows:

$$\chi^2 = \frac{(o_1 - e_1)^2}{e_1} + \frac{(o_2 - e_2)^2}{e_2} + \cdots + \frac{(o_k - e_k)^2}{e_k}$$

Under the assumption that the distributions of percentages of categorical responses are identical, the χ^2 statistic has a sampling distribution which is approximately a ***chi-square (χ^2) distribution.*** The chi-square probability distribution, like the t distribution, is characterized by a quantity, called the ***degrees of freedom associated with the distribution.*** Several chi-square probability distributions with different degrees of freedom are shown in Figure 14.1.

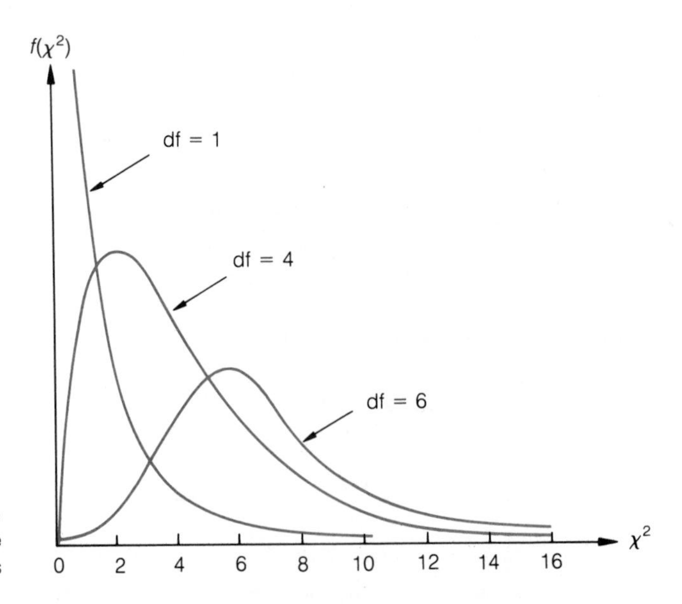

FIGURE 14.1
Several Chi-Square Probability Distributions

Throughout this chapter, we will use the words *chi-square* and the Greek symbol χ^2 interchangeably. Tabulated values of the χ^2 distribution are given in Table 8, Appendix E; part of this table is shown in Figure 14.2. Entries in the table give an upper-tail value of χ^2, call it χ^2_α, such that $P(\chi^2 > \chi^2_\alpha) = \alpha$.

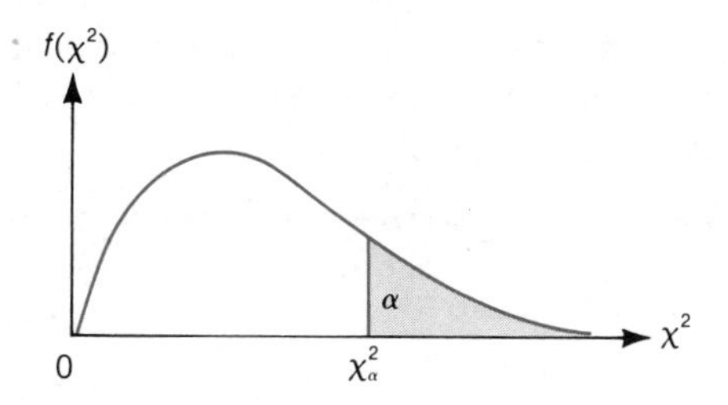

DEGREES OF FREEDOM	$\chi^2_{.100}$	$\chi^2_{.050}$	$\chi^2_{.025}$	$\chi^2_{.010}$	$\chi^2_{.005}$
1	2.70554	3.84146	5.02389	6.63490	7.87944
2	4.60517	5.99147	7.37776	9.21034	10.5966
3	6.25139	7.81473	9.34840	11.3449	12.8381
4	7.77944	9.48773	11.1433	13.2767	14.8602
5	9.23635	11.0705	12.8325	15.0863	16.7496
6	10.6446	12.5916	14.4494	16.8119	18.5476
7	12.0170	14.0671	16.0128	18.4753	20.2777
8	13.3616	15.5073	17.5346	20.0902	21.9550
9	14.6837	16.9190	19.0228	21.6660	23.5893
10	15.9871	18.3070	20.4831	23.2093	25.1882
11	17.2750	19.6751	21.9200	24.7250	26.7569
12	18.5494	21.0261	23.3367	26.2170	28.2995
13	19.8119	22.3621	24.7356	27.6883	29.8194
14	21.0642	23.6848	26.1190	29.1413	31.3193
15	22.3072	24.9958	27.4884	30.5779	32.8013
16	23.5418	26.2962	28.8454	31.9999	34.2672
17	24.7690	27.5871	30.1910	33.4087	35.7185
18	25.9894	28.8693	31.5264	34.8053	37.1564
19	27.2036	30.1435	32.8523	36.1908	38.5822

FIGURE 14.2 Reproduction of Part of Table 8, Appendix E

EXAMPLE 14.3 Find the tabulated value of χ^2 corresponding to 9 degrees of freedom which cuts off an upper-tail area of .05.

Solution The value of χ^2 which we seek appears (shaded) in the partial reproduction of Table 8, Appendix E, given in Figure 14.2. The columns of the table identify the value of α associated with the tabulated value χ^2_α and the rows correspond to the degrees of freedom. For this example, we have df = 9 and $\alpha = .05$. Thus, the tabulated value of χ^2 corresponding to 9 degrees of freedom is

$$\chi^2_{.05} = 16.9190$$

We use the tabulated values of χ^2 given in Table 8, Appendix E, to locate the appropriate rejection region for the test of hypothesis which we carry out in the following section.

EXERCISES

14.1 A company conducted a survey to determine whether the proportion of employees favoring a new pension plan depends upon whether workers hold production, clerical, or management jobs. Four hundred employees were randomly selected for the survey. A summary of their responses is shown in Table 14.3.

TABLE 14.3
Employee Responses, Exercise 14.1

	EMPLOYMENT CATEGORY			TOTALS
	Production	Clerical	Management	
Favor new plan	169	76	26	271
Do not favor new plan	87	35	7	129
TOTALS	256	111	33	400

a. Calculate the percentages of employees in favor of the new pension plan for each of the employee categories.
b. Do these percentages suggest a difference in the proportions for the three employment categories?
c. Why is a statistical test useful in answering part b? [*Note:* You do not need to convert the response counts of Table 14.3 to percentages in order to carry out the statistical test referred to in part c. However, the calculations, part a, become necessary when additional analysis is required. See Section 14.3.]

14.2 Refer to the data in Exercise 14.1.

a. Calculate the number of employees that you would expect to fall in each of the six cells of the contingency table if in fact the percentages who favor the new plan in each employment category are identical.
b. Find the difference between the observed and the (estimated) expected number for each of the six cells.
c. Calculate the value of the chi-square statistic for the contingency table.

14.3 A survey was conducted to determine whether a relationship exists between a new college graduate's expectations of acquiring rewarding employment and the

TABLE 14.4
Graduates' Expectations, Exercise 14.3

	COLLEGE MAJOR				TOTALS
	Social Sciences	Biological Sciences	Physical Sciences	Arts and Humanities	
High expectations for employment	12	27	43	16	98
Modest expectations for employment	36	45	38	27	146
Poor expectations for employment or no opinion	14	6	3	33	56
TOTALS	62	78	84	76	300

graduate's college major. Three hundred graduates were randomly selected from among prospective graduates in the social, biological, and physical sciences, and from the arts and humanities. A summary of their responses is shown in Table 14.4.

a. Calculate the percentage of social sciences graduates in each of the three employment expectation response categories. Calculate these percentages for graduates in the biological and physical sciences and arts and humanities.
b. Do the distributions of percentages for the four college majors appear to differ?
c. Why is a statistical test useful in answering part b?

14.4 Refer to the data in Exercise 14.3.

a. Calculate the number of graduates that you would expect to fall in each of the twelve cells of the contingency table if in fact the distributions of percentages for the four college majors are identical.
b. Find the difference between the observed and the (estimated) expected number for each of the twelve cells.
c. Calculate the value of the chi-square statistic for the contingency table.

14.5 For each of the following combinations of α and degrees of freedom (df), find the value of chi-square, χ^2_α, that places an area α in the upper tail of the chi-square distribution. Sketch the chi-square distribution showing the locations of α and χ^2_α.

a. $\alpha = .05$, df = 7
b. $\alpha = .10$, df = 16
c. $\alpha = .01$, df = 10
d. $\alpha = .025$, df = 8
e. $\alpha = .005$, df = 5

14.6 Find the value of α that corresponds to:

a. A value $\chi^2 = 20.4831$, based on df = 10
b. A value $\chi^2 = 13.2767$, based on df = 4
c. A value $\chi^2 = 26.2962$, based on df = 16

14.7 The advances of medical technology have led to an increased survival rate of infants born with genetic diseases. Since no true cure for genetic disease is currently available, some physicians provide genetic counseling for their patients in order to prevent the birth of genetically defective infants. However, genetic counseling has faced a certain amount of resistance from both physicians and patients. Weitz (1979) conducted a survey of general and family practitioners, pediatricians, and obstetrician-gynecologists in the cities of Phoenix and Tucson, Arizona. In one part of the study, each physician was classified according to religion and opinion on genetic counseling. A summary of the responses for Jewish, Protestant, and Catholic physicians is shown in Table 14.5 on page 500.

a. Scan the data of Table 14.5. Do you believe there is evidence of a difference in the proportion of physicians who strongly support genetic counseling for the three religions?
b. Why is a statistical test useful in answering part a?
c. Calculate the number of physicians you would expect to fall in each of the six cells of the contingency table.

TABLE 14.5
Physicians' Responses, Exercise 14.7

	RELIGION			TOTALS
	Jewish	Protestant	Catholic	
Strongly support genetic counseling	21	36	10	67
Do not strongly support genetic counseling	26	142	52	220
TOTALS	47	178	62	287

Source: Rose Weitz, "Barriers to acceptance of genetic counseling among primary care physicians." *Social Biology,* Fall 1979, *26,* 192.

d. Find the difference between the observed and the (estimated) expected number for each of the six cells.

e. Calculate the value of the chi-square statistic for the contingency table.

14.3 Analysis of the Herring Gull Data

Do the data on the number of dead herring gull young collected during the months July to December reflect actual differences in the distributions of percentages of responses in the time period categories among the populations of all herring gull young that might be observed at the three locations? We will answer this question by performing a complete test of hypothesis on the data of Table 14.2.

EXAMPLE 14.4 Set up the appropriate null and alternative hypotheses for the test.

Solution We have already stated the null hypothesis in Section 14.1. For convenience, we restate it here:

> H_0: The distributions of percentages of dead herring gull young in the categories corresponding to time period are identical for the three locations. Equivalently, we are hypothesizing that the distribution of percentages for one direction of classification (Time Periods) in Table 14.2 is *independent of* the second direction of classification (Location).

The alternative hypothesis can then be phrased:

> H_a: The distributions of percentages of responses in the categories corresponding to time period differ for the three locations. That is, the distributions *depend* on location.

The null and alternative hypotheses for the statistical tests which we have discussed previously were given in terms of values of a target parameter or the differences between two target parameters. Since there are a large number of parameters to consider in this categorical data analysis (12, in fact, one proportion for each cell in Table 14.2), it is more convenient to write H_0 and H_a as above.

EXAMPLE 14.5 Use the data of Table 14.2 to compute the χ^2 test statistic.

Solution The first step in the calculation of the test statistic is to compute the (estimated) expected count, e_i, in each cell of the table. In Example 14.1, we gave the formula for finding expected counts and illustrated its use by computing the expected number of responses in each of the three cells in the top row of Table 14.2. Using the formula

$$e_i = \frac{(R_i)(C_i)}{n}$$

we computed the expected counts for the remaining 9 cells. These values are given in parentheses with the observed counts in Table 14.6.

Once these (estimated) expected cell frequencies have been computed, we substitute their values into the formula for the χ^2 statistic given in the box following Example 14.2 in Section 14.2.

TABLE 14.6 Observed and Estimated Expected (in Parentheses) Counts for the Herring Gull Data

	LOCATION		
	New Jersey	**Netherlands**	**England**
July–August	11	32	70
	(6.888)	(66.586)	(39.525)
September	19	130	89
	(14.508)	(140.244)	(83.248)
October	9	150	39
	(12.070)	(116.673)	(69.257)
November–December	3	94	43
	(8.534)	(82.496)	(48.970)

The computed χ^2 test statistic is

$$\chi^2 = \frac{(o_1 - e_1)^2}{e_1} + \frac{(o_2 - e_2)^2}{e_2} + \cdots + \frac{(o_{12} - e_{12})^2}{e_{12}}$$

$$= \frac{(11 - 6.888)^2}{6.888} + \frac{(32 - 66.586)^2}{66.586} + \cdots + \frac{(43 - 48.970)^2}{48.970}$$

$$= 75.89$$

Is the value 75.89 large enough for us to conclude that the distributions of percentages are different? We answer this question in the following example.

EXAMPLE 14.6 Specify the rejection region for the test and make the proper conclusion. Use a significance level of $\alpha = .05$.

Solution If the computed value of the χ^2 statistic is "too large," there is evidence to reject H_0 and conclude that the differences in the distributions of percentages of responses exhibited by the data in Table 14.2 are due not to sampling variation but to actual differences in the populations. To determine how large χ^2 must be before it is too large to be attributed to chance, we make use of the fact that, under certain conditions (see Section 14.4), the sampling distribution of χ^2 is approximately a chi-square probability distribution if the null hypothesis is true. For a significance level of $\alpha = .05$, we need to find the tabulated value of $\chi^2_{.05}$ in Table 8, Appendix E. If our computed test statistic is larger than this critical value, i.e., if

$$\chi^2 > \chi^2_{.05}$$

then we will reject the null hypothesis. However, we noted in Section 14.2 that the tabulated values of χ^2 in Table 8, Appendix E, depend upon degrees of freedom. In analyses of data reported in the form of Table 14.2, the appropriate degrees of freedom will be $(r-1)(c-1)$, where r is the number of rows and c is the number of columns in the table.

For the herring gull data, we have $r = 4$ rows and $c = 3$ columns; hence, the appropriate number of degrees of freedom for χ^2 is

$$\text{df} = (r-1)(c-1) = (3)(2) = 6$$

The tabulated value of $\chi^2_{.05}$ corresponding to 6 df is 12.5916; thus, the rejection region (shaded in Figure 14.3) is

$$\chi^2 > 12.5916$$

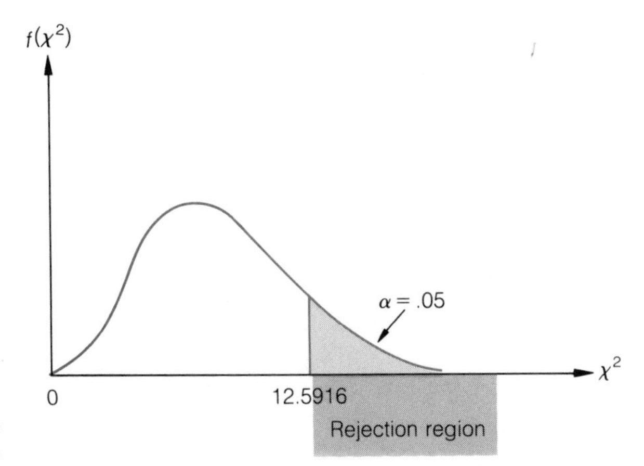

FIGURE 14.3 Rejection Region for the Herring Gull Example

Since the computed value $\chi^2 = 75.89$ exceeds the critical value 12.5916, we reject the null hypothesis in favor of the alternative. At significance level $\alpha = .05$, the data of Table 14.2 indicate that the distributions of the percentages of responses in the categories corresponding to the time periods are different among the three locations. Thus, it appears that the length of parental feeding time varies among the three locations. (Recall that a high frequency of deaths of young gulls in a given time period locates the point at which parental feeding becomes negligible.)

The elements of a χ^2 test for differences in the distributions of percentages of responses from opinion polls, sample surveys, etc., are summarized in the box.

GENERAL FORM OF A CHI-SQUARE TEST FOR INDEPENDENCE OF TWO DIRECTIONS OF CLASSIFICATION

H_0: The two directions of classification in the contingency table are independent

H_a: The two directions of classification in the contingency table are dependent

Test statistic: $\chi^2 = \sum_{i=1}^{k} \frac{(o_i - e_i)^2}{e_i}$

where

k = Number of cells (rc) in the table consisting of r rows and c columns

o_i = Observed number of responses in cell i

e_i = Estimated expected number of responses in cell i

Rejection region: $\chi^2 > \chi^2_\alpha$

where χ^2_α is the tabulated value of the chi-square distribution based on $(r-1)(c-1)$ degrees of freedom such that $P(\chi^2 > \chi^2_\alpha) = \alpha$.

Assumptions: See Section 14.4.

The data of Table 14.2 could also be used to obtain an estimate of the percentage of responses in a specific category of the population or to test hypotheses about the value of a particular percentage. The techniques are identical to those of Chapters 8, 9, and 10 where we considered large-sample inferences about binomial proportions. We illustrate with two examples.

EXAMPLE 14.7 Use the data of Table 14.2 to estimate the proportion of deaths of herring gull young that might be expected to occur during the month of September in the Netherlands. Use a 95% confidence interval.

Solution Let p be the true proportion of dead herring gull young found in the Netherlands that died during the month of September. Since we are now interested in only one population proportion, we may treat the data of Table 14.2 as binomial data. We can think of a dead gull found in the Netherlands as being classified into one of only two categories: (1) died in September; or (2) did not die in September. The proportion p then represents the probability of success in a binomial experiment consisting of $n = 406$ trials (i.e., 406 observations of dead herring gulls in the Netherlands), where a "success" is defined as observing a dead herring gull in September.

Following the procedure of Chapter 8, a 95% confidence interval for p is given by

$$\hat{p} \pm 1.96\sqrt{\frac{\hat{p}\hat{q}}{n}}$$

where $\hat{p}$ is the sample proportion of successes in n trials and $\hat{q} = 1 - \hat{p}$. From Table 14.2, the number of "successes" (i.e., the number of dead gulls in September in the Netherlands) is 130. Thus, our estimate is

$$\hat{p} = \frac{\text{Number of successes in the sample}}{n} = \frac{130}{406} = .32$$

Substituting $\hat{p} = .32$, $\hat{q} = 1 - .32 = .68$ and $n = 406$ into the confidence interval for p, we obtain

$$\begin{aligned}\hat{p} \pm 1.96\sqrt{\frac{\hat{p}\hat{q}}{n}} &= .32 \pm 1.96\sqrt{\frac{(.32)(.68)}{406}} \\ &= .32 \pm (1.96)(.02315) \\ &= .32 \pm .045 \\ &= (.275,\ .365)\end{aligned}$$

We estimate, with 95% confidence, that the percentage of dead herring gulls found in the Netherlands that died in September falls between 27.5% and 36.5%.

EXAMPLE 14.8 Refer to Table 14.2. Test to determine whether the proportion of dead herring gull young found in New Jersey that died in October is larger than the corresponding proportion in England. Use $\alpha = .01$.

Solution Let p_1 and p_2 be the true proportions of dead herring gull young found in New Jersey and England, respectively, that died during the month of October. To determine whether p_1 is larger than p_2, we test the hypotheses:

$$H_0: \ (p_1 - p_2) = 0$$

$$H_a: \ (p_1 - p_2) > 0$$

Again, we recognize that the portion of the data of Table 14.2 which we are interested in may be treated as data from two independent binomial experiments; consequently, the procedure outlined in Chapters 9 and 10 for testing the difference between binomial proportions may be applied. The appropriate test statistic is given by

$$z = \frac{(\hat{p}_1 - \hat{p}_2) - 0}{\sqrt{\dfrac{\hat{p}_1\hat{q}_1}{n_1} + \dfrac{\hat{p}_2\hat{q}_2}{n_2}}}$$

We need to obtain the estimates $\hat{p}_1$ and $\hat{p}_2$ from Table 14.2.

Consider first the sample of 42 dead herring gull young found in New Jersey. We can think of a dead gull as being classified into one of two categories: (1) died in October; or (2) did not die in October. The proportion p_1 then represents the probability of success in a binomial experiment consisting of $n_1 = 42$ trials, where "success" is defined as observing a dead herring gull in October. The sample proportion of successes used to estimate p_1 is obtained from Table 14.2 as follows:

$$\hat{p}_1 = \frac{\substack{\text{Number of young herring gulls which died in October} \\ \text{in the sample of 42 dead gulls found in New Jersey}}}{42}$$

$$= \frac{9}{42} = .214$$

Similarly, p_2 is the probability of success in a binomial experiment consisting of $n_2 = 241$ trials and is estimated by

$$\hat{p}_2 = \frac{\substack{\text{Number of young herring gulls which died in October} \\ \text{in the sample of 241 dead gulls found in England}}}{241}$$

$$= \frac{39}{241} = .162$$

Substituting these values into the test statistic, we have

$$z = \frac{(\hat{p}_1 - \hat{p}_2) - 0}{\sqrt{\frac{\hat{p}_1\hat{q}_1}{n_1} + \frac{\hat{p}_2\hat{q}_2}{n_2}}} = \frac{(.214) - (.162)}{\sqrt{\frac{(.214)(.786)}{42} + \frac{(.162)(.838)}{241}}}$$

$$= \frac{.052}{.0676} = .769$$

The rejection region for this test is given by

$$z > z_\alpha$$

For $\alpha = .01$, $z_\alpha = z_{.01} = 2.33$ (from Table 3, Appendix E). Since the computed value $z = .769$ does not exceed the critical value 2.33, we fail to reject the null hypothesis at $\alpha = .01$. There is insufficient evidence to claim that the proportion of dead herring gulls that died in October is greater for those dead gulls found in New Jersey than for those found in England. However, it is important to remember that the data were collected at the two locations in different years. Consequently, the percentages p_1 and p_2 can be compared only if the environmental conditions as well as food supplies at the two locations were similar at the times the samples were collected.

We conclude this section with another example of a χ^2 test for differences between the distributions of percentages of categorical responses.

EXAMPLE 14.9 In the past few years, many U.S. farmers have gone on strike to protest that the current prices of farm products, chiefly grains, are less than the cost of production. A common, though controversial, goal of many of the strikes is to induce farmers to reduce production, thereby reducing surpluses and boosting prices. Suppose that a survey of 100 randomly selected U.S. farmers was conducted to determine whether a relationship exists between a farmer's decision to participate in a strike and the farmer's opinion concerning the necessity for a 50% cutback in production. Use the survey results, Table 14.7, to determine if the distributions of percentages of responses corresponding to farmers' opinions about the production cutback are different for those farmers who participate and those who do not participate in the strike. Test at significance level $\alpha = .05$.

TABLE 14.7
Farmer Opinion Poll

		PARTICIPATION IN STRIKE		TOTALS
		Yes	No	
50% CUTBACK IN PRODUCTION	Favor	20	8	28
	Undecided	37	2	39
	Opposed	22	11	33
TOTALS		79	21	100

Solution Notice that the only difference between the structure of the farmer opinion poll and that of the herring gull investigation is that there are no restrictions on the row or column totals in Table 14.7. In the herring gull study, the column totals, i.e., the total number of dead gulls found during the months of July–December at each of the three locations, were fixed in the sense that the researcher could not observe more than 42, 406, and 241 dead gulls at New Jersey, the Netherlands, and England, respectively. The column totals were restricted to the numbers of dead gulls at the three locations. In contrast, the column totals of Table 14.7 were not known until after the farmer opinion poll was conducted. Thus, prior to the survey, there were no restrictions on the number of farmers in the sample who participated in the strike. Fortunately, this fact does not affect the analysis. We proceed, then, with the test of hypothesis outlined in the box following Example 14.6.

The null and alternative hypotheses we wish to test are

H_0: The distributions of percentages of responses corresponding to opinion on the cutback in production are identical for striking and nonstriking farmers (i.e., the two directions of classification "participation in strike" and "position on production cutback" are independent).

H_a: The distributions of percentages of responses corresponding to opinion on the cutback in production are different for striking and nonstriking farmers (i.e., the two directions of classification "participation in strike" and "position on production cutback" are dependent).

From the box, we have

Test statistic: $\chi^2 = \sum_{i=1}^{k} \frac{(o_i - e_i)^2}{e_i}$

Rejection region: For $\alpha = .05$ and $(r-1)(c-1) = (3-1)(2-1) = 2$ df, we will reject H_0 if $\chi^2 > \chi^2_{.05}$, where $\chi^2_{.05} = 5.99147$

The first step in the computation of the test statistic is to calculate the (estimated) expected cell frequencies, $e_i = R_iC_i/n$, according to the box. Proceeding as in Example 14.1, we obtain

$$e_1 = \frac{R_1C_1}{n} = \frac{(28)(79)}{100} = 22.12$$

$$e_2 = \frac{R_2C_2}{n} = \frac{(28)(21)}{100} = 5.88$$

and so forth. The observed numbers of responses and the expected counts (in parentheses) are shown in Table 14.8. Substituting the values of Table 14.8 into the expression for χ^2, we obtain

$$\chi^2 = \frac{(20-22.12)^2}{22.12} + \frac{(8-5.88)^2}{5.88} + \cdots + \frac{(11-6.93)^2}{6.93} = 9.915$$

Since the computed value $\chi^2 = 9.915$ exceeds the critical value of 5.99147 and thus falls in the rejection region, we have sufficient evidence, at $\alpha = .05$, to conclude that the distributions of percentages of responses corresponding to farmers' opinions about a 50% cutback in production are different for striking and nonstriking farmers.

TABLE 14.8
Observed and Expected (in Parentheses) Counts for the Farmer Poll

		PARTICIPATION IN STRIKE	
		Yes	No
50% CUTBACK IN PRODUCTION	Favor	20 (22.12)	8 (5.88)
	Undecided	37 (30.81)	2 (8.19)
	Opposed	22 (26.07)	11 (6.93)

EXAMPLE 14.10 Refer to Example 14.9. Find the approximate attained significance level of the test. Interpret your result.

Solution The computed value of the χ^2 test statistic in Exercise 14.9 was found to be $\chi_c^2 = 9.915$. Since the null hypothesis of independence will be rejected for large values of χ^2, the attained significance level of the test is given by the probability

$$P(\chi^2 > \chi_c^2) = P(\chi^2 > 9.915)$$

The distribution of the χ^2 statistic in this example is based on 2 df; hence, to find the attained significance level or p-value for the test, we need to search for the value 9.915 in the row corresponding to 2 df in Table 8, Appendix E. This computed value falls between 9.21034, the critical value in the $\chi^2_{.010}$ column, and 10.5966, the critical value in the $\chi^2_{.005}$ column. Thus, the attained significance level of the test falls between the p-values .005 and .01. By convention, we report the larger of these two p-values, namely .01, as the approximate attained significance level of the test. Our interpretation of this value is that we will reject the null hypothesis of independence for any fixed significance level α larger than or equal to .01.

EXERCISES

14.8 Give the degrees of freedom for a test of independence of the two directions of classification in a contingency table with:

a. $r = 2$ rows and $c = 2$ columns
b. $r = 4$ rows and $c = 2$ columns
c. $r = 3$ rows and $c = 3$ columns
d. $r = 3$ rows and $c = 4$ columns

14.9 The results of the employee pension survey, Exercise 14.1, are reproduced in Table 14.9.

TABLE 14.9
Employee Responses, Exercise 14.9

OPINION	EMPLOYMENT CATEGORY Production	Clerical	Management	TOTALS
Favor new plan	169	76	26	271
Do not favor new plan	87	35	7	129
TOTALS	256	111	33	400

a. State the null and alternative hypotheses of interest to the company.
b. How many degrees of freedom will the chi-square statistic for this contingency table possess?
c. Use the chi-square statistic to test the hypotheses specified in part a. Use $\alpha = .10$. Do the data provide sufficient evidence to indicate that employees' attitudes toward the new pension plan depend upon their employment category?

14.10 Find the approximate attained significance level for the test in part c of Exercise 14.9. Interpret its value.

14.11 A medical researcher conducted an experiment to compare the frequencies of occurrence of side effects to treatment by a new anti-arthritic drug for two groups of patients, those who had received prior treatment with the drug and those who had not. Four hundred patients who had received the new anti-arthritic drug were randomly selected from the records at a large hospital. The numbers of patients falling in the four treatment-response categories are shown in Table 14.10.

TABLE 14.10
Number of Patients Who Incurred Side Effects When Treated by an Anti-Arthritic Drug

	SIDE EFFECTS	NO SIDE EFFECTS	TOTALS
No prior treatment	39	271	310
Prior treatment	7	83	90
TOTALS	46	354	400

a. Suppose the researcher is interested in detecting a difference in the proportions of patients who suffer side effects for the two types of patients (those who have and those who have not received prior treatment with the drug). State the null and alternative hypotheses the researcher wishes to test.
b. How many degrees of freedom will the chi-square statistic for this contingency table possess?

c. Use the chi-square statistic to test the hypotheses specified in part a. Use $\alpha = .10$.

d. Note that you could also test the null hypothesis, part a, using a two-tailed test as described in Section 10.7. Test the null hypothesis (at $\alpha = .10$) using the z-statistic and compare your results with those of the chi-square test, part c.

e. Refer to part d. When analyzing a contingency table, can you always use the method of Section 10.7 in place of the chi-square test? Explain.

14.12 Find the approximate attained significance level for the chi-square test, Exercise 14.11, part c.

14.13 The results of the college graduate survey, Exercise 14.3, are reproduced in Table 14.11.

TABLE 14.11
Graduates' Expectations, Exercise 14.13

	COLLEGE MAJOR				TOTALS
	Social Sciences	Biological Sciences	Physical Sciences	Arts and Humanities	
High expectations for employment	12	27	43	16	98
Modest expectations for employment	36	45	38	27	146
Poor expectations for employment or no opinion	14	6	3	33	56
TOTALS	62	78	84	76	300

a. State the null and alternative hypotheses involved in a test to determine whether the distributions of responses differ among the four types of majors.

b. How many degrees of freedom will the chi-square statistic for this contingency table possess?

c. Use the chi-square statistic to test the hypotheses, part a. Use $\alpha = .05$. Do the data provide sufficient evidence to indicate differences in the patterns of response for the four types of majors?

14.14 Find the approximate attained significance level for the test in part c of Exercise 14.13. Interpret its value.

14.15 Refer to Exercise 14.1. Estimate the proportion of all employees who favor the new pension plan. Use a 90% confidence interval and interpret your result. [*Hint:* See Section 8.4.]

14.16 Refer to Exercise 14.1. Do the data provide sufficient evidence to indicate a difference in preference for the new pension plan between production and clerical employees? Test using $\alpha = .10$. [*Hint:* See Section 10.7.]

14.17 Find the attained significance level for the test in Exercise 14.16. Interpret its value.

14.4 Assumptions: Situations for Which the Chi-Square Test Is Appropriate

We have discussed one method of analyzing data from opinion polls, sample surveys, experiments, etc., which allow for more than two categories for a response. However, as with most statistical procedures, the chi-square test for distributional differences in response percentages will be valid only when certain assumptions are satisfied. These assumptions will be met if the underlying probability distribution of the response data has the properties outlined in the box.

PROPERTIES OF THE UNDERLYING DISTRIBUTION OF CATEGORICAL DATA

ASSUMPTIONS: CHI-SQUARE TEST FOR INDEPENDENCE OF TWO DIRECTIONS OF CLASSIFICATION

1. The experiment consists of n identical trials.
2. There are k possible outcomes to each trial.
3. The probabilities of the k outcomes, denoted by $p_1, p_2, \ldots, p_k$, remain the same from trial to trial, where

$$p_1 + p_2 + \cdots + p_k = 1$$

4. The trials are independent.

Note that the properties in the box closely resemble those of a binomial experiment. (In fact, when $k = 2$, these are the properties of a binomial experiment.) This underlying probability distribution of the response data is simply an extension of the binomial distribution to include more than two possible outcomes.

Because it is widely used, the chi-square test is also one of the most abused statistical procedures. The user should always be certain that the experiment satisfies the boxed assumptions before proceeding with the test. *In addition, the chi-square test should be avoided when the estimated expected cell counts are small, for it is in this instance that the chi-square probability distribution gives a poor approximation to the sampling distribution of the χ^2 statistic.* As a rule of thumb, an estimated expected cell count of at least five will mean that the chi-square distribution can be used to determine an approximate critical value to specify the rejection region.

ADDITIONAL ASSUMPTION FOR THE VALID USE OF THE CHI-SQUARE TEST

5. The (estimated) expected number of responses for each of the k cells should be at least five.

14.5 Summary

Opinion polls, sample surveys, experiments, etc. which allow for more than two categories for a response can be analyzed using the technique outlined in this chapter, namely, the chi-square test for independence of two directions of classification. The appropriate test statistic, called the χ^2 statistic, has a sampling distribution which is (approximately) a chi-square probability distribution and measures the amount of disagreement between the observed number of responses and the expected number of responses in each category.

Caution should be exercised to avoid misuse of the χ^2 procedure. The underlying distribution of the response data should have the properties outlined in the box of Section 14.4. Also, the estimated number of responses in any cell should not be too small.

KEY WORDS

Categorical data
Observed cell counts
Expected cell counts
Chi-square statistic: χ^2
Chi-square probability distribution
Contingency table
Independence of two directions of classification

SUPPLEMENTARY EXERCISES

14.18 For each of the following combinations of α and degrees of freedom (df), find the value of chi-square, χ^2_α, that places an area α in the upper tail of the chi-square distribution. Sketch the chi-square distribution, showing the location of α and χ^2_α for each part of the exercise.

a. $\alpha = .025$, df = 8
b. $\alpha = .05$, df = 5
c. $\alpha = .05$, df = 10
d. $\alpha = .10$, df = 3
e. $\alpha = .01$, df = 2

14.19 Find the values of $\chi^2_{.10}$, $\chi^2_{.05}$, $\chi^2_{.025}$, $\chi^2_{.01}$, and $\chi^2_{.005}$ for df = 2. Draw a rough sketch of a chi-square distribution with 2 df and locate the values of χ^2 along the horizontal axis.

14.20 An archeologist classified artifacts found at two different excavations into one of four periods of civilization. One hundred thirty artifacts were classified at site #1 and 179 at site #2. The data are shown in Table 14.12. Do the data provide

TABLE 14.12
Counts on Artifacts, Exercise 14.20

EXCAVATION SITE	PERIOD 1	2	3	4	TOTALS
1	21	63	29	17	130
2	19	75	71	14	179
TOTALS	40	138	100	31	309

sufficient evidence to indicate that the distributions of artifacts over time periods differ for the two sites? Test using $\alpha = .05$.

14.21 The results of the Arizona physicians' survey, Exercise 14.7, are reproduced in Table 14.13.

TABLE 14.13
Physicians' Responses, Exercise 14.21

	RELIGION			TOTALS
	Jewish	Protestant	Catholic	
Strongly support genetic counseling	21	36	10	67
Do not strongly support genetic counseling	26	142	52	220
TOTALS	47	178	62	287

a. Conduct a test of hypothesis to determine if the two categories of classification, Religion and Support of genetic counseling, are dependent. Use $\alpha = .05$.
b. Calculate the approximate attained significance level for the test, part a. Interpret its value.
c. Estimate the true proportion of Catholic physicians in Arizona who strongly support genetic counseling with a 95% confidence interval.

14.22 A bank conducted a survey to compare the attitudes of young married customers (those having recently opened their first bank accounts) with the attitudes of older, established customers to their new automated teller system. Two hundred customers were randomly selected from each of these categories and asked whether they preferred the automated teller system to the personal service obtained at the bank. A summary of the customers' responses is shown in Table 14.14.

TABLE 14.14
Bank Customers' Responses, Exercise 14.22

	NEW CUSTOMER	ESTABLISHED CUSTOMER	TOTALS
Favor the automatic teller	87	65	152
Favor personal service	113	135	248
TOTALS	200	200	400

a. Do the data provide sufficient evidence to indicate a difference between the proportions of new and established customers who favor use of the automatic teller? Test using $\alpha = .05$. Interpret the results of the test.
b. Conduct this same test using the method of Section 10.7. Compare your result to that obtained in part a.
c. When analyzing a contingency table, can you always use the method of Section 10.7 instead of the chi-square test? Explain.

14.23 Four hundred college students were randomly selected from the student bodies at each of two universities, one located on the west coast and the other in the northeast, and were asked their opinion on the Israeli preemptive bombing of Iraq's atomic reactor in 1981. The responses are shown in Table 14.15. Do the data provide sufficient evidence to indicate a difference in student reactions to the Israeli preemptive bombing at the two universities? Use $\alpha = .05$. Interpret your result.

TABLE 14.15
Opinion Survey, Exercise 14.23

COLLEGE	SUPPORT ACTION	OPPOSE ACTION	NO OPINION	TOTALS
West coast	211	155	34	400
Northeast	249	121	30	400
TOTALS	460	276	64	800

14.24 Along with the technological age comes the problem of workers being replaced by machines. A labor management organization wants to study the problem of workers displaced by automation within three industries. Case reports for 100 workers whose loss of job is directly attributable to technological advances are selected within each industry. For each worker selected, it is determined whether he or she was given another job within the same company, found a job with another company in the same industry, found a job in a new industry, or has been unemployed for longer than 6 months. The results are given in Table 14.16. Does the plight of workers displaced by automation depend on the industry? Test using $\alpha = .01$.

TABLE 14.16
Current Status of Workers Displaced by Automation, Exercise 14.24

		SAME COMPANY	NEW COMPANY (SAME INDUSTRY)	NEW INDUSTRY	UNEMPLOYED
INDUSTRY	A	62	11	20	7
	B	45	8	38	9
	C	68	19	8	5

14.25 Refer to Exercise 14.24. Estimate the difference between the proportions of displaced workers who find work in another industry for industries A and C. Use a 95% confidence interval.

14.26 *Dear enemy recognition* is the term used by naturalists and ecologists for the aggressive behavior of birds, mammals, and ants when their territorial boundaries are violated by one of their own species. Dear enemy recognition is often followed by escalated attacks on the invading animal. Jaeger (1981) explored the possibility that the red-backed salamander employs dear enemy recognition by using chemical signals to distinguish familiar from unfamiliar salamanders. In escalated contests, a salamander will attempt to bite an opponent's snout—an injury which could reduce a salamander's ability to locate prey, mates, and territorial competitors. One part of Jaeger's study focused on a comparison of the proportions of males and females exhibiting wounds in the snout. One hundred forty-four salamanders were collected from a forest, killed, and inspected for scar tissue in the snout. The results are shown in Table 14.17.

TABLE 14.17
Salamander Data, Exercise 14.26

	MALE	FEMALE	TOTALS
Scar tissue in snout	5	12	17
No scar tissue in snout	76	51	127
TOTALS	81	63	144

Source: R. G. Jaeger. "Dear enemy recognition and the costs of aggression between salamanders." *The American Naturalist,* June 1981, *117*, 962–973.

a. Use a chi-square test to determine if there is a difference between the proportions of males and females with scar tissue in the snout. Use $\alpha = .01$.

b. Perform the test in part a using the technique of Section 10.7.

14.27 Blood and Seider (1981) obtained information on 469 stutterers 14 years or younger from speech-language pathologists selected from each of the fifty states. Each of the stutterers had one other accompanying problem and was classified according to therapeutic intervention, as shown in Table 14.18.

TABLE 14.18
Accompanying Problems of Stutterers, Exercise 14.27

ACCOMPANYING PROBLEM	THERAPEUTIC INTERVENTION			TOTALS
	Stuttering therapy only	Both stuttering therapy and other therapy	Other therapy only	
Articulation disorder	10	155	5	170
Language disorder	8	89	7	104
Language disability	6	60	9	75
Others	4	62	54	120
TOTALS	28	366	75	469

Source: G. W. Blood and R. Seider. "The concomitant problems of young stutterers." *Journal of Speech and Hearing Disorders,* February 1981, *46*, 31–33.

a. Use a chi-square test to determine if the distributions of percentages of responses differ among the categories of accompanying problems of young stutterers. Use $\alpha = .05$.

b. Construct a 95% confidence interval for the true proportion of young stutterers with an accompanying language disorder who are receiving stuttering therapy only.

14.28 Time compression, a method of shortening the time required for broadcasting a television commercial, has enabled television advertisers to cut the high costs of television commercials. But can shorter commercials be effective? In order to answer this question, 200 introductory psychology students were randomly divided into three groups. The first group (57 students) was shown a video tape of a television program which included a 30-second commercial; the second group (74 students) was shown the same video tape but with the 24-second time-compressed version of the commercial; and the third group (69 students) was shown a 20-second time-compressed version of the commercial. Two days after viewing the tape, the three groups of students were asked to name the brand that was advertised. The numbers of students recalling the brand name for each of the three groups are given in Table 14.19.

TABLE 14.19
Students' Recall of Brand Name, Exercise 14.28

		TYPE OF COMMERCIAL			TOTALS
		Normal version (30 seconds)	Time-compressed version #1 (24 seconds)	Time-compressed version #2 (20 seconds)	
RECALL OF BRAND NAME	Yes	15	32	10	57
	No	42	42	59	143
TOTALS		57	74	69	200

a. Do the data provide sufficient evidence (at $\alpha = .05$) that the two directions of classification, Type of commercial and Recall of brand name, are dependent? Interpret your results.
b. Construct a 95% confidence interval for the difference between the proportions recalling brand name for viewers of normal and time-compressed (24-second) commercials.

CASE STUDY 14.1 Our Declining Voter Turnout

Political observers have noted with alarm the increase in nonvoting in U.S. presidential elections since 1960.*

In the presidential election of 1960 (Nixon versus Kennedy) 83.7% of the U.S. public reported turning out to cast their vote. However, by 1976 (Ford versus Carter), the reported voter turnout had slipped to 75.2%, a decrease of 8.5%. Political analysts, political science experts, journalists, and American citizens from all walks of life have opinions on the reason for the decline in voter turnout. Their explanations range from political disillusionment, political mistrust, weakening party ties, and a growing voter apathy to income, education, closeness of the election, the changing age distribution of the population, and even the mass media itself. Journalists particularly stress the growing political apathy among Americans and various census studies have indicated that disinterest in politics is the major reason given for nonvoting. However, Stephen D. Shaffer (1981) discounts voter apathy as a factor. In fact, he shows that "people were as interested in politics and election campaigns in 1976 as they were in 1960."

Shaffer attempts to explain decreasing voter turnout by examining each possible explanatory factor (e.g., apathy) in terms of its categories associated with low voter turnout (e.g., somewhat apathetic, very apathetic). He writes: "In order for a factor to account for declining turnout, it must shape the probability of voting in individual elections and its categories associated with low turnout should increase in size over the years to encompass a greater proportion of the voting-age population." Data for Shaffer's research were obtained from the Consortium for Political and Social Research (CPS) National Election Studies of 1960–1976. For various theoretical and methodological reasons, Shaffer excluded data from the eleven southern states of the old Confederacy and data pertaining to those voters under the age of

*S. D. Shaffer. "A multivariate explanation of decreasing turnout in presidential elections, 1960–1976." *American Journal of Political Science,* February 1981, *25*, 68–93. Austin: University of Texas Press.

21. Voters' responses in the various categories of five of the many possible explanatory factors are presented in this case study in Tables 14.20–14.24.*

TABLE 14.20
Variable: Campaign Interest

	NATIONAL ELECTION YEAR					TOTALS
	1960	1964	1968	1972	1976	
Very much interested	521	418	409	251	638	2,237
Somewhat interested	540	415	432	376	753	2,516
Not much interested	280	276	211	196	324	1,287
TOTALS	1,341	1,109	1,052	823	1,715	6,040

TABLE 14.21
Variable: Political Efficacy**

	NATIONAL ELECTION YEAR					TOTALS
	1960	1964	1968	1972	1976	
High	899	637	494	369	693	3,092
Medium	296	269	262	230	490	1,547
Low	146	203	296	224	532	1,401
TOTALS	1,341	1,109	1,052	823	1,715	6,040

TABLE 14.22
Variable: Party Identification

	NATIONAL ELECTION YEAR					TOTALS
	1960	1964	1968	1972	1976	
Strong Democrat	253	258	185	103	223	1,022
Weak Democrat	294	272	247	200	379	1,392
Independent Democrat	100	111	106	86	211	614
Independent	111	86	104	85	232	618
Independent Republican	113	65	92	98	185	553
Weak Republican	207	175	180	133	279	974
Strong Republican	236	135	124	107	182	784
Apolitical	27	7	14	11	24	83
TOTALS	1,341	1,109	1,052	823	1,715	6,040

TABLE 14.23
Variable: Frequency of Reading Newspaper Articles about Campaign

	NATIONAL ELECTION YEAR					TOTALS
	1960	1964	1968	1972	1976	
Regularly	622	438	380	219	456	2,115
Often	174	153	136	104	264	831
Time to time	196	220	214	142	396	1,168
Once in a great while	94	61	71	31	201	458
Never	255	237	251	327	398	1,468
TOTALS	1,341	1,109	1,052	823	1,715	6,040

*Shaffer reports his results in terms of percentages of responses. For the reader's convenience, we have converted cell percentages to cell counts.

**Respondents scored low in political efficacy if they agreed with both of the following statements; "I don't think public officials care much about what people like me think" and "People like me don't have any say about what the government does."

TABLE 14.24
Variable: Age (Years)

	NATIONAL ELECTION YEAR					TOTALS
	1960	1964	1968	1972	1976	
21–24	45	80	85	82	166	458
25–28	123	103	92	78	219	615
29–32	119	103	75	66	146	509
33–36	135	90	79	49	135	488
37–40	137	100	112	55	122	526
41–44	114	109	86	63	96	468
45–48	117	86	109	81	106	499
49–52	95	96	80	59	109	439
53–56	106	70	68	59	113	416
57–60	87	63	65	50	101	366
61–64	66	57	46	41	116	326
65–68	60	45	48	42	79	274
69–72	50	42	35	31	82	240
73–76	50	28	34	31	46	189
77+	37	37	38	36	79	227
TOTALS	1,341	1,109	1,052	823	1,715	6,040

Recall that Shaffer discounts growing apathy as a factor in the declining voter turnout since 1960. However, his analyses of the data suggest a

> combination of four factors is responsible for decreased turnout among all eligible voters—the most important of which is the changing age composition of the electorate. . . . A close second . . . is a growing public feeling of political inefficaciousness since 1960. . . . A third factor is that people have become less reliant on the more intellectually demanding medium of newspapers, which imparts more campaign information and heightens political interest. The fourth factor in explaining decreased turnout is the increase in avowed independents and the decline in intense partisans.

a. Using the procedure outlined in this chapter, analyze the data in Table 14.20, i.e., test the hypothesis that the distributions of percentages of responses corresponding to the categories of the variable Campaign interest differ among the five national election years. Use $\alpha = .05$. Do your results agree with Shaffer's analysis?

b. Repeat part a for the variables Political efficacy (Table 14.21), Party identification (Table 14.22), Frequency of reading newspaper (Table 14.23), and Age (Table 14.24).

c. Because the χ^2 test for independence is a highly abused statistical tool, we emphasize that, in practice, the user should always check the required assumptions *before* applying the test. For the data of Tables 14.20–14.24, carefully check to determine whether each of the assumptions in the box of Section 14.4 is satisfied.

CASE STUDY 14.2 Testing the Effectiveness of an Antiviral Drug

Vira-A is one of the few existing antiviral drugs on the market. Manufactured by Warner-Lambert Co., it is used primarily to treat herpes virus infections of the eye and brain in adults. However, the *Wall Street Journal* (October 8, 1980) reports that Warner-Lambert will seek approval from the Food and Drug Administration (FDA) to use the drug in treating a rare, but usually fatal, infant illness.

Warner-Lambert claims that Vira-A is useful in treating babies suffering from herpes simplex virus infection. The illness, often transmitted from an infected mother, can leave surviving infants with permanent brain damage. In one portion of a study conducted in 18 health centers throughout the country, the drug was given to 24 babies suffering from the disease. Included in the study was a control group of 19 infected babies who were left untreated. The number of infants surviving the illness in each group is given in Table 14.25.

TABLE 14.25

	SURVIVORS	DEATHS	TOTALS
Treated Group (Drug)	15	9	24
Control Group (No Drug)	5	14	19
TOTALS	20	23	43

a. Use a χ^2 test (if appropriate) to determine whether the distributions of the percentages of survivors and deaths are different for infected babies treated with the drug and those left untreated.

b. You can perform the identical analysis of part a by comparing the binomial proportions p_1 and p_2 using the techniques of Chapter 10, where p_1 is the true survival rate of infected babies treated with the drug, and p_2 is the true survival rate of infected babies who are left untreated. Test the hypothesis that p_1 is larger than p_2, using a significance level of $\alpha = .05$.

c. Based *only* on the results of the survivor-rate study shown in this case study, would you back Warner-Lambert's claim that the drug Vira-A is useful in treating babies suffering from herpes simplex virus infection?

CASE STUDY 14.3 Where Grocery Shoppers Shop—High-, Medium-, or Low-Priced Stores

Refer to the case study on the typology of grocery shoppers, Case Study 11.2. Williams et al. (*Journal of Retailing,* Spring 1978) developed the typology based on customers' involvement with the price policies or customer service policies of retail food stores. The four types of shoppers are identified as (1) Apathetic shoppers, (2) Convenience shoppers, (3) Involved shoppers, and (4) Price shoppers.

In addition to the research outlined in Case Study 11.2, Williams et al. also investigated the stores and chains where members of each of the four groups shopped. The analysis necessitated that the stores included in the study (10 primary stores or chain supermarkets) be classified into various categories. Since the price level of grocery stores is one attribute which can be used to differentiate among stores, Williams et al. classified each of the ten stores as either high-, medium-, or low-priced. This classification was accomplished with the help of experts and shoppers, who were asked to evaluate each of the stores on the basis of their perception of the prices charged. Since there were differences in how the experts and shoppers ranked the stores, the combined rankings were used. (The combined rank of a store was simply the rank—high, medium, or low—provided by the majority of sources.)

After the ten stores were classified on price level, counts were made to determine what percentage of each grocery shopper group shopped in either high-priced, medium-priced, low-priced, or nonclassifiable stores. The results are shown in Table 14.26.

TABLE 14.26 Count of Group Respondents Shopping at High-, Medium-, Low- and Nonclassifiable-Priced Stores

		GROCERY SHOPPER GROUP				TOTALS
		Apathetic	Convenience	Price	Involved	
STORE PRICE LEVEL	High	17	38	13	6	74
	Medium	19	55	35	19	128
	Low	14	12	20	3	49
	Nonclassifiable	9	20	13	5	47
TOTALS		59	125	81	33	298

Is there sufficient evidence to indicate that the two classifications "grocery shopper group" and "store price level" are dependent, i.e., are the distributions of percentages of responses corresponding to the "store price level" categories different among the four shopper groups? Use $\alpha = .05$.

REFERENCES

Blood, G. W., & Seider, R. "The concomitant problems of young stutterers." *Journal of Speech and Hearing Disorders,* February 1981, *46,* 31–33.

Burger, J. "On becoming independent in herring gulls: Parent-young conflict." *The American Naturalist,* April 1981, *117.*

Cochran, W. G., "The χ^2 test of goodness of fit." *Annals of Mathematical Statistics,* 1952, *23,* 315–345.

Jaeger, R. G. "Dear enemy recognition and the costs of aggression between salamanders." *The American Naturalist,* June 1981, *117,* 962–973.

McClave, J. T., & Dietrich, F. H. *Statistics.* 2d ed. San Francisco: Dellen, 1982. Chapter 11.

Shaffer, S. D. "A multivariate explanation of decreasing turnout in presidential elections, 1960–1976." *American Journal of Political Science,* February 1981, *25,* 68–93.

Siegel, S. *Nonparametric statistics for the behavioral sciences.* New York: McGraw-Hill, 1956. Chapter 9.

"Warner-Lambert finds antiviral drug useful." *Wall Street Journal,* October 8, 1980, 27.

Weitz, R. "Barriers to acceptance of genetic counseling among primary care physicians." *Social Biology,* Fall 1979, *26,* 189–197.

Williams, R. H., Painter, J. J., & Nicholas, H. R. "A policy-oriented typology of grocery shoppers." *Journal of Retailing,* Spring 1978, *54,* 27–41.

Appendix A Data Set: Starting Salaries of 948 University of Florida Graduates, June 1980 to March 1981

STARTING SALARIES OF 948 UNIVERSITY OF FLORIDA GRADUATES
JUNE 1980 TO MARCH 1981

OBS	DATE	SEX	COLLEGE	MAJOR	SALARY
1	JUN80	M	ENGINEERING	MECHANICAL	$22,900
2	JUN80	F	SCIENCES	CHEMISTRY	$15,400
3	JUN80	F	JOURNALISM	JOURNALISM	$10,400
4	JUN80	M	ENGINEERING	CIVIL	$16,800
5	JUN80	F	SCIENCES	COMPUTER SCIENCE	$18,300
6	JUN80	F	BUSINESS ADMINISTRATION	FINANCE	$15,600
7	JUN80	F	FINE ARTS	GRAPHIC DESIGN	$6,800
8	JUN80	F	SCIENCES	CHEMISTRY	$10,800
9	JUN80	F	NURSING	NURSING	$14,000
10	JUN80	M	ENGINEERING	MECHANICAL	$22,400
11	JUN80	M	BUSINESS ADMINISTRATION	MANAGEMENT	$11,300
12	JUN80	M	FINE ARTS	MUSIC	$11,600
13	JUN80	M	ENGINEERING	MECHANICAL	$26,500
14	JUN80	M	ENGINEERING	CIVIL	$22,000
15	JUN80	F	ARCHITECTURE	BUILDING CONSTRUCTION	$22,800
16	JUN80	M	ARCHITECTURE	BUILDING CONSTRUCTION	$19,000
17	JUN80	M	ENGINEERING	NUCLEAR	$9,000
18	JUN80	F	JOURNALISM	NEWS EDUCATION	$13,100
19	JUN80	M	ENGINEERING	MECHANICAL	$21,100
20	JUN80	F	LIBERAL ARTS	ENGLISH	$6,800
21	JUN80	F	EDUCATION	SPEECH	$12,000
22	JUN80	F	AGRICULTURE	ANIMAL SCIENCE	$15,700
23	JUN80	M	BUSINESS ADMINISTRATION	FINANCE	$13,000
24	JUN80	F	SCIENCES	MATHEMATICS	$6,000
25	JUN80	M	AGRICULTURE	FOOD SCIENCE	$8,500
26	JUN80	F	NURSING	NURSING	$14,800
27	JUN80	M	ENGINEERING	MECHANICAL	$20,100
28	JUN80	M	LAW	LAW	$13,800
29	JUN80	M	BUSINESS ADMINISTRATION	COMPUTER SCIENCE	$17,100
30	JUN80	F	PHYS ED,HEALTH,RECREATION	RECREATION	$4,900
31	JUN80	F	PHARMACY	PHARMACY	$20,400
32	JUN80	M	ENGINEERING	ELECTRICAL	$15,900
33	JUN80	M	PHARMACY	PHARMACY	$25,000
34	JUN80	M	EDUCATION	SOCIAL SCIENCE	$12,900
35	JUN80	F	HEALTH RELATED	MEDICAL TECHNOLOGY	$10,400
36	JUN80	F	NURSING	NURSING	$10,000
37	JUN80	M	ENGINEERING	ELECTRICAL	$17,700
38	JUN80	M	ARCHITECTURE	AEROSPACE ENGINEER	$6,800
39	JUN80	M	ENGINEERING	ELECTRICAL	$18,700
40	JUN80	M	BUSINESS ADMINISTRATION	ACCOUNTING	$10,200
41	JUN80	F	LIBERAL ARTS	SPEECH COMMUNICATIONS	$14,500
42	JUN80	M	ENGINEERING	ELECTRICAL	$18,100
43	JUN80	M	JOURNALISM	BROADCASTING	$6,200
44	JUN80	M	AGRICULTURE	FOOD RESOURCE ECONOMICS	$8,500
45	JUN80	M	ACCOUNTING	ACCOUNTING	$14,500
46	JUN80	M	ENGINEERING	ELECTRICAL	$19,400
47	JUN80	M	AGRICULTURE	AGRONOMY	$10,900
48	JUN80	M	PHARMACY	PHARMACY	$20,800
49	JUN80	M	BUSINESS ADMINISTRATION	ECONOMICS	$12,900
50	JUN80	M	BUSINESS ADMINISTRATION	FINANCE	$13,200
51	JUN80	M	ARCHITECTURE	LANDSCAPING	$8,600
52	JUN80	F	JOURNALISM	PUBLIC RELATIONS	$14,500
53	JUN80	M	LIBERAL ARTS	SPEECH PATHOLOGY	$20,700
54	JUN80	M	BUILDING CONSTRUCTION	BUILDING CONSTRUCTION	$21,700
55	JUN80	M	BUSINESS ADMINISTRATION	FINANCE	$14,500
56	JUN80	F	BUSINESS ADMINISTRATION	FINANCE	$15,400
57	JUN80	M	BUSINESS ADMINISTRATION	ECONOMICS	$22,600
58	JUN80	F	LIBERAL ARTS	POLITICAL SCIENCE	$9,700
59	JUN80	F	HEALTH RELATED	MEDICAL TECHNOLOGY	$12,000
60	JUN80	M	BUSINESS ADMINISTRATION	MANAGEMENT	$8,100
61	JUN80	M	ENGINEERING	MECHANICAL	$23,900
62	JUN80	M	ENGINEERING	CIVIL	$20,600
63	JUN80	F	AGRICULTURE	FORESTRY	$14,600
64	JUN80	F	SCIENCES	COMPUTER SCIENCE	$15,300
65	JUN80	F	ARCHITECTURE	AEROSPACE ENGINEER	$5,000
66	JUN80	M	AGRICULTURE	ANIMAL SCIENCE	$11,000
67	JUN80	F	JOURNALISM	ADVERTISING	$10,800
68	JUN80	F	SCIENCES	COMPUTER SCIENCE	$20,300
69	JUN80	M	BUSINESS ADMINISTRATION	MANAGEMENT	$13,100
70	JUN80	M	ENGINEERING	CIVIL	$23,700
71	JUN80	M	BUSINESS ADMINISTRATION	REAL ESTATE	$12,200
72	JUN80	F	LIBERAL ARTS	POLITICAL SCIENCE	$6,300
73	JUN80	M	ENGINEERING	MECHANICAL	$19,300
74	JUN80	F	SCIENCES	COMPUTER SCIENCE	$11,000
75	JUN80	M	FORESTRY	FORESTRY	$9,000
76	JUN80	F	ACCOUNTING	ACCOUNTING	$14,600
77	JUN80	M	ENGINEERING	INDUSTRIAL	$21,400
78	JUN80	F	JOURNALISM	JOURNALISM	$7,100
79	JUN80	F	BUSINESS ADMINISTRATION	MARKETING	$15,400
80	JUN80	M	AGRICULTURE	MECHANIZED	$11,000
81	JUN80	F	NURSING	NURSING	$13,300
82	JUN80	M	ENGINEERING	ELECTRICAL	$19,100
83	JUN80	M	ACCOUNTING	ACCOUNTING	$13,500
84	JUN80	M	PHARMACY	PHARMACY	$18,600
85	JUN80	M	PHARMACY	PHARMACY	$20,100
86	JUN80	F	SCIENCES	MATHEMATICS	$9,300
87	JUN80	F	HEALTH RELATED	MEDICAL TECHNOLOGY	$16,500
88	JUN80	M	BUSINESS ADMINISTRATION	FINANCE	$14,400
89	JUN80	F	BUSINESS ADMINISTRATION	MARKETING	$7,400
90	JUN80	F	ACCOUNTING	ACCOUNTING	$14,700
91	JUN80	F	PHARMACY	PHARMACY	$19,400
92	JUN80	M	ENGINEERING	ELECTRICAL	$12,800
93	JUN80	F	LIBERAL ARTS	THEATER	$22,700
94	JUN80	M	LIBERAL ARTS	HISTORY	$13,400
95	JUN80	M	ENGINEERING	AEROSPACE	$19,900
96	JUN80	M	ENGINEERING	MECHANICAL	$22,200
97	JUN80	F	BUSINESS ADMINISTRATION	MANAGEMENT	$14,400
98	JUN80	M	FINE ARTS	MUSIC	$12,700
99	JUN80	F	ENGINEERING	CIVIL	$20,800
100	JUN80	M	ENGINEERING	ELECTRICAL	$25,700
101	JUN80	F	BUSINESS ADMINISTRATION	MANAGEMENT	$12,500
102	JUN80	F	PHARMACY	PHARMACY	$21,400
103	JUN80	F	BUSINESS ADMINISTRATION	MARKETING	$16,700
104	JUN80	M	ENGINEERING	INDUSTRIAL	$20,400
105	JUN80	F	NURSING	NURSING	$14,300

STARTING SALARIES OF 948 UNIVERSITY OF FLORIDA GRADUATES
JUNE 1980 TO MARCH 1981

OBS	DATE	SEX	COLLEGE	MAJOR	SALARY
106	JUN80	M	ENGINEERING	ELECTRICAL	$19,000
107	JUN80	F	AGRICULTURE	EXTENDED EDUCATION	$8,700
108	JUN80	M	SCIENCES	ECONOMICS	$15,000
109	JUN80	M	ACCOUNTING	ACCOUNTING	$14,700
110	JUN80	M	FORESTRY	FORESTRY	$12,300
111	JUN80	M	ENGINEERING	MECHANICAL	$17,700
112	JUN80	M	JOURNALISM	JOURNALISM	$10,500
113	JUN80	M	ENGINEERING	CIVIL	$17,100
114	JUN80	M	PHARMACY	PHARMACY	$27,300
115	JUN80	M	ENGINEERING	INDUSTRIAL	$21,800
116	JUN80	M	FORESTRY	FORESTRY	$12,900
117	JUN80	M	AGRICULTURE	FOOD RESOURCE ECONOMICS	$14,800
118	JUN80	F	SCIENCES	ZOOLOGY	$13,900
119	JUN80	M	ACCOUNTING	ACCOUNTING	$16,000
120	JUN80	M	ACCOUNTING	ACCOUNTING	$17,600
121	JUN80	M	AGRICULTURE	FOOD CROP	$10,000
122	JUN80	M	ENGINEERING	ELECTRICAL	$22,300
123	JUN80	F	HEALTH RELATED	MEDICAL TECHNOLOGY	$8,600
124	JUN80	F	EDUCATION	MATHEMATICS	$14,000
125	JUN80	F	PHARMACY	PHARMACY	$26,300
126	JUN80	F	HEALTH RELATED	CLINICAL DIATETICS	$15,000
127	JUN80	M	ENGINEERING	MECHANICAL	$22,600
128	JUN80	F	NURSING	NURSING	$15,400
129	JUN80	M	AGRICULTURE	POULTRY SCIENCE	$38,100
130	JUN80	M	AGRICULTURE	POULTRY SCIENCE	$16,400
131	JUN80	F	SCIENCES	COMPUTER SCIENCE	$20,500
132	JUN80	M	ARCHITECTURE	BUILDING CONSTRUCTION	$25,700
133	JUN80	M	ACCOUNTING	ACCOUNTING	$17,700
134	JUN80	M	ENGINEERING	CIVIL	$16,900
135	JUN80	M	ACCOUNTING	ACCOUNTING	$14,800
136	JUN80	M	ENGINEERING	COMPUTER SCIENCE	$11,800
137	JUN80	F	HEALTH RELATED	CLINICAL DIATETICS	$8,300
138	JUN80	M	ENGINEERING	CIVIL	$14,100
139	JUN80	F	LIBERAL ARTS	POLITICAL SCIENCE	$10,600
140	JUN80	M	ENGINEERING	MECHANICAL	$22,700
141	JUN80	M	LIBERAL ARTS	POLITICAL SCIENCE	$11,400
142	JUN80	M	ENGINEERING	INDUSTRIAL	$19,900
143	JUN80	M	AGRICULTURE	ORNAMENTAL HORTICULTURE	$16,100
144	JUN80	M	PHARMACY	PHARMACY	$17,600
145	JUN80	F	ENGINEERING	INDUSTRIAL	$22,400
146	JUN80	M	ENGINEERING	MECHANICAL	$21,000
147	JUN80	M	AGRICULTURE	FOOD RESOURCE ECONOMICS	$14,900
148	JUN80	M	ACCOUNTING	ACCOUNTING	$22,100
149	JUN80	F	PHARMACY	PHARMACY	$18,100
150	JUN80	M	ACCOUNTING	ACCOUNTING	$17,800
151	JUN80	M	AGRICULTURE	AGRONOMY	$11,300
152	JUN80	M	BUSINESS ADMINISTRATION	MANAGEMENT	$13,500
153	JUN80	M	BUSINESS ADMINISTRATION	COMPUTER SCIENCE	$18,400
154	JUN80	M	PHARMACY	PHARMACY	$20,700
155	JUN80	F	AGRICULTURE	FOOD RESOURCE ECONOMICS	$6,200
156	JUN80	F	LIBERAL ARTS	POLITICAL SCIENCE	$13,800
157	JUN80	M	BUSINESS ADMINISTRATION	MANAGEMENT	$14,900
158	JUN80	M	AGRICULTURE	ANIMAL SCIENCE	$12,500
159	JUN80	M	AGRICULTURE	ANIMAL SCIENCE	$7,700
160	JUN80	M	ARCHITECTURE	AEROSPACE ENGINEER	$14,000
161	JUN80	M	ARCHITECTURE	BUILDING CONSTRUCTION	$19,500
162	JUN80	M	PHARMACY	PHARMACY	$19,700
163	JUN80	M	JOURNALISM	JOURNALISM	$11,000
164	JUN80	M	BUSINESS ADMINISTRATION	MANAGEMENT	$40,400
165	JUN80	M	ENGINEERING	ELECTRICAL	$21,300
166	JUN80	M	SCIENCES	ECONOMICS	$9,400
167	JUN80	M	ENGINEERING	ELECTRICAL	$18,400
168	JUN80	M	ENGINEERING	CIVIL	$14,800
169	JUN80	M	BUSINESS ADMINISTRATION	MANAGEMENT	$10,200
170	JUN80	M	ENGINEERING	ELECTRICAL	$20,400
171	JUN80	F	AGRICULTURE	FOOD SCIENCE	$6,100
172	JUN80	M	ENGINEERING	ELECTRICAL	$14,500
173	JUN80	M	BUSINESS ADMINISTRATION	COMPUTER SCIENCE	$17,100
174	JUN80	M	ENGINEERING	NUCLEAR	$21,600
175	JUN80	F	ACCOUNTING	ACCOUNTING	$12,600
176	JUN80	F	BUSINESS ADMINISTRATION	ACCOUNTING	$14,900
177	JUN80	M	ACCOUNTING	ACCOUNTING	$16,200
178	JUN80	M	BUSINESS ADMINISTRATION	MARKETING	$10,800
179	JUN80	M	BUSINESS ADMINISTRATION	MARKETING	$11,900
180	JUN80	M	PHARMACY	PHARMACY	$19,100
181	JUN80	M	PHARMACY	PHARMACY	$19,500
182	JUN80	M	BUSINESS ADMINISTRATION	COMPUTER SCIENCE	$21,300
183	JUN80	M	ENGINEERING	CIVIL	$19,000
184	JUN80	M	ENGINEERING	MECHANICAL	$17,100
185	JUN80	F	LIBERAL ARTS	POLITICAL SCIENCE	$11,000
186	JUN80	M	BUILDING CONSTRUCTION	BUILDING CONSTRUCTION	$20,700
187	JUN80	F	HEALTH RELATED	MEDICAL TECHNOLOGY	$11,300
188	JUN80	M	ENGINEERING	ELECTRICAL	$22,500
189	JUN80	F	ENGINEERING	COMPUTER SCIENCE	$22,800
190	JUN80	M	PHARMACY	PHARMACY	$20,700
191	JUN80	F	JOURNALISM	ADVERTISING	$13,600
192	JUN80	M	ENGINEERING	NUCLEAR	$19,200
193	JUN80	M	PHARMACY	PHARMACY	$20,400
194	JUN80	M	FORESTRY	WILDLIFE ECOLOGY	$6,800
195	JUN80	F	HEALTH RELATED	MEDICAL TECHNOLOGY	$13,800
196	JUN80	F	PHARMACY	PHARMACY	$11,700
197	JUN80	F	EDUCATION	ELEMENTARY	$8,000
198	JUN80	M	BUSINESS ADMINISTRATION	FINANCE	$14,000
199	JUN80	F	LIBERAL ARTS	POLITICAL SCIENCE	$14,700
200	JUN80	F	BUSINESS ADMINISTRATION	MANAGEMENT	$12,300
201	JUN80	M	ENGINEERING	ELECTRICAL	$19,700
202	JUN80	M	SCIENCES	ECONOMICS	$7,600
203	JUN80	F	ENGINEERING	ELECTRICAL	$12,100
204	JUN80	M	AGRICULTURE	FOOD RESOURCE ECONOMICS	$15,100
205	JUN80	M	PHARMACY	PHARMACY	$9,300
206	JUN80	M	AGRICULTURE	ORNAMENTAL HORTICULTURE	$10,300
207	JUN80	M	ENGINEERING	CIVIL	$26,300
208	JUN80	F	ENGINEERING	MECHANICAL	$21,400
209	JUN80	F	PHARMACY	PHARMACY	$16,100
210	JUN80	M	BUSINESS ADMINISTRATION	MANAGEMENT	$16,600

STARTING SALARIES OF 948 UNIVERSITY OF FLORIDA GRADUATES
JUNE 1980 TO MARCH 1981

OBS	DATE	SEX	COLLEGE	MAJOR	SALARY
211	JUN80	M	AGRICULTURE	FOOD RESOURCE ECONOMICS	$11,500
212	JUN80	M	FORESTRY	FORESTRY	$7,500
213	JUN80	F	NURSING	NURSING	$12,900
214	JUN80	M	PHARMACY	PHARMACY	$19,700
215	JUN80	F	JOURNALISM	JOURNALISM	$8,400
216	JUN80	F	SCIENCES	ECONOMICS	$15,500
217	JUN80	F	PHARMACY	PHARMACY	$19,800
218	JUN80	M	SCIENCES	ECONOMICS	$17,300
219	JUN80	F	AGRICULTURE	ORNAMENTAL HORTICULTURE	$10,400
220	JUN80	F	EDUCATION	EDUCATION	$6,400
221	JUN80	F	PHARMACY	PHARMACY	$24,600
222	JUN80	M	ACCOUNTING	ACCOUNTING	$22,400
223	JUN80	M	ENGINEERING	CIVIL	$19,800
224	JUN80	M	AGRICULTURE	FOOD RESOURCE ECONOMICS	$14,000
225	JUN80	M	ACCOUNTING	ACCOUNTING	$15,200
226	JUN80	F	JOURNALISM	JOURNALISM	$16,400
227	JUN80	M	LIBERAL ARTS	PSYCHOLOGY	$13,500
228	JUN80	M	ENGINEERING	ELECTRICAL	$25,300
229	JUN80	F	ENGINEERING	CIVIL	$20,800
230	JUN80	M	ENGINEERING	ELECTRICAL	$20,300
231	JUN80	M	ARCHITECTURE	AEROSPACE ENGINEER	$14,500
232	JUN80	F	EDUCATION	SPEECH	$10,500
233	JUN80	M	AGRICULTURE	EXTENDED EDUCATION	$15,200
234	JUN80	M	ENGINEERING	CIVIL	$13,500
235	JUN80	M	BUSINESS ADMINISTRATION	FINANCE	$12,200
236	JUN80	M	BUSINESS ADMINISTRATION	FINANCE	$15,500
237	JUN80	M	ENGINEERING	CIVIL	$22,400
238	JUN80	M	BUSINESS ADMINISTRATION	MANAGEMENT	$15,900
239	JUN80	F	SCIENCES	CHEMISTRY	$11,200
240	JUN80	M	ENGINEERING	NUCLEAR	$19,800
241	JUN80	F	JOURNALISM	ADVERTISING	$12,200
242	JUN80	F	JOURNALISM	PUBLIC RELATIONS	$18,700
243	JUN80	F	PHARMACY	PHARMACY	$21,500
244	JUN80	M	PHARMACY	PHARMACY	$23,300
245	JUN80	F	NURSING	NURSING	$13,000
246	JUN80	M	ENGINEERING	CIVIL	$19,800
247	JUN80	M	JOURNALISM	ADVERTISING	$10,300
248	JUN80	F	AGRICULTURE	FOOD RESOURCE ECONOMICS	$14,000
249	JUN80	M	FORESTRY	FORESTRY	$14,800
250	JUN80	M	BUSINESS ADMINISTRATION	ACCOUNTING	$16,600
251	JUN80	F	EDUCATION	MATHEMATICS	$9,700
252	JUN80	F	ENGINEERING	MECHANICAL	$21,500
253	JUN80	M	ACCOUNTING	ACCOUNTING	$15,200
254	JUN80	F	BUSINESS ADMINISTRATION	MARKETING	$19,200
255	JUN80	F	JOURNALISM	JOURNALISM	$11,600
256	JUN80	M	ENGINEERING	AEROSPACE	$18,000
257	JUN80	M	ENGINEERING	CIVIL	$22,400
258	JUN80	M	BUSINESS ADMINISTRATION	FINANCE	$18,900
259	JUN80	M	MEDICINE	MEDICINE	$12,100
260	JUN80	F	HEALTH RELATED	HEALTH SCIENCE	$12,300
261	JUN80	M	BUSINESS ADMINISTRATION	ACCOUNTING	$16,500
262	JUN80	F	NURSING	NURSING	$19,900
263	JUN80	F	SCIENCES	CHEMISTRY	$15,400
264	JUN80	F	BUSINESS ADMINISTRATION	MANAGEMENT	$17,400
265	JUN80	M	BUSINESS ADMINISTRATION	COMPUTER SCIENCE	$20,200
266	JUN80	M	MEDICINE	MEDICINE	$18,700
267	JUN80	F	EDUCATION	SPEECH	$10,900
268	JUN80	F	PHARMACY	PHARMACY	$9,700
269	JUN80	F	FORESTRY	WILDLIFE ECOLOGY	$11,300
270	JUN80	M	ENGINEERING	MECHANICAL	$20,600
271	JUN80	M	BUSINESS ADMINISTRATION	MARKETING	$19,600
272	JUN80	M	LIBERAL ARTS	POLITICAL SCIENCE	$14,100
273	JUN80	M	PHARMACY	PHARMACY	$26,900
274	JUN80	F	AGRICULTURE	FORESTRY	$9,300
275	JUN80	M	SCIENCES	ECONOMICS	$14,700
276	JUN80	M	BUSINESS ADMINISTRATION	MANAGEMENT	$16,200
277	JUN80	F	LIBERAL ARTS	RELIGION	$7,100
278	JUN80	F	PHYS ED,HEALTH,RECREATION	HEALTH EDUCATION	$10,500
279	JUN80	M	ENGINEERING	ELECTRICAL	$17,500
280	JUN80	M	PHARMACY	PHARMACY	$11,500
281	JUN80	M	ENGINEERING	CIVIL	$22,100
282	JUN80	F	AGRICULTURE	POULTRY SCIENCE	$11,900
283	JUN80	M	ENGINEERING	ELECTRICAL	$24,400
284	JUN80	M	ENGINEERING	MECHANICAL	$19,300
285	JUN80	M	BUSINESS ADMINISTRATION	MANAGEMENT	$15,600
286	JUN80	F	SCIENCES	COMPUTER SCIENCE	$20,700
287	JUN80	M	BUSINESS ADMINISTRATION	FINANCE	$14,600
288	JUN80	M	BUSINESS ADMINISTRATION	COMPUTER SCIENCE	$15,700
289	JUN80	F	LIBERAL ARTS	ENGLISH	$11,200
290	JUN80	F	ACCOUNTING	ACCOUNTING	$16,600
291	JUN80	M	ENGINEERING	MECHANICAL	$22,700
292	JUN80	F	JOURNALISM	JOURNALISM	$9,200
293	JUN80	M	SCIENCES	ECONOMICS	$12,100
294	JUN80	F	BUSINESS ADMINISTRATION	MANAGEMENT	$13,900
295	JUN80	F	NURSING	NURSING	$5,800
296	JUN80	M	ENGINEERING	CIVIL	$20,400
297	JUN80	M	BUSINESS ADMINISTRATION	MANAGEMENT	$13,100
298	JUN80	F	ACCOUNTING	ACCOUNTING	$15,400
299	JUN80	M	BUSINESS ADMINISTRATION	MANAGEMENT	$11,500
300	JUN80	F	BUSINESS ADMINISTRATION	ACCOUNTING	$18,500
301	JUN80	F	NURSING	NURSING	$16,700
302	JUN80	F	SCIENCES	COMPUTER SCIENCE	$21,200
303	JUN80	M	BUSINESS ADMINISTRATION	FINANCE	$19,300
304	JUN80	M	PHARMACY	PHARMACY	$19,800
305	JUN80	M	ENGINEERING	ELECTRICAL	$17,600
306	JUN80	M	ENGINEERING	ELECTRICAL	$17,700
307	JUN80	M	ENGINEERING	ELECTRICAL	$26,600
308	JUN80	M	AGRICULTURE	ANIMAL SCIENCE	$21,300
309	JUN80	F	ACCOUNTING	ACCOUNTING	$14,200
310	JUN80	M	FORESTRY	FORESTRY	$9,300
311	JUN80	F	ACCOUNTING	ACCOUNTING	$14,300
312	JUN80	M	ACCOUNTING	ACCOUNTING	$21,300
313	JUN80	F	BUSINESS ADMINISTRATION	FINANCE	$13,800
314	JUN80	M	LIBERAL ARTS	POLITICAL SCIENCE	$11,600
315	JUN80	M	ENGINEERING	NUCLEAR	$12,000

STARTING SALARIES OF 948 UNIVERSITY OF FLORIDA GRADUATES
JUNE 1980 TO MARCH 1981

OBS	DATE	SEX	COLLEGE	MAJOR	SALARY
316	JUN80	M	ENGINEERING	MECHANICAL	$21,900
317	JUN80	F	BUSINESS ADMINISTRATION	MARKETING	$14,800
318	JUN80	M	SCIENCES	GEOLOGY	$32,400
319	JUN80	F	JOURNALISM	ADVERTISING	$13,000
320	JUN80	M	BUSINESS ADMINISTRATION	MANAGEMENT	$11,600
321	JUN80	M	BUSINESS ADMINISTRATION	MANAGEMENT	$21,200
322	JUN80	F	BUSINESS ADMINISTRATION	COMPUTER SCIENCE	$17,300
323	JUN80	M	PHARMACY	PHARMACY	$22,800
324	JUN80	M	PHYS ED,HEALTH,RECREATION	HEALTH EDUCATION	$11,400
325	JUN80	M	ENGINEERING	NUCLEAR	$19,400
326	JUN80	M	BUILDING CONSTRUCTION	BUILDING CONSTRUCTION	$19,600
327	JUN80	F	PHARMACY	PHARMACY	$25,100
328	JUN80	M	ENGINEERING	INDUSTRIAL	$21,600
329	JUN80	M	ENGINEERING	MECHANICAL	$22,000
330	JUN80	M	EDUCATION	MATHEMATICS	$10,000
331	JUN80	M	JOURNALISM	ADVERTISING	$17,100
332	JUN80	F	ACCOUNTING	ACCOUNTING	$13,100
333	JUN80	M	AGRICULTURE	AGRICULTURE	$16,700
334	JUN80	M	ARCHITECTURE	BUILDING CONSTRUCTION	$20,000
335	JUN80	M	ENGINEERING	INDUSTRIAL	$21,400
336	JUN80	M	BUSINESS ADMINISTRATION	MANAGEMENT	$18,700
337	JUN80	M	ENGINEERING	CIVIL	$21,200
338	JUN80	M	ENGINEERING	COMPUTER SCIENCE	$17,900
339	JUN80	M	PHARMACY	PHARMACY	$10,300
340	JUN80	F	HEALTH RELATED	MEDICAL TECHNOLOGY	$12,800
341	JUN80	F	BUSINESS ADMINISTRATION	MARKETING	$11,400
342	JUN80	M	BUSINESS ADMINISTRATION	MANAGEMENT	$11,700
343	JUN80	F	PHARMACY	PHARMACY	$17,000
344	JUN80	F	BUSINESS ADMINISTRATION	ACCOUNTING	$13,800
345	JUN80	M	PHARMACY	PHARMACY	$21,400
346	JUN80	F	AGRICULTURE	EXTENDED EDUCATION	$9,500
347	JUN80	M	ENGINEERING	METALLOGY	$23,400
348	JUN80	F	NURSING	NURSING	$10,800
349	JUN80	M	LIBERAL ARTS	ENGLISH	$18,900
350	JUN80	M	ENGINEERING	MECHANICAL	$22,800
351	JUN80	F	BUSINESS ADMINISTRATION	MARKETING	$11,200
352	JUN80	M	ARCHITECTURE	BUILDING CONSTRUCTION	$19,300
353	JUN80	F	BUSINESS ADMINISTRATION	MANAGEMENT	$9,200
354	JUN80	M	BUSINESS ADMINISTRATION	FINANCE	$12,500
355	JUN80	F	PHARMACY	PHARMACY	$20,800
356	JUN80	M	ACCOUNTING	ACCOUNTING	$14,400
357	JUN80	M	ENGINEERING	ELECTRICAL	$18,900
358	JUN80	M	ENGINEERING	ELECTRICAL	$19,700
359	JUN80	M	PHARMACY	PHARMACY	$18,600
360	JUN80	M	BUSINESS ADMINISTRATION	FINANCE	$13,200
361	JUN80	M	BUSINESS ADMINISTRATION	FINANCE	$15,600
362	JUN80	F	LIBERAL ARTS	HISTORY	$5,400
363	JUN80	F	AGRICULTURE	ORNAMENTAL HORTICULTURE	$8,400
364	JUN80	F	BUSINESS ADMINISTRATION	MARKETING	$15,000
365	JUN80	F	JOURNALISM	PHOTOGRAPHIC	$8,500
366	JUN80	M	BUSINESS ADMINISTRATION	MANAGEMENT	$13,400
367	JUN80	M	ENGINEERING	CIVIL	$13,400
368	AUG80	M	ACCOUNTING	ACCOUNTING	$12,000
369	AUG80	M	AGRICULTURE	FOOD RESOURCE ECONOMICS	$12,000
370	AUG80	M	FORESTRY	WILDLIFE ECOLOGY	$11,200
371	AUG80	F	ACCOUNTING	ACCOUNTING	$16,900
372	AUG80	M	PHYSICIANS ASSISTANT	PHYSICIAN'S ASSISTANT	$21,400
373	AUG80	M	ENGINEERING	MECHANICAL	$22,800
374	AUG80	F	ACCOUNTING	ACCOUNTING	$14,400
375	AUG80	M	ENGINEERING	ELECTRICAL	$18,800
376	AUG80	M	ENGINEERING	MECHANICAL	$19,400
377	AUG80	M	ENGINEERING	ELECTRICAL	$21,200
378	AUG80	F	JOURNALISM	JOURNALISM	$16,700
379	AUG80	F	LIBERAL ARTS	PSYCHOLOGY	$8,900
380	AUG80	F	EDUCATION	MATHEMATICS	$12,800
381	AUG80	F	NURSING	NURSING	$14,000
382	AUG80	M	AGRICULTURE	FOOD RESOURCE ECONOMICS	$12,100
383	AUG80	M	LAW	TAX	$20,300
384	AUG80	F	BUSINESS ADMINISTRATION	MARKETING	$17,100
385	AUG80	M	BUSINESS ADMINISTRATION	MANAGEMENT	$12,000
386	AUG80	F	NURSING	NURSING	$11,800
387	AUG80	M	NURSING	NURSING	$13,100
388	AUG80	M	FORESTRY	FORESTRY	$13,500
389	AUG80	M	ENGINEERING	CIVIL	$21,500
390	AUG80	M	ENGINEERING	AGRICULTURAL	$15,000
391	AUG80	M	ENGINEERING	CIVIL	$22,900
392	AUG80	M	HEALTH RELATED	PHYSICAL THERAPY	$14,800
393	AUG80	F	BUSINESS ADMINISTRATION	MARKETING	$10,600
394	AUG80	M	ARCHITECTURE	AEROSPACE ENGINEER	$21,300
395	AUG80	M	ENGINEERING	CHEMICAL	$21,900
396	AUG80	M	BUSINESS ADMINISTRATION	MARKETING	$13,100
397	AUG80	F	ENGINEERING	NUCLEAR	$17,900
398	AUG80	F	EDUCATION	EDUCATION	$7,800
399	AUG80	F	MEDICINE	PHYSICIAN'S ASSISTANT	$19,200
400	AUG80	M	BUSINESS ADMINISTRATION	MARKETING	$14,300
401	AUG80	M	SCIENCES	ECONOMICS	$15,200
402	AUG80	M	BUSINESS ADMINISTRATION	FINANCE	$20,600
403	AUG80	F	HEALTH RELATED	PHYSICAL THERAPY	$14,300
404	AUG80	M	MEDICINE	PHYSICIAN'S ASSISTANT	$16,800
405	AUG80	F	BUSINESS ADMINISTRATION	MARKETING	$12,300
406	AUG80	M	AGRICULTURE	POULTRY SCIENCE	$14,600
407	AUG80	F	HEALTH RELATED	PHYSICAL THERAPY	$16,300
408	AUG80	F	EDUCATION	MEDICINE	$7,100
409	AUG80	F	HEALTH RELATED	MEDICAL TECHNOLOGY	$13,800
410	AUG80	M	BUILDING CONSTRUCTION	BUILDING CONSTRUCTION	$17,000
411	AUG80	M	BUSINESS ADMINISTRATION	MANAGEMENT	$18,300
412	AUG80	M	EDUCATION	SOCIAL SCIENCE	$5,000
413	AUG80	M	AGRICULTURE	FOOD RESOURCE ECONOMICS	$12,200
414	AUG80	M	BUSINESS ADMINISTRATION	ACCOUNTING	$16,600
415	AUG80	M	SCIENCES	ECONOMICS	$17,200
416	AUG80	F	HEALTH RELATED	MEDICAL TECHNOLOGY	$14,600
417	AUG80	M	ENGINEERING	ELECTRICAL	$18,000
418	AUG80	F	NURSING	NURSING	$13,000
419	AUG80	M	ENGINEERING	ENVIRONMENTAL	$14,900
420	AUG80	F	BUSINESS ADMINISTRATION	MARKETING	$7,000

STARTING SALARIES OF 948 UNIVERSITY OF FLORIDA GRADUATES
JUNE 1980 TO MARCH 1981

OBS	DATE	SEX	COLLEGE	MAJOR	SALARY
421	AUG80	F	HEALTH RELATED	APPLIED HEALTH	$9,800
422	AUG80	M	ENGINEERING	MECHANICAL	$19,800
423	AUG80	M	ENGINEERING	ELECTRICAL	$17,600
424	AUG80	F	NURSING	NURSING	$11,700
425	AUG80	M	AGRICULTURE	FOOD RESOURCE ECONOMICS	$12,300
426	AUG80	F	HEALTH RELATED	MEDICAL TECHNOLOGY	$13,300
427	AUG80	F	NURSING	NURSING	$13,400
428	AUG80	M	JOURNALISM	JOURNALISM	$14,100
429	AUG80	M	ARCHITECTURE	BUILDING CONSTRUCTION	$17,000
430	AUG80	F	HEALTH RELATED	DENTAL HYGIENE	$7,500
431	AUG80	M	ACCOUNTING	ACCOUNTING	$15,900
432	AUG80	F	HEALTH RELATED	PHYSICAL THERAPY	$15,000
433	AUG80	M	ACCOUNTING	ACCOUNTING	$14,700
434	AUG80	M	LAW	LAW	$22,500
435	AUG80	M	ENGINEERING	ELECTRICAL	$18,100
436	AUG80	F	JOURNALISM	PHOTOGRAPHIC	$11,400
437	AUG80	M	ENGINEERING	ELECTRICAL	$20,800
438	AUG80	M	BUSINESS ADMINISTRATION	COMPUTER SCIENCE	$11,700
439	AUG80	F	JOURNALISM	PUBLIC RELATIONS	$16,400
440	AUG80	F	ENGINEERING	AEROSPACE	$21,200
441	AUG80	M	ENGINEERING	INDUSTRIAL	$20,800
442	AUG80	F	HEALTH RELATED	APPLIED HEALTH	$16,300
443	AUG80	M	BUSINESS ADMINISTRATION	MANAGEMENT	$11,000
444	AUG80	M	ENGINEERING	MECHANICAL	$14,800
445	AUG80	F	NURSING	NURSING	$11,700
446	AUG80	M	AGRICULTURE	FOOD RESOURCE ECONOMICS	$16,000
447	AUG80	M	BUILDING CONSTRUCTION	BUILDING CONSTRUCTION	$19,500
448	AUG80	M	ENGINEERING	MECHANICAL	$19,300
449	AUG80	M	PHYS ED,HEALTH,RECREATION	RECREATION	$8,900
450	AUG80	M	BUILDING CONSTRUCTION	BUILDING CONSTRUCTION	$18,100
451	AUG80	M	ENGINEERING	INDUSTRIAL	$20,400
452	AUG80	M	ACCOUNTING	ACCOUNTING	$20,100
453	AUG80	M	ENGINEERING	ENERGY	$18,000
454	AUG80	M	ENGINEERING	ELECTRICAL	$24,000
455	AUG80	M	ENGINEERING	MECHANICAL	$14,400
456	AUG80	M	ENGINEERING	CHEMICAL	$22,700
457	AUG80	F	HEALTH RELATED	PHYSICAL THERAPY	$14,800
458	AUG80	M	SCIENCES	ECONOMICS	$13,400
459	AUG80	M	ENGINEERING	ELECTRICAL	$17,300
460	AUG80	M	AGRICULTURE	PLANT PATHOLOGY	$11,300
461	AUG80	M	ENGINEERING	CHEMICAL	$23,200
462	AUG80	M	JOURNALISM	ADVERTISING	$9,700
463	AUG80	M	ENGINEERING	MECHANICAL	$14,800
464	AUG80	M	BUSINESS ADMINISTRATION	MARKETING	$36,500
465	AUG80	M	JOURNALISM	ADVERTISING	$8,700
466	AUG80	M	BUSINESS ADMINISTRATION	MANAGEMENT	$11,400
467	AUG80	F	ACCOUNTING	ACCOUNTING	$16,100
468	AUG80	F	HEALTH RELATED	MEDICAL TECHNOLOGY	$10,900
469	AUG80	M	MEDICINE	PHYSICIAN'S ASSISTANT	$15,300
470	AUG80	M	ENGINEERING	CIVIL	$22,300
471	AUG80	M	BUSINESS ADMINISTRATION	MARKETING	$10,900
472	AUG80	M	AGRICULTURE	MECHANIZED	$17,900
473	AUG80	M	MEDICINE	PHYSICIAN'S ASSISTANT	$17,000
474	AUG80	M	ACCOUNTING	ACCOUNTING	$12,500
475	AUG80	F	ENGINEERING	INDUSTRIAL	$19,200
476	AUG80	F	ACCOUNTING	ACCOUNTING	$13,800
477	AUG80	M	HEALTH RELATED	PHYSICAL THERAPY	$15,000
478	AUG80	M	BUILDING CONSTRUCTION	BUILDING CONSTRUCTION	$15,200
479	AUG80	M	ENGINEERING	ELECTRICAL	$21,300
480	AUG80	M	MEDICINE	PHYSICIAN'S ASSISTANT	$15,700
481	AUG80	F	ENGINEERING	INDUSTRIAL	$20,200
482	AUG80	F	AGRICULTURE	NUTRITION	$9,000
483	AUG80	M	ENGINEERING	CIVIL	$17,800
484	AUG80	M	BUILDING CONSTRUCTION	BUILDING CONSTRUCTION	$21,400
485	AUG80	M	ENGINEERING	CIVIL	$17,700
486	AUG80	F	HEALTH RELATED	PHYSICAL THERAPY	$13,600
487	AUG80	F	FINE ARTS	GRAPHIC DESIGN	$5,100
488	AUG80	M	ENGINEERING	MECHANICAL	$19,700
489	AUG80	F	EDUCATION	ELEMENTARY	$6,800
490	AUG80	M	BUSINESS ADMINISTRATION	COMPUTER SCIENCE	$16,300
491	AUG80	M	BUSINESS ADMINISTRATION	MANAGEMENT	$13,700
492	AUG80	M	AGRICULTURE	ANIMAL SCIENCE	$7,800
493	AUG80	M	BUSINESS ADMINISTRATION	REAL ESTATE	$12,100
494	AUG80	F	ENGINEERING	ELECTRICAL	$22,100
495	AUG80	F	JOURNALISM	JOURNALISM	$11,200
496	AUG80	F	BUSINESS ADMINISTRATION	COMPUTER SCIENCE	$15,600
497	AUG80	M	ARCHITECTURE	BUILDING CONSTRUCTION	$18,300
498	AUG80	M	BUSINESS ADMINISTRATION	MANAGEMENT	$16,000
499	AUG80	F	NURSING	NURSING	$11,800
500	AUG80	M	ENGINEERING	MECHANICAL	$20,900
501	AUG80	M	MEDICINE	PHYSICIAN'S ASSISTANT	$22,600
502	AUG80	F	HEALTH RELATED	MEDICAL TECHNOLOGY	$14,000
503	AUG80	M	ENGINEERING	CIVIL	$15,600
504	AUG80	M	HEALTH RELATED	SPEECH PATHOLOGY	$7,300
505	AUG80	M	BUSINESS ADMINISTRATION	REAL ESTATE	$5,500
506	AUG80	F	AGRICULTURE	ORNAMENTAL HORTICULTURE	$9,100
507	AUG80	M	ACCOUNTING	ACCOUNTING	$15,100
508	AUG80	M	ARCHITECTURE	BUILDING CONSTRUCTION	$20,200
509	AUG80	F	MEDICINE	MEDICINE	$13,900
510	AUG80	F	EDUCATION	SPEECH	$10,500
511	AUG80	F	JOURNALISM	ADVERTISING	$10,300
512	AUG80	F	JOURNALISM	ADVERTISING	$8,100
513	AUG80	M	BUSINESS ADMINISTRATION	MARKETING	$14,500
514	AUG80	M	BUILDING CONSTRUCTION	BUILDING CONSTRUCTION	$18,700
515	AUG80	M	ENGINEERING	CIVIL	$24,100
516	AUG80	M	LIBERAL ARTS	PSYCHOLOGY	$16,800
517	AUG80	F	NURSING	NURSING	$11,100
518	AUG80	M	AGRICULTURE	EXTENDED EDUCATION	$9,200
519	AUG80	F	ACCOUNTING	ACCOUNTING	$14,000
520	AUG80	M	ENGINEERING	MECHANICAL	$20,600
521	AUG80	M	ARCHITECTURE	AEROSPACE ENGINEER	$9,200
522	AUG80	M	ENGINEERING	ELECTRICAL	$21,500
523	AUG80	M	ENGINEERING	ELECTRICAL	$27,300
524	AUG80	M	LIBERAL ARTS	PSYCHOLOGY	$22,800
525	AUG80	M	ENGINEERING	ELECTRICAL	$16,900

STARTING SALARIES OF 948 UNIVERSITY OF FLORIDA GRADUATES
JUNE 1980 TO MARCH 1981

OBS	DATE	SEX	COLLEGE	MAJOR	SALARY
526	AUG80	M	ARCHITECTURE	AEROSPACE ENGINEER	$7,900
527	AUG80	F	BUSINESS ADMINISTRATION	MARKETING	$12,700
528	AUG80	M	ACCOUNTING	ACCOUNTING	$18,500
529	AUG80	M	ENGINEERING	CIVIL	$21,100
530	AUG80	F	HEALTH RELATED	PHYSICAL THERAPY	$17,500
531	AUG80	M	ACCOUNTING	ACCOUNTING	$12,800
532	AUG80	F	HEALTH RELATED	MEDICAL TECHNOLOGY	$10,500
533	AUG80	M	ACCOUNTING	ACCOUNTING	$18,500
534	AUG80	F	JOURNALISM	PUBLIC RELATIONS	$10,400
535	AUG80	F	MEDICINE	PHYSICIAN'S ASSISTANT	$15,300
536	AUG80	M	AGRICULTURE	FORESTRY	$12,900
537	AUG80	M	ENGINEERING	ELECTRICAL	$18,400
538	AUG80	M	SCIENCES	ECONOMICS	$26,500
539	AUG80	M	ENGINEERING	MECHANICAL	$18,400
540	AUG80	M	BUSINESS ADMINISTRATION	COMPUTER SCIENCE	$19,500
541	AUG80	M	JOURNALISM	JOURNALISM	$9,500
542	AUG80	M	ENGINEERING	COMPUTER SCIENCE	$19,300
543	AUG80	F	NURSING	NURSING	$10,700
544	AUG80	M	ENGINEERING	ELECTRICAL	$23,000
545	AUG80	F	NURSING	NURSING	$12,000
546	AUG80	F	AGRICULTURE	ANIMAL SCIENCE	$5,400
547	AUG80	M	AGRICULTURE	ORNAMENTAL HORTICULTURE	$15,300
548	AUG80	F	NURSING	NURSING	$12,900
549	AUG80	M	ENGINEERING	CIVIL	$15,000
550	AUG80	F	HEALTH RELATED	PHYSICAL THERAPY	$14,900
551	AUG80	F	NURSING	NURSING	$14,700
552	AUG80	F	NURSING	NURSING	$11,800
553	AUG80	M	PHYS ED,HEALTH,RECREATION	RECREATION	$7,300
554	AUG80	M	ENGINEERING	AEROSPACE	$13,700
555	AUG80	F	JOURNALISM	BROADCASTING	$10,700
556	AUG80	M	LAW	CRIMINAL JUSTICE	$17,900
557	AUG80	M	ENGINEERING	CIVIL	$21,600
558	AUG80	M	PHYSICIANS ASSISTANT	PHYSICIAN'S ASSISTANT	$15,300
559	AUG80	F	PHYS ED,HEALTH,RECREATION	RECREATION	$12,600
560	AUG80	F	NURSING	NURSING	$12,900
561	AUG80	F	EDUCATION	SPEECH	$8,000
562	AUG80	M	ACCOUNTING	ACCOUNTING	$15,400
563	AUG80	M	ENGINEERING	ELECTRICAL	$11,100
564	AUG80	F	NURSING	NURSING	$13,300
565	AUG80	M	BUSINESS ADMINISTRATION	COMPUTER SCIENCE	$17,500
566	AUG80	M	BUSINESS ADMINISTRATION	MARKETING	$13,900
567	AUG80	M	ACCOUNTING	ACCOUNTING	$13,000
568	AUG80	M	AGRICULTURE	ORNAMENTAL HORTICULTURE	$10,600
569	AUG80	F	BUILDING CONSTRUCTION	BUILDING CONSTRUCTION	$11,400
570	AUG80	F	NURSING	NURSING	$6,600
571	AUG80	M	ENGINEERING	INDUSTRIAL	$25,700
572	AUG80	M	ENGINEERING	CIVIL	$21,700
573	AUG80	M	ARCHITECTURE	BUILDING CONSTRUCTION	$19,900
574	AUG80	F	NURSING	NURSING	$13,000
575	AUG80	F	ENGINEERING	INDUSTRIAL	$22,000
576	AUG80	F	NURSING	NURSING	$15,100
577	AUG80	F	MEDICINE	MEDICINE	$12,200
578	AUG80	F	BUSINESS ADMINISTRATION	MARKETING	$8,900
579	AUG80	M	ACCOUNTING	ACCOUNTING	$16,700
580	AUG80	F	ARCHITECTURE	AEROSPACE ENGINEER	$10,900
581	AUG80	F	LIBERAL ARTS	PSYCHOLOGY	$11,100
582	AUG80	F	HEALTH RELATED	PHYSICAL THERAPY	$18,500
583	AUG80	F	PHYS ED,HEALTH,RECREATION	RECREATION	$7,600
584	AUG80	M	BUSINESS ADMINISTRATION	ACCOUNTING	$14,300
585	AUG80	F	BUSINESS ADMINISTRATION	BUSINESS ADMINISTRATION	$12,600
586	AUG80	M	JOURNALISM	REPORTING	$8,600
587	AUG80	M	NURSING	NURSING	$12,400
588	AUG80	F	ENGINEERING	COMPUTER SCIENCE	$20,100
589	AUG80	F	JOURNALISM	JOURNALISM	$8,300
590	AUG80	F	ENGINEERING	COMPUTER SCIENCE	$18,600
591	AUG80	F	ENGINEERING	ELECTRICAL	$20,100
592	AUG80	M	BUSINESS ADMINISTRATION	FINANCE	$10,900
593	AUG80	M	NURSING	NURSING	$12,100
594	AUG80	M	ENGINEERING	MECHANICAL	$19,200
595	AUG80	M	BUSINESS ADMINISTRATION	MANAGEMENT	$11,300
596	AUG80	M	ENGINEERING	CHEMICAL	$21,000
597	AUG80	M	JOURNALISM	JOURNALISM	$6,500
598	AUG80	M	ENGINEERING	MECHANICAL	$21,000
599	AUG80	M	ENGINEERING	ELECTRICAL	$23,600
600	AUG80	F	NURSING	NURSING	$12,900
601	AUG80	F	ENGINEERING	CHEMICAL	$21,000
602	AUG80	F	AGRICULTURE	MICROBIOLOGY	$13,200
603	AUG80	M	AGRICULTURE	FOOD RESOURCE ECONOMICS	$14,400
604	AUG80	M	ENGINEERING	AEROSPACE	$18,400
605	DEC80	M	ENGINEERING	MECHANICAL	$22,800
606	DEC80	M	BUSINESS ADMINISTRATION	FINANCE	$12,500
607	DEC80	M	ENGINEERING	ENVIRONMENTAL	$18,600
608	DEC80	M	LIBERAL ARTS	SOCIOLOGY	$9,500
609	DEC80	F	NURSING	NURSING	$12,400
610	DEC80	F	AGRICULTURE	PLANT PATHOLOGY	$11,800
611	DEC80	M	BUILDING CONSTRUCTION	BUILDING CONSTRUCTION	$11,300
612	DEC80	M	AGRICULTURE	EXTENDED EDUCATION	$13,500
613	DEC80	F	PHYS ED,HEALTH,RECREATION	PHYSICAL EDUCATION	$10,400
614	DEC80	M	ENGINEERING	CIVIL	$17,300
615	DEC80	M	NURSING	NURSING	$14,900
616	DEC80	F	NURSING	NURSING	$12,800
617	DEC80	F	PHYS ED,HEALTH,RECREATION	PHYSICAL EDUCATION	$10,400
618	DEC80	M	LIBERAL ARTS	HISTORY	$12,900
619	DEC80	F	BUSINESS ADMINISTRATION	FINANCE	$13,900
620	DEC80	F	NURSING	NURSING	$16,200
621	DEC80	F	SCIENCES	COMPUTER SCIENCE	$19,600
622	DEC80	F	EDUCATION	ELEMENTARY	$14,600
623	DEC80	M	ENGINEERING	MECHANICAL	$23,000
624	DEC80	M	ENGINEERING	ELECTRICAL	$21,300
625	DEC80	M	ENGINEERING	ELECTRICAL	$22,700
626	DEC80	M	ENGINEERING	ENVIRONMENTAL	$21,400
627	DEC80	M	ENGINEERING	INDUSTRIAL	$22,100
628	DEC80	F	AGRICULTURE	EXTENDED EDUCATION	$38,400
629	DEC80	F	HEALTH RELATED	EDUCATION	$11,000
630	DEC80	F	ENGINEERING	ENVIRONMENTAL	$19,800

STARTING SALARIES OF 948 UNIVERSITY OF FLORIDA GRADUATES
JUNE 1980 TO MARCH 1981

OBS	DATE	SEX	COLLEGE	MAJOR	SALARY
631	DEC80	M	ENGINEERING	MECHANICAL	$20,500
632	DEC80	M	BUSINESS ADMINISTRATION	FINANCE	$13,700
633	DEC80	F	LIBERAL ARTS	HISTORY	$11,300
634	DEC80	F	BUSINESS ADMINISTRATION	ACCOUNTING	$15,200
635	DEC80	F	FINE ARTS	PHOTOGRAPHY	$9,800
636	DEC80	M	BUSINESS ADMINISTRATION	FINANCE	$18,500
637	DEC80	M	ENGINEERING	ENVIRONMENTAL	$22,100
638	DEC80	F	LIBERAL ARTS	PSYCHOLOGY	$7,200
639	DEC80	F	ENGINEERING	INDUSTRIAL	$9,700
640	DEC80	F	LIBERAL ARTS	POLITICAL SCIENCE	$15,100
641	DEC80	F	EDUCATION	ELEMENTARY	$13,100
642	DEC80	M	SCIENCES	COMPUTER SCIENCE	$21,600
643	DEC80	F	AGRICULTURE	EXTENDED EDUCATION	$12,400
644	DEC80	M	LIBERAL ARTS	PSYCHOLOGY	$9,100
645	DEC80	M	ENGINEERING	ELECTRICAL	$19,900
646	DEC80	F	ARCHITECTURE	BUILDING CONSTRUCTION	$20,600
647	DEC80	M	ENGINEERING	ENVIRONMENTAL	$22,700
648	DEC80	F	PHYS ED,HEALTH,RECREATION	RECREATION	$12,200
649	DEC80	F	BUSINESS ADMINISTRATION	MARKETING	$17,300
650	DEC80	M	ENGINEERING	COMPUTER SCIENCE	$19,400
651	DEC80	M	ARCHITECTURE	BUILDING CONSTRUCTION	$42,000
652	DEC80	M	FORESTRY	FORESTRY	$14,500
653	DEC80	F	HEALTH RELATED	OCCUPATIONAL THERAPY	$12,100
654	DEC80	M	ENGINEERING	MECHANICAL	$20,400
655	DEC80	F	HEALTH RELATED	OCCUPATIONAL THERAPY	$14,500
656	DEC80	M	BUSINESS ADMINISTRATION	FINANCE	$14,500
657	DEC80	F	LIBERAL ARTS	SOCIOLOGY	$14,900
658	DEC80	F	SCIENCES	BOTANY	$12,500
659	DEC80	M	LIBERAL ARTS	THEATER	$11,600
660	DEC80	F	NURSING	NURSING	$15,600
661	DEC80	F	HEALTH RELATED	PHYSICAL THERAPY	$14,000
662	DEC80	F	NURSING	NURSING	$12,500
663	DEC80	M	BUSINESS ADMINISTRATION	MARKETING	$25,300
664	DEC80	F	NURSING	NURSING	$14,900
665	DEC80	M	ENGINEERING	MECHANICAL	$19,100
666	DEC80	M	ENGINEERING	MECHANICAL	$14,100
667	DEC80	F	PHARMACY	PHARMACY	$7,600
668	DEC80	F	JOURNALISM	PUBLIC RELATIONS	$16,800
669	DEC80	F	HEALTH RELATED	OCCUPATIONAL THERAPY	$11,900
670	DEC80	M	ARCHITECTURE	BUILDING CONSTRUCTION	$17,200
671	DEC80	F	LIBERAL ARTS	SOCIOLOGY	$13,800
672	DEC80	F	PHYS ED,HEALTH,RECREATION	PHYSICAL EDUCATION	$9,300
673	DEC80	M	ENGINEERING	ENVIRONMENTAL	$19,300
674	DEC80	M	ENGINEERING	ELECTRICAL	$19,300
675	DEC80	F	ACCOUNTING	ACCOUNTING	$10,200
676	DEC80	M	ARCHITECTURE	BUILDING CONSTRUCTION	$20,100
677	DEC80	M	BUSINESS ADMINISTRATION	ACCOUNTING	$11,100
678	DEC80	M	AGRICULTURE	ANIMAL SCIENCE	$21,400
679	DEC80	M	ACCOUNTING	ACCOUNTING	$15,900
680	DEC80	F	HEALTH RELATED	OCCUPATIONAL THERAPY	$13,200
681	DEC80	M	ENGINEERING	ELECTRICAL	$21,200
682	DEC80	M	ENGINEERING	ELECTRICAL	$23,500
683	DEC80	F	NURSING	NURSING	$12,700
684	DEC80	F	EDUCATION	SPEECH	$15,500
685	DEC80	F	PHYS ED,HEALTH,RECREATION	PHYSICAL EDUCATION	$7,400
686	DEC80	F	ACCOUNTING	ACCOUNTING	$23,100
687	DEC80	M	ENGINEERING	MECHANICAL	$22,600
688	DEC80	M	LAW	LAW	$22,000
689	DEC80	M	ACCOUNTING	ACCOUNTING	$15,600
690	DEC80	F	JOURNALISM	PUBLIC RELATIONS	$16,100
691	DEC80	F	NURSING	NURSING	$14,500
692	DEC80	M	ACCOUNTING	ACCOUNTING	$16,500
693	DEC80	F	HEALTH RELATED	OCCUPATIONAL THERAPY	$16,500
694	DEC80	M	ARCHITECTURE	BUILDING CONSTRUCTION	$25,300
695	DEC80	M	EDUCATION	ENGLISH	$18,600
696	DEC80	M	EDUCATION	SOCIAL SCIENCE	$12,700
697	DEC90	M	ENGINEERING	ELECTRICAL	$21,700
698	DEC80	M	ACCOUNTING	ACCOUNTING	$18,000
699	DEC80	M	ENGINEERING	CIVIL	$18,100
700	DEC80	M	BUILDING CONSTRUCTION	BUILDING CONSTRUCTION	$16,800
701	DEC80	M	ENGINEERING	METALLOGY	$21,000
702	DEC80	F	BUSINESS ADMINISTRATION	ACCOUNTING	$21,500
703	DEC80	M	BUSINESS ADMINISTRATION	MARKETING	$11,300
704	DEC80	F	HEALTH RELATED	PHYSICAL THERAPY	$17,300
705	DEC80	M	JOURNALISM	NEWS EDUCATION	$13,300
706	DEC80	F	ENGINEERING	MECHANICAL	$20,600
707	DEC80	M	ENGINEERING	ELECTRICAL	$17,800
708	DEC80	M	ENGINEERING	LAND SURVEYING	$11,600
709	DEC80	F	JOURNALISM	MAGAZINES	$8,200
710	DEC80	F	BUSINESS ADMINISTRATION	FINANCE	$12,400
711	DEC80	M	ENGINEERING	INDUSTRIAL	$21,800
712	DEC80	M	BUSINESS ADMINISTRATION	MARKETING	$12,900
713	DEC80	M	ENGINEERING	CIVIL	$20,000
714	DEC80	M	BUSINESS ADMINISTRATION	FINANCE	$13,300
715	DEC80	F	HEALTH RELATED	OCCUPATIONAL THERAPY	$11,700
716	DEC80	M	ENGINEERING	ELECTRICAL	$22,400
717	DEC80	M	ENGINEERING	ELECTRICAL	$20,500
718	DEC80	F	EDUCATION	SPEECH	$7,600
719	DEC80	F	HEALTH RELATED	OCCUPATIONAL THERAPY	$8,000
720	DEC80	M	ENGINEERING	INDUSTRIAL	$13,500
721	DEC80	F	NURSING	NURSING	$10,200
722	DEC80	F	ENGINEERING	COMPUTER SCIENCE	$22,100
723	DEC80	F	HEALTH RELATED	OCCUPATIONAL THERAPY	$10,200
724	DEC80	M	BUSINESS ADMINISTRATION	MARKETING	$10,000
725	DEC80	F	BUSINESS ADMINISTRATION	MANAGEMENT	$11,600
726	DEC80	M	ENGINEERING	CIVIL	$15,700
727	DEC80	F	JOURNALISM	BROADCASTING	$7,700
728	DEC80	M	ACCOUNTING	ACCOUNTING	$16,700
729	DEC80	F	LIBERAL ARTS	CRIMINAL JUSTICE	$11,100
730	DEC80	F	HEALTH RELATED	OCCUPATIONAL THERAPY	$14,000
731	DEC80	M	AGRICULTURE	EXTENDED EDUCATION	$7,900
732	DEC80	M	BUSINESS ADMINISTRATION	COMPUTER SCIENCE	$11,900
733	DEC80	F	HEALTH RELATED	PHYSICAL THERAPY	$15,200
734	DEC80	M	AGRICULTURE	ENTOMOLOGY	$17,000
735	DEC80	F	JOURNALISM	PUBLIC RELATIONS	$12,500

STARTING SALARIES OF 948 UNIVERSITY OF FLORIDA GRADUATES
JUNE 1980 TO MARCH 1981

OBS	DATE	SEX	COLLEGE	MAJOR	SALARY
736	DEC80	M	ARCHITECTURE	AEROSPACE ENGINEER	$14,100
737	DEC80	F	LIBERAL ARTS	PSYCHOLOGY	$8,000
738	DEC80	F	NURSING	NURSING	$13,500
739	DEC80	F	PHARMACY	PHARMACY	$23,600
740	DEC80	M	EDUCATION	ELEMENTARY	$9,300
741	DEC80	M	BUSINESS ADMINISTRATION	ECONOMICS	$13,000
742	DEC80	M	ACCOUNTING	ACCOUNTING	$13,000
743	DEC80	F	PHARMACY	PHARMACY	$8,000
744	DEC80	M	SCIENCES	COMPUTER SCIENCE	$19,300
745	DEC80	M	ACCOUNTING	ACCOUNTING	$18,000
746	DEC80	F	EDUCATION	SPEECH COMMUNICATIONS	$11,200
747	DEC80	F	HEALTH RELATED	OCCUPATIONAL THERAPY	$17,200
748	DEC80	F	HEALTH RELATED	OCCUPATIONAL THERAPY	$10,900
749	DEC80	F	HEALTH RELATED	OCCUPATIONAL THERAPY	$15,200
750	DEC80	M	ENGINEERING	CHEMICAL	$25,100
751	DEC80	M	LIBERAL ARTS	CRIMINAL JUSTICE	$12,400
752	DEC80	M	LAW	LAW	$8,700
753	DEC80	M	SCIENCES	ECONOMICS	$12,100
754	DEC80	M	ENGINEERING	ELECTRICAL	$22,600
755	DEC80	F	JOURNALISM	PUBLIC RELATIONS	$16,900
756	DEC80	F	JOURNALISM	JOURNALISM	$10,200
757	DEC80	F	HEALTH RELATED	OCCUPATIONAL THERAPY	$17,200
758	DEC80	M	ENGINEERING	ENVIRONMENTAL	$18,900
759	DEC80	F	HEALTH RELATED	PHYSICAL THERAPY	$17,700
760	DEC80	M	BUSINESS ADMINISTRATION	FINANCE	$15,200
761	DEC80	F	ACCOUNTING	ACCOUNTING	$16,600
762	DEC80	M	ENGINEERING	MECHANICAL	$22,600
763	DEC80	M	ARCHITECTURE	BUILDING CONSTRUCTION	$21,600
764	DEC80	M	BUSINESS ADMINISTRATION	FINANCE	$17,300
765	DEC80	M	ENGINEERING	ELECTRICAL	$19,500
766	DEC80	M	SCIENCES	GEOGRAPHY	$9,400
767	DEC80	F	LIBERAL ARTS	SPEECH	$12,000
768	DEC80	M	SCIENCES	ECONOMICS	$19,500
769	DEC80	M	LIBERAL ARTS	POLITICAL SCIENCE	$14,000
770	DEC80	F	NURSING	NURSING	$15,600
771	DEC80	M	LIBERAL ARTS	HISTORY	$16,500
772	DEC80	M	LIBERAL ARTS	POLITICAL SCIENCE	$9,500
773	DEC80	M	AGRICULTURE	FOOD CROP	$12,400
774	DEC80	M	BUSINESS ADMINISTRATION	FINANCE	$11,000
775	DEC80	M	BUSINESS ADMINISTRATION	COMPUTER SCIENCE	$16,900
776	DEC80	M	LIBERAL ARTS	PSYCHOLOGY	$11,800
777	DEC80	M	BUSINESS ADMINISTRATION	MANAGEMENT	$11,500
778	MAR81	M	PHARMACY	PHARMACY	$11,400
779	MAR81	M	ENGINEERING	ELECTRICAL	$21,900
780	MAR81	F	PHARMACY	PHARMACY	$10,900
781	MAR81	M	LIBERAL ARTS	CRIMINAL JUSTICE	$13,500
782	MAR81	F	PHARMACY	PHARMACY	$20,900
783	MAR81	F	EDUCATION	ELEMENTARY	$7,300
784	MAR81	F	ARCHITECTURE	INTERIOR DESIGN	$10,500
785	MAR81	M	BUSINESS ADMINISTRATION	FINANCE	$14,500
786	MAR81	F	AGRICULTURE	ORNAMENTAL HORTICULTURE	$17,100
787	MAR81	M	PHARMACY	PHARMACY	$25,700
788	MAR81	F	EDUCATION	ELEMENTARY	$9,500
789	MAR81	F	LIBERAL ARTS	ENGLISH	$8,700
790	MAR81	M	LIBERAL ARTS	PSYCHOLOGY	$14,700
791	MAR81	F	AGRICULTURE	PLANT SCIENCE	$10,400
792	MAR81	F	ARCHITECTURE	INTERIOR DESIGN	$11,300
793	MAR81	M	ARCHITECTURE	INTERIOR DESIGN	$13,900
794	MAR81	M	ENGINEERING	COMPUTER SCIENCE	$21,400
795	MAR81	M	ENGINEERING	ELECTRICAL	$20,100
796	MAR81	F	ENGINEERING	MECHANICAL	$22,300
797	MAR81	M	ENGINEERING	CIVIL	$22,800
798	MAR81	M	BUSINESS ADMINISTRATION	MANAGEMENT	$17,300
799	MAR81	M	ENGINEERING	ELECTRICAL	$24,000
800	MAR81	F	ENGINEERING	ENVIRONMENTAL	$16,600
801	MAR81	M	ENGINEERING	CHEMICAL	$24,500
802	MAR81	F	NURSING	NURSING	$8,500
803	MAR81	F	NURSING	NURSING	$15,200
804	MAR81	F	NURSING	NURSING	$14,700
805	MAR81	M	BUSINESS ADMINISTRATION	FINANCE	$20,200
806	MAR81	F	JOURNALISM	BROADCASTING	$9,400
807	MAR81	F	EDUCATION	ELEMENTARY	$8,200
808	MAR81	M	ENGINEERING	ELECTRICAL	$21,500
809	MAR81	M	ENGINEERING	ELECTRICAL	$22,700
810	MAR81	F	ENGINEERING	ELECTRICAL	$18,200
811	MAR81	F	NURSING	NURSING	$18,100
812	MAR81	F	BUSINESS ADMINISTRATION	COMPUTER SCIENCE	$21,100
813	MAR81	F	JOURNALISM	PUBLIC RELATIONS	$10,500
814	MAR81	F	NURSING	NURSING	$20,400
815	MAR81	F	PHARMACY	PHARMACY	$22,300
816	MAR81	M	BUSINESS ADMINISTRATION	FINANCE	$18,400
817	MAR81	M	ARCHITECTURE	LANDSCAPING	$9,300
818	MAR81	F	NURSING	NURSING	$11,400
819	MAR81	F	HEALTH RELATED	PHYSICAL THERAPY	$17,400
820	MAR81	M	BUSINESS ADMINISTRATION	FINANCE	$14,800
821	MAR81	M	ENGINEERING	ELECTRICAL	$20,300
822	MAR81	F	SCIENCES	COMPUTER SCIENCE	$33,400
823	MAR81	M	SCIENCES	GEOLOGY	$11,300
824	MAR81	F	EDUCATION	SPEECH	$7,300
825	MAR81	F	NURSING	NURSING	$16,500
826	MAR81	M	PHYS ED.HEALTH.RECREATION	RECREATION	$8,200
827	MAR81	M	AGRICULTURE	VEGETABLE CROP	$10,400
828	MAR81	F	PHARMACY	PHARMACY	$24,500
829	MAR81	M	AGRICULTURE	ANIMAL SCIENCE	$7,500
830	MAR81	F	BUSINESS ADMINISTRATION	FINANCE	$14,700
831	MAR81	M	PHARMACY	PHARMACY	$8,700
832	MAR81	F	PHARMACY	PHARMACY	$17,700
833	MAR81	F	ENGINEERING	MECHANICAL	$22,600
834	MAR81	M	ENGINEERING	MECHANICAL	$23,400
835	MAR81	M	ENGINEERING	COMPUTER SCIENCE	$17,500
836	MAR81	M	AGRICULTURE	EXTENDED EDUCATION	$7,400
837	MAR81	F	NURSING	NURSING	$14,700
838	MAR81	M	ENGINEERING	CIVIL	$26,200
839	MAR81	M	BUSINESS ADMINISTRATION	COMPUTER SCIENCE	$17,300
840	MAR81	M	ENGINEERING	CIVIL	$18,000

STARTING SALARIES OF 948 UNIVERSITY OF FLORIDA GRADUATES
JUNE 1980 TO MARCH 1981

OBS	DATE	SEX	COLLEGE	MAJOR	SALARY
841	MAR81	F	PHARMACY	PHARMACY	$20,000
842	MAR81	F	AGRICULTURE	FOOD RESOURCE ECONOMICS	$14,500
843	MAR81	M	ENGINEERING	ELECTRICAL	$24,000
844	MAR81	M	BUSINESS ADMINISTRATION	FINANCE	$39,400
845	MAR81	M	AGRICULTURE	MECHANIZED	$12,600
846	MAR81	M	BUILDING CONSTRUCTION	BUILDING CONSTRUCTION	$16,900
847	MAR81	M	ACCOUNTING	ACCOUNTING	$16,400
848	MAR81	F	PHARMACY	PHARMACY	$24,300
849	MAR81	M	BUSINESS ADMINISTRATION	MARKETING	$36,300
850	MAR81	M	ENGINEERING	MECHANICAL	$20,500
851	MAR81	F	AGRICULTURE	ORNAMENTAL HORTICULTURE	$8,700
852	MAR81	M	ENGINEERING	ELECTRICAL	$22,500
853	MAR81	M	ENGINEERING	ELECTRICAL	$28,200
854	MAR81	F	EDUCATION	SPEECH	$13,600
855	MAR81	F	AGRICULTURE	MICROBIOLOGY	$9,900
856	MAR81	M	ENGINEERING	ELECTRICAL	$21,900
857	MAR81	F	SCIENCES	COMPUTER SCIENCE	$19,500
858	MAR81	F	ENGINEERING	ENVIRONMENTAL	$13,400
859	MAR81	M	ENGINEERING	ELECTRICAL	$26,200
860	MAR81	M	BUSINESS ADMINISTRATION	MANAGEMENT	$9,200
861	MAR81	M	ENGINEERING	INDUSTRIAL	$27,200
862	MAR81	M	ACCOUNTING	ACCOUNTING	$19,700
863	MAR81	M	ACCOUNTING	ACCOUNTING	$16,400
864	MAR81	M	LIBERAL ARTS	PSYCHOLOGY	$10,600
865	MAR81	F	NURSING	NURSING	$16,000
866	MAR81	F	BUSINESS ADMINISTRATION	FINANCE	$16,300
867	MAR81	M	ENGINEERING	ENVIRONMENTAL	$21,600
868	MAR81	M	ENGINEERING	CIVIL	$13,700
869	MAR81	F	EDUCATION	ELEMENTARY	$9,300
870	MAR81	M	ARCHITECTURE	BUILDING CONSTRUCTION	$23,800
871	MAR81	F	NURSING	NURSING	$14,800
872	MAR81	M	BUSINESS ADMINISTRATION	MANAGEMENT	$9,400
873	MAR81	F	BUSINESS ADMINISTRATION	ACCOUNTING	$16,400
874	MAR81	F	NURSING	NURSING	$20,400
875	MAR81	M	PHARMACY	PHARMACY	$9,900
876	MAR81	M	BUSINESS ADMINISTRATION	REAL ESTATE	$14,800
877	MAR81	F	PHARMACY	PHARMACY	$7,700
878	MAR81	M	ARCHITECTURE	AEROSPACE ENGINEER	$10,800
879	MAR81	F	ENGINEERING	ENVIRONMENTAL	$19,800
880	MAR81	M	SCIENCES	COMPUTER SCIENCE	$13,900
881	MAR81	M	ENGINEERING	ELECTRICAL	$11,100
882	MAR81	M	BUSINESS ADMINISTRATION	COMPUTER SCIENCE	$13,200
883	MAR81	M	ENGINEERING	INDUSTRIAL	$25,600
884	MAR81	F	EDUCATION	SPEECH	$9,200
885	MAR81	F	ACCOUNTING	ACCOUNTING	$17,600
886	MAR81	M	ENGINEERING	ELECTRICAL	$23,500
887	MAR81	F	SCIENCES	ANTHROPOLOGY	$10,600
888	MAR81	M	ENGINEERING	INDUSTRIAL	$14,600
889	MAR81	F	PHYS ED,HEALTH,RECREATION	RECREATION	$10,900
890	MAR81	M	JOURNALISM	MAGAZINES	$10,700
891	MAR81	M	ENGINEERING	ELECTRICAL	$28,100
892	MAR81	M	ENGINEERING	ELECTRICAL	$23,000
893	MAR81	M	BUSINESS ADMINISTRATION	FINANCE	$19,100
894	MAR81	M	ENGINEERING	COMPUTER SCIENCE	$20,600
895	MAR81	F	NURSING	NURSING	$15,300
896	MAR81	F	SCIENCES	COMPUTER SCIENCE	$22,600
897	MAR81	F	AGRICULTURE	FOOD RESOURCE ECONOMICS	$7,600
898	MAR81	F	NURSING	NURSING	$13,600
899	MAR81	M	ACCOUNTING	ACCOUNTING	$16,100
900	MAR81	M	ENGINEERING	ELECTRICAL	$20,200
901	MAR81	M	JOURNALISM	BROADCASTING	$15,500
902	MAR81	M	ACCOUNTING	ACCOUNTING	$16,400
903	MAR81	F	LIBERAL ARTS	PSYCHOLOGY	$15,500
904	MAR81	M	ENGINEERING	ELECTRICAL	$22,700
905	MAR81	F	PHARMACY	PHARMACY	$25,000
906	MAR81	M	ENGINEERING	CHEMICAL	$25,100
907	MAR81	F	ACCOUNTING	ACCOUNTING	$19,200
908	MAR81	M	BUSINESS ADMINISTRATION	ACCOUNTING	$17,000
909	MAR81	F	PHARMACY	PHARMACY	$13,300
910	MAR81	M	ARCHITECTURE	AEROSPACE ENGINEER	$8,500
911	MAR81	F	NURSING	NURSING	$14,400
912	MAR81	M	PHARMACY	PHARMACY	$28,500
913	MAR81	M	ENGINEERING	ELECTRICAL	$23,400
914	MAR81	F	SCIENCES	MATHEMATICS	$20,800
915	MAR81	M	ACCOUNTING	ACCOUNTING	$18,700
916	MAR81	M	ENGINEERING	LAND SURVEYING	$17,400
917	MAR81	M	ACCOUNTING	ACCOUNTING	$18,400
918	MAR81	F	PHARMACY	PHARMACY	$15,000
919	MAR81	M	ENGINEERING	LAND SURVEYING	$18,800
920	MAR81	M	ENGINEERING	CIVIL	$22,400
921	MAR81	M	ENGINEERING	ELECTRICAL	$21,500
922	MAR81	F	BUSINESS ADMINISTRATION	ACCOUNTING	$15,800
923	MAR81	M	ENGINEERING	ELECTRICAL	$20,900
924	MAR81	F	HEALTH RELATED	PHYSICAL THERAPY	$15,400
925	MAR81	M	ENGINEERING	ELECTRICAL	$20,900
926	MAR81	F	EDUCATION	HISTORY	$11,400
927	MAR81	M	ACCOUNTING	ACCOUNTING	$15,500
928	MAR81	M	ENGINEERING	ELECTRICAL	$22,800
929	MAR81	M	PHARMACY	PHARMACY	$23,900
930	MAR81	F	EDUCATION	ELEMENTARY	$7,800
931	MAR81	M	ENGINEERING	CIVIL	$12,000
932	MAR81	F	JOURNALISM	BROADCASTING	$10,800
933	MAR81	M	ENGINEERING	ELECTRICAL	$21,000
934	MAR81	M	ENGINEERING	CIVIL	$21,700
935	MAR81	M	PHARMACY	PHARMACY	$19,500
936	MAR81	F	EDUCATION	SPEECH	$8,600
937	MAR81	M	ENGINEERING	CHEMICAL	$22,400
938	MAR81	M	BUSINESS ADMINISTRATION	FINANCE	$20,000
939	MAR81	M	ENGINEERING	MECHANICAL	$28,600
940	MAR81	F	NURSING	NURSING	$13,600
941	MAR81	F	AGRICULTURE	AGRONOMY	$13,700
942	MAR81	F	PHARMACY	PHARMACY	$28,700
943	MAR81	M	ENGINEERING	INDUSTRIAL	$19,800
944	MAR81	F	PHARMACY	PHARMACY	$11,600
945	MAR81	M	ENGINEERING	MECHANICAL	$21,800
946	MAR81	M	ENGINEERING	ELECTRICAL	$20,700
947	MAR81	F	NURSING	NURSING	$11,700
948	MAR81	M	ARCHITECTURE	BUILDING CONSTRUCTION	$17,400

Appendix B Data Set: Starting Salaries of University of Florida Graduates in Five Colleges

STARTING SALARIES OF UNIVERSITY OF FLORIDA GRADUATES
IN FIVE COLLEGES

COLLEGE OF BUSINESS ADMINISTRATION

OBS	SEX	SALARY	OBS	SEX	SALARY	OBS	SEX	SALARY
1	F	$15,600	2	M	$11,300	3	M	$13,000
4	M	$17,100	5	M	$10,200	6	M	$12,900
7	M	$13,200	8	M	$14,500	9	F	$15,400
10	M	$22,600	11	M	$8,100	12	M	$13,100
13	M	$12,200	14	F	$15,400	15	M	$14,400
16	F	$7,400	17	F	$14,400	18	F	$12,500
19	F	$16,700	20	M	$13,500	21	M	$18,400
22	M	$14,900	23	M	$40,400	24	M	$10,200
25	M	$17,100	26	F	$14,900	27	M	$10,800
28	M	$11,900	29	M	$21,300	30	M	$14,000
31	F	$12,300	32	M	$16,600	33	M	$12,200
34	M	$15,500	35	M	$15,900	36	M	$16,600
37	F	$19,200	38	M	$18,900	39	M	$16,500
40	F	$17,400	41	M	$20,200	42	M	$19,600
43	M	$16,200	44	M	$15,600	45	M	$14,600
46	M	$15,700	47	F	$13,900	48	M	$13,100
49	M	$11,500	50	F	$18,500	51	M	$19,300
52	F	$13,800	53	F	$14,800	54	M	$11,600
55	M	$21,200	56	F	$17,300	57	M	$18,700
58	F	$11,400	59	M	$11,700	60	F	$13,800
61	F	$11,200	62	F	$9,200	63	M	$12,500
64	M	$13,200	65	M	$15,600	66	F	$15,000
67	M	$13,400	68	F	$17,100	69	M	$12,000
70	F	$10,600	71	M	$13,100	72	M	$14,300
73	M	$20,600	74	F	$12,300	75	M	$18,300
76	M	$16,600	77	F	$7,000	78	M	$11,700
79	M	$11,000	80	M	$36,500	81	M	$11,400
82	M	$10,900	83	M	$16,300	84	M	$13,700
85	M	$12,100	86	F	$15,600	87	M	$16,000
88	M	$5,500	89	M	$14,500	90	F	$12,700
91	M	$19,500	92	M	$17,500	93	M	$13,900
94	F	$8,900	95	M	$14,300	96	F	$12,600
97	M	$10,900	98	M	$11,300	99	M	$12,500
100	F	$13,900	101	M	$13,700	102	F	$15,200
103	M	$18,500	104	F	$17,300	105	M	$14,500
106	M	$25,300	107	M	$11,100	108	F	$21,500
109	M	$11,300	110	F	$12,400	111	M	$12,900
112	M	$13,300	113	M	$10,000	114	F	$11,600
115	M	$11,900	116	M	$13,000	117	M	$15,200
118	M	$17,300	119	M	$11,000	120	M	$16,900
121	M	$11,500	122	M	$14,500	123	M	$17,300
124	M	$20,200	125	F	$21,100	126	M	$18,400
127	M	$14,800	128	F	$14,700	129	M	$17,300
130	M	$39,400	131	M	$36,300	132	M	$9,200
133	F	$16,300	134	M	$9,400	135	F	$16,400
136	M	$14,800	137	M	$13,200	138	M	$19,100
139	M	$17,000	140	F	$15,800	141	M	$20,000

COLLEGE OF EDUCATION

OBS	SEX	SALARY	OBS	SEX	SALARY	OBS	SEX	SALARY
142	F	$12,000	143	M	$12,900	144	F	$14,000
145	F	$8,000	146	F	$6,400	147	F	$10,500
148	F	$9,700	149	F	$10,900	150	M	$10,000
151	F	$12,800	152	F	$7,800	153	F	$7,100
154	M	$5,000	155	F	$6,800	156	F	$10,500
157	F	$8,000	158	F	$14,600	159	F	$13,100
160	F	$15,500	161	M	$18,600	162	M	$12,700
163	F	$7,600	164	M	$9,300	165	F	$11,200
166	F	$7,300	167	F	$9,500	168	F	$8,200
169	F	$7,300	170	F	$13,600	171	F	$9,300
172	F	$9,200	173	F	$11,400	174	F	$7,800
175	F	$8,600						

COLLEGE OF ENGINEERING

OBS	SEX	SALARY	OBS	SEX	SALARY	OBS	SEX	SALARY
			176	M	$22,900	177	M	$16,800
178	M	$22,400	179	M	$26,500	180	M	$22,000
181	M	$9,000	182	M	$21,100	183	M	$20,100
184	M	$15,900	185	M	$17,700	186	M	$18,700
187	M	$18,100	188	M	$19,400	189	M	$23,900
190	M	$20,600	191	M	$23,700	192	M	$19,300
193	M	$21,400	194	M	$19,100	195	M	$12,800
196	M	$19,900	197	M	$22,200	198	F	$20,800
199	M	$25,700	200	M	$20,400	201	M	$19,000
202	M	$17,700	203	M	$17,100	204	M	$21,800
205	M	$22,300	206	M	$22,600	207	M	$16,900
208	M	$11,800	209	M	$14,100	210	M	$22,700
211	M	$19,900	212	F	$22,400	213	M	$21,000
214	M	$21,300	215	M	$18,400	216	M	$14,800
217	M	$20,400	218	M	$14,500	219	M	$21,600
220	M	$19,000	221	M	$17,100	222	M	$22,500
223	F	$22,800	224	M	$19,200	225	M	$19,700
226	F	$12,100	227	M	$26,300	228	F	$21,400
229	M	$19,800	230	M	$25,300	231	F	$20,800
232	M	$20,300	233	M	$13,500	234	M	$22,400
235	M	$19,800	236	M	$19,800	237	F	$21,500
238	M	$18,000	239	M	$22,400	240	M	$20,600
241	M	$17,500	242	M	$22,100	243	M	$24,400
244	M	$19,300	245	M	$22,700	246	M	$20,400
247	M	$17,600	248	M	$17,700	249	M	$26,600
250	M	$12,000	251	M	$21,900	252	M	$19,400
253	M	$21,600	254	M	$22,000	255	M	$21,400

STARTING SALARIES OF UNIVERSITY OF FLORIDA GRADUATES IN FIVE COLLEGES

COLLEGE OF ENGINEERING (CONTINUED)

OBS	SEX	SALARY	OBS	SEX	SALARY	OBS	SEX	SALARY
256	M	$21,200	257	M	$17,900	258	M	$23,400
259	M	$22,800	260	M	$18,900	261	M	$19,700
262	M	$13,400	263	M	$22,800	264	M	$18,800
265	M	$19,400	266	M	$21,200	267	M	$21,500
268	M	$15,000	269	M	$22,900	270	M	$21,900
271	F	$17,900	272	M	$18,000	273	M	$14,900
274	M	$19,800	275	M	$17,600	276	M	$18,100
277	M	$20,800	278	F	$21,200	279	M	$20,800
280	M	$14,800	281	M	$19,300	282	M	$20,400
283	M	$18,000	284	M	$24,000	285	M	$14,400
286	M	$22,700	287	M	$17,300	288	M	$23,200
289	M	$14,800	290	M	$22,300	291	F	$19,200
292	M	$21,300	293	F	$20,200	294	M	$17,800
295	M	$17,700	296	M	$19,700	297	F	$22,100
298	M	$20,900	299	M	$15,600	300	M	$24,100
301	M	$20,600	302	M	$21,500	303	M	$27,300
304	M	$16,900	305	M	$21,100	306	M	$18,400
307	M	$18,400	308	M	$19,300	309	M	$23,000
310	M	$15,000	311	M	$13,700	312	M	$21,600
313	M	$11,100	314	M	$25,700	315	M	$21,700
316	F	$22,000	317	F	$20,100	318	F	$18,600
319	F	$20,100	320	M	$19,200	321	M	$21,000
322	M	$21,000	323	M	$23,600	324	F	$21,000
325	M	$18,400	326	M	$22,800	327	M	$18,600
328	M	$17,300	329	M	$23,000	330	M	$21,300
331	M	$22,700	332	M	$21,400	333	M	$22,100
334	F	$19,800	335	M	$20,500	336	M	$22,100
337	F	$9,700	338	M	$19,900	339	M	$22,700
340	M	$19,400	341	M	$20,400	342	M	$19,100
343	M	$14,100	344	M	$19,300	345	M	$19,300
346	M	$21,200	347	M	$23,500	348	M	$22,600
349	M	$21,700	350	M	$18,100	351	M	$21,000
352	F	$20,600	353	M	$17,800	354	M	$11,600
355	M	$21,800	356	M	$20,000	357	M	$22,400
358	M	$20,500	359	M	$13,500	360	F	$22,100
361	M	$15,700	362	M	$25,100	363	M	$22,600
364	M	$18,900	365	M	$22,600	366	M	$19,500
367	M	$21,900	368	M	$21,400	369	M	$20,100
370	F	$22,300	371	M	$22,800	372	M	$24,000
373	F	$16,600	374	M	$24,500	375	M	$21,500
376	M	$22,700	377	F	$18,200	378	M	$20,300
379	F	$22,600	380	M	$23,400	381	M	$17,500
382	M	$26,200	383	M	$18,000	384	M	$24,000
385	M	$20,500	386	M	$22,500	387	M	$28,200
388	M	$21,900	389	F	$13,400	390	M	$26,200
391	M	$27,200	392	M	$21,600	393	M	$13,700
394	F	$19,800	395	M	$11,100	396	M	$25,600
397	M	$23,500	398	M	$14,600	399	M	$28,100
400	M	$23,000	401	M	$20,600	402	M	$20,200
403	M	$22,700	404	M	$25,100	405	M	$23,400
406	M	$17,400	407	M	$18,800	408	M	$22,400
409	M	$21,500	410	M	$20,900	411	M	$20,900
412	M	$22,800	413	M	$12,000	414	M	$21,000
415	M	$21,700	416	M	$22,400	417	M	$28,600
418	M	$19,800	419	M	$21,800	420	M	$20,700

COLLEGE OF LIBERAL ARTS

OBS	SEX	SALARY	OBS	SEX	SALARY	OBS	SEX	SALARY
421	F	$6,800	422	F	$14,500	423	M	$20,700
424	F	$9,700	425	F	$6,300	426	F	$22,700
427	M	$13,400	428	F	$10,600	429	M	$11,400
430	F	$13,800	431	F	$11,000	432	F	$14,700
433	M	$13,500	434	M	$14,100	435	F	$7,100
436	F	$11,200	437	M	$11,600	438	M	$18,900
439	F	$5,400	440	F	$8,900	441	M	$16,800
442	M	$22,800	443	F	$11,100	444	M	$9,500
445	M	$12,900	446	F	$11,300	447	F	$7,200
448	F	$15,100	449	M	$9,100	450	F	$14,900
451	M	$11,600	452	F	$13,800	453	F	$11,100
454	F	$8,000	455	M	$12,400	456	F	$12,000
457	M	$14,000	458	M	$16,500	459	M	$9,500
460	M	$11,800	461	M	$13,500	462	F	$8,700
463	M	$14,700	464	M	$10,600	465	F	$15,500

COLLEGE OF SCIENCES

OBS	SEX	SALARY	OBS	SEX	SALARY	OBS	SEX	SALARY
466	F	$15,400	467	F	$18,300	468	F	$10,800
469	F	$6,000	470	F	$15,300	471	F	$20,300
472	F	$11,000	473	F	$9,300	474	M	$15,000
475	F	$13,900	476	F	$20,500	477	M	$9,400
478	M	$7,600	479	F	$15,500	480	M	$17,300
481	F	$11,200	482	F	$15,400	483	M	$14,700
484	F	$20,700	485	M	$12,100	486	F	$21,200
487	M	$32,400	488	M	$15,200	489	M	$17,200
490	M	$13,400	491	M	$26,500	492	F	$19,600
493	M	$21,600	494	F	$12,500	495	M	$19,300
496	M	$12,100	497	M	$9,400	498	M	$19,500
499	F	$33,400	500	M	$11,300	501	F	$19,500
502	M	$13,900	503	F	$10,600	504	F	$22,600
505	F	$20,800						

Appendix C Data Set: Supermarket Customer Checkout Times for Mechanical and Automated Checking

CUSTOMER CHECKOUT TIMES(SECONDS)
AT SUPERMARKETS WITH MECHANICAL AND AUTOMATED CHECKERS

SUPERMARKET A (MECHANICAL)

OBS	CHKTIME	OBS	CHKTIME	OBS	CHKTIME	OBS	CHKTIME	OBS	CHKTIME	OBS	CHKTIME	OBS	CHKTIME	OBS	CHKTIME	OBS	CHKTIME	OBS	CHKTIME
1	62	51	112	101	115	151	40	201	77	251	56	301	144	351	100	401	137	451	38
2	133	52	132	102	147	152	65	202	64	252	87	302	75	352	85	402	145	452	27
3	190	53	93	103	63	153	147	203	36	253	112	303	85	353	185	403	80	453	100
4	25	54	70	104	130	154	125	204	131	254	135	304	138	354	95	404	132	454	96
5	80	55	270	105	135	155	112	205	123	255	145	305	75	355	112	405	75	455	70
6	57	56	130	106	40	156	192	206	48	256	90	306	67	356	204	406	52	456	87
7	55	57	85	107	105	157	190	207	72	257	124	307	56	357	108	407	65	457	58
8	135	58	70	108	65	158	135	208	75	258	137	308	90	358	96	408	55	458	106
9	150	59	50	109	35	159	140	209	138	259	116	309	168	359	62	409	100	459	112
10	100	60	145	110	70	160	95	210	70	260	148	310	100	360	58	410	85	460	49
11	110	61	35	111	152	161	76	211	87	261	160	311	95	361	75	411	170	461	147
12	152	62	75	112	177	162	138	212	58	262	97	312	185	362	125	412	65	462	92
13	27	63	55	113	113	163	157	213	113	263	172	313	147	363	200	413	76	463	72
14	96	64	85	114	85	164	34	214	160	264	49	314	132	364	70	414	48	464	65
15	145	65	60	115	205	165	96	215	145	265	53	315	58	365	144	415	25	465	110
16	102	66	65	116	170	166	74	216	75	266	113	316	21	366	149	416	78	466	145
17	35	67	95	117	15	167	36	217	73	267	160	317	62	367	70	417	130	467	160
18	147	68	125	118	27	168	152	218	178	268	84	318	108	368	130	418	115	468	96
19	202	69	160	119	52	169	148	219	60	269	48	319	105	369	116	419	95	469	84
20	38	70	85	120	20	170	73	220	206	270	80	320	18	370	145	420	138	470	106
21	53	71	70	121	63	171	57	221	84	271	102	321	80	371	85	421	112	471	70
22	33	72	25	122	47	172	53	222	81	272	125	322	170	372	46	422	49	472	55
23	77	73	80	123	84	173	197	223	156	273	165	323	84	373	138	423	96	473	86
24	90	74	72	124	132	174	222	224	139	274	90	324	96	374	70	424	205	474	112
25	38	75	140	125	70	175	70	225	111	275	130	325	110	375	85	425	88	475	130
26	275	76	35	126	80	176	147	226	57	276	200	326	70	376	135	426	88	476	192
27	142	77	157	127	220	177	106	227	165	277	95	327	85	377	85	427	172	477	100
28	146	78	65	128	70	178	117	228	65	278	80	328	130	378	72	428	112	478	114
29	77	79	124	129	64	179	94	229	137	279	138	329	144	379	60	429	106	479	58
30	32	80	170	130	91	180	58	230	125	280	85	330	125	380	52	430	70	480	78
31	137	81	160	131	27	181	15	231	64	281	172	331	115	381	147	431	155	481	147
32	132	82	75	132	125	182	137	232	26	282	135	332	28	382	155	432	110	482	116
33	47	83	145	133	95	183	132	233	72	283	90	333	83	383	64	433	87	483	148
34	112	84	100	134	75	184	107	234	220	284	90	334	166	384	108	434	202	484	85
35	145	85	97	135	65	185	35	235	37	285	98	335	72	385	92	435	106	485	115
36	327	86	120	136	56	186	106	236	85	286	118	336	87	386	116	436	78	486	166
37	130	87	25	137	130	187	205	237	80	287	154	337	176	387	85	437	118	487	105
38	45	88	133	138	85	188	15	238	128	288	60	338	94	388	114	438	176	488	75
39	72	89	52	139	27	189	27	239	65	289	70	339	96	389	115	439	144	489	162
40	93	90	50	140	35	190	75	240	95	290	155	340	58	390	130	440	75	490	78
41	5	91	80	141	162	191	85	241	95	291	146	341	84	391	136	441	48	491	220
42	80	92	67	142	117	192	125	242	138	292	200	342	85	392	97	442	85	492	115
43	215	93	120	143	72	193	129	243	127	293	71	343	170	393	110	443	70	493	73
44	186	94	127	144	63	194	60	244	83	294	115	344	84	394	70	444	146	494	162
45	85	95	42	145	105	195	84	245	45	295	98	345	87	395	65	445	82	495	138
46	70	96	180	146	155	196	95	246	96	296	45	346	143	396	152	446	100	496	175
47	77	97	70	147	77	197	132	247	220	297	65	347	310	397	100	447	75	497	110
48	100	98	80	148	208	198	185	248	98	298	60	348	75	398	75	448	70	498	84
49	72	99	245	149	130	199	76	249	165	299	146	349	172	399	42	449	144	499	95
50	194	100	98	150	112	200	69	250	154	300	132	350	135	400	156	450	137	500	85

CUSTOMER CHECKOUT TIMES(SECONDS)
AT SUPERMARKETS WITH MECHANICAL AND AUTOMATED CHECKERS

SUPERMARKET B (AUTOMATED)

OBS	CHKTIME	OBS	CHKTIME	OBS	CHKTIME	OBS	CHKTIME	OBS	CHKTIME	OBS	CHKTIME	OBS	CHKTIME	OBS	CHKTIME	OBS	CHKTIME	OBS	CHKTIME
501	18	551	25	601	25	651	40	701	30	751	20	801	4	851	35	901	30	951	100
502	37	552	7	602	12	652	47	702	15	752	20	802	45	852	30	902	50	952	25
503	63	553	30	603	45	653	13	703	215	753	135	803	4	853	25	903	45	953	80
504	6	554	8	604	26	654	5	704	150	754	40	804	40	854	15	904	2	954	20
505	116	555	73	605	48	655	35	705	120	755	57	805	15	855	40	905	13	955	35
506	53	556	10	606	165	656	120	706	40	756	145	806	45	856	70	906	57	956	70
507	65	557	39	607	44	657	44	707	43	757	7	807	50	857	17	907	26	957	10
508	35	558	61	608	38	658	45	708	35	758	159	808	15	858	10	908	23	958	5
509	229	559	126	609	8	659	26	709	40	759	62	809	5	859	5	909	2	959	60
510	67	560	8	610	35	660	38	710	25	760	15	810	10	860	35	910	100	960	120
511	13	561	44	611	22	661	40	711	22	761	115	811	45	861	55	911	23	961	23
512	63	562	120	612	17	662	50	712	40	762	292	812	105	862	80	912	52	962	7
513	20	563	80	613	78	663	25	713	24	763	85	813	40	863	70	913	45	963	13
514	37	564	36	614	15	664	26	714	340	764	33	814	10	864	40	914	95	964	15
515	77	565	17	615	113	665	25	715	350	765	34	815	30	865	45	915	30	965	40
516	69	566	156	616	17	666	70	716	107	766	99	816	15	866	15	916	50	966	20
517	16	567	31	617	25	667	20	717	80	767	103	817	30	867	120	917	8	967	38
518	12	568	16	618	30	668	25	718	110	768	155	818	110	868	12	918	37	968	13
519	98	569	47	619	20	669	110	719	180	769	45	819	30	869	90	919	40	969	25
520	12	570	36	620	6	670	49	720	75	770	50	820	5	870	30	920	30	970	30
521	57	571	56	621	32	671	180	721	155	771	57	821	10	871	3	921	70	971	4
522	84	572	17	622	7	672	90	722	53	772	30	822	10	872	100	922	80	972	3
523	65	573	23	623	7	673	111	723	55	773	103	823	5	873	8	923	7	973	26
524	124	574	60	624	18	674	69	724	170	774	30	824	8	874	55	924	80	974	60
525	53	575	22	625	20	675	75	725	20	775	135	825	10	875	15	925	100	975	65
526	18	576	39	626	15	676	40	726	10	776	30	826	50	876	50	926	60	976	30
527	12	577	93	627	10	677	39	727	20	777	130	827	100	877	25	927	75	977	4
528	57	578	20	628	63	678	40	728	123	778	60	828	18	878	2	928	45	978	70
529	84	579	96	629	30	679	60	729	90	779	25	829	10	879	55	929	60	979	50
530	65	580	237	630	36	680	85	730	40	780	353	830	70	880	30	930	35	980	87
531	124	581	245	631	108	681	20	731	63	781	110	831	40	881	35	931	17	981	30
532	53	582	10	632	45	682	24	732	25	782	25	832	20	882	80	932	12	982	13
533	18	583	52	633	33	683	91	733	15	783	10	833	15	883	25	933	20	983	135
534	12	584	33	634	10	684	50	734	57	784	9	834	20	884	42	934	125	984	6
535	18	585	10	635	6	685	63	735	30	785	35	835	15	885	21	935	18	985	35
536	12	586	27	636	32	686	40	736	40	786	35	836	5	886	48	936	3	986	130
537	15	587	12	637	50	687	30	737	35	787	10	837	8	887	90	937	10	987	75
538	18	588	74	638	143	688	20	738	50	788	20	838	10	888	30	938	20	988	35
539	7	589	101	639	16	689	120	739	16	789	15	839	50	889	100	939	50	989	40
540	27	590	46	640	135	690	29	740	120	790	5	840	70	890	7	940	60	990	55
541	34	591	142	641	81	691	30	741	23	791	20	841	3	891	10	941	20	991	65
542	15	592	100	642	28	692	50	742	50	792	8	842	110	892	40	942	35	992	97
543	51	593	7	643	27	693	82	743	23	793	15	843	10	893	5	943	40	993	65
544	14	594	73	644	13	694	100	744	40	794	5	844	95	894	35	944	80	994	45
545	25	595	25	645	24	695	120	745	140	795	20	845	30	895	60	945	30	995	70
546	43	596	66	646	4	696	40	746	15	796	8	846	10	896	50	946	65	996	85
547	216	597	12	647	39	697	55	747	20	797	10	847	5	897	7	947	35	997	10
548	119	598	25	648	76	698	110	748	43	798	5	848	45	898	20	948	10	998	27
549	45	599	66	649	71	699	40	749	228	799	35	849	85	899	5	949	100	999	20
550	15	600	12	650	6	700	53	750	127	800	50	850	30	900	10	950	35	1000	25

Appendix D Data Set: Results of DDT Analyses on Fish Samples Collected Summer 1980, Tennessee River, Alabama

RESULTS OF DDT ANALYSES
ON FISH SAMPLES COLLECTED SUMMER 1980
TENNESSEE RIVER, ALABAMA

OBS	LOCATION	SPECIES	LENGTH	WEIGHT	DDT
1	FCM5	CHANNEL CATFISH	42.5	732	10.00
2	FCM5	CHANNEL CATFISH	44.0	795	16.00
3	FCM5	CHANNEL CATFISH	41.5	547	23.00
4	FCM5	CHANNEL CATFISH	39.0	465	21.00
5	FCM5	CHANNEL CATFISH	50.5	1252	50.00
6	FCM5	CHANNEL CATFISH	52.0	1255	150.00
7	LCM3	CHANNEL CATFISH	40.5	741	28.00
8	LCM3	CHANNEL CATFISH	48.0	1151	7.70
9	LCM3	CHANNEL CATFISH	48.0	1186	2.00
10	LCM3	CHANNEL CATFISH	43.5	754	19.00
11	LCM3	CHANNEL CATFISH	40.5	679	16.00
12	LCM3	CHANNEL CATFISH	47.5	985	5.40
13	SCM1	CHANNEL CATFISH	44.5	1133	2.60
14	SCM1	CHANNEL CATFISH	46.0	1139	3.10
15	SCM1	CHANNEL CATFISH	48.0	1186	3.50
16	SCM1	CHANNEL CATFISH	45.0	984	9.10
17	SCM1	CHANNEL CATFISH	43.0	965	7.80
18	SCM1	CHANNEL CATFISH	45.0	1084	4.10
19	TRM275	CHANNEL CATFISH	48.0	986	8.40
20	TRM275	CHANNEL CATFISH	45.0	1023	15.00
21	TRM275	CHANNEL CATFISH	49.0	1266	25.00
22	TRM275	CHANNEL CATFISH	50.0	1086	5.60
23	TRM275	CHANNEL CATFISH	46.0	1044	4.60
24	TRM275	CHANNEL CATFISH	52.0	1770	8.20
25	TRM280	CHANNEL CATFISH	48.0	1048	6.10
26	TRM280	CHANNEL CATFISH	51.0	1641	13.00
27	TRM280	CHANNEL CATFISH	48.5	1331	6.00
28	TRM280	CHANNEL CATFISH	51.0	1728	6.60
29	TRM280	CHANNEL CATFISH	44.0	917	5.50
30	TRM280	CHANNEL CATFISH	51.0	1398	11.00
31	TRM280	SMALL MOUTH BUFFALO	49.0	1763	4.50
32	TRM280	SMALL MOUTH BUFFALO	46.0	1459	4.20
33	TRM280	SMALL MOUTH BUFFALO	52.0	2302	3.00
34	TRM280	SMALL MOUTH BUFFALO	46.0	1614	2.30
35	TRM280	SMALL MOUTH BUFFALO	46.0	1444	2.50
36	TRM280	SMALL MOUTH BUFFALO	48.0	2006	6.80
37	TRM285	CHANNEL CATFISH	44.0	936	19.00
38	TRM285	CHANNEL CATFISH	42.0	1058	7.20
39	TRM285	CHANNEL CATFISH	42.5	800	6.00
40	TRM285	CHANNEL CATFISH	45.5	1087	10.00
41	TRM285	CHANNEL CATFISH	48.0	1329	12.00
42	TRM285	CHANNEL CATFISH	44.0	897	2.80
43	TRM285	LARGE MOUTH BASS	28.5	778	0.48
44	TRM285	LARGE MOUTH BASS	26.0	532	0.18
45	TRM285	LARGE MOUTH BASS	25.5	441	0.34
46	TRM285	LARGE MOUTH BASS	25.0	544	0.11
47	TRM285	LARGE MOUTH BASS	23.0	393	0.22
48	TRM285	LARGE MOUTH BASS	28.0	733	0.80
49	TRM290	CHANNEL CATFISH	41.0	961	8.70
50	TRM290	CHANNEL CATFISH	44.0	886	22.00
51	TRM290	CHANNEL CATFISH	41.0	678	13.00
52	TRM290	CHANNEL CATFISH	42.0	1011	3.50
53	TRM290	CHANNEL CATFISH	42.5	947	9.30
54	TRM290	CHANNEL CATFISH	44.0	989	21.00
55	TRM290	SMALL MOUTH BUFFALO	43.5	1291	3.40
56	TRM290	SMALL MOUTH BUFFALO	46.5	1186	13.00
57	TRM290	SMALL MOUTH BUFFALO	43.0	1293	5.60
58	TRM290	SMALL MOUTH BUFFALO	47.0	1709	12.00
59	TRM290	SMALL MOUTH BUFFALO	46.0	1425	21.00
60	TRM290	SMALL MOUTH BUFFALO	41.0	1176	8.00
61	TRM295	CHANNEL CATFISH	36.0	980	12.00
62	TRM295	CHANNEL CATFISH	47.5	1176	6.00
63	TRM295	CHANNEL CATFISH	41.5	989	4.70
64	TRM295	CHANNEL CATFISH	49.5	1084	31.00
65	TRM295	CHANNEL CATFISH	46.0	1115	5.20
66	TRM295	CHANNEL CATFISH	46.5	724	27.00
67	TRM300	CHANNEL CATFISH	36.0	847	18.00
68	TRM300	CHANNEL CATFISH	37.0	876	7.50
69	TRM300	CHANNEL CATFISH	35.0	844	3.00
70	TRM300	CHANNEL CATFISH	36.0	908	13.00
71	TRM300	CHANNEL CATFISH	48.0	1358	7.30
72	TRM300	CHANNEL CATFISH	49.0	1019	15.00
73	TRM300	SMALL MOUTH BUFFALO	35.5	1300	1.30
74	TRM300	SMALL MOUTH BUFFALO	46.0	1365	4.80
75	TRM300	SMALL MOUTH BUFFALO	45.0	1437	5.10
76	TRM300	SMALL MOUTH BUFFALO	44.5	1460	5.10
77	TRM300	SMALL MOUTH BUFFALO	49.0	1671	4.00
78	TRM300	SMALL MOUTH BUFFALO	47.5	1717	10.00
79	TRM305	CHANNEL CATFISH	35.0	613	12.00
80	TRM305	CHANNEL CATFISH	51.0	353	22.00
81	TRM305	CHANNEL CATFISH	42.5	909	10.00
82	TRM305	CHANNEL CATFISH	38.0	886	11.00
83	TRM305	CHANNEL CATFISH	41.0	890	17.00
84	TRM305	CHANNEL CATFISH	47.0	1031	9.70
85	TRM310	CHANNEL CATFISH	45.0	1083	12.00
86	TRM310	CHANNEL CATFISH	45.5	864	4.70
87	TRM310	CHANNEL CATFISH	45.0	886	6.00
88	TRM310	CHANNEL CATFISH	45.0	965	3.80
89	TRM310	CHANNEL CATFISH	39.0	537	17.00
90	TRM310	CHANNEL CATFISH	40.5	630	12.00
91	TRM310	SMALL MOUTH BUFFALO	46.0	1486	1.40
92	TRM310	SMALL MOUTH BUFFALO	47.0	1743	6.10
93	TRM310	SMALL MOUTH BUFFALO	48.5	2061	2.80
94	TRM310	SMALL MOUTH BUFFALO	48.0	1707	4.80
95	TRM310	SMALL MOUTH BUFFALO	38.0	862	5.70
96	TRM310	SMALL MOUTH BUFFALO	38.5	911	3.30
97	TRM315	CHANNEL CATFISH	29.5	476	3.30
98	TRM315	CHANNEL CATFISH	42.0	743	3.70
99	TRM315	CHANNEL CATFISH	47.5	1128	9.90
100	TRM315	CHANNEL CATFISH	43.5	848	6.80
101	TRM315	CHANNEL CATFISH	47.5	1091	13.00
102	TRM315	CHANNEL CATFISH	43.5	715	8.80
103	TRM320	CHANNEL CATFISH	47.5	983	57.00
104	TRM320	CHANNEL CATFISH	51.5	1251	96.00
105	TRM320	CHANNEL CATFISH	49.5	1255	360.00
106	TRM320	CHANNEL CATFISH	47.0	1152	130.00

RESULTS OF DDT ANALYSES
ON FISH SAMPLES COLLECTED SUMMER 1980
TENNESSEE RIVER, ALABAMA

OBS	LOCATION	SPECIES	LENGTH	WEIGHT	DDT
107	TRM320	CHANNEL CATFISH	47.5	1085	13.00
108	TRM320	CHANNEL CATFISH	47.0	1118	61.00
109	TRM320	SMALL MOUTH BUFFALO	36.0	1285	12.00
110	TRM320	SMALL MOUTH BUFFALO	34.5	1178	33.00
111	TRM320	SMALL MOUTH BUFFALO	44.5	1492	48.00
112	TRM320	SMALL MOUTH BUFFALO	46.0	1524	10.00
113	TRM320	SMALL MOUTH BUFFALO	46.0	1473	44.00
114	TRM320	SMALL MOUTH BUFFALO	32.5	520	0.43
115	TRM325	CHANNEL CATFISH	46.0	863	1100.00
116	TRM325	CHANNEL CATFISH	40.0	549	9.40
117	TRM325	CHANNEL CATFISH	43.5	810	4.10
118	TRM325	CHANNEL CATFISH	46.5	908	2.80
119	TRM325	CHANNEL CATFISH	43.0	804	0.74
120	TRM325	CHANNEL CATFISH	47.5	1179	14.00
121	TRM330	CHANNEL CATFISH	32.0	556	22.00
122	TRM330	CHANNEL CATFISH	40.5	659	9.10
123	TRM330	CHANNEL CATFISH	51.5	1229	140.00
124	TRM330	CHANNEL CATFISH	48.0	1050	4.20
125	TRM330	CHANNEL CATFISH	47.0	952	12.00
126	TRM330	CHANNEL CATFISH	41.0	826	2.00
127	TRM330	SMALL MOUTH BUFFALO	33.5	599	0.30
128	TRM330	SMALL MOUTH BUFFALO	47.0	1704	1.20
129	TRM340	CHANNEL CATFISH	50.0	1207	7.10
130	TRM340	CHANNEL CATFISH	45.0	911	180.00
131	TRM340	CHANNEL CATFISH	49.0	1498	1.50
132	TRM340	CHANNEL CATFISH	49.5	1496	2.40
133	TRM340	CHANNEL CATFISH	50.0	1142	4.30
134	TRM340	CHANNEL CATFISH	45.0	879	3.90
135	TRM340	SMALL MOUTH BUFFALO	32.5	525	0.99
136	TRM340	SMALL MOUTH BUFFALO	38.0	806	0.45
137	TRM340	SMALL MOUTH BUFFALO	38.5	694	2.50
138	TRM340	SMALL MOUTH BUFFALO	36.0	643	0.25
139	TRM345	LARGE MOUTH BASS	26.5	514	0.58
140	TRM345	LARGE MOUTH BASS	23.5	358	2.00
141	TRM345	LARGE MOUTH BASS	30.0	856	2.20
142	TRM345	LARGE MOUTH BASS	29.0	793	7.40
143	TRM345	LARGE MOUTH BASS	17.5	173	0.35
144	TRM345	LARGE MOUTH BASS	36.0	1433	1.90

Appendix E Statistical Tables

Contents

TABLE 1 Binomial Probabilities

TABULATED VALUES ARE P(X)

N=5

X	0.01	0.05	0.1	0.2	0.3	0.4	0.5	0.6	0.7	0.8	0.9	0.95	0.99	X
0	.9510	.7738	.5905	.3277	.1681	.0778	.0313	.0102	.0024	.0003	.0000	.0000	.0000	0
1	.0480	.2036	.3280	.4096	.3601	.2592	.1563	.0768	.0283	.0064	.0005	.0000	.0000	1
2	.0010	.0214	.0729	.2048	.3087	.3456	.3125	.2304	.1323	.0512	.0081	.0011	.0000	2
3	.0000	.0011	.0081	.0512	.1323	.2304	.3125	.3456	.3087	.2048	.0729	.0214	.0010	3
4	.0000	.0000	.0004	.0064	.0283	.0768	.1563	.2592	.3601	.4096	.3280	.2036	.0480	4
5	.0000	.0000	.0000	.0003	.0024	.0102	.0313	.0778	.1681	.3277	.5905	.7738	.9510	5

TABULATED VALUES ARE P(X)

N=6

X	0.01	0.05	0.1	0.2	0.3	0.4	0.5	0.6	0.7	0.8	0.9	0.95	0.99	X
0	.9415	.7351	.5314	.2621	.1176	.0467	.0156	.0041	.0007	.0001	.0000	.0000	.0000	0
1	.0571	.2321	.3543	.3932	.3025	.1866	.0938	.0369	.0102	.0015	.0001	.0000	.0000	1
2	.0014	.0305	.0984	.2458	.3241	.3110	.2344	.1382	.0595	.0154	.0012	.0001	.0000	2
3	.0000	.0021	.0146	.0819	.1852	.2765	.3125	.2765	.1852	.0819	.0146	.0021	.0000	3
4	.0000	.0001	.0012	.0154	.0595	.1382	.2344	.3110	.3241	.2458	.0984	.0305	.0014	4
5	.0000	.0000	.0001	.0015	.0102	.0369	.0938	.1866	.3025	.3932	.3543	.2321	.0571	5
6	.0000	.0000	.0000	.0001	.0007	.0041	.0156	.0467	.1176	.2621	.5314	.7351	.9415	6

TABULATED VALUES ARE P(X)

N=7

X	0.01	0.05	0.1	0.2	0.3	0.4	0.5	0.6	0.7	0.8	0.9	0.95	0.99	X
0	.9321	.6983	.4783	.2097	.0824	.0280	.0078	.0016	.0002	.0000	.0000	.0000	.0000	0
1	.0659	.2573	.3720	.3670	.2471	.1306	.0547	.0172	.0036	.0004	.0000	.0000	.0000	1
2	.0020	.0406	.1240	.2753	.3177	.2613	.1641	.0774	.0250	.0043	.0002	.0000	.0000	2
3	.0000	.0036	.0230	.1147	.2269	.2903	.2734	.1935	.0972	.0287	.0026	.0002	.0000	3
4	.0000	.0002	.0026	.0287	.0972	.1935	.2734	.2903	.2269	.1147	.0230	.0036	.0000	4
5	.0000	.0000	.0002	.0043	.0250	.0774	.1641	.2613	.3177	.2753	.1240	.0406	.0020	5
6	.0000	.0000	.0000	.0004	.0036	.0172	.0547	.1306	.2471	.3670	.3720	.2573	.0659	6
7	.0000	.0000	.0000	.0000	.0002	.0016	.0078	.0280	.0824	.2097	.4783	.6983	.9321	7

TABULATED VALUES ARE P(X)

N=8

X	0.01	0.05	0.1	0.2	0.3	0.4	0.5	0.6	0.7	0.8	0.9	0.95	0.99	X
0	.9227	.6634	.4305	.1678	.0576	.0168	.0039	.0007	.0001	.0000	.0000	.0000	.0000	0
1	.0746	.2793	.3826	.3355	.1977	.0896	.0313	.0079	.0012	.0001	.0000	.0000	.0000	1
2	.0026	.0515	.1488	.2936	.2965	.2090	.1094	.0413	.0100	.0011	.0000	.0000	.0000	2
3	.0001	.0054	.0331	.1468	.2541	.2787	.2187	.1239	.0467	.0092	.0004	.0000	.0000	3
4	.0000	.0004	.0046	.0459	.1361	.2322	.2734	.2322	.1361	.0459	.0046	.0004	.0000	4
5	.0000	.0000	.0004	.0092	.0467	.1239	.2187	.2787	.2541	.1468	.0331	.0054	.0001	5
6	.0000	.0000	.0000	.0011	.0100	.0413	.1094	.2090	.2965	.2936	.1488	.0515	.0026	6
7	.0000	.0000	.0000	.0001	.0012	.0079	.0313	.0896	.1977	.3355	.3826	.2793	.0746	7
8	.0000	.0000	.0000	.0000	.0001	.0007	.0039	.0168	.0576	.1678	.4305	.6634	.9227	8

TABULATED VALUES ARE P(X)

N=9

X	P: 0.01	0.05	0.1	0.2	0.3	0.4	0.5	0.6	0.7	0.8	0.9	0.95	0.99	X
0	.9135	.6302	.3874	.1342	.0404	.0101	.0020	.0003	.0000	.0000	.0000	.0000	.0000	0
1	.0830	.2985	.3874	.3020	.1556	.0605	.0176	.0035	.0004	.0000	.0000	.0000	.0000	1
2	.0034	.0629	.1722	.3020	.2668	.1612	.0703	.0212	.0039	.0003	.0000	.0000	.0000	2
3	.0001	.0077	.0446	.1762	.2668	.2508	.1641	.0743	.0210	.0028	.0001	.0000	.0000	3
4	.0000	.0006	.0074	.0661	.1715	.2508	.2461	.1672	.0735	.0165	.0008	.0000	.0000	4
5	.0000	.0000	.0008	.0165	.0735	.1672	.2461	.2508	.1715	.0661	.0074	.0006	.0000	5
6	.0000	.0000	.0001	.0028	.0210	.0743	.1641	.2508	.2668	.1762	.0446	.0077	.0001	6
7	.0000	.0000	.0000	.0003	.0039	.0212	.0703	.1612	.2668	.3020	.1722	.0629	.0034	7
8	.0000	.0000	.0000	.0000	.0004	.0035	.0176	.0605	.1556	.3020	.3874	.2985	.0830	8
9	.0000	.0000	.0000	.0000	.0000	.0003	.0020	.0101	.0404	.1342	.3874	.6302	.9135	9

TABULATED VALUES ARE P(X)

N=10

X	P: 0.01	0.05	0.1	0.2	0.3	0.4	0.5	0.6	0.7	0.8	0.9	0.95	0.99	X
0	.9044	.5987	.3487	.1074	.0282	.0060	.0010	.0001	.0000	.0000	.0000	.0000	.0000	0
1	.0914	.3151	.3874	.2684	.1211	.0403	.0098	.0016	.0001	.0000	.0000	.0000	.0000	1
2	.0042	.0746	.1937	.3020	.2335	.1209	.0439	.0106	.0014	.0001	.0000	.0000	.0000	2
3	.0001	.0105	.0574	.2013	.2668	.2150	.1172	.0425	.0090	.0008	.0000	.0000	.0000	3
4	.0000	.0010	.0112	.0881	.2001	.2508	.2051	.1115	.0368	.0055	.0001	.0000	.0000	4
5	.0000	.0001	.0015	.0264	.1029	.2007	.2461	.2007	.1029	.0264	.0015	.0001	.0000	5
6	.0000	.0000	.0001	.0055	.0368	.1115	.2051	.2508	.2001	.0881	.0112	.0010	.0000	6
7	.0000	.0000	.0000	.0008	.0090	.0425	.1172	.2150	.2668	.2013	.0574	.0105	.0001	7
8	.0000	.0000	.0000	.0001	.0014	.0106	.0439	.1209	.2335	.3020	.1937	.0746	.0042	8
9	.0000	.0000	.0000	.0000	.0001	.0016	.0098	.0403	.1211	.2684	.3874	.3151	.0914	9
10	.0000	.0000	.0000	.0000	.0000	.0001	.0010	.0060	.0282	.1074	.3487	.5987	.9044	10

TABULATED VALUES ARE P(X)

N=15

X	P: 0.01	0.05	0.1	0.2	0.3	0.4	0.5	0.6	0.7	0.8	0.9	0.95	0.99	X
0	.8601	.4633	.2059	.0352	.0047	.0005	.0000	.0000	.0000	.0000	.0000	.0000	.0000	0
1	.1303	.3658	.3432	.1319	.0305	.0047	.0005	.0000	.0000	.0000	.0000	.0000	.0000	1
2	.0092	.1348	.2669	.2309	.0916	.0219	.0032	.0003	.0000	.0000	.0000	.0000	.0000	2
3	.0004	.0307	.1285	.2501	.1700	.0634	.0139	.0016	.0001	.0000	.0000	.0000	.0000	3
4	.0000	.0049	.0428	.1876	.2186	.1268	.0417	.0074	.0006	.0000	.0000	.0000	.0000	4
5	.0000	.0006	.0105	.1032	.2061	.1859	.0916	.0245	.0030	.0001	.0000	.0000	.0000	5
6	.0000	.0000	.0019	.0430	.1472	.2066	.1527	.0612	.0116	.0007	.0000	.0000	.0000	6
7	.0000	.0000	.0003	.0138	.0811	.1771	.1964	.1181	.0348	.0035	.0000	.0000	.0000	7
8	.0000	.0000	.0000	.0035	.0348	.1181	.1964	.1771	.0811	.0138	.0003	.0000	.0000	8
9	.0000	.0000	.0000	.0007	.0116	.0612	.1527	.2066	.1472	.0430	.0019	.0000	.0000	9
10	.0000	.0000	.0000	.0001	.0030	.0245	.0916	.1859	.2061	.1032	.0105	.0006	.0000	10
11	.0000	.0000	.0000	.0000	.0006	.0074	.0417	.1268	.2186	.1876	.0428	.0049	.0000	11
12	.0000	.0000	.0000	.0000	.0001	.0016	.0139	.0634	.1700	.2501	.1285	.0307	.0004	12
13	.0000	.0000	.0000	.0000	.0000	.0003	.0032	.0219	.0916	.2309	.2669	.1348	.0092	13
14	.0000	.0000	.0000	.0000	.0000	.0000	.0005	.0047	.0305	.1319	.3432	.3658	.1303	14
15	.0000	.0000	.0000	.0000	.0000	.0000	.0000	.0005	.0047	.0352	.2059	.4633	.8601	15

TABLE 1 *(continued)*

TABULATED VALUES ARE P(X)

N=20

X	0.01	0.05	0.1	0.2	0.3	0.4	0.5	0.6	0.7	0.8	0.9	0.95	0.99	X
0	.8179	.3585	.1216	.0115	.0008	.0000	.0000	.0000	.0000	.0000	.0000	.0000	.0000	0
1	.1652	.3774	.2702	.0576	.0068	.0005	.0000	.0000	.0000	.0000	.0000	.0000	.0000	1
2	.0159	.1887	.2852	.1369	.0278	.0031	.0002	.0000	.0000	.0000	.0000	.0000	.0000	2
3	.0010	.0596	.1901	.2054	.0716	.0123	.0011	.0000	.0000	.0000	.0000	.0000	.0000	3
4	.0000	.0133	.0898	.2182	.1304	.0350	.0046	.0003	.0000	.0000	.0000	.0000	.0000	4
5	.0000	.0022	.0319	.1746	.1789	.0746	.0148	.0013	.0000	.0000	.0000	.0000	.0000	5
6	.0000	.0003	.0089	.1091	.1916	.1244	.0370	.0049	.0002	.0000	.0000	.0000	.0000	6
7	.0000	.0000	.0020	.0545	.1643	.1659	.0739	.0146	.0010	.0000	.0000	.0000	.0000	7
8	.0000	.0000	.0004	.0222	.1144	.1797	.1201	.0355	.0039	.0001	.0000	.0000	.0000	8
9	.0000	.0000	.0001	.0074	.0654	.1597	.1602	.0710	.0120	.0005	.0000	.0000	.0000	9
10	.0000	.0000	.0000	.0020	.0308	.1171	.1762	.1171	.0308	.0020	.0000	.0000	.0000	10
11	.0000	.0000	.0000	.0005	.0120	.0710	.1602	.1597	.0654	.0074	.0001	.0000	.0000	11
12	.0000	.0000	.0000	.0001	.0039	.0355	.1201	.1797	.1144	.0222	.0004	.0000	.0000	12
13	.0000	.0000	.0000	.0000	.0010	.0146	.0739	.1659	.1643	.0545	.0020	.0000	.0000	13
14	.0000	.0000	.0000	.0000	.0002	.0049	.0370	.1244	.1916	.1091	.0089	.0003	.0000	14
15	.0000	.0000	.0000	.0000	.0000	.0013	.0148	.0746	.1789	.1746	.0319	.0022	.0000	15
16	.0000	.0000	.0000	.0000	.0000	.0003	.0046	.0350	.1304	.2182	.0898	.0133	.0000	16
17	.0000	.0000	.0000	.0000	.0000	.0000	.0011	.0123	.0716	.2054	.1901	.0596	.0010	17
18	.0000	.0000	.0000	.0000	.0000	.0000	.0002	.0031	.0278	.1369	.2852	.1887	.0159	18
19	.0000	.0000	.0000	.0000	.0000	.0000	.0000	.0005	.0068	.0576	.2702	.3774	.1652	19
20	.0000	.0000	.0000	.0000	.0000	.0000	.0000	.0000	.0008	.0115	.1216	.3585	.8179	20

TABULATED VALUES ARE P(X)

N=25

X	0.01	0.05	0.1	0.2	0.3	0.4	0.5	0.6	0.7	0.8	0.9	0.95	0.99	X
0	.7778	.2774	.0718	.0038	.0001	.0000	.0000	.0000	.0000	.0000	.0000	.0000	.0000	0
1	.1964	.3650	.1994	.0236	.0014	.0000	.0000	.0000	.0000	.0000	.0000	.0000	.0000	1
2	.0238	.2305	.2659	.0708	.0074	.0004	.0000	.0000	.0000	.0000	.0000	.0000	.0000	2
3	.0018	.0930	.2265	.1358	.0243	.0019	.0001	.0000	.0000	.0000	.0000	.0000	.0000	3
4	.0001	.0269	.1384	.1867	.0572	.0071	.0004	.0000	.0000	.0000	.0000	.0000	.0000	4
5	.0000	.0060	.0646	.1960	.1030	.0199	.0016	.0000	.0000	.0000	.0000	.0000	.0000	5
6	.0000	.0010	.0239	.1633	.1472	.0442	.0053	.0002	.0000	.0000	.0000	.0000	.0000	6
7	.0000	.0001	.0072	.1108	.1712	.0800	.0143	.0009	.0000	.0000	.0000	.0000	.0000	7
8	.0000	.0000	.0018	.0623	.1651	.1200	.0322	.0031	.0001	.0000	.0000	.0000	.0000	8
9	.0000	.0000	.0004	.0294	.1336	.1511	.0609	.0088	.0004	.0000	.0000	.0000	.0000	9
10	.0000	.0000	.0001	.0118	.0916	.1612	.0974	.0212	.0013	.0000	.0000	.0000	.0000	10
11	.0000	.0000	.0000	.0040	.0536	.1465	.1328	.0434	.0042	.0001	.0000	.0000	.0000	11
12	.0000	.0000	.0000	.0012	.0268	.1140	.1550	.0760	.0115	.0003	.0000	.0000	.0000	12
13	.0000	.0000	.0000	.0003	.0115	.0760	.1550	.1140	.0268	.0012	.0000	.0000	.0000	13
14	.0000	.0000	.0000	.0001	.0042	.0434	.1328	.1465	.0536	.0040	.0000	.0000	.0000	14
15	.0000	.0000	.0000	.0000	.0013	.0212	.0974	.1612	.0916	.0118	.0001	.0000	.0000	15
16	.0000	.0000	.0000	.0000	.0004	.0088	.0609	.1511	.1336	.0294	.0004	.0000	.0000	16
17	.0000	.0000	.0000	.0000	.0001	.0031	.0322	.1200	.1651	.0623	.0018	.0000	.0000	17
18	.0000	.0000	.0000	.0000	.0000	.0009	.0143	.0800	.1712	.1108	.0072	.0001	.0000	18
19	.0000	.0000	.0000	.0000	.0000	.0002	.0053	.0442	.1472	.1633	.0239	.0010	.0000	19
20	.0000	.0000	.0000	.0000	.0000	.0000	.0016	.0199	.1030	.1960	.0646	.0060	.0000	20
21	.0000	.0000	.0000	.0000	.0000	.0000	.0004	.0071	.0572	.1867	.1384	.0269	.0001	21
22	.0000	.0000	.0000	.0000	.0000	.0000	.0001	.0019	.0243	.1358	.2265	.0930	.0018	22
23	.0000	.0000	.0000	.0000	.0000	.0000	.0000	.0004	.0074	.0708	.2659	.2305	.0238	23
24	.0000	.0000	.0000	.0000	.0000	.0000	.0000	.0000	.0014	.0236	.1994	.3650	.1964	24
25	.0000	.0000	.0000	.0000	.0000	.0000	.0000	.0000	.0001	.0038	.0718	.2774	.7778	25

TABLE 2 Cumulative Binomial Probabilities

N=5

x	0.01	0.05	0.1	0.2	0.3	0.4	0.5	0.6	0.7	0.8	0.9	0.95	0.99	x
0	.9510	.7738	.5905	.3277	.1681	.0778	.0313	.0102	.0024	.0003	.0000	.0000	.0000	0
1	.9990	.9774	.9185	.7373	.5282	.3370	.1875	.0870	.0308	.0067	.0005	.0000	.0000	1
2	1.0000	.9988	.9914	.9421	.8369	.6826	.5000	.3174	.1631	.0579	.0086	.0012	.0000	2
3	1.0000	1.0000	.9995	.9933	.9692	.9130	.8125	.6630	.4718	.2627	.0815	.0226	.0010	3
4	1.0000	1.0000	1.0000	.9997	.9976	.9898	.9687	.9222	.8319	.6723	.4095	.2262	.0490	4

N=6

x	0.01	0.05	0.1	0.2	0.3	0.4	0.5	0.6	0.7	0.8	0.9	0.95	0.99	x
0	.9415	.7351	.5314	.2621	.1176	.0467	.0156	.0041	.0007	.0001	.0000	.0000	.0000	0
1	.9985	.9672	.8857	.6554	.4202	.2333	.1094	.0410	.0109	.0016	.0001	.0000	.0000	1
2	1.0000	.9978	.9841	.9011	.7443	.5443	.3437	.1792	.0705	.0170	.0013	.0001	.0000	2
3	1.0000	.9999	.9987	.9830	.9295	.8208	.6562	.4557	.2557	.0989	.0158	.0022	.0000	3
4	1.0000	1.0000	.9999	.9984	.9891	.9590	.8906	.7667	.5798	.3446	.1143	.0328	.0015	4
5	1.0000	1.0000	1.0000	.9999	.9993	.9959	.9844	.9533	.8824	.7379	.4686	.2649	.0585	5

N=7

x	0.01	0.05	0.1	0.2	0.3	0.4	0.5	0.6	0.7	0.8	0.9	0.95	0.99	x
0	.9321	.6983	.4783	.2097	.0824	.0280	.0078	.0016	.0002	.0000	.0000	.0000	.0000	0
1	.9980	.9556	.8503	.5767	.3294	.1586	.0625	.0188	.0038	.0004	.0000	.0000	.0000	1
2	1.0000	.9962	.9743	.8520	.6471	.4199	.2266	.0963	.0288	.0047	.0002	.0000	.0000	2
3	1.0000	.9998	.9973	.9667	.8740	.7102	.5000	.2898	.1260	.0333	.0027	.0002	.0000	3
4	1.0000	1.0000	.9998	.9953	.9712	.9037	.7734	.5801	.3529	.1480	.0257	.0038	.0000	4
5	1.0000	1.0000	1.0000	.9996	.9962	.9812	.9375	.8414	.6706	.4233	.1497	.0444	.0020	5
6	1.0000	1.0000	1.0000	1.0000	.9998	.9984	.9922	.9720	.9176	.7903	.5217	.3017	.0679	6

N=8

x	0.01	0.05	0.1	0.2	0.3	0.4	0.5	0.6	0.7	0.8	0.9	0.95	0.99	x
0	.9227	.6634	.4305	.1678	.0576	.0168	.0039	.0007	.0001	.0000	.0000	.0000	.0000	0
1	.9973	.9428	.8131	.5033	.2553	.1064	.0352	.0085	.0013	.0001	.0000	.0000	.0000	1
2	.9999	.9942	.9619	.7969	.5518	.3154	.1445	.0498	.0113	.0012	.0000	.0000	.0000	2
3	1.0000	.9996	.9950	.9437	.8059	.5941	.3633	.1737	.0580	.0104	.0004	.0000	.0000	3
4	1.0000	1.0000	.9996	.9896	.9420	.8263	.6367	.4059	.1941	.0563	.0050	.0004	.0000	4
5	1.0000	1.0000	1.0000	.9988	.9887	.9502	.8555	.6846	.4482	.2031	.0381	.0058	.0001	5
6	1.0000	1.0000	1.0000	.9999	.9987	.9915	.9648	.8936	.7447	.4967	.1869	.0572	.0027	6
7	1.0000	1.0000	1.0000	1.0000	.9999	.9993	.9961	.9832	.9424	.8322	.5695	.3366	.0773	7

N=9

x	0.01	0.05	0.1	0.2	0.3	0.4	0.5	0.6	0.7	0.8	0.9	0.95	0.99	x
0	.9135	.6302	.3874	.1342	.0404	.0101	.0020	.0003	.0000	.0000	.0000	.0000	.0000	0
1	.9966	.9288	.7748	.4362	.1960	.0705	.0195	.0038	.0004	.0000	.0000	.0000	.0000	1
2	.9999	.9916	.9470	.7382	.4628	.2318	.0898	.0250	.0043	.0003	.0000	.0000	.0000	2
3	1.0000	.9994	.9917	.9144	.7297	.4826	.2539	.0994	.0253	.0031	.0001	.0000	.0000	3
4	1.0000	1.0000	.9991	.9804	.9012	.7334	.5000	.2666	.0988	.0196	.0009	.0000	.0000	4
5	1.0000	1.0000	.9999	.9969	.9747	.9006	.7461	.5174	.2703	.0856	.0083	.0006	.0000	5
6	1.0000	1.0000	1.0000	.9997	.9957	.9750	.9102	.7682	.5372	.2618	.0530	.0084	.0001	6
7	1.0000	1.0000	1.0000	1.0000	.9996	.9962	.9805	.9295	.8040	.5638	.2252	.0712	.0034	7
8	1.0000	1.0000	1.0000	1.0000	1.0000	.9997	.9980	.9899	.9596	.8658	.6126	.3698	.0865	8

TABLE 2 *(continued)*

N=10

P

X	0.01	0.05	0.1	0.2	0.3	0.4	0.5	0.6	0.7	0.8	0.9	0.95	0.99	X
0	.9044	.5987	.3487	.1074	.0282	.0060	.0010	.0001	.0000	.0000	.0000	.0000	.0000	0
1	.9957	.9139	.7361	.3758	.1493	.0464	.0107	.0017	.0001	.0000	.0000	.0000	.0000	1
2	.9999	.9885	.9298	.6778	.3828	.1673	.0547	.0123	.0016	.0001	.0000	.0000	.0000	2
3	1.0000	.9990	.9872	.8791	.6496	.3823	.1719	.0548	.0106	.0009	.0000	.0000	.0000	3
4	1.0000	.9999	.9984	.9672	.8497	.6331	.3770	.1662	.0473	.0064	.0001	.0000	.0000	4
5	1.0000	1.0000	.9999	.9936	.9527	.8338	.6230	.3669	.1503	.0328	.0016	.0001	.0000	5
6	1.0000	1.0000	1.0000	.9991	.9894	.9452	.8281	.6177	.3504	.1209	.0128	.0010	.0000	6
7	1.0000	1.0000	1.0000	.9999	.9984	.9877	.9453	.8327	.6172	.3222	.0702	.0115	.0001	7
8	1.0000	1.0000	1.0000	1.0000	.9999	.9983	.9893	.9536	.8507	.6242	.2639	.0861	.0043	8
9	1.0000	1.0000	1.0000	1.0000	1.0000	.9999	.9990	.9940	.9718	.8926	.6513	.4013	.0956	9

N=15

P

X	0.01	0.05	0.1	0.2	0.3	0.4	0.5	0.6	0.7	0.8	0.9	0.95	0.99	X
0	.8601	.4633	.2059	.0352	.0047	.0005	.0000	.0000	.0000	.0000	.0000	.0000	.0000	0
1	.9904	.8290	.5490	.1671	.0353	.0052	.0005	.0000	.0000	.0000	.0000	.0000	.0000	1
2	.9996	.9638	.8159	.3980	.1268	.0271	.0037	.0003	.0000	.0000	.0000	.0000	.0000	2
3	1.0000	.9945	.9444	.6482	.2969	.0905	.0176	.0019	.0001	.0000	.0000	.0000	.0000	3
4	1.0000	.9994	.9873	.8358	.5155	.2173	.0592	.0093	.0007	.0000	.0000	.0000	.0000	4
5	1.0000	.9999	.9978	.9389	.7216	.4032	.1509	.0338	.0037	.0001	.0000	.0000	.0000	5
6	1.0000	1.0000	.9997	.9819	.8689	.6098	.3036	.0950	.0152	.0008	.0000	.0000	.0000	6
7	1.0000	1.0000	1.0000	.9958	.9500	.7869	.5000	.2131	.0500	.0042	.0000	.0000	.0000	7
8	1.0000	1.0000	1.0000	.9992	.9848	.9050	.6964	.3902	.1311	.0181	.0003	.0000	.0000	8
9	1.0000	1.0000	1.0000	.9999	.9963	.9662	.8491	.5968	.2784	.0611	.0022	.0001	.0000	9
10	1.0000	1.0000	1.0000	1.0000	.9993	.9907	.9408	.7827	.4845	.1642	.0127	.0006	.0000	10
11	1.0000	1.0000	1.0000	1.0000	.9999	.9981	.9824	.9095	.7031	.3518	.0556	.0055	.0000	11
12	1.0000	1.0000	1.0000	1.0000	1.0000	.9997	.9963	.9729	.8732	.6020	.1841	.0362	.0004	12
13	1.0000	1.0000	1.0000	1.0000	1.0000	1.0000	.9995	.9948	.9647	.8329	.4510	.1710	.0096	13
14	1.0000	1.0000	1.0000	1.0000	1.0000	1.0000	1.0000	.9995	.9953	.9648	.7941	.5367	.1399	14

N=20

P

X	0.01	0.05	0.1	0.2	0.3	0.4	0.5	0.6	0.7	0.8	0.9	0.95	0.99	X
0	.8179	.3585	.1216	.0115	.0008	.0000	.0000	.0000	.0000	.0000	.0000	.0000	.0000	0
1	.9831	.7358	.3917	.0692	.0076	.0005	.0000	.0000	.0000	.0000	.0000	.0000	.0000	1
2	.9990	.9245	.6769	.2061	.0355	.0036	.0002	.0000	.0000	.0000	.0000	.0000	.0000	2
3	1.0000	.9841	.8670	.4114	.1071	.0160	.0013	.0000	.0000	.0000	.0000	.0000	.0000	3
4	1.0000	.9974	.9568	.6296	.2375	.0510	.0059	.0003	.0000	.0000	.0000	.0000	.0000	4
5	1.0000	.9997	.9887	.8042	.4164	.1256	.0207	.0016	.0000	.0000	.0000	.0000	.0000	5
6	1.0000	1.0000	.9976	.9133	.6080	.2500	.0577	.0065	.0003	.0000	.0000	.0000	.0000	6
7	1.0000	1.0000	.9996	.9679	.7723	.4159	.1316	.0210	.0013	.0000	.0000	.0000	.0000	7
8	1.0000	1.0000	.9999	.9900	.8867	.5956	.2517	.0565	.0051	.0001	.0000	.0000	.0000	8
9	1.0000	1.0000	1.0000	.9974	.9520	.7553	.4119	.1275	.0171	.0006	.0000	.0000	.0000	9
10	1.0000	1.0000	1.0000	.9994	.9829	.8725	.5881	.2447	.0480	.0026	.0000	.0000	.0000	10
11	1.0000	1.0000	1.0000	.9999	.9949	.9435	.7483	.4044	.1133	.0100	.0001	.0000	.0000	11
12	1.0000	1.0000	1.0000	1.0000	.9987	.9790	.8684	.5841	.2277	.0321	.0004	.0000	.0000	12
13	1.0000	1.0000	1.0000	1.0000	.9997	.9935	.9423	.7500	.3920	.0867	.0024	.0000	.0000	13
14	1.0000	1.0000	1.0000	1.0000	1.0000	.9984	.9793	.8744	.5836	.1958	.0113	.0003	.0000	14
15	1.0000	1.0000	1.0000	1.0000	1.0000	.9997	.9941	.9490	.7625	.3704	.0432	.0026	.0000	15
16	1.0000	1.0000	1.0000	1.0000	1.0000	1.0000	.9987	.9840	.8929	.5886	.1330	.0159	.0000	16
17	1.0000	1.0000	1.0000	1.0000	1.0000	1.0000	.9998	.9964	.9645	.7939	.3231	.0755	.0010	17
18	1.0000	1.0000	1.0000	1.0000	1.0000	1.0000	1.0000	.9995	.9924	.9308	.6083	.2642	.0169	18
19	1.0000	1.0000	1.0000	1.0000	1.0000	1.0000	1.0000	1.0000	.9992	.9885	.8784	.6415	.1821	19

N=25

p

x	0.01	0.05	0.1	0.2	0.3	0.4	0.5	0.6	0.7	0.8	0.9	0.95	0.99	x
0	.7778	.2774	.0718	.0038	.0001	.0000	.0000	.0000	.0000	.0000	.0000	.0000	.0000	0
1	.9742	.6424	.2712	.0274	.0016	.0001	.0000	.0000	.0000	.0000	.0000	.0000	.0000	1
2	.9980	.8729	.5371	.0982	.0090	.0004	.0000	.0000	.0000	.0000	.0000	.0000	.0000	2
3	.9999	.9659	.7636	.2340	.0332	.0024	.0001	.0000	.0000	.0000	.0000	.0000	.0000	3
4	1.0000	.9928	.9020	.4207	.0905	.0095	.0005	.0000	.0000	.0000	.0000	.0000	.0000	4
5	1.0000	.9988	.9666	.6167	.1935	.0294	.0020	.0001	.0000	.0000	.0000	.0000	.0000	5
6	1.0000	.9998	.9905	.7800	.3407	.0736	.0073	.0003	.0000	.0000	.0000	.0000	.0000	6
7	1.0000	1.0000	.9977	.8909	.5118	.1536	.0216	.0012	.0000	.0000	.0000	.0000	.0000	7
8	1.0000	1.0000	.9995	.9532	.6769	.2735	.0539	.0043	.0001	.0000	.0000	.0000	.0000	8
9	1.0000	1.0000	.9999	.9827	.8106	.4246	.1148	.0132	.0005	.0000	.0000	.0000	.0000	9
10	1.0000	1.0000	1.0000	.9944	.9022	.5858	.2122	.0344	.0018	.0000	.0000	.0000	.0000	10
11	1.0000	1.0000	1.0000	.9985	.9558	.7323	.3450	.0778	.0060	.0001	.0000	.0000	.0000	11
12	1.0000	1.0000	1.0000	.9996	.9825	.8462	.5000	.1538	.0175	.0004	.0000	.0000	.0000	12
13	1.0000	1.0000	1.0000	.9999	.9940	.9222	.6550	.2677	.0442	.0015	.0000	.0000	.0000	13
14	1.0000	1.0000	1.0000	1.0000	.9982	.9656	.7878	.4142	.0978	.0056	.0000	.0000	.0000	14
15	1.0000	1.0000	1.0000	1.0000	.9995	.9868	.8852	.5754	.1894	.0173	.0001	.0000	.0000	15
16	1.0000	1.0000	1.0000	1.0000	.9999	.9957	.9461	.7265	.3231	.0468	.0005	.0000	.0000	16
17	1.0000	1.0000	1.0000	1.0000	1.0000	.9988	.9784	.8464	.4882	.1091	.0023	.0000	.0000	17
18	1.0000	1.0000	1.0000	1.0000	1.0000	.9997	.9927	.9264	.6593	.2200	.0095	.0002	.0000	18
19	1.0000	1.0000	1.0000	1.0000	1.0000	.9999	.9980	.9706	.8065	.3833	.0334	.0012	.0000	19
20	1.0000	1.0000	1.0000	1.0000	1.0000	1.0000	.9995	.9905	.9095	.5793	.0980	.0072	.0000	20
21	1.0000	1.0000	1.0000	1.0000	1.0000	1.0000	.9999	.9976	.9668	.7660	.2364	.0341	.0001	21
22	1.0000	1.0000	1.0000	1.0000	1.0000	1.0000	1.0000	.9996	.9910	.9018	.4629	.1271	.0020	22
23	1.0000	1.0000	1.0000	1.0000	1.0000	1.0000	1.0000	.9999	.9984	.9726	.7288	.3576	.0258	23
24	1.0000	1.0000	1.0000	1.0000	1.0000	1.0000	1.0000	1.0000	.9999	.9962	.9282	.7226	.2222	24

TABLE 3
Normal Curve Areas

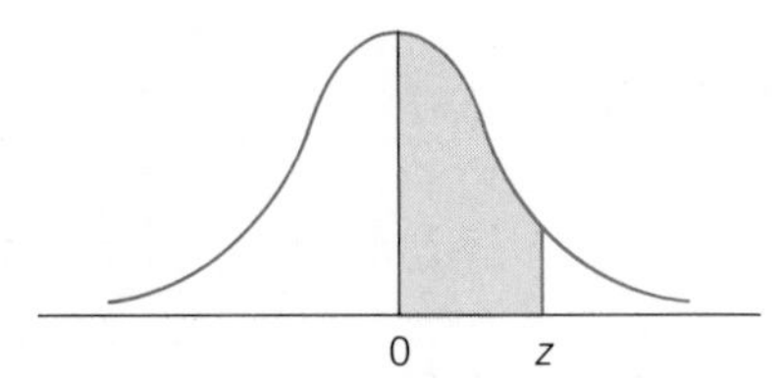

z	.00	.01	.02	.03	.04	.05	.06	.07	.08	.09
0.0	.0000	.0040	.0080	.0120	.0160	.0199	.0239	.0279	.0319	.0359
0.1	.0398	.0438	.0478	.0517	.0557	.0596	.0636	.0675	.0714	.0753
0.2	.0793	.0832	.0871	.0910	.0948	.0987	.1026	.1064	.1103	.1141
0.3	.1179	.1217	.1255	.1293	.1331	.1368	.1406	.1443	.1480	.1517
0.4	.1554	.1591	.1628	.1664	.1700	.1736	.1772	.1808	.1844	.1879
0.5	.1915	.1950	.1985	.2019	.2054	.2088	.2123	.2157	.2190	.2224
0.6	.2257	.2291	.2324	.2357	.2389	.2422	.2454	.2486	.2517	.2549
0.7	.2580	.2611	.2642	.2673	.2704	.2734	.2764	.2794	.2823	.2852
0.8	.2881	.2910	.2939	.2967	.2995	.3023	.3051	.3078	.3106	.3133
0.9	.3159	.3186	.3212	.3238	.3264	.3289	.3315	.3340	.3365	.3389
1.0	.3413	.3438	.3461	.3485	.3508	.3531	.3554	.3577	.3599	.3621
1.1	.3643	.3665	.3686	.3708	.3729	.3749	.3770	.3790	.3810	.3830
1.2	.3849	.3869	.3888	.3907	.3925	.3944	.3962	.3980	.3997	.4015
1.3	.4032	.4049	.4066	.4082	.4099	.4115	.4131	.4147	.4162	.4177
1.4	.4192	.4207	.4222	.4236	.4251	.4265	.4279	.4292	.4306	.4319
1.5	.4332	.4345	.4357	.4370	.4382	.4394	.4406	.4418	.4429	.4441
1.6	.4452	.4463	.4474	.4484	.4495	.4505	.4515	.4525	.4535	.4545
1.7	.4554	.4564	.4573	.4582	.4591	.4599	.4608	.4616	.4625	.4633
1.8	.4641	.4649	.4656	.4664	.4671	.4678	.4686	.4693	.4699	.4706
1.9	.4713	.4719	.4726	.4732	.4738	.4744	.4750	.4756	.4761	.4767
2.0	.4772	.4778	.4783	.4788	.4793	.4798	.4803	.4808	.4812	.4817
2.1	.4821	.4826	.4830	.4834	.4838	.4842	.4846	.4850	.4854	.4857
2.2	.4861	.4864	.4868	.4871	.4875	.4878	.4881	.4884	.4887	.4890
2.3	.4893	.4896	.4898	.4901	.4904	.4906	.4909	.4911	.4913	.4916
2.4	.4918	.4920	.4922	.4925	.4927	.4929	.4931	.4932	.4934	.4936
2.5	.4938	.4940	.4941	.4943	.4945	.4946	.4948	.4949	.4951	.4952
2.6	.4953	.4955	.4956	.4957	.4959	.4960	.4961	.4962	.4963	.4964
2.7	.4965	.4966	.4967	.4968	.4969	.4970	.4971	.4972	.4973	.4974
2.8	.4974	.4975	.4976	.4977	.4977	.4978	.4979	.4979	.4980	.4981
2.9	.4981	.4982	.4982	.4983	.4984	.4984	.4985	.4985	.4986	.4986
3.0	.4987	.4987	.4987	.4988	.4988	.4989	.4989	.4989	.4990	.4990

Source: Abridged from Table I of A. Hald, *Statistical Tables and Formulas* (New York: John Wiley & Sons, Inc.), 1952. Reproduced by permission of A. Hald and the publisher.

TABLE 4
Critical Values for Student's t

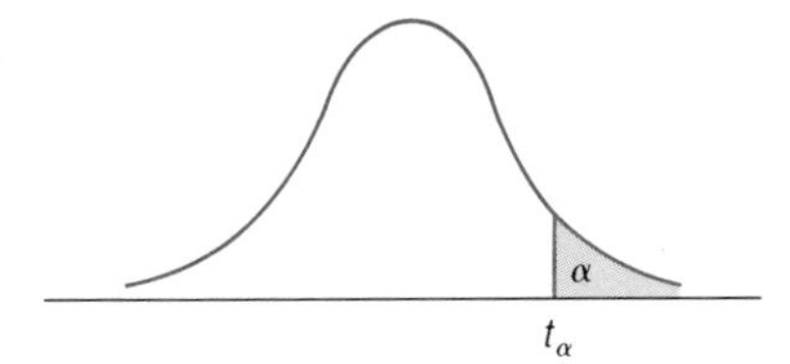

DEGREES OF FREEDOM	$t_{.100}$	$t_{.050}$	$t_{.025}$	$t_{.010}$	$t_{.005}$
1	3.078	6.314	12.706	31.821	63.657
2	1.886	2.920	4.303	6.965	9.925
3	1.638	2.353	3.182	4.541	5.841
4	1.533	2.132	2.776	3.747	4.604
5	1.476	2.015	2.571	3.365	4.032
6	1.440	1.943	2.447	3.143	3.707
7	1.415	1.895	2.365	2.998	3.499
8	1.397	1.860	2.306	2.896	3.355
9	1.383	1.833	2.262	2.821	3.250
10	1.372	1.812	2.228	2.764	3.169
11	1.363	1.796	2.201	2.718	3.106
12	1.356	1.782	2.179	2.681	3.055
13	1.350	1.771	2.160	2.650	3.012
14	1.345	1.761	2.145	2.624	2.977
15	1.341	1.753	2.131	2.602	2.947
16	1.337	1.746	2.120	2.583	2.921
17	1.333	1.740	2.110	2.567	2.898
18	1.330	1.734	2.101	2.552	2.878
19	1.328	1.729	2.093	2.539	2.861
20	1.325	1.725	2.086	2.528	2.845
21	1.323	1.721	2.080	2.518	2.831
22	1.321	1.717	2.074	2.508	2.819
23	1.319	1.714	2.069	2.500	2.807
24	1.318	1.711	2.064	2.492	2.797
25	1.316	1.708	2.060	2.485	2.787
26	1.315	1.706	2.056	2.479	2.779
27	1.314	1.703	2.052	2.473	2.771
28	1.313	1.701	2.048	2.467	2.763
29	1.311	1.699	2.045	2.462	2.756
∞	1.282	1.645	1.960	2.326	2.576

Source: From M. Merrington, "Table of Percentage Points of the t-Distribution," *Biometrika,* 1941, *32,* 300. Reproduced by permission of E. S. Pearson.

TABLE 5 Critical Values for the F Statistic: $F_{.10}$

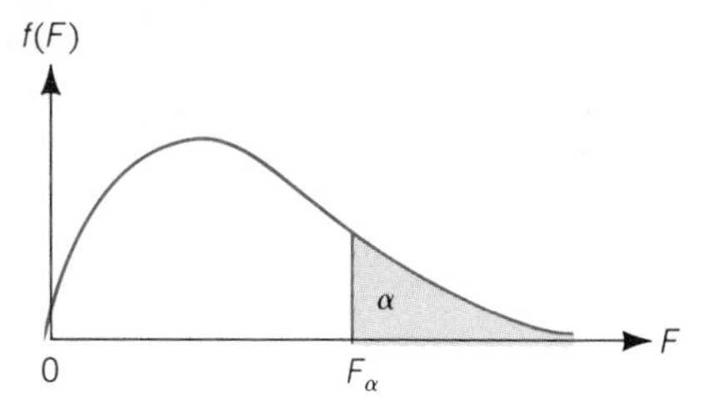

ν_2 \ ν_1	NUMERATOR DEGREES OF FREEDOM 1	2	3	4	5	6	7	8	9
1	39.86	49.50	53.59	55.83	57.24	58.20	58.91	59.44	59.86
2	8.53	9.00	9.16	9.24	9.29	9.33	9.35	9.37	9.38
3	5.54	5.46	5.39	5.34	5.31	5.28	5.27	5.25	5.24
4	4.54	4.32	4.19	4.11	4.05	4.01	3.98	3.95	3.94
5	4.06	3.78	3.62	3.52	3.45	3.40	3.37	3.34	3.32
6	3.78	3.46	3.29	3.18	3.11	3.05	3.01	2.98	2.96
7	3.59	3.26	3.07	2.96	2.88	2.83	2.78	2.75	2.72
8	3.46	3.11	2.92	2.81	2.73	2.67	2.62	2.59	2.56
9	3.36	3.01	2.81	2.69	2.61	2.55	2.51	2.47	2.44
10	3.29	2.92	2.73	2.61	2.52	2.46	2.41	2.38	2.35
11	3.23	2.86	2.66	2.54	2.45	2.39	2.34	2.30	2.27
12	3.18	2.81	2.61	2.48	2.39	2.33	2.28	2.24	2.21
13	3.14	2.76	2.56	2.43	2.35	2.28	2.23	2.20	2.16
14	3.10	2.73	2.52	2.39	2.31	2.24	2.19	2.15	2.12
15	3.07	2.70	2.49	2.36	2.27	2.21	2.16	2.12	2.09
16	3.05	2.67	2.46	2.33	2.24	2.18	2.13	2.09	2.06
17	3.03	2.64	2.44	2.31	2.22	2.15	2.10	2.06	2.03
18	3.01	2.62	2.42	2.29	2.20	2.13	2.08	2.04	2.00
19	2.99	2.61	2.40	2.27	2.18	2.11	2.06	2.02	1.98
20	2.97	2.59	2.38	2.25	2.16	2.09	2.04	2.00	1.96
21	2.96	2.57	2.36	2.23	2.14	2.08	2.02	1.98	1.95
22	2.95	2.56	2.35	2.22	2.13	2.06	2.01	1.97	1.93
23	2.94	2.55	2.34	2.21	2.11	2.05	1.99	1.95	1.92
24	2.93	2.54	2.33	2.19	2.10	2.04	1.98	1.94	1.91
25	2.92	2.53	2.32	2.18	2.09	2.02	1.97	1.93	1.89
26	2.91	2.52	2.31	2.17	2.08	2.01	1.96	1.92	1.88
27	2.90	2.51	2.30	2.17	2.07	2.00	1.95	1.91	1.87
28	2.89	2.50	2.29	2.16	2.06	2.00	1.94	1.90	1.87
29	2.89	2.50	2.28	2.15	2.06	1.99	1.93	1.89	1.86
30	2.88	2.49	2.28	2.14	2.05	1.98	1.93	1.88	1.85
40	2.84	2.44	2.23	2.09	2.00	1.93	1.87	1.83	1.79
60	2.79	2.39	2.18	2.04	1.95	1.87	1.82	1.77	1.74
120	2.75	2.35	2.13	1.99	1.90	1.82	1.77	1.72	1.68
∞	2.71	2.30	2.08	1.94	1.85	1.77	1.72	1.67	1.63

(Row labels: DENOMINATOR DEGREES OF FREEDOM)

Source: From M. Merrington and C. M. Thompson, "Tables of Percentage Points of the Inverted Beta (*F*)-Distribution," *Biometrika*, 1943, *33*, 73–88. Reproduced by permission of the *Biometrika* Trustees.

ν_2 \ ν_1	NUMERATOR DEGREES OF FREEDOM									
	10	12	15	20	24	30	40	60	120	∞
1	60.19	60.71	61.22	61.74	62.00	62.26	62.53	62.79	63.06	63.33
2	9.39	9.41	9.42	9.44	9.45	9.46	9.47	9.47	9.48	9.49
3	5.23	5.22	5.20	5.18	5.18	5.17	5.16	5.15	5.14	5.13
4	3.92	3.90	3.87	3.84	3.83	3.82	3.80	3.79	3.78	3.76
5	3.30	3.27	3.24	3.21	3.19	3.17	3.16	3.14	3.12	3.10
6	2.94	2.90	2.87	2.84	2.82	2.80	2.78	2.76	2.74	2.72
7	2.70	2.67	2.63	2.59	2.58	2.56	2.54	2.51	2.49	2.47
8	2.54	2.50	2.46	2.42	2.40	2.38	2.36	2.34	2.32	2.29
9	2.42	2.38	2.34	2.30	2.28	2.25	2.23	2.21	2.18	2.16
10	2.32	2.28	2.24	2.20	2.18	2.16	2.13	2.11	2.08	2.06
11	2.25	2.21	2.17	2.12	2.10	2.08	2.05	2.03	2.00	1.97
12	2.19	2.15	2.10	2.06	2.04	2.01	1.99	1.96	1.93	1.90
13	2.14	2.10	2.05	2.01	1.98	1.96	1.93	1.90	1.88	1.85
14	2.10	2.05	2.01	1.96	1.94	1.91	1.89	1.86	1.83	1.80
15	2.06	2.02	1.97	1.92	1.90	1.87	1.85	1.82	1.79	1.76
16	2.03	1.99	1.94	1.89	1.87	1.84	1.81	1.78	1.75	1.72
17	2.00	1.96	1.91	1.86	1.84	1.81	1.78	1.75	1.72	1.69
18	1.98	1.93	1.89	1.84	1.81	1.78	1.75	1.72	1.69	1.66
19	1.96	1.91	1.86	1.81	1.79	1.76	1.73	1.70	1.67	1.63
20	1.94	1.89	1.84	1.79	1.77	1.74	1.71	1.68	1.64	1.61
21	1.92	1.87	1.83	1.78	1.75	1.72	1.69	1.66	1.62	1.59
22	1.90	1.86	1.81	1.76	1.73	1.70	1.67	1.64	1.60	1.57
23	1.89	1.84	1.80	1.74	1.72	1.69	1.66	1.62	1.59	1.55
24	1.88	1.83	1.78	1.73	1.70	1.67	1.64	1.61	1.57	1.53
25	1.87	1.82	1.77	1.72	1.69	1.66	1.63	1.59	1.56	1.52
26	1.86	1.81	1.76	1.71	1.68	1.65	1.61	1.58	1.54	1.50
27	1.85	1.80	1.75	1.70	1.67	1.64	1.60	1.57	1.53	1.49
28	1.84	1.79	1.74	1.69	1.66	1.63	1.59	1.56	1.52	1.48
29	1.83	1.78	1.73	1.68	1.65	1.62	1.58	1.55	1.51	1.47
30	1.82	1.77	1.72	1.67	1.64	1.61	1.57	1.54	1.50	1.46
40	1.76	1.71	1.66	1.61	1.57	1.54	1.51	1.47	1.42	1.38
60	1.71	1.66	1.60	1.54	1.51	1.48	1.44	1.40	1.35	1.29
120	1.65	1.60	1.55	1.48	1.45	1.41	1.37	1.32	1.26	1.19
∞	1.60	1.55	1.49	1.42	1.38	1.34	1.30	1.24	1.17	1.00

DENOMINATOR DEGREES OF FREEDOM

TABLE 6 Critical Values for the F Statistic: $F_{.05}$

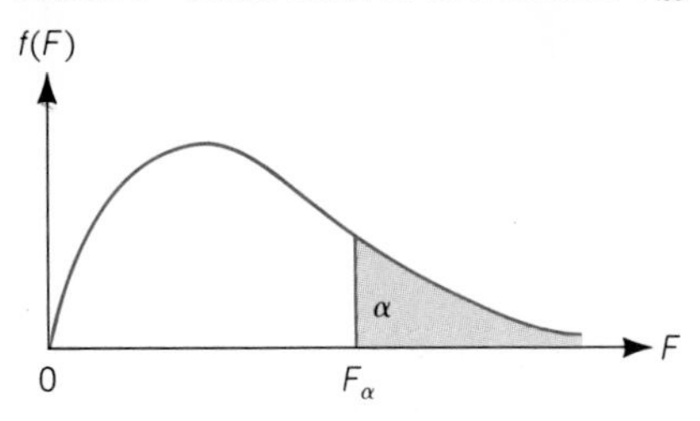

ν_2 \ ν_1 (NUMERATOR DEGREES OF FREEDOM)	1	2	3	4	5	6	7	8	9
1	161.4	199.5	215.7	224.6	230.2	234.0	236.8	238.9	240.5
2	18.51	19.00	19.16	19.25	19.30	19.33	19.35	19.37	19.38
3	10.13	9.55	9.28	9.12	9.01	8.94	8.89	8.85	8.81
4	7.71	6.94	6.59	6.39	6.26	6.16	6.09	6.04	6.00
5	6.61	5.79	5.41	5.19	5.05	4.95	4.88	4.82	4.77
6	5.99	5.14	4.76	4.53	4.39	4.28	4.21	4.15	4.10
7	5.59	4.74	4.35	4.12	3.97	3.87	3.79	3.73	3.68
8	5.32	4.46	4.07	3.84	3.69	3.58	3.50	3.44	3.39
9	5.12	4.26	3.86	3.63	3.48	3.37	3.29	3.23	3.18
10	4.96	4.10	3.71	3.48	3.33	3.22	3.14	3.07	3.02
11	4.84	3.98	3.59	3.36	3.20	3.09	3.01	2.95	2.90
12	4.75	3.89	3.49	3.26	3.11	3.00	2.91	2.85	2.80
13	4.67	3.81	3.41	3.18	3.03	2.92	2.83	2.77	2.71
14	4.60	3.74	3.34	3.11	2.96	2.85	2.76	2.70	2.65
15	4.54	3.68	3.29	3.06	2.90	2.79	2.71	2.64	2.59
16	4.49	3.63	3.24	3.01	2.85	2.74	2.66	2.59	2.54
17	4.45	3.59	3.20	2.96	2.81	2.70	2.61	2.55	2.49
18	4.41	3.55	3.16	2.93	2.77	2.66	2.58	2.51	2.46
19	4.38	3.52	3.13	2.90	2.74	2.63	2.54	2.48	2.42
20	4.35	3.49	3.10	2.87	2.71	2.60	2.51	2.45	2.39
21	4.32	3.47	3.07	2.84	2.68	2.57	2.49	2.42	2.37
22	4.30	3.44	3.05	2.82	2.66	2.55	2.46	2.40	2.34
23	4.28	3.42	3.03	2.80	2.64	2.53	2.44	2.37	2.32
24	4.26	3.40	3.01	2.78	2.62	2.51	2.42	2.36	2.30
25	4.24	3.39	2.99	2.76	2.60	2.49	2.40	2.34	2.28
26	4.23	3.37	2.98	2.74	2.59	2.47	2.39	2.32	2.27
27	4.21	3.35	2.96	2.73	2.57	2.46	2.37	2.31	2.25
28	4.20	3.34	2.95	2.71	2.56	2.45	2.36	2.29	2.24
29	4.18	3.33	2.93	2.70	2.55	2.43	2.35	2.28	2.22
30	4.17	3.32	2.92	2.69	2.53	2.42	2.33	2.27	2.21
40	4.08	3.23	2.84	2.61	2.45	2.34	2.25	2.18	2.12
60	4.00	3.15	2.76	2.53	2.37	2.25	2.17	2.10	2.04
120	3.92	3.07	2.68	2.45	2.29	2.17	2.09	2.02	1.96
∞	3.84	3.00	2.60	2.37	2.21	2.10	2.01	1.94	1.88

DENOMINATOR DEGREES OF FREEDOM (ν_2)

Source: From M. Merrington and C. M. Thompson, "Tables of Percentage Points of the Inverted Beta (*F*)-Distribution," *Biometrika*, 1943, *33*, 73–88. Reproduced by permission of the *Biometrika* Trustees.

ν_2 \ ν_1 (DENOMINATOR DEGREES OF FREEDOM)	NUMERATOR DEGREES OF FREEDOM 10	12	15	20	24	30	40	60	120	∞
1	241.9	243.9	245.9	248.0	249.1	250.1	251.1	252.2	253.3	254.3
2	19.40	19.41	19.43	19.45	19.45	19.46	19.47	19.48	19.49	19.50
3	8.79	8.74	8.70	8.66	8.64	8.62	8.59	8.57	8.55	8.53
4	5.96	5.91	5.86	5.80	5.77	5.75	5.72	5.69	5.66	5.63
5	4.74	4.68	4.62	4.56	4.53	4.50	4.46	4.43	4.40	4.36
6	4.06	4.00	3.94	3.87	3.84	3.81	3.77	3.74	3.70	3.67
7	3.64	3.57	3.51	3.44	3.41	3.38	3.34	3.30	3.27	3.23
8	3.35	3.28	3.22	3.15	3.12	3.08	3.04	3.01	2.97	2.93
9	3.14	3.07	3.01	2.94	2.90	2.86	2.83	2.79	2.75	2.71
10	2.98	2.91	2.85	2.77	2.74	2.70	2.66	2.62	2.58	2.54
11	2.85	2.79	2.72	2.65	2.61	2.57	2.53	2.49	2.45	2.40
12	2.75	2.69	2.62	2.54	2.51	2.47	2.43	2.38	2.34	2.30
13	2.67	2.60	2.53	2.46	2.42	2.38	2.34	2.30	2.25	2.21
14	2.60	2.53	2.46	2.39	2.35	2.31	2.27	2.22	2.18	2.13
15	2.54	2.48	2.40	2.33	2.29	2.25	2.20	2.16	2.11	2.07
16	2.49	2.42	2.35	2.28	2.24	2.19	2.15	2.11	2.06	2.01
17	2.45	2.38	2.31	2.23	2.19	2.15	2.10	2.06	2.01	1.96
18	2.41	2.34	2.27	2.19	2.15	2.11	2.06	2.02	1.97	1.92
19	2.38	2.31	2.23	2.16	2.11	2.07	2.03	1.98	1.93	1.88
20	2.35	2.28	2.20	2.12	2.08	2.04	1.99	1.95	1.90	1.84
21	2.32	2.25	2.18	2.10	2.05	2.01	1.96	1.92	1.87	1.81
22	2.30	2.23	2.15	2.07	2.03	1.98	1.94	1.89	1.84	1.78
23	2.27	2.20	2.13	2.05	2.01	1.96	1.91	1.86	1.81	1.76
24	2.25	2.18	2.11	2.03	1.98	1.94	1.89	1.84	1.79	1.73
25	2.24	2.16	2.09	2.01	1.96	1.92	1.87	1.82	1.77	1.71
26	2.22	2.15	2.07	1.99	1.95	1.90	1.85	1.80	1.75	1.69
27	2.20	2.13	2.06	1.97	1.93	1.88	1.84	1.79	1.73	1.67
28	2.19	2.12	2.04	1.96	1.91	1.87	1.82	1.77	1.71	1.65
29	2.18	2.10	2.03	1.94	1.90	1.85	1.81	1.75	1.70	1.64
30	2.16	2.09	2.01	1.93	1.89	1.84	1.79	1.74	1.68	1.62
40	2.08	2.00	1.92	1.84	1.79	1.74	1.69	1.64	1.58	1.51
60	1.99	1.92	1.84	1.75	1.70	1.65	1.59	1.53	1.47	1.39
120	1.91	1.83	1.75	1.66	1.61	1.55	1.50	1.43	1.35	1.25
∞	1.83	1.75	1.67	1.57	1.52	1.46	1.39	1.32	1.22	1.00

TABLE 7 Critical Values for the F Statistic: $F_{.025}$

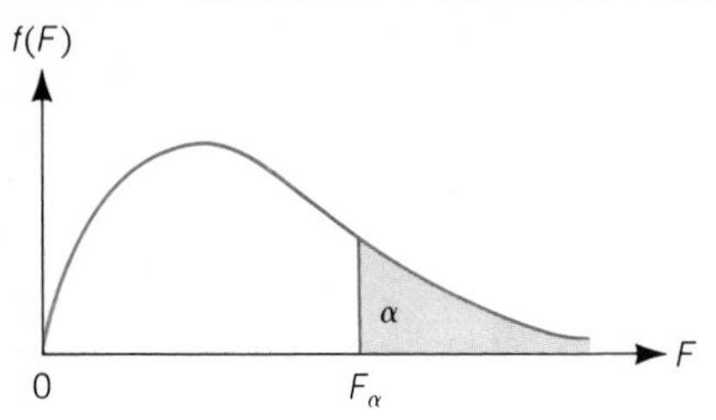

ν_2 \ ν_1 (NUMERATOR DEGREES OF FREEDOM)	1	2	3	4	5	6	7	8	9
1	647.8	799.5	864.2	899.6	921.8	937.1	948.2	956.7	963.3
2	38.51	39.00	39.17	39.25	39.30	39.33	39.36	39.37	39.39
3	17.44	16.04	15.44	15.10	14.88	14.73	14.62	14.54	14.47
4	12.22	10.65	9.98	9.60	9.36	9.20	9.07	8.98	8.90
5	10.01	8.43	7.76	7.39	7.15	6.98	6.85	6.76	6.68
6	8.81	7.26	6.60	6.23	5.99	5.82	5.70	5.60	5.52
7	8.07	6.54	5.89	5.52	5.29	5.12	4.99	4.90	4.82
8	7.57	6.06	5.42	5.05	4.82	4.65	4.53	4.43	4.36
9	7.21	5.71	5.08	4.72	4.48	4.32	4.20	4.10	4.03
10	6.94	5.46	4.83	4.47	4.24	4.07	3.95	3.85	3.78
11	6.72	5.26	4.63	4.28	4.04	3.88	3.76	3.66	3.59
12	6.55	5.10	4.47	4.12	3.89	3.73	3.61	3.51	3.44
13	6.41	4.97	4.35	4.00	3.77	3.60	3.48	3.39	3.31
14	6.30	4.86	4.24	3.89	3.66	3.50	3.38	3.29	3.21
15	6.20	4.77	4.15	3.80	3.58	3.41	3.29	3.20	3.12
16	6.12	4.69	4.08	3.73	3.50	3.34	3.22	3.12	3.05
17	6.04	4.62	4.01	3.66	3.44	3.28	3.16	3.06	2.98
18	5.98	4.56	3.95	3.61	3.38	3.22	3.10	3.01	2.93
19	5.92	4.51	3.90	3.56	3.33	3.17	3.05	2.96	2.88
20	5.87	4.46	3.86	3.51	3.29	3.13	3.01	2.91	2.84
21	5.83	4.42	3.82	3.48	3.25	3.09	2.97	2.87	2.80
22	5.79	4.38	3.78	3.44	3.22	3.05	2.93	2.84	2.76
23	5.75	4.35	3.75	3.41	3.18	3.02	2.90	2.81	2.73
24	5.72	4.32	3.72	3.38	3.15	2.99	2.87	2.78	2.70
25	5.69	4.29	3.69	3.35	3.13	2.97	2.85	2.75	2.68
26	5.66	4.27	3.67	3.33	3.10	2.94	2.82	2.73	2.65
27	5.63	4.24	3.65	3.31	3.08	2.92	2.80	2.71	2.63
28	5.61	4.22	3.63	3.29	3.06	2.90	2.78	2.69	2.61
29	5.59	4.20	3.61	3.27	3.04	2.88	2.76	2.67	2.59
30	5.57	4.18	3.59	3.25	3.03	2.87	2.75	2.65	2.57
40	5.42	4.05	3.46	3.13	2.90	2.74	2.62	2.53	2.45
60	5.29	3.93	3.34	3.01	2.79	2.63	2.51	2.41	2.33
120	5.15	3.80	3.23	2.89	2.67	2.52	2.39	2.30	2.22
∞	5.02	3.69	3.12	2.79	2.57	2.41	2.29	2.19	2.11

(Row labels ν_2: DENOMINATOR DEGREES OF FREEDOM)

Source: From M. Merrington and C. M. Thompson, "Tables of Percentage Points of the Inverted Beta (F)-Distribution," *Biometrika*, 1943, *33*, 73–88. Reproduced by permission of the *Biometrika* Trustees.

DENOMINATOR DEGREES OF FREEDOM (ν_2); NUMERATOR DEGREES OF FREEDOM (ν_1)

ν_2 \ ν_1	10	12	15	20	24	30	40	60	120	∞
1	968.6	976.7	984.9	993.1	997.2	1001	1006	1010	1014	1018
2	39.40	39.41	39.43	39.45	39.46	39.46	39.47	39.48	39.49	39.50
3	14.42	14.34	14.25	14.17	14.12	14.08	14.04	13.99	13.95	13.90
4	8.84	8.75	8.66	8.56	8.51	8.46	8.41	8.36	8.31	8.26
5	6.62	6.52	6.43	6.33	6.28	6.23	6.18	6.12	6.07	6.02
6	5.46	5.37	5.27	5.17	5.12	5.07	5.01	4.96	4.90	4.85
7	4.76	4.67	4.57	4.47	4.42	4.36	4.31	4.25	4.20	4.14
8	4.30	4.20	4.10	4.00	3.95	3.89	3.84	3.78	3.73	3.67
9	3.96	3.87	3.77	3.67	3.61	3.56	3.51	3.45	3.39	3.33
10	3.72	3.62	3.52	3.42	3.37	3.31	3.26	3.20	3.14	3.08
11	3.53	3.43	3.33	3.23	3.17	3.12	3.06	3.00	2.94	2.88
12	3.37	3.28	3.18	3.07	3.02	2.96	2.91	2.85	2.79	2.72
13	3.25	3.15	3.05	2.95	2.89	2.84	2.78	2.72	2.66	2.60
14	3.15	3.05	2.95	2.84	2.79	2.73	2.67	2.61	2.55	2.49
15	3.06	2.96	2.86	2.76	2.70	2.64	2.59	2.52	2.46	2.40
16	2.99	2.89	2.79	2.68	2.63	2.57	2.51	2.45	2.38	2.32
17	2.92	2.82	2.72	2.62	2.56	2.50	2.44	2.38	2.32	2.25
18	2.87	2.77	2.67	2.56	2.50	2.44	2.38	2.32	2.26	2.19
19	2.82	2.72	2.62	2.51	2.45	2.39	2.33	2.27	2.20	2.13
20	2.77	2.68	2.57	2.46	2.41	2.35	2.29	2.22	2.16	2.09
21	2.73	2.64	2.53	2.42	2.37	2.31	2.25	2.18	2.11	2.04
22	2.70	2.60	2.50	2.39	2.33	2.27	2.21	2.14	2.08	2.00
23	2.67	2.57	2.47	2.36	2.30	2.24	2.18	2.11	2.04	1.97
24	2.64	2.54	2.44	2.33	2.27	2.21	2.15	2.08	2.01	1.94
25	2.61	2.51	2.41	2.30	2.24	2.18	2.12	2.05	1.98	1.91
26	2.59	2.49	2.39	2.28	2.22	2.16	2.09	2.03	1.95	1.88
27	2.57	2.47	2.36	2.25	2.19	2.13	2.07	2.00	1.93	1.85
28	2.55	2.45	2.34	2.23	2.17	2.11	2.05	1.98	1.91	1.83
29	2.53	2.43	2.32	2.21	2.15	2.09	2.03	1.96	1.89	1.81
30	2.51	2.41	2.31	2.20	2.14	2.07	2.01	1.94	1.87	1.79
40	2.39	2.29	2.18	2.07	2.01	1.94	1.88	1.80	1.72	1.64
60	2.27	2.17	2.06	1.94	1.88	1.82	1.74	1.67	1.58	1.48
120	2.16	2.05	1.94	1.82	1.76	1.69	1.61	1.53	1.43	1.31
∞	2.05	1.94	1.83	1.71	1.64	1.57	1.48	1.39	1.27	1.00

TABLE 8
Critical Values for the χ^2 Statistic

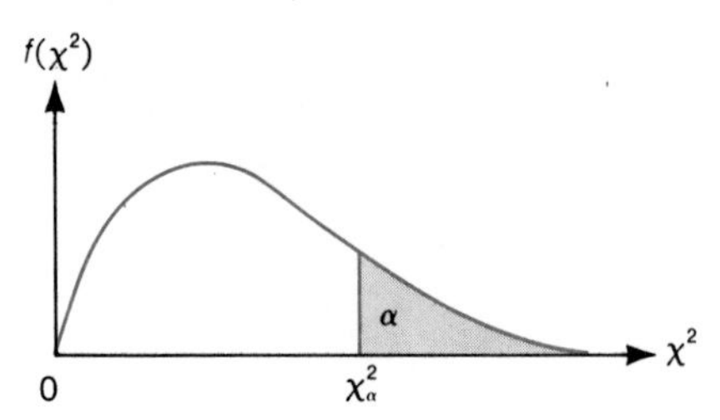

DEGREES OF FREEDOM	$\chi^2_{.995}$	$\chi^2_{.990}$	$\chi^2_{.975}$	$\chi^2_{.950}$	$\chi^2_{.900}$
1	0.0000393	0.0001571	0.0009821	0.0039321	0.0157908
2	0.0100251	0.0201007	0.0506356	0.102587	0.210720
3	0.0717212	0.114832	0.215795	0.351846	0.584375
4	0.206990	0.297110	0.484419	0.710721	1.063623
5	0.411740	0.554300	0.831211	1.145476	1.61031
6	0.675727	0.872085	1.237347	1.63539	2.20413
7	0.989265	1.239043	1.68987	2.16735	2.83311
8	1.344419	1.646482	2.17973	2.73264	3.48954
9	1.734926	2.087912	2.70039	3.32511	4.16816
10	2.15585	2.55821	3.24697	3.94030	4.86518
11	2.60321	3.05347	3.81575	4.57481	5.57779
12	3.07382	3.57056	4.40379	5.22603	6.30380
13	3.56503	4.10691	5.00874	5.89186	7.04150
14	4.07468	4.66043	5.62872	6.57063	7.78953
15	4.60094	5.22935	6.26214	7.26094	8.54675
16	5.14224	5.81221	6.90766	7.96164	9.31223
17	5.69724	6.40776	7.56418	8.67176	10.0852
18	6.26481	7.01491	8.23075	9.39046	10.8649
19	6.84398	7.63273	8.90655	10.1170	11.6509
20	7.43386	8.26040	9.59083	10.8508	12.4426
21	8.03366	8.89720	10.28293	11.5913	13.2396
22	8.64272	9.54249	10.9823	12.3380	14.0415
23	9.26042	10.19567	11.6885	13.0905	14.8479
24	9.88623	10.8564	12.4011	13.8484	15.6587
25	10.5197	11.5240	13.1197	14.6114	16.4734
26	11.1603	12.1981	13.8439	15.3791	17.2919
27	11.8076	12.8786	14.5733	16.1513	18.1138
28	12.4613	13.5648	15.3079	16.9279	18.9392
29	13.1211	14.2565	16.0471	17.7083	19.7677
30	13.7867	14.9535	16.7908	18.4926	20.5992
40	20.7065	22.1643	24.4331	26.5093	29.0505
50	27.9907	29.7067	32.3574	34.7642	37.6886
60	35.5346	37.4848	40.4817	43.1879	46.4589
70	43.2752	45.4418	48.7576	51.7393	55.3290
80	51.1720	53.5400	57.1532	60.3915	64.2778
90	59.1963	61.7541	65.6466	69.1260	73.2912
100	67.3276	70.0648	74.2219	77.9295	82.3581

Source: From C. M. Thompson, "Tables of the Percentage Points of the χ^2-Distribution," *Biometrika*, 1941, *32*, 188–189. Reproduced by permission of E. S. Pearson.

DEGREES OF FREEDOM	$\chi^2_{.100}$	$\chi^2_{.050}$	$\chi^2_{.025}$	$\chi^2_{.010}$	$\chi^2_{.005}$
1	2.70554	3.84146	5.02389	6.63490	7.87944
2	4.60517	5.99147	7.37776	9.21034	10.5966
3	6.25139	7.81473	9.34840	11.3449	12.8381
4	7.77944	9.48773	11.1433	13.2767	14.8602
5	9.23635	11.0705	12.8325	15.0863	16.7496
6	10.6446	12.5916	14.4494	16.8119	18.5476
7	12.0170	14.0671	16.0128	18.4753	20.2777
8	13.3616	15.5073	17.5346	20.0902	21.9550
9	14.6837	16.9190	19.0228	21.6660	23.5893
10	15.9871	18.3070	20.4831	23.2093	25.1882
11	17.2750	19.6751	21.9200	24.7250	26.7569
12	18.5494	21.0261	23.3367	26.2170	28.2995
13	19.8119	22.3621	24.7356	27.6883	29.8194
14	21.0642	23.6848	26.1190	29.1413	31.3193
15	22.3072	24.9958	27.4884	30.5779	32.8013
16	23.5418	26.2962	28.8454	31.9999	34.2672
17	24.7690	27.5871	30.1910	33.4087	35.7185
18	25.9894	28.8693	31.5264	34.8053	37.1564
19	27.2036	30.1435	32.8523	36.1908	38.5822
20	28.4120	31.4104	34.1696	37.5662	39.9968
21	29.6151	32.6705	35.4789	38.9321	41.4010
22	30.8133	33.9244	36.7807	40.2894	42.7956
23	32.0069	35.1725	38.0757	41.6384	44.1813
24	33.1963	36.4151	39.3641	42.9798	45.5585
25	34.3816	37.6525	40.6465	44.3141	46.9278
26	35.5631	38.8852	41.9232	45.6417	48.2899
27	36.7412	40.1133	43.1944	46.9630	49.6449
28	37.9159	41.3372	44.4607	48.2782	50.9933
29	39.0875	42.5569	45.7222	49.5879	52.3356
30	40.2560	43.7729	46.9792	50.8922	53.6720
40	51.8050	55.7585	59.3417	63.6907	66.7659
50	63.1671	67.5048	71.4202	76.1539	79.4900
60	74.3970	79.0819	83.2976	88.3794	91.9517
70	85.5271	90.5312	95.0231	100.425	104.215
80	96.5782	101.879	106.629	112.329	116.321
90	107.565	113.145	118.136	124.116	128.299
100	118.498	124.342	129.561	135.807	140.169

TABLE 9 Random Numbers

COLUMN / ROW	1	2	3	4	5	6	7	8	9	10	11	12	13	14
1	10480	15011	01536	02011	81647	91646	69179	14194	62590	36207	20969	99570	91291	90700
2	22368	46573	25595	85393	30995	89198	27982	53402	93965	34095	52666	19174	39615	99505
3	24130	48360	22527	97265	76393	64809	15179	24830	49340	32081	30680	19655	63348	58629
4	42167	93093	06243	61680	07856	16376	39440	53537	71341	57004	00849	74917	97758	16379
5	37570	39975	81837	16656	06121	91782	60468	81305	49684	60672	14110	06927	01263	54613
6	77921	06907	11008	42751	27756	53498	18602	70659	90655	15053	21916	81825	44394	42880
7	99562	72905	56420	69994	98872	31016	71194	18738	44013	48840	63213	21069	10634	12952
8	96301	91977	05463	07972	18876	20922	94595	56869	69014	60045	18425	84903	42508	32307
9	89579	14342	63661	10281	17453	18103	57740	84378	25331	12566	58678	44947	05585	56941
10	85475	36857	53342	53988	53060	59533	38867	62300	08158	17983	16439	11458	18593	64952
11	28918	69578	88231	33276	70997	79936	56865	05859	90106	31595	01547	85590	91610	78188
12	63553	40961	48235	03427	49626	69445	18663	72695	52180	20847	12234	90511	33703	90322
13	09429	93969	52636	92737	88974	33488	36320	17617	30015	08272	84115	27156	30613	74952
14	10365	61129	87529	85689	48237	52267	67689	93394	01511	26358	85104	20285	29975	89868
15	07119	97336	71048	08178	77233	13916	47564	81056	97735	85977	29372	74461	28551	90707
16	51085	12765	51821	51259	77452	16308	60756	92144	49442	53900	70960	63990	75601	40719
17	02368	21382	52404	60268	89368	19885	55322	44819	01188	65255	64835	44919	05944	55157
18	01011	54092	33362	94904	31273	04146	18594	29852	71585	85030	51132	01915	92747	64951
19	52162	53916	46369	58586	23216	14513	83149	98736	23495	64350	94738	17752	35156	35749
20	07056	97628	33787	09998	42698	06691	76988	13602	51851	46104	88916	19509	25625	58104
21	48663	91245	85828	14346	09172	30168	90229	04734	59193	22178	30421	61666	99904	32812
22	54164	58492	22421	74103	47070	25306	76468	26384	58151	06646	21524	15227	96909	44592
23	32639	32363	05597	24200	13363	38005	94342	28728	35806	06912	17012	64161	18296	22851
24	29334	27001	87637	87308	58731	00256	45834	15398	46557	41135	10367	07684	36188	18510
25	02488	33062	28834	07351	19731	92420	60952	61280	50001	67658	32586	86679	50720	94953
26	81525	72295	04839	96423	24878	82651	66566	14778	76797	14780	13300	87074	79666	95725
27	29676	20591	68086	26432	46901	20849	89768	81536	86645	12659	92259	57102	80428	25280

28	00742	57392	39064	66432	84673	40027	32832	61362	98947	96067	64760	64584	96096	98253
29	05366	04213	25669	26422	44407	44048	37937	63904	45766	66134	75470	66520	34693	90449
30	91921	26418	64117	94305	26766	25940	39972	22209	71500	64568	91402	42416	07844	69618
31	00582	04711	87917	77341	42206	35126	74087	99547	81817	42607	43808	76655	62028	76630
32	00725	69884	62797	56170	86324	88072	76222	36086	84637	93161	76038	65855	77919	88006
33	69011	65795	95876	55293	18988	27354	26575	08625	40801	59920	29841	80150	12777	48501
34	25976	57948	29888	88604	67917	48708	18912	82271	65424	69774	33611	54262	85963	03547
35	09763	83473	73577	12908	30883	18317	28290	35797	05998	41688	34952	37888	38917	88050
36	91576	42595	27958	30134	04024	86385	29880	99730	55536	84855	29080	09250	79656	73211
37	17955	56349	90999	49127	20044	59931	06115	20542	18059	02008	73708	83517	36103	42791
38	46503	18584	18845	49618	02304	51038	20655	58727	28168	15475	56942	53389	20562	87338
39	92157	89634	94824	78171	84610	82834	09922	25417	44137	48413	25555	21246	35509	20468
40	14577	62765	35605	81263	39667	47358	56873	56307	61607	49518	89656	20103	77490	18062
41	98427	07523	33362	64270	01638	92477	66969	98420	04880	45585	46565	04102	46880	45709
42	34914	63976	88720	82765	34476	17032	87589	40836	32427	70002	70663	88863	77775	69348
43	70060	28277	39475	46473	23219	53416	94970	25832	69975	94884	19661	72828	00102	66794
44	53976	54914	06990	67245	68350	82948	11398	42878	80287	88267	47363	46634	06541	97809
45	76072	29515	40980	07391	58745	25774	22987	80059	39911	96189	41151	14222	60697	59583
46	90725	52210	83974	29992	65831	38857	50490	83765	55657	14361	31720	57375	56228	41546
47	64364	67412	33339	31926	14883	24413	59744	92351	97473	89286	35931	04110	23726	51900
48	08962	00358	31662	25388	61642	34072	81249	35648	56891	69352	48373	45578	78547	81788
49	95012	68379	93526	70765	10592	04542	76463	54328	02349	17247	28865	14777	62730	92277
50	15664	10493	20492	38391	91132	21999	59516	81652	27195	48223	46751	22923	32261	85653
51	16408	81899	04153	53381	79401	21438	83035	92350	36693	31238	59649	91754	72772	02338
52	18629	81953	05520	91962	04739	13092	97662	24822	94730	06496	35090	04822	86774	98289
53	73115	35101	47498	87637	99016	71060	88824	71013	18735	20286	23153	72924	35165	43040
54	57491	16703	23167	49323	45021	33132	12544	41035	80780	45393	44812	12515	98931	91202
55	30405	83946	23792	14422	15059	45799	22716	19792	09983	74353	68668	30429	70735	25499
56	16631	35006	85900	98275	32388	52390	16815	69298	82732	38480	73817	32523	41961	44437
57	96773	20206	42559	78985	05300	22164	24369	54224	35083	19687	11052	91491	60383	19746
58	38935	64202	14349	82674	66523	44133	00697	35552	35970	19124	63318	29686	03387	59846
59	31624	76384	17403	53363	44167	64486	64758	75366	76554	31601	12614	33072	60332	92325
60	78919	19474	23632	27889	47914	02584	37680	20801	72152	39339	34806	08930	85001	87820
61	03931	33309	57047	74211	63445	17361	62825	39908	05607	91284	68833	25570	38818	46920
62	74426	33278	43972	10119	89917	15665	52872	73823	73144	88662	88970	74492	51805	99378
63	09066	00903	20795	95452	92648	45454	09552	88815	16553	51125	79375	97596	16296	66092
64	42238	12426	87025	14267	20979	04508	64535	31355	86064	29472	47689	05974	52468	16834
65	16153	08002	26504	41744	81959	65642	74240	56302	00033	67107	77510	70625	28725	34191
66	21457	40742	29820	96783	29400	21840	15035	34537	33310	06116	95240	15957	16572	06004

TABLE 9 *(continued)*

ROW \ COLUMN	1	2	3	4	5	6	7	8	9	10	11	12	13	14
67	21581	57802	02050	89728	17937	37621	47075	42080	97403	48626	68995	43805	33386	21597
68	55612	78095	83197	33732	05810	24813	86902	60397	16489	03264	88525	42786	05269	92532
69	44657	66999	99324	51281	84463	60563	79312	93454	68876	25471	93911	25650	12682	73572
70	91340	84979	46949	81973	37949	61023	43997	15263	80644	43942	89203	71795	99533	50501
71	91227	21199	31935	27022	84067	05462	35216	14486	29891	68607	41867	14951	91696	85065
72	50001	38140	66321	19924	72163	09538	12151	06878	91903	18749	34405	56087	82790	70925
73	65390	05224	72958	28609	81406	39147	25549	48542	42627	45233	57202	94617	23772	07896
74	27504	96131	83944	41575	10573	08619	64482	73923	36152	05184	94142	25299	84387	34925
75	37169	94851	39117	89632	00959	16487	65536	49071	39782	17095	02330	74301	00275	48280
76	11508	70225	51111	38351	19444	66499	71945	05422	13442	78675	84081	66938	93654	59894
77	37449	30362	06694	54690	04052	53115	62757	95348	78662	11163	81651	50245	34971	52924
78	46515	70331	85922	38329	57015	15765	97161	17869	45349	61796	66345	81073	49106	79860
79	30986	81223	42416	58353	21532	30502	32305	86482	05174	07901	54339	58861	74818	46942
80	63798	64995	46583	09785	44160	78128	83991	42865	92520	83531	80377	35909	81250	54238
81	82486	84846	99254	67632	43218	50076	21361	64816	51202	88124	41870	52689	51275	83556
82	21885	32906	92431	09060	64297	51674	64126	62570	26123	05155	59194	52799	28225	85762
83	60336	98782	07408	53458	13564	59089	26445	29789	85205	41001	12535	12133	14645	23541
84	43937	46891	24010	25560	86355	33941	25786	54990	71899	15475	95434	98227	21824	19585
85	97656	63175	89303	16275	07100	92063	21942	18611	47348	20203	18534	03862	78095	50136
86	03299	01221	05418	38982	55758	92237	26759	86367	21216	98442	08303	56613	91511	75928
87	79626	06486	03574	17668	07785	76020	79924	25651	83325	88428	85076	72811	22717	50585
88	85636	68335	47539	03129	65651	11977	02510	26113	99447	68645	34327	15152	55230	93448
89	18039	14367	61337	06177	12143	46609	32989	74014	64708	00533	35398	58408	13261	47908
90	08362	15656	60627	36478	65648	16764	53412	09013	07832	41574	17639	82163	60859	75567
91	79556	29068	04142	16268	15387	12856	66227	38358	22478	73373	88732	09443	82558	05250
92	92608	82674	27072	32534	17075	27698	98204	63863	11951	34648	88022	56148	34925	57031
93	23982	25835	40055	67006	12293	02753	14827	23235	35071	99704	37543	11601	35503	85171
94	09915	96306	05908	97901	28395	14186	00821	80703	70426	75647	76310	88717	37890	40129
95	59037	33300	26695	62247	69927	76123	50842	43834	86654	70959	79725	93872	28117	19233
96	42488	78077	69882	61657	34136	79180	97526	43092	04098	73571	80799	76536	71255	64239
97	46764	86273	63003	93017	31204	36692	40202	35275	57306	55543	53203	18098	47625	88684
98	03237	45430	55417	63282	90816	17349	88298	90183	36600	78406	06216	95787	42579	90730
99	86591	81482	52667	61582	14972	90053	89534	76036	49199	43716	97548	04379	46370	28672
100	38534	01715	94964	87288	65680	43772	39560	12918	86537	62738	19636	51132	25739	56947

Source: Abridged from W. H. Beyer, Ed., *Handbook of Tables for Probability and Statistics*, 2d ed. (Cleveland: The Chemical Rubber Company), 1968. Reproduced by permission of the publisher.

TABLE 10
Critical Values of the Sample Coefficient of Correlation, r

SAMPLE SIZE n	$r_{.050}$	$r_{.025}$	$r_{.010}$	$r_{.005}$
3	.988	.969	.951	.988
4	.900	.950	.980	.900
5	.805	.878	.934	.959
6	.729	.811	.882	.917
7	.669	.754	.833	.875
8	.621	.707	.789	.834
9	.582	.666	.750	.798
10	.549	.632	.715	.765
11	.521	.602	.685	.735
12	.497	.576	.658	.708
13	.476	.553	.634	.684
14	.457	.532	.612	.661
15	.441	.514	.592	.641
16	.426	.497	.574	.623
17	.412	.482	.558	.606
18	.400	.468	.543	.590
19	.389	.456	.529	.575
20	.378	.444	.516	.561
21	.369	.433	.503	.549
22	.360	.423	.492	.537
27	.323	.381	.445	.487
32	.296	.349	.409	.449
37	.275	.325	.381	.418
42	.257	.304	.358	.393
47	.243	.288	.338	.372
52	.231	.273	.322	.354
62	.211	.250	.295	.325
72	.195	.232	.274	.302
82	.183	.217	.257	.283
92	.173	.205	.242	.267
102	.164	.195	.230	.254

Source: From *Biometrika Tables for Statisticians, Vol I,* "Percentage points for the distribution of the correlation coefficient, r, when $\rho = 0$." Editors, E. S. Pearson and H. O. Hartley, 1966, p. 146.

Answers to Selected Exercises

Chapter 2

2.1a. Quantitative **b.** Qualitative **c.** Qualitative **d.** Quantitative
2.2a. Quantitative **b.** Qualitative **c.** Quantitative
2.4a. Qualitative **b.** Qualitative **c.** Quantitative **d.** Quantitative **e.** Qualitative
2.6b. 10 **c.** .16

***2.9a.**

CATEGORY	1981	1982
Interest on debt	12.1%	12.2%
Health	10.0%	10.1%
Energy	1.3%	1.6%
Environment	2.1%	1.9%
Veterans' Benefits	3.4%	3.3%
Transportation	3.6%	2.9%
Education	4.8%	4.7%
Defense	24.3%	24.9%
Social Security	35.0%	34.5%
Other	3.4%	3.9%

*Numbers rounded and based on 1981 budget of 662.8 and 1982 budget of 739.3 (billions of $).

c. Defense **2.17** ≈39%

2.18a. Quantitative **b.** Qualitative **c.** Quantitative **d.** Qualitative **e.** Qualitative **f.** Quantitative **g.** Quantitative **h.** Qualitative
2.19b. .80 **2.20a.** Qualitative **2.22b.** Cola **c.** 37.8% **2.23a.** Bar graph **b.** ≈18%
2.26a. Quantitative **2.27a.** Bar graph **c.** 0% **2.28a.** 22.7% **b.** 71.4% **c.** 3.3%
2.30a. Qualitative **b.** Bar graph or pie chart **d.** ≈32.7% **e.** ≈12.1%
2.31a. Quantitative **b.** Frequency distribution **c.** ≈28%
d. Yes; no diameters recorded in the interval (.9985, .9995)
2.32a. Quantitative

Chapter 3

3.1a. 12 **b.** 40 **c.** 7 **d.** 21 **3.2a.** 33 **b.** 175 **c.** 20 **d.** 71 **3.3a.** 11.2 **b.** 12
3.4a. 19.43 **b.** 20 **3.5a.** 6 **b.** 50 **c.** 42.8 **3.6** Mean = 4.6, Median = 4

3.7 Mean = 4.33, Median = 4.5 **3.8** Mean = 15,400, Median = 15,450
3.9 Modal class: 14,800 – 17,800; Mode = 16,050 **3.10** Larger, yes
3.12b. Mean = 81.15, Median = 83, Mode = 83 **3.13** Range = 9, St. dev. = 3.51, Var. = 12.3
3.14 Range = 6, St. dev. = 2.16, Var. = 4.67 **3.15** Var. = 3.66, St. dev. = 1.91
3.16 4.64 ± 1.91 (proportion = .72), 4.64 ± 3.82 (proportion = .96), 4.64 ± 5.73 (proportion = 1.00)
3.18 $\bar{x} = 380.4$, $s = 217.59$, 380.4 ± 217.59 (proportion = .657), 380.4 ± 435.18 (proportion = .971), 380.4 ± 652.77 (proportion = 1.00)
3.19a. $\bar{x} = 14.85$, $s = 5.12$ **b.** 14.85 ± 5.12, 14.85 ± 10.24, 14.85 ± 15.36 **c.** Yes
3.20b. ≈60–80% **c.** ≈95% **3.21b.** ≈95% **c.** ≈60–80%
3.22 70% of the starting salaries lie below $19,000 and 30% lie above $19,000
3.23a. −1.25 **b.** 1.10; negative z-values lie to the left or below the mean
3.24a. $28.75 **b.** −1.12 **c.** −.38, skewed right **3.25b.** −1.72 **c.** No
3.26a. 1.0 **b.** Possibly, .44 is within 2 standard deviations of the mean ($z = 1.73$)
3.27b. 3.05 **c.** No; most likely; otherwise we have observed a rare event **3.28b.** $\bar{x} = 377.5$, $s = 207.89$
3.29a. $\bar{x} = 16$, $s = 6.56$ **3.30** $\bar{x} = 2285$, $s = 1329.20$ **3.31** $\bar{x} = 3.07$, $s^2 = 51.80$, $s = 7.20$
3.32a. 37, 341, 1369 **b.** 470, 86,138, 220,900 **c.** −10, 58, 100 **d.** 89, 1,651, 7,921
3.33a. 6.17, 7.5, 8 **b.** 117.5, 111.5, all 4 numbers are the mode **c.** −1.43, −2, bimodal (−2 and −3)
d. 17.8, 17, 17
3.34a. 14, 22.6, 4.75 **b.** 247, 10,304.3, 101.5 **c.** 8, 7.29,, 2.70 **d.** 11, 16.7, 4.09
3.35a. .76, .836, .914 **b.** 4.15, 15.32, 3.91 **c.** 2, 9.67, 3.11
3.36a. 5, 35.2, 5.93 **b.** 3, 140, 11.8 **c.** 2, 17.8, 4.21 **d.** 1000, 407,679.33, 638.5
3.38 Standard deviation **3.39** Percentiles, z-scores
3.40a. 1.4, 1, bimodal (0 and 1), 4, 1.82, 1.35 **b.** Range, variance, standard deviation; mean, median, mode
3.41 $\bar{x} = 2.2$, $s = 2.38$, $s^2 = 5.68$
3.42b. $\bar{x} = 98.67$, $s = 18.17$ **c.** 98.67 ± 18.17, 98.67 ± 36.34, 98.67 ± 54.51 **d.** Yes
3.43 $\bar{x} = 67.88$, $s^2 = 382.0$, $s = 19.5$ **3.45a.** Skewed right **b.** Median **c.** Mean
3.47a. 29.45 **b.** 30 **c.** 30 **3.48** Range = 50, Var. = 87.02, St. dev. = 9.33
3.49 29.45 ± 9.33 (proportion = .70), 29.45 ± 18.66 (proportion = .95), 29.45 ± 27.99 (proportion = 1.00) **3.50** 18
3.51b. ≈95% **c.** ≈20–40% **3.52a.** 1, −.5, .67, −.33, −.67 **b.** Highest: #1, Lowest: #5
3.53b. ≈95% **c.** 50% **d.** no **3.55a.** ≈95% **b.** Almost 100% **3.56a.** −3.43 **b.** Yes
3.57a. 36.36 **b.** 12.39 **3.58a.** 9.07, 5.65 **b.** 9.07 ± 11.30; 54
3.59b. ≈60–80% **c.** $z = -1.25$ **d.** Greater than
3.60a. Approx. 60–80% within 5.0 ± 1.5, approx. 95% within 5.0 ± 3.0, almost all within 5.0 ± 4.5 **b.** No
3.61a. Almost 100% **b.** $z = 2.0$, ≈97.5% had scores below 76 **3.62a.** 2.15–2.35 **b.** $\bar{x} = 2.28$, $s = .40$
3.63a. $\bar{x} = 410.347$, $s = 143.886$, $s^2 = 20{,}703{,}161$ (in thousands) **b.** Median = 375,700, Range = 488,900
c. $z = -.12$ **d.** Boeing
e. 410.347 ± 143.886 (proportion = .87), 410.347 ± 287.772 (proportion = .93), 410.347 ± 431.658 (proportion = 1.00)
3.64b. ≈95% **c.** No **3.66a.** 22.2, 54.92, 7.41 **c.** $z = -.30$
3.67a. Approx. 60–80% within 8.5 ± 2.0, approx. 95% within 8.5 ± 4.0, almost all within 8.5 ± 6.0
b. Yes, on approx. 95% of the nonholiday weekends the number of no shows will fall between 4.5 and 12.5

Chapter 4

4.1 A and C are mutually exclusive **4.2** A and B are mutually exclusive **4.3** All 3 pairs are mutually exclusive
4.4 No pair is mutually exclusive **4.5a.** WES, WSE, EWS, ESW, SWE, SEW **b.** Yes **4.7** $2/365 = .0054795$
4.8 .08 **4.9a.** .3 **4.11a.** 1/4 **b.** 1/4 **c.** 3/4 **d.** Observe at most one head; 3/4
4.12a. 1/36 **b.** 3/36 **c.** 2/36 **4.14** 1/6 **4.16a.** 10 **4.17a.** 3/10 **b.** No **4.18** 32
4.19 8/32 = 1/4 **4.20b.** 1/16 **c.** Either public prefers A or we have observed a rare event **4.21** 1/36
4.22 1/8000 = .000125 **4.23a.** .2 **b.** 10/49 **c.** 10/46 **4.24a.** .00000016 **b.** Claim is probably false

4.25a. 8/36 **b.** 4/36 **c.** 8/36 **d.** 4/36 **4.26a.** .01 **b.** .5 **c.** .005
4.27a. .63 **b.** .57 **c.** .21 **d.** .41 **e.** 0 **f.** .79 **g.** 41/63 **4.28** .99
4.29b. .15 **d.** .90 **4.30a.** 73/503 **b.** 81/1000 **c.** 590/1000 **d.** 491/1000 **4.31** No
4.32 (B and C) and (A and B) are mutually exclusive **4.33** (A and B) and (B and C) are mutually exclusive
4.34b. ¼ **c.** ½ **4.35** No **4.36** ½
4.37a. .15 **b.** .80 **c.** .60 **d.** None are mutually exclusive
4.38a. 410/500 **b.** 125/500 **c.** 375/500 **d.** 205/500 **e.** 1 **f.** 61/500 **g.** 0 **h.** 10/125 **i.** 80/90
j. No **k.** Yes
4.39a. .5 **b.** At least .1 **4.40** .045 **4.41a.** .01 **b.** .99 **c.** .0001 **d.** Most likely false
4.42 2,598,960 **4.43a.** .30 **b.** .90 **4.44a.** 20 **c.** 4/20 **d.** 16/20 **e.** 4/10
4.45a. .81 **b.** .1296 **4.46a.** 23/50 **b.** 24/50 **c.** 12/50 **d.** 6/50 **e.** 15/27
4.47a. .40 **b.** .12 **c.** .70 **d.** 18/40
4.48a. .216 **b.** .064 **c.** Probably not; highly unlikely to observe three incorrect predictions if claim is true
4.49b. 5/9 **c.** ⅓ **4.50a.** .5 **b.** .1 **4.51a.** ⅙ **b.** 1/11 **c.** 1/66 **d.** Yes; a good chance
4.52a. 120 **b.** .0083 **c.** .9917 **d.** Yes **4.53a.** .06 **b.** .94
4.54a. .729 **b.** .081 **c.** .001 **d.** .81 **e.** 0 **f.** 0 **g.** No **4.55a.** 10 ways **b.** 4/10
4.56a. .31 **b.** .87 **c.** .27 **4.57a.** .23 **b.** .13 **c.** .94 **d.** 5/6 **e.** .48
4.58a. 24/36 **b.** 5/36 **c.** 6/36 **4.59b.** No **4.60a.** .60 **b.** .70 **c.** .55

Chapter 5

5.1 Yes **5.2** No **5.3** No **5.4** Yes **5.5** Yes **5.6** Yes **5.7a.** Yes **b.** $p > ½$ **c.** $p = ½$
5.8 .0625, .25, .375, .25, .0625 **5.9** .4096, .4096, .1536, .0256, .0016 **5.11a.** .3125 **b.** .6875
5.12a. .8192 **b.** .1808 **c.** Complementary **d.** Probabilities sum to 1
5.13a. .008 **b.** .384 **c.** .896 **5.14a.** .8 **c.** $x = 0$ **d.** $x = 2$ **e.** x is at least 2
5.15b. .6 **c.** .48 **5.16b.** .4 **c.** .48 **5.17a.** .3241 **b.** .3125 **c.** .0001
5.18a. .3487 **b.** .3874 **c.** .1937 **5.19a.** .1172 **b.** .2051 **c.** .2461
5.22a. .8369 **b.** .9976 **c.** 1 **5.23a.** .9983 **b.** .9877 **c.** .0017 **5.24a.** .9298 **b.** .0702
5.25a. .8125 **b.** .1875 **c.** .1875 **d.** .5000 **5.26a.** .3585 **b.** .7358 **c.** .7358 **d.** .2642
5.27a. .9706 **b.** .0043 **c.** .9868 **5.28a.** .8725 **b.** .5955 **c.** .0510
5.29a. 1.5, 1.16 **b.** 7.5, 1.94 **5.30** ≈95% **5.31** .944 **b.** .964 **c.** Yes
5.32a. 99, .995 **b.** 80, 4 **c.** 50, 5 **d.** 20, 4 **e.** 1, .995
5.34a. 891, 2.98 **b.** 720, 12 **c.** 450, 15 **d.** 180, 12 **e.** 9, 2.98
5.36a. 120 **b.** 120 ± 20.78 **c.** Yes, yes **5.37a.** 1960 **b.** 6.26 **c.** 1960 ± 12.52 **d.** Yes
5.39a. .50 **b.** 500, 15.8 **c.** 500 ± 31.6 **5.40** .729, .243, .027, .001 **5.41** .343, .441, .189, .027
5.42 .125, .375, .375, .125 **5.43** .027, .189, .441, .343 **5.44** .001, .027, .243, .729
5.45a. .2734 **b.** .3633 **c.** .0352
5.46a. .0712 **b.** Yes **c.** Yes, claim is most likely false **d.** .2252, no, no
5.47a. .0523 **b.** Possibly; observed event is rare if $p = .1$ **5.48** Yes, $p = .4$, $n = 50$ **5.49** Yes
5.50 No **5.51a.** Yes **b.** No **c.** No **5.52a.** .0282 **b.** .9984 **c.** .0001 **5.53** .0677
5.55a. 3, 1.225 **b.** 3 ± 2.45 **5.56a.** .5793 **b.** .1091 **c.** .0982
5.57a. 36 **b.** 5.23 **c.** 36 ± 10.46 **d.** Claim is false; p is less than .24
5.58a. Yes **b.** .0715 **c.** .9285 **d.** .3199 **5.59** No

Chapter 6

6.1a. .3849 **b.** .4319 **c.** .1844 **d.** .4147 **e.** .0918
6.2a. .4750 **b.** .4750 **c.** .95 **d.** .2912 **e.** .1075

6.3a. .0934 **b.** .5 **c.** .9115 **d.** .9066 **e.** .8164
6.4a. .0869 **b.** .0099 **c.** .0099 **d.** .8965 **6.5a.** .25 **b.** .92 **c.** 1.28 **d.** 1.65 **e.** 1.96
6.6a. 1.13 **b.** 2.33 **c.** .67 **d.** .84 **e.** 1.00
6.7a. 1.96 **b.** 1.65 **c.** 2.58 **d.** 2.33 **e.** 1.28
6.8a. .75 **b.** −1.00 **c.** −1.625 **d.** 2.00 **e.** −2.00
6.9a. .75 **b.** 1.46 **c.** 1.82 **d.** .38 **e.** .99 **6.10a.** 1.28 **b.** 1.04 **c.** .84 **d.** .67
6.11a. −1.28 **b.** −1.04 **c.** −.52 **d.** .00 **6.12a.** .3446 **b.** .0082 **c.** .6294
6.13a. .0838 **b.** ≈1 **c.** .1793 **6.14a.** .0436 **b.** .1271 **c.** .6141
6.15a. .50 **b.** .1056 **c.** .1056 **6.16** .7580 **6.17** .0075 **6.18** .2743 **6.19** .0294, no
6.20 4.58 minutes **6.21** $360.60 **6.22a.** .1721 **b.** .2372 **c.** .8257 **d.** .9441 **e.** .7794
6.23a. 0.0 **b.** 1.18 **c.** .13 **d.** .67 **6.24a.** −1.65 **b.** −1.96 **c.** −.99 **d.** −.77
6.25a. .0401 **b.** .1056 **6.26a.** .0099 **b.** .1359 **c.** No, probability of this event is only .0013
6.27a. .9808 **b.** .0032 **c.** 10.21 oz. **6.28a.** .7642 **b.** .2037 **c.** 65,825 miles
6.29a. .6736 **b.** .3264 **c.** .50 **d.** .3264 **6.30** 19.08 minutes **6.31a.** .9236 **b.** .5
6.32a. .0548 **b.** 97 days **6.33a.** .3192 **b.** 8.22 minutes **6.34** 47.6 ± 20.6

Chapter 7

7.3a. 45 **7.4a.** 20 **c.** No **7.5a.** 35 **7.9** Method B, method A **7.13** Statistic B
7.14a. Approx. normal, $\mu_{\bar{x}} = 3.25$, $\sigma_{\bar{x}} = .07$ **b.** 0.00 **c.** .2368
7.15a. Approx. normal, $\mu_{\bar{x}} = 9.8$, $\sigma_{\bar{x}} = .078$ **b.** .8997 **c.** No, $z = -3.21$
7.17a. Approx. normal, $\mu_{\bar{x}} = 22{,}500$, $\sigma_{\bar{x}} = 1095.44$ **b.** .9774 **7.18a.** It will be larger **b.** ≈1, yes
7.19a. .0062 **b.** Yes **7.20a.** Approx. normal, $\mu_{\bar{x}} = 4.1$, $\sigma_{\bar{x}} = .83$ **b.** .0057 **c.** .0019 **d.** No
7.21a. .0018 **b.** Yes **7.22a.** 6 **7.23a.** 15
7.24a. Approx. normal, $\mu_{\bar{x}} = 170$, $\sigma_{\bar{x}} = 15$ **b.** .0690 **c.** .2514
7.25a. Approx. normal, $\mu_{\bar{x}} = 90$, $\sigma_{\bar{x}} = .52$ **b.** ≈0 **c.** Yes
7.26a. $\mu_{\bar{x}} = 7500$, $\sigma_{\bar{x}} = 40$; no **b.** $\mu_{\bar{x}} = 7500$, $\sigma_{\bar{x}} = 8$; yes **7.28a.** 35 **c.** No
7.29a. Approx. normal, $\mu_{\bar{x}} = 200$, $\sigma_{\bar{x}} = 4$ **b.** .1587 **c.** .9938 **d.** .1974
7.30a. (1.452, 2.548) **b.** .0026
7.31a. Approx. normal **b.** 80,000 **c.** 4518.5 **d.** .0040 **e.** No
7.32a. Approx. normal, $\mu_{\bar{x}} = 75$, $\sigma_{\bar{x}} = 1.67$ **b.** .0082 **7.33a.** .5752 **b.** .1335 **c.** 5.2 ± 1.44
7.34b. .0013 **c.** Probably wrong, yes
7.35a. Approx. normal, $\mu_{\bar{x}_{25}} = 17$, $\sigma_{\bar{x}_{25}} = 2$ **b.** Approx. normal, $\mu_{\bar{x}_{100}} = 17$, $\sigma_{\bar{x}_{100}} = 1$ **c.** $P(15 < \bar{x}_{100} < 19)$
d. .6826, .9544
7.36a. .0179 **b.** Yes **7.37a.** .0139 **b.** Yes **7.38** .0062 **7.39** .6915

Chapter 8

8.2a. 1.44 **b.** 1.96 **c.** 2.24 **8.4** 2.5 ± .54 **8.5** Increase *n*, decrease confidence coefficient
8.6 20 ± 1.028 **8.7** 37.1 ± .464 **8.8a.** 3.2 ± .082 **c.** Increase *n* or decrease confidence coefficient
8.10a. $425 **b.** 425 ± 18.21 **c.** .97 **d.** No **8.11a.** 2.898 **b.** 2.262 **c.** 1.761
8.13a. 54.8 ± 12.74 mph **8.15b.** 10.1 ± 3.85 bushels **8.16b.** $41.75 ± $2.65 **c.** No **8.17** 27.8 ± 2.50
8.18a. 105.7 ± 7.78 **8.19** 16.05% ± .585% **8.21a.** .24 ± .08 **c.** Increase **8.22** .043 ± .045
8.23 .612 ± .062 **8.24** .3 ± .195 **8.25a.** .45 ± .126 **8.26a.** .086 **b.** .086 ± .0155 **8.27** .69 ± .03
8.28 −61°F ± 11.53°F **8.29a.** 2.9 ± 2.15 **b.** Yes **8.30a.** 214 ± 70.16
8.31a. −61 **b.** −61 ± 10.65 **d.** Yes **8.32a.** −4.6 ± 4.06 **b.** Yes **8.34a.** −61°F ± 25.65°F

8.36 1.6 ± 2.42 **8.37** -3.75 ± 2.89 **8.38a.** -4.8 ± 23.38 **c.** No **8.40** $-.027 \pm .024$
8.41a. $.22 \pm .142$ **b.** Yes **8.42a.** $-.227 \pm .218$ **b.** Japan **c.** Decrease
8.43a. $.18 \pm .148$ **b.** Yes, ration A **8.44** $-.03 \pm .054$ **8.45a.** $-.046 \pm .133$ **b.** Yes
8.48 $.74 \pm .024$ **8.49a.** 7.5 ± 1.11 **b.** 3.7 ± 1.59 **8.50** $3.8 \pm .27$
8.51a. 29.6 ± 43.91 **b.** No **c.** Decrease **8.52** $.107 \pm .042$ **8.53a.** $-.06 \pm .182$ **b.** No
8.54 27.5 mm ± 1.42 mm **8.55a.** 7.5 ± 10.23 hrs. **b.** No **8.56a.** 47.17 ± 3.74 **b.** No
8.57 $-.08 \pm .098$ **8.58** $.615 \pm .033$ **8.59a.** -25 ± 26.56 **b.** No **8.60a.** $.255 \pm .036$
8.61a. −150°F ± 18.91°F **b.** Yes **8.62** 112 ± 134.85, no **8.63** 95% confidence interval: $11.6 \pm .54$
8.64 $-.54 \pm .22$ **8.65a.** 86,890 ± 27,668 **b.** Claim is false, since the interval includes only positive differences

Chapter 9

9.2 H_0: $\mu = \$35.00$, H_a: $\mu > \$35.00$ **9.3** H_0: $\mu = .04$, H_a: $\mu < .04$
9.4 H_0: $(\mu_1 - \mu_2) = 0$, H_a: $(\mu_1 - \mu_2) \neq 0$ **9.5** H_0: $(p_1 - p_2) = 0$, H_a: $(p_1 - p_2) > 0$
9.6 H_0: $p = .90$, H_a: $p > .90$ **9.7** 9.2, 9.3, 9.5, 9.6 are one-tailed, 9.4 is two-tailed
9.12c. Type II error, Type I error **9.13** $z < -2.33$ or $z > 2.33$ **9.14** $z > 1.645$
9.15a. $\alpha = .025$ **b.** $\alpha = .05$ **c.** $\alpha = .01$
9.16a. $z < -2.33$ **b.** $z < -2.05$ **c.** $z < -1.645$ **d.** $z < -1.28$
9.17a. $z = -2.65$ **b.** $z < -2.33$ **c.** Reject H_0 at $\alpha = .01$
9.18a. $z < -1.96$ or $z > 1.96$ **b.** $z = .81$ **c.** Do not reject H_0 at $\alpha = .05$ **9.21** Reject H_0, do not reject H_0
9.22 Decreases **9.25a.** $z < -2.58$ or $z > 2.58$ **b.** $z < -2.33$ or $z > 2.33$ **c.** $z < -2.05$ or $z > 2.05$
9.26a. .005 **b.** .0985 **c.** .10 **9.27** H_0: $\mu = 22$, H_a: $\mu < 22$ **9.28** H_0: $(p_1 - p_2) = 0$, H_a: $(p_1 - p_2) \neq 0$
9.29 H_0: $(\mu_1 - \mu_2) = 0$, H_a: $(\mu_1 - \mu_2) > 0$ **9.30** H_0: $p = 1/6$, H_a: $p \neq 1/6$
9.31a. $z = -3.13$ **b.** $z < -1.645$ or $z > 1.645$ **c.** Reject H_0 **9.32a.** $z > 2.33$ **b.** $z = 1.57$
9.33a. $z < -2.33$ or $z > 2.33$ **b.** $z = 6.56$ **c.** Yes

Chapter 10

10.1 Yes, $z = -8.65$ **10.3** Yes, $z = 1.94$ **10.4** No, $z = 1.43$ **10.5** Yes, $z = -7.07$ **10.7** No, $t = -1.29$
10.8a. No, $t = .87$ **10.9** No, $t = 1.34$ **10.10** Yes, $t = -2.96$ **10.12** $z = 2.04$, reject H_0: $p = .40$
10.13a. $z = 21.14$, reject H_0: $p = .5$ **10.14** No, $z = -1.27$ **10.15** $z = -.97$, do not reject H_0: $p = .95$
10.16 $z = .44$, do not reject H_0: $p = .85$ **10.17** Yes, $z = -10.37$ **10.18** Yes, $z = 6.83$
10.19 Yes, $z = -1.83$ **10.20** $z = -.46$, do not reject H_0 **10.21** $t = -3.47$, reject H_0
10.22 $t = -1.42$, reject H_0 **10.23** Yes, $t = 3.19$ **10.24** $t = 2.97$, reject H_0 **10.25** Yes, $z = -4.40$
10.26 No, $z = 1.32$ **10.27** Yes, $z = -2.70$ **10.28** Yes, $z = -2.08$
10.29a. .0250 **b.** .05 **c.** .0038 **d.** .1056 **10.30a.** .3124 **b.** .0178 **c.** 0 **d.** .147
10.31 ≈ 0 **10.32** .0281, do not reject H_0 **10.33** .3228 **10.34** .0376, do not reject H_0
10.35 Yes, $z = 3.54$ **10.36** No, $z = -1.92$ **10.37** No, $t = -.17$ **10.38** Yes, $z = 2.31$
10.39 $z = 2.95$, reject H_0 **10.40** Yes, $z = 11.14$ **10.41** Yes, $z = 30.93$ **10.42** $z = -3.56$, reject H_0
10.43 No, $z = 1.33$ **10.44** No, $t = 1.58$ **10.45** $z = -2.60$, reject H_0
10.46 $z = 1.39$, p-value = .0838, do not reject H_0 (at $\alpha = .05$) **10.47** $t = 1.10$, do not reject H_0
10.48 Yes (at $\alpha = .05$), $z = 1.74$, p-value = .0409 **10.49** $t = -.46$, do not reject H_0 **10.50** No, $z = -1.35$
10.51 No, $t = .35$ **10.52** No, $z = .92$, p-value = .1788 **10.53** $t = 1.11$, do not reject H_0
10.54 $z = 8.94$, reject H_0 **10.55** $z = 4.03$, reject H_0 **10.56** Yes, $z = 6.67$, p-value ≈ 0 **10.57** Yes, $z = -2.47$
10.58 $z = .57$, do not reject H_0 **10.59** H_0: $(p_1 - p_2) = .5$, H_a: $(p_1 - p_2) > .5$ **10.60** $z = -.13$, do not reject H_0

Chapter 11

11.1a. 2.40 **b.** 3.35 **c.** 1.65 **d.** 5.86 **11.2a.** 3.18 **b.** 2.62 **c.** 2.10
11.3a. MST = 3.111 **b.** MSE = 1.405 **c.** 2 **d.** 7 **e.** $F = 2.21$ **f.** Reject H_0 if $F > 4.74$
g. Do not reject H_0: $\mu_1 = \mu_2 = \mu_3$
11.4a. H_0: $\mu_1 = \mu_2 = \mu_3 = \mu_4$, H_a: At least 2 means are different **b.** MST = 375.6 **c.** MSE = 140.775
d. $F = 2.67$ **e.** 3 **f.** 16 **g.** Reject H_0 if $F > 4.08$ **h.** No, do not reject H_0
11.5 Yes, reject H_0, $F = 9.50$ **11.6** Reject H_0, $F = 3.63$ **11.7** Yes, reject H_0, $F = 9.46$
11.8 Do not reject H_0, $F = 1.62$ **11.9** Reject H_0, $F = 13.00$ **11.10** -15 ± 15.91
11.11a. 34.3 ± 4.97 **b.** 6.25 ± 7.03; no evidence of a difference in means, 0 is included in the interval
11.12a. 90% CI for μ_1: 154.125 ± 11.013, 90% CI for μ_2: 129.75 ± 11.013, 90% CI for μ_3: 141.5 ± 11.013
b. 12.625 ± 15.576
11.13 -21.8 ± 10.15 **11.14** 3.167 ± 1.98
11.15a. 99% CI for $(\mu_1 - \mu_2)$: $-.97 \pm .702$, 99% CI for $(\mu_1 - \mu_3)$: $.148 \pm .702$, 99% CI for $(\mu_2 - \mu_3)$: $1.118 \pm .702$
b. Firm 2 (Theory Y)

11.16

SOURCE	df	SS	MS	F
City	3	1126.8	375.60	2.67
Error	16	2252.4	140.775	
Total	19	3379.2		

11.17

SOURCE	df	SS	MS	F
Hair	3	1566.5686	522.189	9.46
Error	13	717.6667	55.205	
Total	16	2284.2353		

11.18

SOURCE	df	SS	MS	F
Word	2	24.7778	12.3889	1.62
Error	15	114.8333	7.6556	
Total	17	139.6111		

11.19 SST = 0.0992333, MST = 0.04961667, df(drugs) = 2, SSE = 0.0450625, MSE = 0.00214583, df(error) = 21, SS(Total) = 0.14429583, $F = 23.12$; Reject H_0: $\mu_1 = \mu_2 = \mu_3$, p-value = .0001
11.20a. H_0: $\mu_R = \mu_L = \mu_E$, H_a: At least 2 means differ **b.** SST = 574.89, MST = 287.45, $F = 10.70$, SSE = 4944.27, MSE = 26.87 **c.** Yes, reject H_0, p-value = .0001
11.21a. SST = 7.015, MST = 2.338, $F = 24.83$, SSE = 1.883, MSE = 0.094
b. Yes, reject H_0: $\mu_1 = \mu_2 = \mu_3 = \mu_4$, p-value = .0001
11.22 Reject H_0: $\mu_1 = \mu_2 = \mu_3 = \mu_4$, $F = 3.77$, p-value = .028 **11.23a.** 2 **b.** 42 **11.24a.** 7 **b.** 24

11.25

SOURCE	df	SS	MS	F
Treatments	9	136.8	15.2	2.27
Error	30	200.7	6.69	
Total	39	337.5		

11.26

SOURCE	df	SS	MS	F
Treatments	1	28	28.0	14.0
Error	16	32	2.0	
Total	17	60		

11.27a.

SOURCE	df	SS	MS	F
Plan	3	117.642	39.214	7.79
Error	13	65.417	5.032	
Total	16	183.059		

b. Yes, reject H_0: $\mu_1 = \mu_2 = \mu_3 = \mu_4$; $F_{.025} = 4.35$ **c.** 95% CI for $(\mu_1 - \mu_3)$: -5.65 ± 3.25

11.28a.

SOURCE	df	SS	MS	F
Depth	4	21.5430	5.3857	7.38
Error	15	10.9425	.7295	
Total	19	32.4855		

b. Yes, reject H_0, $F = 7.38$ **c.** 0.225 ± 3.497

11.29a.

SOURCE	df	SS	MS	F
Day	2	76.778	38.389	0.47
Error	15	1233.500	82.233	
Total	17	1310.278		

b. No, do not reject H_0: $\mu_8 = \mu_{10} = \mu_{12}$, $F_{.10} = 2.70$ **c.** 79.333 ± 6.490 **d.** 3.167 ± 9.178

11.30

SOURCE	df	SS	MS	F
Group	9	37.101	4.122	4.57
Error	174	157.023	0.902	
Total	183	194.125		

b. Yes, reject H_0; $F_{.05} = 1.88$

11.31a.

SOURCE	df	SS	MS	F
Treatment	5	55.546867	11.1094	5.49
Error	12	24.286733	2.0239	
Total	17	79.833600		

b. Yes, reject H_0: $\mu_A = \mu_B = \cdots = \mu_F$; $F_{.05} = 3.11$ **c.** 5.12 ± 2.531 **d.** 12.443 ± 1.790

11.32a.

SOURCE	df	SS	MS	F
Food levels	4	784.56	196.14	3.32
Error	111	6562.32	59.12	
Total	115	7346.88		

b. Yes, reject H_0: $\mu_1 = \mu_2 = \cdots = \mu_5$; $F_{.025} = 2.89$

11.33a.

SOURCE	df	SS	MS	F
Watch	2	2.0720533	1.03602667	1.33
Error	12	9.3410400	0.77842000	
Total	14	11.4130933		

b. No, do not reject H_0; $F_{.025} = 5.10$ **c.** $.116 \pm .994$
11.34a. SST = 23454.45, MST = 11727.225, $F = 1.22$, SSE = 1360549.04, MSE = 9649.284
b. No, do not reject H_0: $\mu_1 = \mu_2 = \mu_3$; p-value = .2997

Chapter 12

12.1 Negative **12.2** Positive **12.3** Yes, negative **12.4** Yes, positive
12.6a. Yes, positive **b.** $r = .971$ **c.** Yes, $r_{.025} = .878$
12.7a. Yes, positive **b.** $r = .963$ **c.** Yes, $r_{.025} = .754$
12.8a. Yes, negative **b.** $r = -.929$ **c.** Yes, $-r_{.025} = -.632$
12.9a. Possibly; positive **b.** $r = .342$ **c.** No, $r_{.025} = .754$ **12.10a.** Possibly; negative **b.** $r = -.367$
c. No, $-r_{.025} = -.811$
12.11a. Negative **b.** $r = -.914$ **c.** Yes, $-r_{.025} = -.576$
12.12a. 3.5 **b.** 5.5 **d.** 4.5 **e.** Points fall exactly on the line
12.13a. 1.5 **b.** 2 **c.** Increase by 2 **d.** Decrease by 2 **e.** 1.5
12.14a. −.5 **b.** −2.5 **d.** −1.5 **e.** Points fall exactly on the line
12.15a. 1.5 **b.** −2 **c.** Decrease by 2 **d.** 1.5 **e.** Same y-intercepts, different slopes
12.17a. $\beta_0 = 1, \beta_1 = 3$ **b.** $\beta_0 = 1, \beta_1 = -3$ **c.** $\beta_0 = -1, \beta_1 = \frac{1}{2}$ **d.** $\beta_0 = -1, \beta_1 = -3$
e. $\beta_0 = 2, \beta_1 = -\frac{1}{2}$ **f.** $\beta_0 = -1.5, \beta_1 = 1$ **g.** $\beta_0 = 0, \beta_1 = 3$ **h.** $\beta_0 = 0, \beta_1 = -2$
12.18b. $\hat{y} = 0.35 + 1.05x$ **12.19b.** $\hat{y} = 4.986 - 1.934x$ **12.20b.** $\hat{y} = 3.343 + 0.576x$
12.21b. $\hat{y} = 1.255 - .398x$ **12.22b.** $\hat{y} = 124.9 - 0.2x$ **d.** 28.9 **12.23b.** $\hat{y} = 18.89 + .0109x$ **d.** 22.37
12.24b. $\hat{y} = 126.98 - 0.2739x$ **d.** 61.23 **12.25b.** $\hat{y} = 69.65 - 2.285x$ **d.** 17.095
12.26a. SSE = 0.675, $s^2 = 0.225$ **b.** 3 **c.** $t = 7.00$, reject H_0 **d.** 0.0060 **e.** $1.05 \pm .353$
12.27a. SSE = 1.816, $s^2 = 0.363$ **b.** 5 **c.** $t = 7.99$, reject H_0 **d.** 0.0005 **e.** $.576 \pm .142$
12.28a. SSE = 78.9, $s^2 = 9.8625$ **b.** 8 **c.** Yes, $t = -7.12$, reject H_0
12.29a. SSE = 95.589, $s^2 = 23.897$ **b.** 4 **c.** No; do not reject H_0, $t = -0.79$
12.30a. $\hat{y} = 20.599 + 5.638x$ **b.** SSE = 562.997, $s^2 = 140.749$ **c.** 5.638 ± 1.448
12.31 Yes; reject H_0: $\beta_1 > 0$, $t = 7.76$ **12.33** $r^2 = .942$ **12.34** $r^2 = .927$ **12.35** $r^2 = .864$
12.36 $r^2 = .135$ **12.37a.** $1.4 \pm .499$ **b.** 1.4 ± 1.223 **c.** Prediction interval is wider
12.38a. $2.767 \pm .481$ **b.** 2.767 ± 1.306 **c.** Prediction interval is wider
12.39a. 22.37 ± 4.26 **b.** 22.37 ± 1.54
12.40a. $\hat{y} = 6.514 + 10.829x$ **b.** $12.904 \pm .445$ **c.** $12.904 \pm .886$
12.41a. $\hat{\beta}_0 = 0, \hat{\beta}_1 = 0.857$ **c.** Yes; reject H_0: $\beta_1 = 0$, $t = 6.71$
12.42a. Positive; as x increases, y increases **b.** $r = 0.958$ **c.** $r^2 = 0.918$
d. Yes **e.** $r = .958$, hence reject H_0; yes
12.43a. $1.714 \pm .619$ **b.** 1.714 ± 1.297 **12.44a.** $\hat{\beta}_0 = 2, \hat{\beta}_1 = -1.2$ **c.** Yes, reject H_0: $\beta_1 = 0$, $t = -5.20$
12.45a. Negative; as x increases, y decreases **b.** $r = -.949$ **c.** $r^2 = 0.900$ **d.** Yes
e. $r = -.949$, hence reject H_0; yes
12.46a. $-.04 \pm 1.331$ **b.** $-.04 \pm 2.714$ **12.47a.** $r = 0.843$ **b.** No; $r_{.025} = .878$, do not reject H_0
12.48a. $\hat{y} = -13.622 - 0.0533x$ **c.** $r = -.923$, $r^2 = .852$ **d.** $.779 \pm .150$ **e.** $.779 \pm .474$
12.49a. $r = -.194$ **b.** No; $-r_{.025} = -.707$ **c.** $\hat{y} = 70.837 - 0.2817x$
d. No; do not reject H_0: $\beta_1 = 0$, $t = -0.48$

12.50a. $\hat{y} = 17.8774 + 1.2907x$ **c.** $r = .9993$, $r^2 = .9987$ **d.** $28.203 \pm .244$ **e.** $26.912 \pm .642$
12.51a. H_0: $\beta_1 = 0$, H_a: $\beta_1 \neq 0$ **b.** Do not reject H_0, at $\alpha = .05$
12.52a. Yes, reject H_0: $\beta_1 = 0$ **b.** Yes, reject H_0: $\beta_1 = 0$ **d.** 8.29 **e.** 21.15

Chapter 13

13.1b. For $x = 30, 31, 32, 33$: linear; for $x = 33, 34, 35, 36$: linear; for all the data: quadratic
13.2a. $y = \beta_0 + \beta_1 x + \text{Random error}$ **b.** $y = \beta_0 + \beta_1 x + \beta_2 x^2 + \text{Random error}$
c. Yes; reject H_0: $\beta_2 = 0$, $t = -3.33$
13.3a. Yes; reject H_0: $\beta_1 = \beta_2 = \beta_3 = 0$, $F = 15.40$ **b.** No; do not reject H_0: $\beta_3 = 0$, $t = -.753$
13.4a. $\hat{y} = 0.4564705 + 0.00078505x_1 + 0.23737262x_2 - 0.00003809x_1x_2$ **b.** $\text{SSE} = 2.71515039$, $s^2 = 0.16969690$
13.5a. Yes; $F = 84.86$, p-value $= .0001$ **13.6** (7.323, 9.449) **13.7** Reject H_0: $\beta_1 = \beta_2 = 0$, $F_{.05} = 6.94$
13.8b. Yes; reject H_0: $\beta_1 = \beta_2 = 0$, $F = 39.51$ **c.** 3.38 **13.9a.** No, $F = 1.28$ **b.** No
13.10a. 13.6812 **b.** Yes
13.11a. $\hat{y} = 69.75354 - 10.09196x_1 - 5.334766x_2$ **b.** $R^2 = .95917$ **c.** Yes, $F = 46.99$
d. Do not reject H_0: $\beta_1 = 0$, $t = \sqrt{F} = -2.23$; reject H_0: $\beta_2 = 0$, $t = \sqrt{F} = -8.48$
13.12a. Yes; reject H_0: $\beta_1 = \beta_2 = 0$, $F = 618.96$ **b.** Yes; reject H_0: $\beta_1 = \beta_2 = 0$, $F = 1540.32$
13.13b. Yes; reject H_0: $\beta_1 = \beta_2 = \beta_3 = \beta_4 = 0$, $F_{.025} = 3.01$
c. Do not reject H_0: $\beta_1 = 0$, $t = -1.64$; reject H_0: $\beta_2 = 0$, $t = -9.62$; reject H_0: $\beta_3 = 0$, $t = 5.59$; reject H_0: $\beta_4 = 0$, $t = 3.59$
13.14a. $\hat{y} = -1180 + 6808x$ **b.** Yes; reject H_0: $\beta_1 = 0$, $t = 19.00$ **c.** $r^2 = .984$
13.15a. $\hat{y} = -3.426 - 0.040x_1 + 0.048x_2 + 0.0018x_3 - 0.0771x_4$ **b.** $R^2 = .94883$
c. Do not reject H_0, p-value $= .3979$ **d.** (1.2271, 1.4666)
13.16 $\hat{y} = 0.5149 + 0.000542x_1 - 0.000353x_2 + 0.001988x_3$; $R^2 = .6988$, $F = 4.64$

Chapter 14

14.1a. Production: 66.02%; clerical: 68.47%; management: 78.79% **b.** Yes

14.2a.

Expected number	Production	Clerical	Management
Favor	173.4	75.2	22.4
Don't favor	82.6	35.8	10.6

b.

Observed–Expected	Production	Clerical	Management
Favor	−4.4	0.8	3.6
Don't favor	4.4	−0.8	−3.6

c. $\chi^2 = 2.219$

14.3a.

Percentage in each expectation category	Social Sciences	Biological Sciences	Physical Sciences	Arts and Humanities
High	19.36	34.62	51.19	21.05
Modest	58.06	57.69	45.24	35.53
Poor	22.58	7.69	3.57	43.42

b. Yes

14.4a.

Expected number	Social Sciences	Biological Sciences	Physical Sciences	Arts and Humanities
High	20.3	25.5	27.4	24.8
Moderate	30.2	38.0	40.9	37.0
Poor	11.6	14.6	15.7	14.2

b.

Observed–Expected	Social Sciences	Biological Sciences	Physical Sciences	Arts and Humanities
High	−8.3	1.5	15.6	−8.8
Moderate	5.8	7.0	−2.9	−10.0
Poor	2.4	−8.6	−12.7	18.8

c. $\chi^2 = 61.49$

14.5a. 14.0671 **b.** 23.5418 **c.** 23.2093 **d.** 17.5436 **e.** 16.7496
14.6a. .025 **b.** .01 **c.** .05
14.7a. Yes **c.**

Expected number	Jewish	Protestant	Catholic
Strongly support	11.0	41.6	14.5
Do not support	36.0	136.4	47.5

d.

Observed–Expected	Jewish	Protestant	Catholic
Strongly support	10.0	−5.6	−4.5
Do not support	−10.0	5.6	4.5

e. $\chi^2 = 14.728$

14.8a. 1 **b.** 3 **c.** 4 **d.** 6
14.9a. H_0: The two directions of classification Employment category and Opinion are independent; H_a: The two directions of classification are dependent **b.** 2 **c.** No; $\chi^2 = 2.219$, do not reject H_0
14.10 p-value $> .10$ (p-value $= .3298$)
14.11a. H_0: The two directions of classification Treatment and Side effects are independent; H_a: The two directions of classification are dependent
b. 1 **c.** Do not reject H_0, $\chi^2 = 1.581$ **d.** Do not reject H_0, $z = 1.415$ **e.** No
14.12 p-value $> .10$ (p-value $= .2086$)
14.13a. H_0: The two directions of classification College major and Employment expectations are independent; H_a: The two directions of classification are dependent **b.** 6 **c.** Yes; reject H_0, $\chi^2 = 61.49$
14.14 p-value $< .005$ (p-value $= .0001$) **14.15** $.6775 \pm .0384$
14.16 No; $z = -0.46$, do not reject H_0: $(p_1 - p_2) = 0$ **14.17** p-value $= .6456$
14.18a. 17.5346 **b.** 11.0705 **c.** 18.3070 **d.** 6.25139 **e.** 9.21034
14.19 $\chi^2_{.10} = 4.60517$; $\chi^2_{.025} = 7.37776$; $\chi^2_{.005} = 10.5966$; $\chi^2_{.05} = 5.99147$; $\chi^2_{.01} = 9.21034$
14.20 Yes; reject H_0, $\chi^2 = 11.595$ **14.21a.** Reject H_0, $\chi^2 = 14.728$ **b.** p-value $< .005$ (p-value $= .0006$)
c. $.161 \pm .091$
14.22a. Yes; $\chi^2 = 5.136$, $.025 < p\text{-value} < .01$ (p-value $= .0235$) **b.** $z = 2.28$, reject H_0: $(p_1 - p_2) = 0$ **c.** No
14.23 Yes; reject H_0, $\chi^2 = 7.578$ **14.24** Yes; $\chi^2 = 31.855$ **14.25** $.12 \pm .095$
14.26a. Do not reject H_0, $\chi^2 = 5.642$ ($\chi^2_{.01} = 6.6349$) **b.** Do not reject H_0, $z = -2.29$ ($-z_{.005} = -2.58$)
14.27a. Reject H_0, $\chi^2 = 105.258$ **b.** $.0769 \pm .0512$ **14.28a.** Yes; reject H_0, $\chi^2 = 14.671$ **b.** $-.169 \pm .161$
14.29a. Yes; reject H_0, $\chi^2 = 55.123$ (Warning: the assumptions required for a valid test may be violated)
b. $-.00183 \pm .242$

Index

E

F

G

H

I

L

M

N

O

P

Q

R

S

T

This book was designed by Janet Bollow. The text was typeset in Helvetica Light by the staff of Jonathan Peck, Typographer. Printing and binding were done by Halliday Lithograph.

The project was managed by Elizabeth Powers of the staff of Jonathan Peck, Typographer, in cooperation with Susan Reiland.

Technical art was prepared by Reese Thornton.

Normal Curve Areas

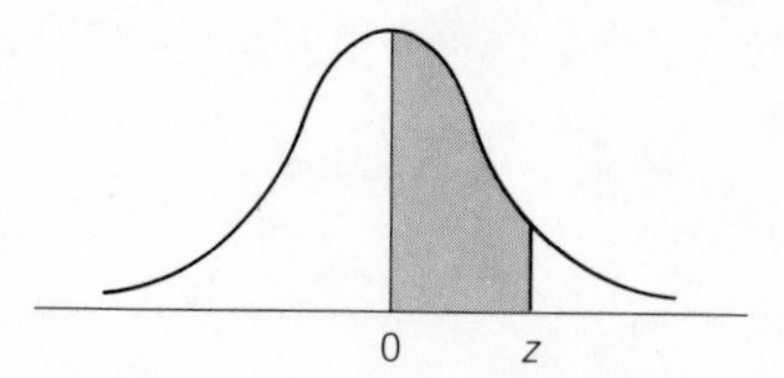

z	.00	.01	.02	.03	.04	.05	.06	.07	.08	.09
0.0	.0000	.0040	.0080	.0120	.0160	.0199	.0239	.0279	.0319	.0359
0.1	.0398	.0438	.0478	.0517	.0557	.0596	.0636	.0675	.0714	.0753
0.2	.0793	.0832	.0871	.0910	.0948	.0987	.1026	.1064	.1103	.1141
0.3	.1179	.1217	.1255	.1293	.1331	.1368	.1406	.1443	.1480	.1517
0.4	.1554	.1591	.1628	.1664	.1700	.1736	.1772	.1808	.1844	.1879
0.5	.1915	.1950	.1985	.2019	.2054	.2088	.2123	.2157	.2190	.2224
0.6	.2257	.2291	.2324	.2357	.2389	.2422	.2454	.2486	.2517	.2549
0.7	.2580	.2611	.2642	.2673	.2704	.2734	.2764	.2794	.2823	.2852
0.8	.2881	.2910	.2939	.2967	.2995	.3023	.3051	.3078	.3106	.3133
0.9	.3159	.3186	.3212	.3238	.3264	.3289	.3315	.3340	.3365	.3389
1.0	.3413	.3438	.3461	.3485	.3508	.3531	.3554	.3577	.3599	.3621
1.1	.3643	.3665	.3686	.3708	.3729	.3749	.3770	.3790	.3810	.3830
1.2	.3849	.3869	.3888	.3907	.3925	.3944	.3962	.3980	.3997	.4015
1.3	.4032	.4049	.4066	.4082	.4099	.4115	.4131	.4147	.4162	.4177
1.4	.4192	.4207	.4222	.4236	.4251	.4265	.4279	.4292	.4306	.4319
1.5	.4332	.4345	.4357	.4370	.4382	.4394	.4406	.4418	.4429	.4441
1.6	.4452	.4463	.4474	.4484	.4495	.4505	.4515	.4525	.4535	.4545
1.7	.4554	.4564	.4573	.4582	.4591	.4599	.4608	.4616	.4625	.4633
1.8	.4641	.4649	.4656	.4664	.4671	.4678	.4686	.4693	.4699	.4706
1.9	.4713	.4719	.4726	.4732	.4738	.4744	.4750	.4756	.4761	.4767
2.0	.4772	.4778	.4783	.4788	.4793	.4798	.4803	.4808	.4812	.4817
2.1	.4821	.4826	.4830	.4834	.4838	.4842	.4846	.4850	.4854	.4857
2.2	.4861	.4864	.4868	.4871	.4875	.4878	.4881	.4884	.4887	.4890
2.3	.4893	.4896	.4898	.4901	.4904	.4906	.4909	.4911	.4913	.4916
2.4	.4918	.4920	.4922	.4925	.4927	.4929	.4931	.4932	.4934	.4936
2.5	.4938	.4940	.4941	.4943	.4945	.4946	.4948	.4949	.4951	.4952
2.6	.4953	.4955	.4956	.4957	.4959	.4960	.4961	.4962	.4963	.4964
2.7	.4965	.4966	.4967	.4968	.4969	.4970	.4971	.4972	.4973	.4974
2.8	.4974	.4975	.4976	.4977	.4977	.4978	.4979	.4979	.4980	.4981
2.9	.4981	.4982	.4982	.4983	.4984	.4984	.4985	.4985	.4986	.4986
3.0	.4987	.4987	.4987	.4988	.4988	.4989	.4989	.4989	.4990	.4990